T0135068

# Lecture Notes in Networks and Systems

## Volume 424

The series "Lecture Notes in Networks and Systems" publishes the latest developments in Networks and Systems—quickly, informally and with high quality. Original research reported in proceedings and post-proceedings represents the core of LNNS.

Volumes published in LNNS embrace all aspects and subfields of, as well as new challenges in, Networks and Systems.

The series contains proceedings and edited volumes in systems and networks, spanning the areas of Cyber-Physical Systems, Autonomous Systems, Sensor Networks, Control Systems, Energy Systems, Automotive Systems, Biological Systems, Vehicular Networking and Connected Vehicles, Aerospace Systems, Automation, Manufacturing, Smart Grids, Nonlinear Systems, Power Systems, Robotics, Social Systems, Economic Systems and other. Of particular value to both the contributors and the readership are the short publication timeframe and the world-wide distribution and exposure which enable both a wide and rapid dissemination of research output.

The series covers the theory, applications, and perspectives on the state of the art and future developments relevant to systems and networks, decision making, control, complex processes and related areas, as embedded in the fields of interdisciplinary and applied sciences, engineering, computer science, physics, economics, social, and life sciences, as well as the paradigms and methodologies behind them.

Indexed by SCOPUS, INSPEC, WTI Frankfurt eG, zbMATH, SCImago.

All books published in the series are submitted for consideration in Web of Science.

For proposals from Asia please contact Aninda Bose (aninda.bose@springer.com).

More information about this series at https://link.springer.com/bookseries/15179

Andrei Tchernykh · Anatoly Alikhanov ·
Mikhail Babenko · Irina Samoylenko
Editors

# Mathematics and its Applications in New Computer Systems

MANCS-2021

 Springer

*Editors*
Andrei Tchernykh 🆔
Director of the Problem-Oriented Cloud
Computing Environment
International Laboratory
CICESE Research Center
Ensenada, Baja California, Mexico

Mikhail Babenko
Faculty of Mathematics and Computer
Science
North-Caucasus Federal University
Stavropol, Russia

Anatoly Alikhanov
Faculty of Mathematics
and Computer Science
North-Caucasus Federal University
Stavropol, Russia

Irina Samoylenko
Department of Informational Systems
Stavropol State Agrarian University
Stavropol, Russia

ISSN 2367-3370         ISSN 2367-3389   (electronic)
Lecture Notes in Networks and Systems
ISBN 978-3-030-97019-2        ISBN 978-3-030-97020-8   (eBook)
https://doi.org/10.1007/978-3-030-97020-8

This Springer imprint is published by the registered company Springer Nature Switzerland AG
The registered company address is: Gewerbestrasse 11, 6330 Cham, Switzerland

# Preface

This volume contains the proceedings of International Conference on Mathematics and its Applications in New Computer Systems. The conference was held by North-Caucasus Centre for Mathematical Research, North-Caucasus Federal University.

The papers are focused on four sections: numerical methods in scientific computing; mathematical solutions to cryptography issues; data analysis and modular computing; and mathematical education. All presented papers include significant research achievements.

Numerical methods field contains such contributions as the application of a generalized multiscale finite element method in numerical simulation of poroelasticity problems in fractured media; a scheme for solving the Cauchy problem for a system of nonlinear differential equations with nonlinear boundary conditions taking into account the energy exchange between the parts of the vibrating object on the left and right parts of the moving boundary. Additionally, authors describe the simulation method of a control propulsion-steering system in tele-controlled uninhabited vehicles, in which movement, direction change and positioning are ensured by four thrusters located in the plan rectangular bluff platform. The synthesis of a numerical method for compensating distortions in the control system of underwater tele-controlled uninhabited vehicles consists of a multiengine system, simultaneously performing the functions of a propulsion unit and a steering device. Modular redundant codes using a non-positional number system in residue number system make it possible to simultaneously translate data from modular representation into positional representation and correct distorted information, which significantly reduces the time spent on detecting and correcting errors.

Mathematical solutions to cryptography issues are presented by a mixed approach based on both arithmetic calculations and a lookup table (LUT) implementation, a reliable method of data transmission over wireless sensor networks and a model for protecting information resources in the cloud. Researchers analyze integro-differential equation that simulates small deformations of an elastically supported console; a finite element method for integro-differential models of a singularly loaded bar; hyperchaotic mappings and their construction based on attractors, as well as studies of image noise characteristics using the investigated attractors and their performance; orthogonal transformations to improve the efficiency of using neural networks and using wavelet transforms in convolutional networks. The possibilities of recurrent quantitative analysis recover the bitstream encrypted with the chaotic parameter modulation. Compaction of subsiding soils by deep explosions is explored. The analysis of the experimental data obtained as a result of the explosions shows that mathematical modeling of this process should use differential calculus methods.

The field of data analysis and modular computing includes such significant contributions, as a discrete model of the Hamming neural network, which makes it possible to simplify the implementation of computations significantly, and a natural gradient descent algorithm with the Dirichlet distribution, which includes step-size adaptation and has an advantage of natural gradient descent over stochastic gradient descent and Adam algorithm. Authors propose a reverse converter from the balanced residue number system (RNS) with low-cost moduli set to the binary number system (BNS); a neural network classification system of pigmented skin lesions according to ten diagnostically significant categories, the highest accuracy rate of which was achieved using the AlexNet convolutional neural network architecture and amounted to 80.15%; a scheme for detecting a moving object based on the Kalman filter and a face manipulation detection technique that can figure out whether a target video contains manipulated faces in it or not. The idea to use a role-based application level decentralization mechanisms against attacks exploiting network decentralization vulnerabilities is presented. A comparative analysis of state-of-the-art approaches to the circuits design for DWT with Cohen–Daubechies–Feauveau 9/7 wavelet is performed. Evaluation results using the "unit-gate" model shows that the optimized direct implementation of DWT exceeds the lifting scheme by 4.59 times in computational speed and requires 9.55% fewer hardware costs when implemented on modern digital devices such as field-programmable gate arrays. A comparison of the direct method of implementation of two-dimensional filtering and the method of implementation of two-dimensional filters by Winograd is demonstrated. A deep learning-based method for breast tumor classification from breast ultrasound tumor images achieves state-of-the-art classification accuracy of 98.72% on the test data set with only 40% top-ranked extracted features by the TL model. An approach for image denoising reduces the noise components level such as Gaussian and salt-and-pepper when performing a discrete wavelet transform (DWT) of images. Application of spline wavelet robust bent codes is based on bent functions in communication systems with code-division multiple access (CDMA). The method of forming a neural control system for underwater robotic complexes in conditions of impossibility of physical expansion of channels is developed.

In addition, the volume contains study in the field of education. The results of testing first-year students of the North-Caucasus Federal University in the international program for assessing the quality of education Program for International Student Assessment (PISA) were obtained by mathematical statistics. Relevant issues of modern trends in the use of distance technologies in the educational space and the transformation of higher education were examined on the example of the North-Caucasus Federal University, including the context of the epidemiological situation caused by the spread of the coronavirus COVID-19 in the Russian Federation. Testing advanced TIMSS implemented for graduate bachelor students in mathematical areas of training is shown.

The target audience of the conference proceedings includes postgraduates, lecturers at institutions of higher education and researchers who study mathematics and its applications in computer systems. Based on the conclusions and results offered, representatives of the targeted audience are likely to find essential knowledge and suggestions for future research.

# Contents

# Comparison of Approaches to the Circuits Design for DWT with CDF 9/7 Wavelet

Pavel Lyakhov⑩, Nikolay Nagornov⑩, and Maxim Bergerman⑩

**Abstract** Discrete wavelet transform (DWT) is widely used in modern science and technology to solve a wide range of signal and image processing problems and digital communications. The high growth rates of quantitative and qualitative characteristics of digital information lead to the need to improve information processing methods and increase the efficiency of their implementation. Specialized hardware circuits are used to solve this problem since they can significantly enhance the characteristics of DWT implementation devices. Calculations for DWT are organized using multiple approaches that differ in the priorities of resource consumption. This paper proposes a comparative analysis of state-of-the-art approaches to the circuits design for DWT with Cohen-Daubechies-Feauveau 9/7 wavelet. The evaluation results using the "unit-gate" model showed that the optimized direct implementation of DWT exceeds the lifting scheme by 4.59 times in computational speed and requires 9.55% fewer hardware costs when implemented on modern digital devices such as Field-Programmable Gate Arrays.

**Keywords** Discrete wavelet transform · Circuits design · CDF 9/7 · Unit-gate model · Lifting scheme

## 1 Introduction

Various transforms are used to process digital information nowadays. The discrete Fourier transform (DFT) [1], Hadamard transform (HT) [2], and discrete wavelet transform (DWT) [3] are the most common of them. Both DFT and HT are widely used in the frequency domain, but the domain characteristics disappeared. The time position and the degree of intensity after signal DFT or HT cannot be determined.

P. Lyakhov
North Caucasus Center for Mathematical Research, North-Caucasus Federal University, 355017 Stavropol, Russia

N. Nagornov (✉) · M. Bergerman
North-Caucasus Federal University, Stavropol, Russia
e-mail: sparta1392@mail.ru

© The Author(s), under exclusive license to Springer Nature Switzerland AG 2022
A. Tchernykh et al. (eds.), *Mathematics and its Applications in New Computer Systems*,
Lecture Notes in Networks and Systems 424,
https://doi.org/10.1007/978-3-030-97020-8_1

1

Local features of the signal time domain are lost in this case. DWT does not have this disadvantage since it allows obtaining both frequency and time information about a signal. DWT uses a filter bank that translates the signal from a time representation into a time–frequency domain. Signal processing is performed by convolution with a pair of low-pass and high-pass wavelet filters of filter bank that highlight main and detailed information, respectively.

The convolution operation has high computational complexity, and its multiple executions in DWT require significant device resources [4]. Signal processing time is the main one. At the same time, the quantitative and qualitative characteristics of digital information are constantly improving. There is a need to improve the efficiency of digital processing devices for one- and multidimensional signals [5]. The development of new and improvement of existing methods of digital processing data is actively being carried out to solve this problem. One of the main approaches to improving the performance of digital devices is based on the use of specialized hardware accelerators such as Field-Programmable Gate Array (FPGA) [6], Application-Specific Integrated Circuit (ASIC) [7], System-on-a-Chip (SoC) [8].

Hardware accelerators are widely used for many digital signal and image processing tasks such as denoising and compression. DWT allows you to solve these problems with high efficiency, using various approaches to the hardware implementation of computations. There are two state-of-the-art approaches to the circuits design for wavelet data processing [9]: the direct implementation (DI) and the lifting scheme (LS). DI aims to achieve maximum processing speed. LS is used when small device area and low power consumption are more priority.

The purpose of this article is to analyze the computational complexity of various approaches to the circuits design for DWT and compare both the time and hardware costs of their implementation on modern digital circuits.

## 2   Materials and Methods

### 2.1   *Theoretical Analysis of Generations Discrete Wavelet Transform*

DWT is performed by signal or image processing using a filter bank that transforms the input signal into a time–frequency form. DI is named first generation DWT and uses low-pass $L = \left(l_1, l_2, ..., l_{k_1-1}, l_{k_1}\right)$ and high-pass $H = \left(h_1, h_2, ..., h_{k_2-1}, h_{k_2}\right)$ filters with $k_1$ and $k_2$ coefficients respectively as follows [9]:

$$X^{[L]}(n) = \sum_{i=1}^{k_1} X(n - i + 1) \cdot l_i, \; X^{[H]}(n) = \sum_{i=1}^{k_2} X(n - i + 1) \cdot h_i,$$

where $X(n) = (x_1, x_2, ..., x_{n-1}, x_n)$ is the input signal; $X^{[L]}(n)$ and $X^{[H]}(n)$ are the input signal's low- and high-frequency components, respectively. DI scheme with downsampling $\downarrow 2$ is shown in Fig. 1a. LS is named second generation DWT and uses predict $P$ and update $U$ steps (functions) instead of low-pass and high-pass filters, as shown in Fig. 1b. The $P$ function predicts the values in the high-frequency part from the low-frequency values. The $U$ function is needed to correct the low-frequency part using changed high-frequency values. LS can have several predict and update steps. The wavelet used for transformation determines the steps number and the values of the functions.

Symmetric biorthogonal wavelets are widely used in circuits design for DWT. Cohen-Daubechies-Feauveau (CDF) 9/7 wavelet is one of the main. CDF 9/7 wavelet has 9 low-pass filter coefficients and 7 high-pass filter coefficients and is used to process one-dimensional and multidimensional (images and video data) signals. In particular, the JPEG 2000 compression standard uses CDF 9/7 for lossy compression [9]. Coefficients of CDF 9/7 wavelet filters for DI are presented in Table 1.

LS for CDF 9/7 wavelet containing 2 prediction and update steps is presented in Fig. 2, where $K = 1.230174$ and uses the functions presented in Table 2.

The number of filter coefficients determines wavelet order. The higher the order of the wavelet and steps in the lifting scheme, the greater the ability to extract various information. The wavelet CDF 9/7 has enough order to successfully solve most practical problems of processing one- and multidimensional signals. However, calculations using this wavelet are associated with an error since its coefficients must be pre-rounded to be represented in the device memory with a preselected accuracy.

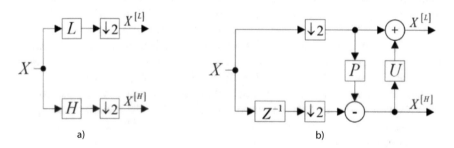

a)                           b)

**Fig. 1** The schemes of the discrete wavelet transform: a) direct implementation; b) lifting scheme

**Table 1** Coefficients of CDF 9/7 wavelet filters

| Coefficient number | Wavelet CDF 9/7 | |
|---|---|---|
| | Low-pass | High-pass |
| 0 | 0.602949 | 1.115087 |
| −1, 1 | 0.266864 | −0.591272 |
| −2, 2 | −0.078223 | −0.057544 |
| −3, 3 | −0.016864 | 0.091272 |
| −4, 4 | 0.026749 | – |

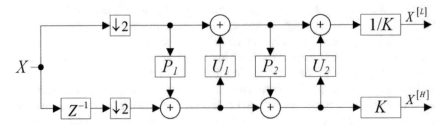

**Fig. 2** The lifting scheme for CDF 9/7 wavelet

**Table 2** Functions of the lifting scheme for CDF 9/7 wavelet

| Function | Wavelet CDF 9/7 |
|----------|-----------------|
| $P_1$ | $-1.586134(x_i + x_{i+1})$ |
| $U_1$ | $-0.052980(x_{i-1} + x_i)$ |
| $P_2$ | $0.882911(x_i + x_{i+1})$ |
| $U_2$ | $-0.443507(x_{i-1} + x_i)$ |

The symmetry of this wavelet is an important advantage concerning conventional orthogonal wavelets (such as Daubechies wavelets, symlets, and coiflets), due to which it has become widespread.

Consider in more detail what advantages the use of symmetric biorthogonal wavelet CDF 9/7 gives compared to conventional orthogonal wavelets at their implementation in modern circuits design. The computation scheme for calculating low- and high-frequency signal values at processing it using DI of DWT with CDF 9/7 wavelet is shown in Fig. 3a. These calculations are performed using adders and multipliers. The main computational load falls on the multipliers since their complexity is quadratic, while for adders, it is linear. Thus, the time and hardware costs for signal processing are primarily determined by the number of multiplications.

The calculations implementation according to the scheme in Fig. 3a requires high resource costs due to the need for multiple uses of the multiplication operation. Biorthogonality of the used wavelet with symmetric filter coefficients allows optimizing computations, reducing the number of multiplication operations due to preliminary summation of the signal values multiplied by the same wavelet filter coefficients. The described calculations are performed according to the scheme shown in Fig. 3b. The figure shows that the number of multipliers in one iteration of the optimized DI (ODI) using wavelet CDF 9/7 decreases from 16 to 9, significantly reducing hardware costs when implementing DWT for signal processing.

LS, unlike DI and ODI, initially takes into account the symmetry of wavelet filters. But besides this, the second generation DWT also considers the relationship between the coefficients of the low-pass and high-pass filters, allowing them to be combined and presented as a group of mathematical functions (predict and update steps). Also, the lifting scheme separates the computations between iterations, allowing the same intermediate results for several adjacent iterations. The computation scheme for

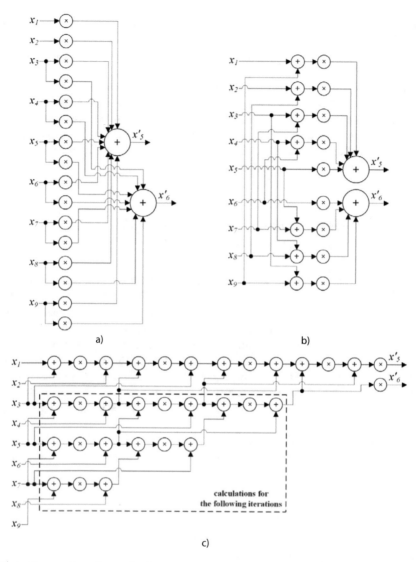

a)  b)

c)

**Fig. 3** Computation schemes for discrete wavelet transform with CDF 9/7 wavelet using various approaches: a) direct implementation; b) optimized direct implementation; c) lifting scheme

calculating low- and high-frequency signal values at processing it using LS of DWT with CDF 9/7 wavelet is shown in Fig. 3c. The figure shows that a significant part of the calculations required for a single iteration is the basis for subsequent iterations and does not require accounting for their costs within the current iteration. Thus, the organization of computations using LS allows reducing the number of multiplications needed further. The number of multipliers in one iteration of computations using LS decreases from 9 to 6. However, the number of successive multiplications in one iteration of calculations increases from 1 to 5. LS leads to the need to organize sequential calculations using values obtained from previous or subsequent iterations. In contrast, DI implements computations in parallel, requiring significantly less time spent on signal wavelet processing than LS.

We can conclude that using symmetric biorthogonal wavelets should significantly reduce hardware costs for signal wavelet processing using various methods on modern microelectronic devices. LS should allow achieving an even greater reduction in hardware costs but at the expense of a significant increase in time costs. An estimate of the time and hardware costs for all considered approaches to the DWT implementation is presented further.

## 2.2 Comparison of Direct Implementations and Lifting Scheme Using the Unit-Gate Model

The unit-gate model (UGM) is a logical device theoretical evaluation technique, taking into account the computational complexity of its constituent elements [10]. This model calculates the hardware and time costs required to implement various methods and algorithms for circuits design. UGM is measured in the unit gate: gate area for device area estimating and gate delay for device delay estimating. Logic two-input gates AND, OR, NAND, and OR occupies 1 gate area and gate delay for each, XOR and XNOR – 2 for each. Operation NOT takes 0 gate area and 0 gate delay. The sum of all elements estimates the device area. The device delay is calculated as the maximum gate delay value from all possible paths from the device input to its output.

Consider a full adder (FA), the scheme of which is shown in Fig. 4. FA consists of 2 XOR elements, 2 AND elements, and 1 OR element. The device area, according to UGM, is:

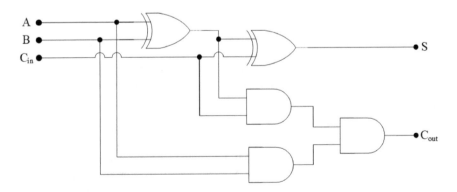

**Fig. 4** The schemes of the full adder

$$U_{area}(FA) = 2 \cdot U_{area}(XOR) + 2 \cdot U_{area}(AND) + U_{area}(OR) = 2 \cdot 2 + 2 \cdot 1 + 1 = 7.$$

Calculate the adder delay. Calculate all possible passes from the input data of the device to the output for this. The adder has 3 inputs: number $A$, number $B$, and carry $C_{in}$; and 2 outputs: sum $S$ and carry $C_{out}$. The number of possible values for device latency calculating is 8:

1. $U_{delay}(A \rightarrow S) = 2 \cdot U_{delay}(XOR) = 2 \cdot 2 = 4$.
2. $U_{delay}(A \rightarrow C_{out}) = U_{delay}(XOR) + U_{delay}(AND) + U_{delay}(OR) = 2 + 1 + 1 = 4$.
3. $U_{delay}(A \rightarrow C_{out}) = U_{delay}(AND) + U_{delay}(OR) = 1 + 1 = 2$.
4. $U_{delay}(B \rightarrow S) = 2 \cdot U_{delay}(XOR) = 2 \cdot 2 = 4$.
5. $U_{delay}(B \rightarrow C_{out}) = U_{delay}(XOR) + U_{delay}(AND) + U_{delay}(OR) = 2 + 1 + 1 = 4$.
6. $U_{delay}(B \rightarrow C_{out}) = U_{delay}(AND) + U_{delay}(OR) = 1 + 1 = 2$.
7. $U_{delay}(C_{in} \rightarrow S) = U_{delay}(XOR) = 2$.
8. $U_{delay}(C_{in} \rightarrow C_{out}) = U_{delay}(AND) + U_{delay}(OR) = 1 + 1 = 2$.

Thus, the longest delay of FA is:

$$U_{delay}(FA) = \max \left\{ \begin{array}{l} U_{delay}(A \rightarrow S), U_{delay}(A \rightarrow C_{out}), U_{delay}(B \rightarrow S), \\ U_{delay}(B \rightarrow C_{out}), U_{delay}(C_{in} \rightarrow S), U_{delay}(C_{in} \rightarrow C_{out}) \end{array} \right\} = 4.$$

The parallel adder architectures Carry-Save adder (CSA) [11] and Kogge-Stone adder (KSA) [12] were used in this research. The area and delay values according to UGM for these elements are:

$$U_{area}(CSA) = 7k, \quad U_{delay}(CSA) = 4,$$
$$U_{area}(KSA) = 3k \log_2 k + 3k + 1, \quad U_{delay}(CSA) = 2 \log_2 k + 4,$$

where $k$ – adder bitness. Compare DI, ODI, and LS using UGM with CSA and KSA. The signal values arriving at the inputs of all circuits are 8 bits wide. Filter coefficients are represented in device memory with 16-bit precision. The first summation of two 8-bit signal values in ODI and LS is performed using KSA, and the result is 9-bit wide. Multiplications for all circuits are performed by generating partial multiplications. All results of these and subsequent calculations are presented with 16-bit precision. The processed signal values are scaled to an 8-bit representation. The results of logical devices evaluating the considered approaches to the DWT implementation and their discussion are presented further.

**Table 3** The evaluation results of various approaches to the circuits design for discrete wavelet transform with CDF 9/7 wavelet using "unit-gate" model

| Approach | Delay | Area |
|---|---|---|
| Direct implementation | 56 | 16,434 |
| Optimized direct implementation | 58 | 10,722 |
| Lifting scheme | 266 | 11,854 |

## 3  Results and Discussion

The evaluation results of analyzed approaches (DI, ODI, LS) to the circuits design for DWT with CDF 9/7 using UGM are presented in Table 3. ODI shows better device area results than DI by reducing the number of multipliers from 16 to 9. However, adding adders at the initial stage of calculations leads to a slight increase in device latency. LS loses to both approaches in latency by almost 5 times due to the organization of sequential computations, forcing each multiplication to be performed using the results of the previous computations. At the same time, LS focused on minimizing the area of the device, loses in this indicator to ODI. The 8-bit and 16-bit numbers arrive at the input of multipliers with DI. The 9-bit and 16-bit numbers are received at the input of multipliers with ODI. Two 16-bit numbers are input to the input of all multipliers (except first) in LS. Thus, multipliers in LS have a higher computational complexity than other approaches, which leads to an increase in the area of the device. As a result, LS is inferior to ODI in both device latency and device area.

## 4  Conclusion

There are two state-of-the-art approaches to the circuits design for DWT: DI and LS. A theoretical analysis of the computational complexity of both generations DWT and their evaluation with a symmetric biorthogonal CDF 9/7 wavelet using a UGM is carried out in this paper. The results showed that the symmetry of the wavelet makes it possible to reduce the number of multipliers due to preliminary summation of the signal values and to optimize DI scheme. ODI of DWT exceeds LS by 4.59 times in computational speed and requires 9.55% fewer hardware costs. Thus, it is advisable to circuits design for wavelet processing of signals and images using symmetric biorthogonal wavelets and optimized DI of DWT.

**Acknowledgements** The research in section 1 and subsection 2.1 was supported by the Russian Science Foundation (project no. 21-71-00017). The research in the remaining subsection and sections was supported by North-Caucasus Center for Mathematical Research under agreement №. 075-02-2021-1749 with the Ministry of Science and Higher Education of the Russian Federation.

# References

1. Rosa BMG, Yang GZ (2020) Bladder volume monitoring using electrical impedance tomography with simultaneous multi-tone tissue stimulation and DFT-based impedance calculation inside an FPGA. IEEE Trans Biomed Circuits Syst 14(4):775–786
2. Hahamovich E, Rosenthal A (2020) Ultrasound detection arrays via coded hadamard apertures. IEEE Trans Ultrason Ferroelectr Freq Control 67(10):2095–2102
3. Cao L, Li H, Zhang Y, Zhang L, Xu L (2020) Hierarchical method for cataract grading based on retinal images using improved Haar wavelet. Inf. Fusion 53:196–208
4. Rossinelli D, Fourestey G, Schmidt F, Busse B, Kurtcuoglu V (2021) High-throughput lossy-to-lossless 3D image compression. IEEE Trans Med Imaging 40(2):607–620
5. Smistad E, Østvik A, Pedersen A (2019) High performance neural network inference, streaming, and visualization of medical images using FAST. IEEE Access 7:136310–136321
6. Ravi M, Sewa A, Shashidhar TG, Sanagapati SSS (2019) FPGA as a hardware accelerator for computation intensive maximum likelihood expectation maximization medical image reconstruction algorithm. IEEE Access 7:111727–111735
7. Janjic J, Tan M, Daeichin V, Noothout E, Chen C, Chen Z, Chang Z-Y, Beurskens RHSH, van Soest G, van der Steen AFW, Verweij MD, Pertijs MAP, de Jong N (2018) A 2-D ultrasound transducer with front-end ASIC and low cable count for 3-D forward-looking intravascular imaging: performance and characterization. IEEE Trans Ultrason Ferroelectr Freq Control 65(10):1832–1844
8. Zhai X, et al (2019) Zynq SoC based acceleration of the lattice Boltzmann method. Concurr Comput Pract Exp 31:e5184
9. Gonzalez RC, Woods RE (2018) Digital image processing. 4th edn. Pearson Education Limited, Harlow
10. Zimmermann R (1997) Binary adder architectures for cell-based VLSI and their synthesis. Hartung-Gorre, Zürich
11. Parhami B (2010) Computer arithmetic: algorithms and hardware designs. Oxford University Press, London
12. Kogge PM, Stone HS (1973) A parallel algorithm for the efficient solution of a general class of recurrence equations. IEEE Trans Comput C-22(8):786–793

# Hamming Neural Network in Discrete Form

**Aleksey Shaposhnikov**[ID]**, Anzor Orazaev**[ID]**, Egor Eremenko**[ID]**, and Danil Malakhov**[ID]

**Abstract** Hamming artificial neural network is used to solve problems of classification of binary input vectors. Its work is based on procedures aimed at choosing, as a solution to the classification problem, one of the reference images closest to the noisy input image supplied to the network input and assigning this image to the corresponding class. At the same time, difficulties were identified in implementing calculations of the Hamming neural network paradigm. One of the ways to simplify computations in the considered neural network paradigm is to transition to integer computations since integer multiplication is computed several times faster than real data type. To simplify the computation model, a Hamming neural network in discrete form was proposed in this paper. A discrete model of the Hamming neural network is proposed, making it possible to simplify the implementation of computations significantly. These results can be applied to solve various problems, including computer vision applications.

**Keywords** Hamming neural network · Neural networks · Discretization · Integer form · Pattern recognition

## 1 Introduction

Recognizing and classifying systems of similar behavior is often a necessary function in performance measurement and modern control and data acquisition applications. Usually, it can be compared with the type of pattern recognition offered by the Hamming neural network [1]. Opportunities for teaching the network opened its application in various fields of engineering sciences, including the use of field-programmable gate arrays (FPGA) [2]. Analog neural networks are more efficient

A. Shaposhnikov · E. Eremenko · D. Malakhov
Department of Mathematical Modeling, North-Caucasus Federal University, Stavropol, Russia
e-mail: ashaposhnikov@ncfu.ru

A. Orazaev (✉)
North-Caucasus Center for Mathematical Research, North-Caucasus Federal University, Stavropol, Russia
e-mail: anz.orazaev95@gmail.com

© The Author(s), under exclusive license to Springer Nature Switzerland AG 2022          11
A. Tchernykh et al. (eds.), *Mathematics and its Applications in New Computer Systems*,
Lecture Notes in Networks and Systems 424,
https://doi.org/10.1007/978-3-030-97020-8_2

in equipment size, power and speed, but suffer from noise sensitivity. Thus, most of the neural network hardware is implemented using analog devices. Hamming's neural network is used to solve many problems. In [3], the network is used to recognize handwritten numbers in real time. In paper [4], a neural network is used to analyze surface parameters in production. In work [5], the authors propose using the Hamming neural network to analyze Li-ion batteries. Hamming neural networks are also used to assess the proximity of binary vectors, which expands the scope of neural networks for solving problems of recognition and classification of input information using proximity functions and more subtle signs of the proximity of discrete objects with binary coding [6].

In this paper, the Hamming neural network sampling will be proposed, and an example will demonstrate the correctness of its operation. The rest of the article is structured in the following way. Section 2 provides basic theoretical information about the Hamming neural network. In Sect. 3, a method for its discretization is proposed. In Sect. 4, conclusions are drawn, and ways of further research on this topic are outlined.

## 2   Background on Hamming Neural Network

Hamming's neural network was positioned as a specialized heteroassociative storage device [7]. The network is used to correlate a binary vector with one of the reference images (each class has its image) or decide that the vector does not correspond to any of the patterns. The network chooses one of the reference images minimizing the Hamming distance to the input vector. Hamming distance is the number of distinct bits between two patterns [8]. For example, for two vectors $X_1 = (1, 1, 0)$ and $X_2 = (1, 0, 1)$ the Hamming distance is equal to 2.

Figure 1 shows a Hamming neural network containing $m$ inputs and consists of two layers. The first and second layers each have $k$ neurons, where $k$ is the number of samples. The neurons of the first layer each have m synapses connected to the network inputs (forming a fictitious zero layer). The neurons of the second layer are interconnected by inhibitory (negative feedback) synaptic connections. A single positive feedback synapse for each neuron is connected to its axon.

The network must select a sample with a minimum Hamming distance to the unknown input signal, which will activate only one network output corresponding to that sample [9]. A binary vector is fed to the network's input, and the result is a unitary code, where the number of the activated output corresponds to the number of the reference image.

At the initialization stage, the weights of the first layer and the threshold of the activation function are assigned the following values:

$$w_{ij} = x_i^{(j)}/2 \qquad (1)$$

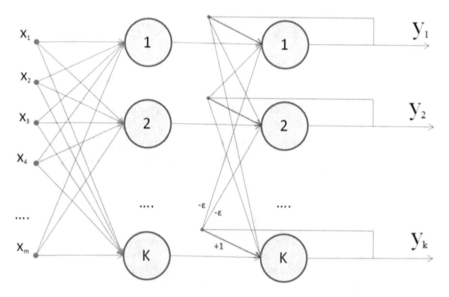

**Fig. 1** Structural scheme of the Hamming network

where $x_i^{(j)}$ is $i$-th element of $j$-th pattern $i = 0...m - 1$ is network input number, $j = 0...k - 1$ is the reference image number.

The weight coefficients of inhibitory synapses in the second layer are taken equal to some value $0 < \varepsilon < 1/k$. The synapse of a neuron is associated with its axon weights $+1$.

The algorithm for the functioning of the Hamming network is as follows:

1. An unknown vector $X = \{x_i : i = 0...m - 1\}$ is fed to the network inputs, based on which the states of the neurons of the first layer are calculated (the superscript in brackets indicates the layer number):

$$y_j^{(1)} = s_j^{(1)} = \sum_{i=0}^{m-1} w_{ij} x_i + T_j \tag{2}$$

$$T_j = m/2 \tag{3}$$

After that, the values of the axons of the second layer are initialized with the obtained values:

$$y_j(2) = y_j(1) \tag{4}$$

2. Calculate the new states of the neurons of the second layer:

$$s_j^{(2)}(p+1) = y_j(p) - \varepsilon \sum_{j=0}^{k-1} y_j^{(2)}(p) \qquad (5)$$

where $p$ is iteration number, and the values of their axons:

$$y_j^{(2)}(p+1) = f[s_j^{(2)}(p+1)] \qquad (6)$$

The activation function $y = f(s)$ has the form of a threshold Fig. 2, and the value of $Q$ must be large enough so that any possible values of the argument do not lead to saturation.

$$y = \begin{cases} 0, s < 0 \\ s, 0 \le s < Q \\ Q, s \ge Q \end{cases} \qquad (7)$$

3.  Check if the outputs of the neurons of the second layer have changed during the last iteration. If yes, go to step 2. Otherwise, end.

    It can be seen from the evaluation of the algorithm that the role of the first layer is rather arbitrary: having used the values of its weight coefficients once at step 1, the network no longer refers to it, therefore the first layer can be completely excluded from the network (replaced by a matrix of weight coefficients) [10].

**Example 1:** Let as reference images in the network be written as $X_1 = (1, -1, 1)$ and $X_2 = (-1, 1, -1)$. Then the matrix of weight coefficients of the first layer of neurons using (1) will be equal to $W^{(1)} = \begin{pmatrix} \frac{1}{2} & -\frac{1}{2} & \frac{1}{2} \\ -\frac{1}{2} & \frac{1}{2} & -\frac{1}{2} \end{pmatrix}$. Bias taking into account (3) $T = 3/2$. $\varepsilon = 1/3$. Let's submit to the network input $X = (1, 1, 1)$. The functioning of the network can be summarized in Table 1.

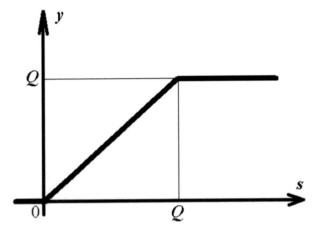

**Fig. 2** Activation function

**Table 1** Hamming neural network functioning

| Iteration number, $p$ | 1 | 2 | 3 |
|---|---|---|---|
| Network output $y_1$ | 2 | 1,7 | 1,6 |
| Network output $y_2$ | 1 | 0,3 | 0 |

The numerical values of the output values of the network during operation are rounded to tenths, which is done to simplify the understanding of the operation of the network. At the fourth iteration, the state of the network will be repeated, so the iterative process will be stopped. The network output indicates that the first image is recognized because Hamming distance to it $d_1 = 1$, and to the second image $d_1 = 2$.

The main difficulty in implementing neural computations in the well-known model of the Hamming paradigm neural network lies in the implementation of multiplications of real data. The number of multiplications for recognizing one pattern will be equal to $k^2(p + 1)$, where $p$ is the number of required iterations. Since $k$ is the number of recognizable patterns, the total number of multiplications can be quite large.

## 3  Hamming Neural Network in Discrete Form

The work aims to simplify the calculations of the Hamming neural network paradigm. One of the ways to simplify computations in the considered neural network paradigm is to transition to integer computations since integer multiplication is computed several times faster than real data type.

The algorithm for the functioning of the network will not change. The network weights of the first layer, determined by the formula (1), can be represented as

$$w_{ij} = x_i^{(j)} \tag{8}$$

The displacement in formula (3) can be taken as

$$T_j = m \tag{9}$$

The weights of the inhibitory synapses of the Hamming discrete neural network in the second layer are taken equal to $\varepsilon = -1$, and the synapse of a neuron associated with its axon will have weight $k$. This will change the network topology. The structure of Hamming's discrete neural network paradigm is shown in Fig. 3. The activation function threshold defined by (7) must be large enough

$$Q = k^3 \tag{10}$$

Let us give an example of the functioning of the proposed Hamming discrete neural network.

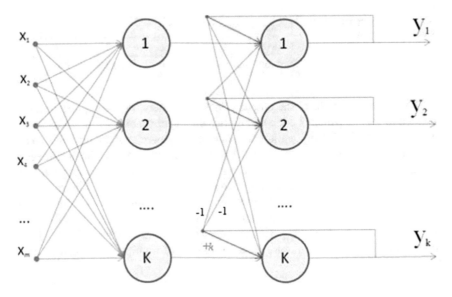

**Fig. 3** Block diagram of Hamming discrete neural network

**Table 2** Functioning of the Hamming discrete neural network

| Iteration number, $p$ | 1 | 2 | 3 |
|---|---|---|---|
| Network output $y_1$ | 2 | 1,7 | 1,6 |
| Network output $y_2$ | 1 | 0,3 | 0 |

**Example 2.** As reference images, we take the initial data of Example 1. Then the matrix of weight coefficients of the first layer of neurons using (7) will be equal to $W^{(1)} = \begin{pmatrix} 1 & -1 & 1 \\ -1 & 1 & -1 \end{pmatrix}$. Bias considering (8) T = 3. The matrix of weights of the second layer of neurons will be equal to $W^{(2)} = \begin{pmatrix} 2 & -1 \\ -1 & 2 \end{pmatrix}$. Let's give the network input $X = (1, 1, 1)$. The saturation threshold for the initial data is an example of the activation function (9) equality $Q = 8$. The operation of the network is summarized in Table 2.

At the fourth iteration, the state of the network will repeat itself so that the iterative process will be stopped.

## 4   Conclusion

This article discusses the well-known Hamming neural network paradigm. The difficulties in the implementation of neurocomputations of this paradigm are revealed. A discrete model of the Hamming neural network is proposed, making it possible to simplify the implementation of computations significantly. These results can be applied to solve various problems, including computer vision applications. In the future, an interesting direction of research is the hardware implementation of the Hamming neural network in a discrete form on the FPGA.

**Acknowledgements** The authors would like to thank the North Caucasus Federal University for supporting the contest of projects competition of scientific groups and individual scientists of the North Caucasus Federal University. The work is supported by the North-Caucasus Center for Mathematical Research under agreement № 075-02-2021-1749 with the Ministry of Science and Higher Education of the Russian Federation and by Russian Foundation for Basic Research project 19-07-00130.

# References

1. Lippmann R (1987) An introduction to computing with neural nets. IEEE ASSP Mag 4(2):4–22
2. Himavathi S, Anitha D, Muthuramalingam A (2007) Feedforward neural network implementation in FPGA using layer multiplexing for effective resource utilization. IEEE Trans Neural Netw 18(3):880–888
3. Knerr S, Personnaz L, Dreyfus G (1992) Handwritten digit recognition by neural networks with single-layer training. IEEE Trans Neural Netw 3(6):962–968
4. Lipinski D, Tomkowski R, Kacalak W (2018) Application of the Hamming network to the classification of surfaces after abrasive machining. J Mach Eng 18(4):114–126
5. Kim J, Lee S, Cho BH (2011) Discrimination of Li-ion batteries based on Hamming network using discharging–charging voltage pattern recognition for improved state-of-charge estimation. J Power Sources 196(4):2227–2240
6. Dmitrienko VD, Zakovorotniy AY, Leonov SY (2020). Neural networks for determining affinity functions. In: 2020 International congress on human-computer interaction, optimization and robotic applications (HORA). IEEE, pp 1–5
7. Dayhoff JE (1990) Neural network architectures: an introduction. Van Nostrand Reinhold Co.
8. Gupta M, Jin L, Homma N (2004) Static and dynamic neural networks: from fundamentals to advanced theory. Wiley, Hoboken
9. Fausett LV (2006) Fundamentals of neural networks: architectures, algorithms and applications. Pearson Education India

# Improving Extreme Search with Natural Gradient Descent Using Dirichlet Distribution

Ruslan I. Abdulkadirov(ID) and Pavel A. Lyakhov(ID)

**Abstract**  Natural gradient descent is an optimization algorithm, which is proposed to replace stochastic gradient descent and its modifications. The most precious ability of this algorithm is to reach the extreme with little number of iterations and required accuracy, which has high value in machine learning and statistics. The goal of this article is to propose a natural gradient descent algorithm with the Dirichlet distribution, which includes step-size adaptation. We will prove experimentally advantage of natural gradient descent over stochastic gradient descent and Adam algorithm. Additionally, the calculating of the Fisher information matrix of Dirichlet distribution will be shown.

**Keywords**  Natural gradient descent · Adam algorithm · Kullback-Leibler divergence · Fisher information matrix · Dirichlet distribution

## 1  Introduction

The problem of increasing the accuracy and minimizing the number of iterations (epochs) in machine learning, optimization methods, and approximation theory remains actual. Especially this is actual in neural networks, where prediction depends on accuracy.

Very often algorithms such as gradient and stochastic gradient descents (GD and SGD, respectively) are applied for optimization of the loss function. Later appeared modification of SGD with momentum such as AdaGrad in [1] and [2], RMSprop [3], ADADELTA [4] and Adam algorithm in [5]. But these methods sometimes converge slowly and accuracy doesn't achieve the required quality. Even step-size adaptation

R. I. Abdulkadirov (✉) · P. A. Lyakhov
North-Caucasus Center for Mathematical Research, North-Caucasus Federal University, Stavropol 355017, Russian Federation
e-mail: ruslanabdulkadirovstavropol@gmail.com

P. A. Lyakhov
Department of Mathematical Modeling, North-Caucasus Federal University, Stavropol 355017, Russian Federation

© The Author(s), under exclusive license to Springer Nature Switzerland AG 2022
A. Tchernykh et al. (eds.), *Mathematics and its Applications in New Computer Systems*, Lecture Notes in Networks and Systems 424,
https://doi.org/10.1007/978-3-030-97020-8_3

from [6] doesn't ensure the extreme reaching in minimum time. But researching the loss function from the geometrical point of view, especially Riemannian, can perform the process of minimization and improve the quality of recognition.

The idea about including metric properties in GD came from Riemannian geometry. In [7] was defined Riemannian gradient flow, that accelerates the optimization process. The Riemannian gradient flow is defined on a smooth manifold with a metric tensor. In information geometry gradient flow is defined on the manifold of probability distributions and called Fisher information matrix. In this case gradient flow is replaced by a natural gradient from [8]. The natural gradient in information geometry is defined as the product between the Fisher information matrix and the gradient of the loss function. For calculating the Fisher matrix will be used the Kullback-Leibler divergence (K-L divergence). Optimization algorithm, which applies natural gradient, is called natural gradient descent (NGD). Remarkable that changing the length of step and value of gradient for NGD is not necessary for increasing the accuracy, selecting appropriate parameters for distributions suffices.

In this article, we propose the algorithm of NGD based on Dirichlet distribution. According to experimental results, we will prove that using Dirichlet distribution gives higher accuracy and less number of iteration than GD with step-size adaptation and Adam algorithm.

The remaining of the paper is organized as follows. Section 2 presents preliminaries, which consist of gradient descent with step-size adaptation, modification of Adam algorithm, NGD with definitions of K-L divergence and Fisher information matrix. Section 3 represents Dirichlet distribution, calculating of proper Fisher matrix, and proposing Algorithm 3. In Sect. 4 shows test functions with related graphs of convergence and table, which shows the number of iterations for every algorithm. In Sect. 5 reported discussions and suggestions for developing the natural gradient descent in further.

## 2 Preliminaries

### 2.1 Gradient Descent with Step-Size Adaptation

Consider optimizing a smooth objective $f : \Omega \rightarrow \mathbb{R}$ over closed convex set $\Omega \in \mathbb{R}$. The problem of minimization is to calculate $min_{x \in \Omega}(f(x))$. This algorithm is still actual in machine learning and optimization methods.

The gradient descent with appropriate step-size adaptation in [6] has an advantages in rate and accuracy. It is possible that Adam algorithm doesn't have the same accuracy and convergence with respect to gradient descent. Especially, this is delighted in case of adjusting step-size in both algorithms. Gradient descent with step-size adaptation is defined as Algorithm 1.

---

**Algorithm 1** Monotonous gradient descent with stepsize adaptation

---

**Input:** starting point $x \in \mathbb{R}^n$, a scalar function $f(x)$ and its gradient $\nabla f(x)$, initial stepsize $\alpha$
**Output:** some $x$ minimizing $f$
1: initialize $f_x = f(x) \in \mathbb{R}$, $g = \nabla f(x)^T \in \mathbb{R}^n$
2: **while** $|y - x| < \theta$ for 10 iterations in sequence **do**
3:     $y \leftarrow x - \alpha g/|g|, f_y \leftarrow f(y)$
4:     **if** $f_y < f_x$ **then**
5:        $x \leftarrow y, f_x \leftarrow f_y, g \leftarrow \nabla f(x)^T, a \leftarrow 1.2a$
6:     **else**
7:        $a \leftarrow 0.5a$
8:     **end if**
9: **end while**
10: **return** $x$

---

**Algorithm 2** Adam algorithm

---

**Input:** starting point $x \in \mathbb{R}^n$, a scalar function $f(x)$, its gradient $\nabla f(x)$, initial step-size $\alpha$, exponential decay rates $\beta_1$, $\beta_2$ and $\epsilon$
**Output:** some $x$ minimizing $f$
1: initialize $f_x = f(x) \in \mathbb{R}$, $g = \nabla f(x)^T \in \mathbb{R}^n$, $m_0 = 0$, $v_0 = 0$
2: **for** i from 0 to n - 1 **do**
3:     $g_i \leftarrow \nabla f(x_i); m_i \leftarrow \beta_1 m_{i-1} + (1 - \beta_1)g_i; v_i \leftarrow \beta_2 v_{i-1} + (1 - \beta_2)g_i^2$
4:     $\hat{m}_i \leftarrow m_i/(1 - \beta_1^i); \hat{v}_i \leftarrow v_i/(1 - \beta_1^i); y \leftarrow x - \alpha \cdot \hat{m}_i/(\sqrt{\hat{v}_i} + \epsilon)$
5:     **if** $f_y < f_x$ **then**
6:        $x \leftarrow y, f_x \leftarrow f_y, a \leftarrow 1.2a$
7:     **else**
8:        $a \leftarrow 0.5a$
9:     **end if**
10: **end for**
11: **return** $x$

---

Note that simple GD in general case does not achieve the extreme with required accuracy. Sometimes it does not converge, because the constant length of step $\eta$ can redirect the gradient in case of functions with several local extremes.

## 2.2 Adam Algorithm

The Adam algorithm computes individual adaptive learning rates for different parameters from estimates of first and second moments of the gradients, the name Adam is derived from adaptive moment estimation. For accelerating extreme search it is necessary to include step-size adaptation. Because Adam algorithm in [5] on practice has the same problems as simple gradient descent. But adjusting of the step allows to minimize the number of iterations (epochs) and to maximize the accuracy. In Algorithm 2 we represent the pseudo-code of Adam method with step-size adaptation.

In machine learning the most preferred optimization method is realized by Adam algorithm, which increase the percent of accuracy in every epoch. But this algorithm doesn't engage the curvature of the function $n - 1$-surface, for $n \geq 2$.

## 2.3  Natural Gradient Descent and KL-divergence

Natural Gradient Descent in [7] is obtained as the forward Euler discrediting with step-size $\eta$ of the gradient flow:

$$x^{(k+1)} = x^{(k)} - \eta_k F(x^{(k)})^{-1} \nabla f(x^{(k)}), \tag{1}$$

where $x^{(0)} = x_0$.

The main part of natural gradient descent is a Fisher matrix, that can be calculated on manifold of probability distributions. Suppose we optimize the $f(\theta)$. Let $p(x; \theta)$ is some family of probability distributions over $x$ parametrized by a vector of real numbers $\theta$. Let's provide the Kullback-Leibler divergence [7, 9] as

$$KL(p(x; \theta_t)||p(x; \theta_t + \delta\theta)) = \int p(x; \theta_t) \log \frac{p(x; \theta_t)}{p(x; \theta_t + \delta\theta)} dx. \tag{2}$$

Then the KL-divergence can be represented as

$$KL(p(x; \theta_t)||p(x; \theta_t + \delta\theta)) = -\frac{1}{2}\delta\theta^T \mathbb{E}\left[\nabla \log p(x; \theta_t)\nabla \log p(x; \theta_t)^T\right]\delta\theta, \tag{3}$$

where $F(\theta_t) = -\mathbb{E}\left[\nabla \log p(x; \theta_t)\nabla \log p(x; \theta_t)^T\right]$ is a Fisher information matrix, which is a Riemannian structure on manifold of probability distributions.

Remark, calculation of the Fisher matrix becomes difficult in case of high dimension. This problem can be solved with using of various approximations such as approaching methods in [10], Gauss-Newton and neglecting cross-unit terms in [11], Kronecker-factored Approximate Curvature in [12], factorized natural gradient in [13] that are designed to facilitate computations.

## 3  Natural Gradient Descent with Dirichlet Distribution

The Dirichlet distribution [14] of order $K \geq 2$ with parameters $\alpha_1, ..., \alpha_K > 0$ has a probability density function with respect to Lebesgue measure on the Euclidean space $\mathbb{R}^{K-1}$ given by

$$f(x_1, ..., x_n; \alpha_1, ..., \alpha_K) = \frac{1}{B(\alpha)} \prod_{i=1}^{K} x_i^{\alpha_i - 1}, \quad B(\alpha) = \frac{\prod_i \Gamma(\alpha_i)}{\Gamma(\sum_i \alpha_i)}, \tag{4}$$

Now let's calculate the logarithm of (4):

$$\log f(x_1, ..., x_n; \alpha_1, ..., \alpha_K) = \log \left[ \frac{\Gamma(\sum_i \alpha_i)}{\prod_i \Gamma(\alpha_i)} \prod_{i=1}^{K} x_i^{\alpha_i - 1} \right]$$

$$= \log \Gamma(\sum_{i=1}^{K} \alpha_i) - \sum_{i=1}^{K} \log \Gamma(\alpha_i) + \sum_{i=1}^{K} (\alpha_i - 1) \log x_i.$$

Second order partial derivative of $f$ with respect to $\alpha$:

$$\frac{\partial^2}{\partial a_j \partial a_k} \log f = \psi' \left( \sum_{i=1}^{K} \alpha_i \right), \quad \frac{\partial^2}{\partial a_j^2} \log f = \psi' \left( \sum_{i=1}^{K} \alpha_i \right) - \psi'(\alpha_j).$$

Therefore, we can find the Fisher matrix

$$F_{Dirichlet}(\alpha) = \begin{pmatrix} \psi'(\alpha_1) - \psi'(\sum_i \alpha_i) \ ... & -\psi'(\sum_i \alpha_i) \\ ... & ... & ... \\ -\psi'(\sum_i \alpha_i) & ... \ \psi'(\alpha_K) - \psi'(\sum_i \alpha_i) \end{pmatrix}. \tag{5}$$

After calculating of Fisher information matrix with Dirichlet distribution, let's demonstrate the work of NGD with step-size adaptation in Algorithm 3.

---

**Algorithm 3** Natural Gradient Descent with Dirichlet distribution

---

**Input:** starting point $x \in \mathbb{R}^n$, a scalar function $f(x)$ and its gradient $\nabla f(x)$, initial step-size $\alpha$
**Output:** some $x$ minimizing $f$
1: initialize $f_x = f(x) \in \mathbb{R}$, $g = \nabla f(x)^T \in \mathbb{R}^n$ and Fisher matrix $F_{Dirichlet}$ from (5)
2: **for** i from 0 to n - 1 **do**
3:    $y \leftarrow x - \alpha F_{Dirichlet}^{-1} g/|g|$, $f_y \leftarrow f(y)$
4:    **if** $f_y < f_x$ **then**
5:      $x \leftarrow y$
6:      $f_x \leftarrow f_y$
7:      $g \leftarrow \nabla f(x)^T$
8:      $a \leftarrow 1.2a$
9:    **else**
10:      $a \leftarrow 0.5a$
11:    **end if**
12: **end for**
13: **return** $x$

---

Remark that for Algorithm 3 it is not necessary to decrease the length of steps or numerical value of gradient for improving final values of extremes, because Fisher matrix contains parameters without items of vector $x$, that allows avoid additional computations in loop. Including curvature properties by Fisher matrix natural gradient achieve extreme faster. If we compare Algorithm 3 with 1 and 2 then it implies that natural gradient descent a little bit slower than gradient descent, but converge in less numbers of iterations. And with respect to Adam algorithm natural gradient descent faster and converge better.

# 4 Experimental Results

The behavior of Algorithms 1–3, realized by Python 3.8.10, will be observed in experiments. Convex and smooth functions with related domains $\Omega \in \mathbb{R}^4$ will be selected for solving of the optimization problem.

Initial points and proper parameters will be provided in each experiment. It is necessary to choose the appropriate distribution parameters for every experimental function. Later we will represent functions with related graphs of the rate of convergence, where are drawn 3 curves (blue - GD, orange - Adam algorithm, green - NGD with Dirichlet distribution). Parameter step-size $\eta = 0.001$ is initial for every algorithms, exponential decay rates $\beta_1 = 0.88$, $\beta_2 = 0.9$ for Adam algorithm are the same in every experiment.

Rayden Function with initial point $x = (-14, 10, 16.5, -7.51)$ and $\alpha = (4, 4, 4, 4)$.

$$f(x) = \sum_{i=1}^{4}(\exp(x_i - x_i)) \tag{6}$$

In Fig. 1 shown that NGD with Dirichlet distributions converges faster than Adam algorithm and achieves minimal value, which is equal to 4.0000000013. For Adam algorithm minimal value is 4.0000000028. GD with step-size adaptation gives 4.0000000027.

Generalized Rosenbrock Function with starting point $x = (15.2, 10, 0, 0)$ and $\alpha = (2, 2, 2, 3)$.

$$f(x) = \sum_{i=1}^{3}\left[100(x_{i+1} - x_i^2)^2 + (1 - x_i)^2\right] \tag{7}$$

In Fig. 2 shown that NGD with Dirichlet distributions has the fastest convergence and achieves minimal value, which is equal to 0.010949. For Adam algorithm minimal value is 0.0139083. GD with step-size adaptation reaches 0.143245.

Extended Trigonometric Function with starting point $x = (5, 1.36, 3, -1)$ and $\alpha = (0.5, 0.5, 0.5, 0.9)$.

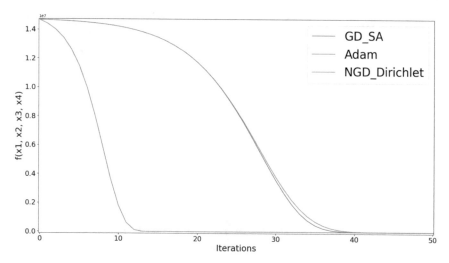

**Fig. 1** Convergence of extreme search methods on Rayden function.

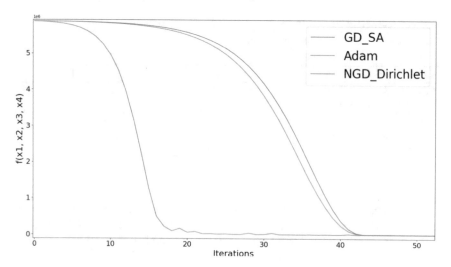

**Fig. 2** Convergence of extreme search methods on generalized Rosenbrock function.

$$f(x) = \sum_{i=1}^{4} \left[ \left( 4 - \sum_{j=1}^{4} \cos x_i \right) + i \cos x_i - \sin x_i \right]^2 \tag{8}$$

In Fig. 3 shown that NGD with Dirichlet distributions reaches the minimum at $1.57389e^{-10}$. For Adam algorithm minimal value is $2.24709e^{-09}$. GD with step-size adaptation shows $5.70724e^{-10}$.

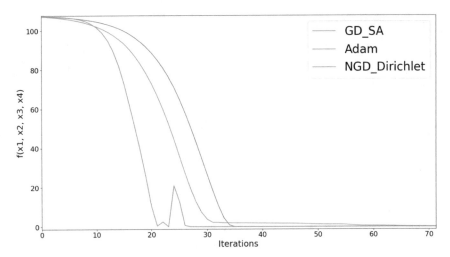

**Fig. 3** Convergence of extreme search methods on extended trigonometric function.

**Table 1** Rate of convergence of extreme search methods on test functions

| Test function | Number of iterations | | |
|---|---|---|---|
| | GD with SA | Adam | NGD dirichlet |
| Rayden function | 37 | 40 | 11 |
| Generalized Rosenbrock function | 42 | 41 | 20 |
| Extended Trigonometric function | 34 | 60 | 26 |

According on the results of experiments we can see that NGD with Dirichlet distribution is the precisest and the fastest with respect to GD with step-size adaptation and Adam algorithm. In Table 1 we represent the number of iterations, which is consumed to converge in minimum.

Remark that parameters of distribution were selected were selected according to initial point and kind of functions in every experiment.

## 5 Discussion

According to the results of experiments presented in Table 1, we can conclude, that NGD with Dirichlet distribution gives value of the extreme with the highest accuracy and least number of iterations than GD with step-size adaptation and Adam algorithm. Moreover, it is not necessary to change the values of the initial step and gradient, because selecting appropriate parameters for Fisher information matrix suffices. But the main disadvantage is the computation of the natural gradient, but it can be solved by applying approximations from [12]. In further exploring natural gradient descent it will be right to compare this algorithm with mirror descent from [15]. Also we need to explore NGD with generalized and multinomial Dirichlet distributions from [16] and [17], respectively.

In conclusion, we can say that natural gradient descent with Dirichlet distribution and step-size adaptation can replace gradient descent and Adam algorithm in various applications, especially in machine learning. Its advantage was proved form theoretical and experimental points of view, which is sufficient fact for developing and modification of this method.

**Acknowledgments** The authors would like to thank the North-Caucasus Federal University for supporting in the contest of projects competition of scientific groups and individual scientists of the North-Caucasus Federal University. The work is supported by North-Caucasus Center for Mathematical Research under agreement No 075-02-2021-1749 with the Ministry of Science and Higher Education of the Russian Federation.

# References

1. Ward R, Wu X, Bottou L (2020) AdaGrad stepsizes: sharp convergence over nonconvex landscapes. J Mach Learn Res 21:1–30
2. Duchi J, Hazan E, Singer Y (2011) Adaptive subgradient methods for online learning and stochastic optimization. J Mach Learn Res 12:2121–2159
3. Hinton G, Srivastava N, Swersky K (2012) Lecture 6D – a separate, adaptive learning rate for each connection. Slides of lecture neural networks for machine learning
4. Zeiler MD (2012) ADADELTA: an adaptive learning rate method. preprint, arXiv:1212.5701v1
5. Kingma DP, Bai JL (2015) Adam: a method for stochastic optimization. In: ICLR
6. Lyakhov P, Abdulkadirov R (2021) Accelerating extreme search based on natural gradient descent with beta distribution. In: International conference engineering and telecommunication (En&T 2021), pp 1–5
7. Martens J (2020) New insights and perspectives on the natural gradient method. J Mach Learn Res 21:1–76
8. Gunasekar S, Woodworth B, Srebro N (2012) Mirrorless mirror descent: a natural derivation of mirror descent. In: Proceedings of the 24th international conference on artificial intelligence and statistics, PMLR, vol 130, pp 2305–2313
9. Amari SI (1998) Natural gradient works efficiently in learning. Neural Comput 10(2):251–276
10. Le Roux N, Schmidt MW, Bach F (2012) A stochastic gradient method with an exponential convergence rate for finite training sets. In: Advances in neural information processing systems (NIPS), pp 2672–2680
11. Ollivier Y (2015) Riemannian metrics for neural networks i: feedforward networks. Inf Inference 4(2):108–153
12. Grosse R, Salakhudinov R (2015) Scaling up natural gradient by sparsely factorizing the inverse fisher matrix. In: Proceedings of the 32nd international conference on machine learning (ICML 2015), pp 2304–2313
13. Martens J, Grosse R (2015) Optimizing neural networks with Kronecker-factored approximate curvature. In: Proceedings of the 32nd international conference on machine learning (ICML)

14. Kotz S, Balakrishnan N, Johnson NL (2000) Continuous multivariate distributions. Volume 1: models and applications. Wiley, New York
15. Miyashita M, Yano S, Kondo T (2018) Mirror descent search and its acceleration. Robot Auton Syst 106:107–116
16. Hankin RKS (2010) A generalization of the dirichlet distribution. J Stat Softw 33(11):1–18
17. Elkan C (2006) Clustering documents with an exponential-family approximation of the Dirichlet compound multinomial distribution. In: Proceedings of the 23rd international conference on machine learning

# CRTf-Based Reverse Converter for RNS with Low-Cost Modules $\{2^n, 2^n - 1, 2^{n+1} - 1\}$

**Maxim Bergerman**📵, **Pavel Lyakhov**📵, **Nataliya Semyonova**📵,
**Danil Bogaevskiy**📵, **and Dmitry Kaplun**📵

**Abstract** In this paper, we propose a reverse converter from the balanced Residue Number System (RNS) with low-cost moduli set $\{2^n, 2^n-1, 2^{(n+1)}-1\}$ to the Binary Number System (BNS). The proposed method is based on the Chinese remainder theorem with fractions (CRTf) for the reverse conversion device. Constant coefficients are calculated in detail, where the bit width depends only on the value of n. Proposed a device architecture using parallel adders to speed up calculations. Hardware modeling of a reverse converter from balanced RNS with moduli set $\{2^n, 2^n-1, 2^{(n+1)}-1\}$ to the positional number system (PNS) was carried out using the field-programmable gate arrays (FPGA) in the VHDL. The results show that the proposed method achieves a faster performance by an average of 8.5% and lower hardware costs by an average of 10.5% compared to the state-of-the-art method. The proposed development can find wide applications in both digital signal and image processing based on RNS.

**Keywords** Residue number system (RNS) · Chinese remainder theorem (CRT) · Reverse conversion · Chinese remainder theorem with fractions (CRTf) · RNS balance · Hardware design

M. Bergerman (✉) · N. Semyonova
Department of Mathematical Modeling, North-Caucasus Federal University, Stavropol, Russia
e-mail: maxx07051997@inbox.ru

P. Lyakhov
North-Caucasus Center for Mathematical Research, North-Caucasus Federal University, 355017 Stavropol, Russia

D. Bogaevskiy · D. Kaplun
Department of Automation and Control Processes, Saint Petersburg Electrotechnical University "LETI", Saint Petersburg, Russia
e-mail: dvbogaevskiy@etu.ru

D. Kaplun
e-mail: dikaplun@etu.ru

© The Author(s), under exclusive license to Springer Nature Switzerland AG 2022          29
A. Tchernykh et al. (eds.), *Mathematics and its Applications in New Computer Systems*,
Lecture Notes in Networks and Systems 424,
https://doi.org/10.1007/978-3-030-97020-8_4

# 1  Introduction

Currently, high-speed computing is very much in demand in modern computer architectures. One of the ways to solve this problem is to use parallel arithmetic operations. The use of the Residue Number System (RNS) allows such calculations to be
performed. Currently, specialized processors are being actively developed that use
RNS to solve problems of digital signal processing [1], cryptography [2], etc. Many
researchers around the world are researching RNS and using RNS in the design of
electronic devices.

The advantage of this number system is the parallel implementation of arithmetic operations: addition, subtraction, and multiplication. Also, the RNS allows
converting data to a very low-bit format, which can be very important for systems
with low power consumption.

However, despite the effective use of multiplication, addition, and subtraction
operations, performing several other operations based on the RNS is very difficult.
Operations of this kind include division and detection of the sign of a number. The
listed operations are a significant limiting factor affecting. First of all, the speed
of devices slows down in such operations. It is necessary to perform the reverse
conversion from the RNS to BNS to solve these disadvantages.

To achieve a high speed of the reverse conversion from RNS to BNS, the
researchers in their work use such sets of modules as $\{2^k - 1, 2^n - 1, 2^n + 1\}$
[3],$\{2^{n+k}, 2^n - 1, 2^{n+1} - 1\}$ [4], $\{2^n, 2^n - 1, 2^{n+1} - 1\}$ [5], $\{2^n - 1, 2^n, 2^n + 1\}$
[6], etc. In this work, an RNS with three low-cost modules $\{2^n, 2^n - 1, 2^{n+1} - 1\}$
will be used with the "measure" of RNS balance and analyze the results of hardware
simulation. The calculations of the constants CRTf for the RNS $\{2^n, 2^n - 1, 2^{n+1} - 1\}$
have also been performed. The architecture of a reverse converter for RNS with a
set of modules $\{2^n, 2^n - 1, 2^{n+1} - 1\}$ was proposed. The simulation was carried
out on FPGA using the VHDL language, where the quality of the proposed reverse
conversion was confirmed.

# 2  Preliminaries

## 2.1  Background on Residue Number System

Residue number system (RNS) is a non-positional number system, where numbers
are represented based on pairwise coprime numbers, called RNS modules $m =
\{m_1, m_2, \ldots, m_n\}$, possessing the property $GCD(m_i, m_j) = 1$ for all $i \neq j$. The
dynamic range *(DR)* of the system is the product of all RNS modules $M = \prod_{i=1}^{n} m_i$,
which determines all possible values in the RNS. The operations of addition, subtraction, and multiplication in the RNS are performed for each remainder, respectively,
independent of each other, it is enough to add, subtract and multiply the corresponding
remainder-digits:

$$A \pm B = \left( \left| a_1 \pm b_1 \right|_{m_1}, \ \ldots, \ \left| a_n \pm b_n \right|_{m_n} \right), \tag{1}$$

$$A \cdot B = \left( \left| a_1 \cdot b_1 \right|_{m_1}, \ \ldots, \ \left| a_n \cdot b_n \right|_{m_n} \right). \tag{2}$$

The traditional way to recover the original number $X$ from the remainders $\{x_1, x_2, \ldots, x_n\}$ is based on the Chinese Remainder Theorem (CRT). Let the modules $\{m_1, m_2, \ldots, m_n\}$ be mutually simple, and they form the DR of the system $M = m_1 \cdot m_2 \cdot \ldots \cdot m_n$. Then any integer $0 \le X \le M$ can be uniquely represented in the RNS as a set of modules $\{x_1, x_2, \ldots, x_n\}$. To recover the number from the RNS, use the formula

$$X = \left| \sum_{i=0}^{n} \left| \left| M_i^{-1} \right|_{m_i} x_i \right|_{m_i} M_i \right|_M, \tag{3}$$

where $M_i = \frac{M}{m_i}$ and $\left| M_i^{-1} \right|_{m_i}$ is the multiplicative inverse $M_i$ for modulo $m_i$.

## 2.2 RNS to BNS Conversion Using CRTf

Most reverse conversion research aims to simplify this operation by reducing the calculations modulo $M$ to simpler operations. One such method is CRT with fractions (CRTf). The essence of the method is to modify formula (3) so that it can be used to estimate the position of a number in a numeric string without performing complex arithmetic operations.

Divide both sides of (3) by $M$:

$$\frac{X}{M} = \left| \sum_{i=1}^{n} x_i \frac{\left| M_i^{-1} \right|_{m_i} M_i}{M} \right|_1 \tag{4}$$

where $|*|_1$ denotes the fractional part of a number. With the notation $k_i = \frac{\left| M_i^{-1} \right|_{m_i} M_i}{M}$, expression (4) can be written in the compact form

$$\frac{X}{M} = \left| \sum_{i=1}^{n} x_i \cdot k_i \right|_1 \tag{5}$$

The values $k_1, k_2, \ldots, k_n$ are rational numbers within the interval $[0, 1)$. In addition, the values depend on the RNS moduli only, thereby being the RNS constants. Formula (5) contains calculations with fractional numbers of arbitrary length and, therefore, cannot be used in hardware implementation. To solve this problem, you can switch to integer arithmetic with relative values. More specifically, it is necessary to multiply each constant from formula (5) by $2^N$ and round the result up:

$$k_i^* = \left\lceil 2^N \frac{\left|M_i^{-1}\right|_{m_i} M_i}{M} \right\rceil \qquad (6)$$

If the value of the coefficient $k_i^*$ in the BNS contains more ones than zeros, it is necessary to convert it to inverse form, depending on the module.

For calculations on modulo $2^\gamma$, negative numbers are represented in additional code, that is

$$\left|(-X) \cdot (-C)\right|_{2^\gamma} = \left|\overline{X} \cdot \left(\overline{C} + 1 - 2^\varepsilon\right) + \Delta_{COR}\right|_{2^\gamma}, \qquad (7)$$

where the correction factor is calculated as follows:

$$\Delta_{COR} = \left|\left(1 - 2^\delta\right) \cdot \left(\overline{C} + 1 - 2^\varepsilon\right)\right|_{2^\gamma}, \qquad (8)$$

where $\delta$ is the bit width of $X$, $\varepsilon$ is the bit width of $k_i^*$, and $\gamma$ is the bit width of $N$.

For calculations on modulo $2^\gamma - 1$, negative numbers are represented in the reverse code, i.e.

$$\left|(-X) \cdot (-C)\right|_{2^\gamma - 1} = \left|\overline{X} \cdot \left(\overline{C} + 1 - 2^\varepsilon\right) + \Delta_{COR}\right|_{2^\gamma - 1}, \qquad (9)$$

where the correction coefficient is calculated using the formula:

$$\Delta_{COR} = \left|\left(1 - 2^\delta\right) \cdot \left(\overline{C} + 1 - 2^\varepsilon\right)\right|_{2^\gamma - 1}. \qquad (10)$$

The sum of the mod $2^N$ products of the remainders by such constants will exceed the relative value $2^N \frac{X}{M}$. Then expression (5) will be written as

$$X^* = \left\lfloor 2^N \frac{X}{M} \right\rfloor = \left|\sum_{i=1}^{n} x_i k_i^*\right|_{2^N}. \qquad (11)$$

As was established in [7], the exact RNS-to-BNS conversion can be guaranteed by choosing

$$N = \left\lceil \log_2(M\theta) \right\rceil - 1, \qquad (12)$$

where $\theta = -n + \sum_{i=1}^{n} m_i$.

The last step to calculate the value $X$ using the value $\left\lfloor 2^N \frac{X}{M} \right\rfloor$ is multiplication by the modulus $M$. The algorithm yields the most significant bits starting from the $(N + 1)$–th one. Thus,

$$X = \frac{X^* M}{2^N}. \qquad (13)$$

During the hardware implementation, the division operation in (13) is ignored since the most significant bits starting from the $(N + 1)$–th one are outputted. In the software implementation, this operation is equivalent to the right $N$-bit shift.

## 2.3   A Measure of RNS Balance with 3 Modules

The concept of "measure of RNS balance" was proposed in [8]. Let modules determine RNS $m_1, m_2, \ldots, m_n$ with bit widths $b_1, b_2, \ldots, b_n$. Let us denote by

$$\bar{b} = \frac{1}{n} \sum_{i=1}^{n} b_i \tag{14}$$

average value of the bit capacity of RNS modules.

Then the value of the RNS balance measure will be equal to.

$$\beta = \frac{1}{n} \sum_{i=1}^{n} (b_i - \bar{b}), \text{ where } i \neq j. \tag{15}$$

In other words, $\beta$ is the variance of the bit width of the RNS modules. The formula can also find it.

$$\beta = \frac{1}{n^2} \sum_{i=1}^{n-1} \sum_{j=1}^{n} (b_i - b_j)^2, \text{ where } i \neq j. \tag{16}$$

If all bit widths of RNS modules are equal, i.e. $b_1 = b_2 = \cdots = b_{n-1} = b_n$, then such an RNS will be called perfectly balanced with $\beta = 0$.

Consider a set of three modules. If modules determine the RNS $m_1, m_2, m_3$ with digits $b_1, b_2, b_3$, then when two digits coincide $b_1 = b_2 \neq b_3$, the value of the measure of balance $\beta$ will be equal to $\frac{2}{9}$, if $b_3 = b_1 - 1$ or $b_3 = b_1 + 1$.

During the research, it turned out that for RNS sets with three modules, 2 sets are the most balanced:

1.   RNS with modules $\{2^n, 2^n - 1, 2^{n-1} - 1\}$;
2.   RNS with modules $\{2^n, 2^n - 1, 2^{n+1} - 1\}$.

In [9], the first set of RNS modules was considered. In this work, consider a reverse conversion from balance RNS with a set of modules $\{2^n, 2^n - 1, 2^{n+1} - 1\}$ into a BNS using CRTf.

## 3   CRTf Method for RNS with Balanced Modules $\{2^n, 2^n - 1, 2^{n+1} - 1\}$

Calculate the dynamic range $M$ and values $M_i$ for RNS with a set of modules $\{2^n, 2^n - 1, 2^{n+1} - 1\}$:

$$M = 2^n \cdot (2^n - 1) \cdot (2^{n+1} - 1) = 2^{3 \cdot n + 1} - 2^{2 \cdot n} - 2^{2 \cdot n + 1} + 2^n,$$

$$M_1 = (2^n - 1) \cdot (2^{n+1} - 1) = 2^{2 \cdot n + 1} - 2^n - 2^{n+1} + 1,$$

$$M_2 = 2^n \cdot (2^{n+1} - 1) = 2^{2 \cdot n + 1} - 2^n,$$

$$M_3 = 2^n \cdot (2^n - 1) = 2^{2n} - 2^n.$$

Let's find the values $\theta$ and $N$:

$$\theta = 2^n + (2^n - 1) + (2^{n+1} - 1) - 3 = 2 \cdot 2^n + 2^{n+1} - 5$$
$$= 2^{n+1} + 2^{n+1} - 2^2 - 2^0 = 2^{n+2} - 2^2 - 2^0;$$

$$N = \log_2((2^{3 \cdot n + 1} - 2^{2 \cdot n} - 2^{2 \cdot n + 1} + 2^n) \cdot (2^{n+2} - 2^2 - 2^0)) - 1$$
$$= 4 \cdot n + 3 - 1 = 4 \cdot n + 2,$$

except when $n = 2$. Then for such a case, $N = 4n + 1$. The values of the multiplicative inverse $M_i^{-1}$ are equal: $M_1^{-1} = 1$; $M_2^{-1} = 1$; $M_3^{-1} = 2^{n+1} - 5$.

Then calculate the values of the coefficients $k_1, k_2, k_3$:

1.   $k_1 = \dfrac{M_1 \cdot |M_1^{-1}|_{m_1}}{M} = \dfrac{|M_1^{-1}|_{m_1}}{m_1} = \dfrac{1}{2^n}$

In binary representation $k_1$ has the form: $k_1 = \dfrac{1}{2^n} = 0.\underbrace{00...01}_{n-bits}00_2$.

2.   $k_2 = \dfrac{M_2 \cdot |M_2^{-1}|_{m_2}}{M} = \dfrac{|M_2^{-1}|_{m_2}}{m_2} = \dfrac{1}{2^n - 1}$.

The number $\dfrac{1}{2^n - 1}$ can be represented as an infinitely decreasing geometric progression:

$$\frac{1}{2^n - 1} = \sum_{i=0}^{\infty} (2^{-n})^{i+1},$$

where the value of $k_2$ can be represented in binary form: $k_2 = 0.\underbrace{(00...01}_{n-bits})_2$.

The period of the infinite fraction is denoted in brackets.

3.   $k_3 = \dfrac{M_3 \cdot |M_3^{-1}|_{m_3}}{M} = \dfrac{|M_3^{-1}|_{m_3}}{m_3} = \dfrac{2^{n+1} - 5}{2^{n+1} - 1}$.

Let us show that the number $\frac{2^{n+1}-5}{2^{n+1}-1}$ can also be represented in the form of an infinitely decreasing geometric progression and find such a progression, the sum of which is equal to the number

$$\frac{1}{2^{n+1}-1} = \frac{2^{-(n+1)}}{1-2^{-(n+1)}} = \sum_{i=0}^{\infty} 2^{-(n+1)}(2^{-(n+1)})^i = \sum_{i=0}^{\infty} (2^{-(n+1)})^{i+1}.$$

Let's represent numbers $\frac{1}{2^{n+1}-1}$ and $2^{n+1}-5$ in the BNS:

$$\frac{1}{2^{n+1}-1} = \frac{2^{-(n+1)}}{1-2^{-(n+1)}} = 0.\underbrace{00...01}_{(n+1)-bits}\ \underbrace{00...01}_{(n+1)-bits}\ ...\ \underbrace{00...01}_{(n+1)-bits}\ ..._2 = 0.(\underbrace{00...01}_{(n+1)-bits})_2,$$

$$2^{n+1}-5 = \underbrace{11...1011}_{(n+1)-bits}{}_2.$$

where the period of the infinite fraction is denoted in brackets. The result of the product of these numbers is $\frac{2^{n+1}-5}{2^{n+1}-1} = 0.\underbrace{11...1011}_{(n+1)-bits}\ \underbrace{11...1011}_{(n+1)-bits}\ ...\ \underbrace{11...1011}_{(n+1)-bits}\ ..._2 =$ $0.(\underbrace{11...1011}_{(n+1)-bits})_2$, where the period of the infinite fraction is denoted in brackets.

This article will consider the cases when $N = 4n + 2$, i.e., where $n > 2$. To obtain $k_i^*$ it is enough to shift the number $k_i$ by $(4n + 2)$-bit width left and rounded up like in formula (6):

$$k_1^* = \lceil 2^N \cdot k_1 \rceil = \left\lceil 2^N \cdot \frac{1}{2^n} \right\rceil = \left\lceil 2^{4n+2} \cdot \frac{1}{2^n} \right\rceil = \lceil 2^{3n+2} \rceil = 2^{3n+2}$$

In binary representation, the value of $k_1^*$ is: $k_1^* = \underbrace{10...000}_{(3n+3)-bits}$. The value $k_1^*$ has a bit width $(3n + 3)$-bits. Find the value $k_2^*$: $k_2^* = \lceil 2^N k_2 \rceil = \lceil \frac{2^{4n+2}}{2^n-1} \rceil$. Since the number $k_2^*$ is taken rounded up we add one to this number in BNS and get:

$$k_2^* = \underbrace{00...001}_{n-bits}\underbrace{00...001}_{n-bits}\underbrace{00...001}_{n-bits}\underbrace{00...001}_{n-bits}\underbrace{00.0...001}_{n-bits}...$$

Remove the zero bits in the most significant bits and rounded up to an integer value, add one to this number: $k_2^* = \underbrace{1}_{1\,bit}\underbrace{00...001}_{n-bits}\underbrace{00...001}_{n-bits}\underbrace{00...001}_{n-bits}\underbrace{01}_{2\,bits}$. The value $k_2^*$ has a bit width $(3n + 3)$-bits. Find the value $k_3^*$: $k_3^* = \lceil 2^N k_3 \rceil = \lceil 2^{4n+2} \cdot \frac{2^{n+1}-5}{2^{n+1}-1} \rceil$. Since the number $k_3^*$ taken rounded up to an integer value, add one to this number and get:

$$k_3^* = \underbrace{11...1011}_{(n+1)-bits} \underbrace{11...1011}_{(n+1)-bits} \underbrace{11...1011}_{(n+1)-bits} \underbrace{11...111}_{(n-1)-bits} \, .$$

Since this number contains more ones than zeros, we represent it in negative form and find correction factor by the formulas (9) and (10): $\Delta_{COR} = \left|(1 - 2^{n+1})(\overline{k_3^*} + 1 - 2^{4n+2})\right| = 1$. Find the inverse of $k_3^*$ to get a number with fewer ones: $\overline{k_3^*} = \underbrace{00...0100}_{(n+1)-bits} \underbrace{00...0100}_{(n+1)-bits} \underbrace{00...0100}_{(n+1)-bits} \underbrace{00...000}_{(n-1)-bits}$. Remove the zero bits in the most significant bits: $\overline{k_3^*} = \underbrace{100}_{3\,bits} \underbrace{00...0100}_{(n+1)-bits} \underbrace{00...0100}_{(n+1)-bits} \underbrace{00...01}_{(n-1)-bits}$. The resulting $\overline{k_3^*}$ has a bit width $(3n + 4)$-bits and $\Delta_{COR} = 1_{10} = 1_2$ is 1 bit.

## 4 Reverse Converter for RNS with a Set of Modules $\{2^n, 2^n - 1, 2^{n+1} - 1\}$

Let us consider in more detail the operation of the reverse converter for RNS with modules $\{2^n, 2^n - 1, 2^{n+1} - 1\}$. Let's divide it into two steps. At the first step, we will find the value of $X^*$, obtained by taking modulo $N = 4n + 2$ the sums of partial products. After applying the compression technique, we get 3 terms, regardless of the bit width, which is equal $w_1, w_2, w_3$ (Fig. 1). Let us operate the addition of the obtained terms using Carry-Save adder (CSA) taken on modulo $4n + 2$. As a result, we get two terms, which we then sum up using the Kogge-Stone's prefix adder (KSA) and get the value $X^*$. The second step is to obtain the original number using the formula (13). First, we multiply the number $X$ by the dynamic range $M$, form partial products (Fig. 2), use CSA until there are 2 terms left at the output. Then summarize with KSA. The final step is to divide by the value using the CSA and KSA tree and then divide that result by $2^N$. In computing, this operation is replaced by shifting the number to the right by $N$ bits. The complete circuit of the proposed reverse converter is shown in Fig. 3.

| | 4n+1 | 4n | 4n-1 | ... | 3n+3 | 3n+2 | 3n+1 | ... | n+2 | n+1 | n | n-1 | ... | 2 | 1 | 0 |
|---|---|---|---|---|---|---|---|---|---|---|---|---|---|---|---|---|
| $W_1 =$ | $x_{1(n-1)}$ | $x_{1(n-2)}$ | $x_{1(n-3)}$ | ... | $x_{11}$ | $x_{10}$ | $x_{2(n-1)}$ | ... | $x_{20}$ | $x_{2(n-1)}$ | $x_{2(n-2)}$ | $x_{2(n-1)}$ | ... | $x_{22}$ | $x_{21}$ | $x_{20}$ |
| $W_2 =$ | $x_{2(n-1)}$ | $x_{2(n-2)}$ | $x_{2(n-3)}$ | ... | $x_{21}$ | $x_{20}$ | $x_{3(n-1)}$ | ... | $x_{31}$ | $x_{30}$ | $x_{3n}$ | $x_{2(n-3)}$ | ... | $x_{20}$ | $x_{31}$ | $x_{30}$ |
| $W_3 =$ | $x_{3(n-2)}$ | $x_{3(n-3)}$ | $x_{3(n-4)}$ | ... | $x_{30}$ | $x_{3n}$ | | | | | | $x_{3(n-1)}$ | ... | $x_{32}$ | | 1 |

**Fig. 1** Grid of terms of partial products $\left|\sum_{i=1}^{n} x_i k_i\right|_{2^N}$ after applying the compression technique for balanced RNS with modules $\{2^n, 2^n - 1, 2^{n+1} - 1\}$

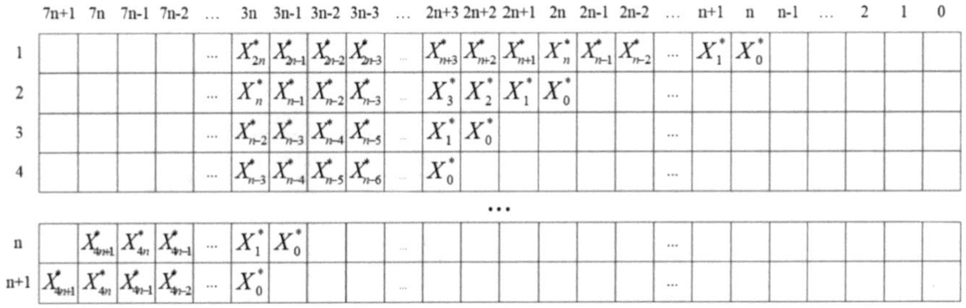

| | 7n+1 | 7n | 7n-1 | 7n-2 | ... | 3n | 3n-1 | 3n-2 | 3n-3 | ... | 2n+3 | 2n+2 | 2n+1 | 2n | 2n-1 | 2n-2 | ... | n+1 | n | n-1 | ... | 2 | 1 | 0 |
|---|---|---|---|---|---|---|---|---|---|---|---|---|---|---|---|---|---|---|---|---|---|---|---|---|
| 1 | | | | | ... | $X_{2n}^*$ | $X_{2n-1}^*$ | $X_{2n-2}^*$ | $X_{2n-3}^*$ | ... | $X_{n+3}^*$ | $X_{n+2}^*$ | $X_{n+1}^*$ | $X_n^*$ | $X_{n-1}^*$ | $X_{n-2}^*$ | ... | $X_1^*$ | $X_0^*$ | | | | | |
| 2 | | | | | ... | $X_n^*$ | $X_{n-1}^*$ | $X_{n-2}^*$ | $X_{n-3}^*$ | | $X_3^*$ | $X_2^*$ | $X_1^*$ | $X_0^*$ | | | ... | | | | | | | |
| 3 | | | | | ... | $X_{n-2}^*$ | $X_{n-3}^*$ | $X_{n-4}^*$ | $X_{n-5}^*$ | | $X_1^*$ | $X_0^*$ | | | | | ... | | | | | | | |
| 4 | | | | | ... | $X_{n-3}^*$ | $X_{n-4}^*$ | $X_{n-5}^*$ | $X_{n-6}^*$ | | $X_0^*$ | | | | | | ... | | | | | | | |
| | | | | | | | | | ... | | | | | | | | | | | | | | | |
| n | | $X_{4n+1}^*$ | $X_{4n}^*$ | $X_{4n-1}^*$ | ... | $X_1^*$ | $X_0^*$ | | | | | | | | | | ... | | | | | | | |
| n+1 | $X_{4n+1}^*$ | $X_{4n}^*$ | $X_{4n-1}^*$ | $X_{4n-2}^*$ | ... | $X_0^*$ | | | | | | | | | | | ... | | | | | | | |

**Fig. 2**  Grid of terms of partial products $X^*M$ for balanced RNS with modules $\{2^n, 2^n - 1, 2^{n+1} - 1\}$

**Fig. 3**  Architecture of the proposed reverse converter from balanced RNS with modules $\{2^n, 2^n - 1, 2^{n+1} - 1\}$ to BNS

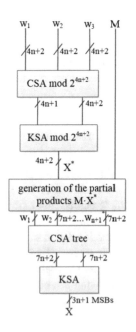

# 5  Hardware Modeling of Reverse Converter for RNS $\{2^n, 2^n - 1, 2^{n+1} - 1\}$

Hardware modeling was implemented in Xilinx ISE Design Suite 14.7 using VHDL. The device has been programmed on a Xilinx xc7k480t-3ffg901 board. We compare the results of the proposed method with the work [10], which also CRTf and used the same set of modules. The obtained simulation results $\{2^n, 2^n - 1, 2^{n+1} - 1\}$ are presented in Table 1.

Based on the analysis of Table 1, we can conclude that the proposed method of reverse conversion from RNS to BNS wins in terms of equipment hardware costs and execution speed. If described as a percentage, then $n = 5$ the proposed method is faster than [10] by 8.5 and 12.8% fewer equipment costs. For $n = 6$, the proposed

**Table 1** Results of hardware simulation of devices for reverse conversion from RNS to BNS

| n, bits | [10] | | | Proposed | | |
|---|---|---|---|---|---|---|
| | Delay, ns | Slices, units | LUT Slices, units | Delay, ns | Slices, units | LUT Slices, units |
| 5 | 17.641 | 125 | 280 | **16.157** | **114** | **246** |
| 6 | 18.509 | 170 | 373 | **16.386** | **141** | **325** |
| 7 | 18.550 | **171** | 427 | **17.521** | 176 | **379** |
| 8 | 21.735 | 305 | 577 | **19.400** | **268** | **524** |
| 9 | 24.003 | 382 | 773 | **22.142** | **336** | **712** |

method is faster by 11.5% and fewer hardware costs by 12.9%. For $n = 7$, the method [10] is inferior to the proposed one in terms of time by 5.5 and 11.2% for equipment. For $n = 8$ the proposed method wins in terms of time by 10.7 and 9.2% for equipment. And, for $n = 9$ the proposed method, better the method [10] in terms of conversion execution time by 7.8 and by 7.9% in terms of hardware costs.

## 6 Conclusions

In this article, the architecture of a reverse converter from balanced RNS with a set of modules $\{2^n, 2^n - 1, 2^{n+1} - 1\}$ to BNS was proposed, hardware simulation was implemented and compared with the existing method. According to the results of comparing the method [8] with the proposed method, it is seen that the proposed method is better, starting from the execution time, this indicator is by 5.5–11.5% and hardware costs by 7.9–12.9%. The proposed architecture of reverse converter can be applied in digital signal processing, digital image processing, cryptography, etc. An interesting direction for further research is the development of such systems using the proposed reverse converter and RNS with well-balanced three low-cost modules.

**Acknowledgements** The work of P.A. Lyakhov was supported by the Presidential Council for grants (project no. МК-3918.2021.1.6). The work of D.I. Kaplun was supported by the Russian Science Foundation, project no.19-19-00566. The work of M.V. Bergerman is supported by North-Caucasus Center for Mathematical Research under agreement №. 075-02-2021-1749 with the Ministry of Science and Higher Education of the Russian Federation.

## References

1. Parhami B (2010) Computer arithmetic Oxford University Press, 2nd edn. New York, Oxford University Press
2. Kumar CS, Prathiba A, Bhaskaran VSK (2016) Implementation of RNS and LNS based addition and subtraction units for cryptography. In: International conference on VLSI systems, architectures, technology and applications (VLSI-SATA), pp 1–5

3. Hiasat A (2019) A reverse converter for three-moduli set $(2^k, 2^n-1, 2^n+1)$, $k < n$. In: IEEE Jordan international joint conference on electrical engineering and information technology (JEEIT), pp 548–553
4. Hiasat A, Sousa L, Anta AF (2019) On the design of RNS inter-modulo processing units for the arithmetic-friendly moduli sets $\{2^{n+k}, 2^{n-1}, 2^{n+1}-1\}$. Comput J 62(2):292–300
5. Hiasat A (2017) An efficient reverse converter for the three-moduli set $(2^{n+1}-1, 2^n, 2^{n-1})$. IEEE Trans Circ Syst II Express Briefs 64(8):962–966
6. Muralidharan R, Chang C (2012) Area-power efficient modulo 2n–1 and modulo 2n+1 multipliers for 2n–1, 2n, 2n+1 based RNS. IEEE Trans Circuits Syst I Regul Pap 59(10):2263–2274
7. Chervyakov NI, Lyakhov PA, Deryabin MA, Nagornov NN, Valueva MV, Valuev GV (2020) Residue number system-based solution for reducing the hardware cost of a convolutional neural network. Neurocomputing 407:439–453
8. Boyvalenkov P, Chervyakov NI, Lyakhov P, Semyonova N, Nazarov A, Valueva M, Boyvalenkov G, Bogaevskiy D, Kaplun D (2020) Classification of moduli sets for residue number system with special diagonal functions. IEEE Access 8:156104–156116
9. Lyakhov P, Bergerman M, Semyonova N, Kaplun D Voznesensky A (2021) Design reverse converter for balanced rns with three low-cost modules. In: 2021 10th mediterranean conference on embedded computing (MECO), pp 1–7
10. Chervyakov NI, Molahosseini AS, Lyakhov PA, Babenko MG, Deryabin MA (2017) Residue-to-binary conversion for general moduli sets based on approximate Chinese remainder theorem. Int J Comput Math 94:1833–1849

# Neural Network Classification of Dermatoscopic Images of Pigmented Skin Lesions

Pavel A. Lyakhov⬛, Ulyana A. Lyakhova⬛, and Valentina A. Baboshina⬛

**Abstract** Today, skin cancer can be regarded as one of the leading causes of death in humans. Skin cancer is the most common type of malignant neoplasm in the body. Rapid and highly accurate diagnosis of malignant skin lesions can reduce the risk of mortality in patients. The paper proposes a neural network classification system of pigmented skin lesions according to 10 diagnostically significant categories. Modeling was carried out using the MATLAB R2020b software package on clinical dermatoscopic images from the international open archive ISIC Melanoma Project. The main convolutional neural network architectures used were SqueezeNet, AlexNet, GoogLeNet, and ResNet101, pre-trained on the ImageNet set of natural images. The highest accuracy rate was achieved using the AlexNet convolutional neural network architecture and amounted to 80.15%. The use of the proposed neural network system for the recognition and classification of dermatoscopic images of pigmented lesions by specialists will improve the accuracy and efficiency of the analysis compared to the methods of visual diagnostics. Timely diagnosis will allow starting treatment at an earlier stage of the disease, which directly affects the percentage of survival and recovery of patients.

**Keywords** Machine learning · Deep learning · Convolutional neural networks · Image classification · Skin cancer · Melanoma · Pigmented skin neoplasms

## 1 Introduction

Nowadays, skin cancer is one of the most common types of malignant lesions in the body [1] and one of the leading causes of death in the world [2]. Dermatoscopy is the most common visual diagnosis method [3]. This method can be effectively used

P. A. Lyakhov · U. A. Lyakhova
Department of Mathematical Modeling, North-Caucasus Federal University, Stavropol, Russia
e-mail: ulaliakhova@ncfu.ru

P. A. Lyakhov · V. A. Baboshina (✉)
North-Caucasus Center for Mathematical Research, North-Caucasus Federal University,
Stavropol, Russia
e-mail: vaababoshina@ncfu.ru

© The Author(s), under exclusive license to Springer Nature Switzerland AG 2022      41
A. Tchernykh et al. (eds.), *Mathematics and its Applications in New Computer Systems*,
Lecture Notes in Networks and Systems 424,
https://doi.org/10.1007/978-3-030-97020-8_5

only by qualified specialists since it is based on visual acuity and the experience of a practicing physician [4]. Early diagnosis of skin cancer has a direct impact on patient survival and recovery rates.

Over the past decade, with the development of artificial intelligence, algorithms of computer analysis of images by their effectiveness were equal and even surpassed human capabilities in some cases. The recognition of visual images is widely used in many fields such as text and speech recognition, semantic search, expert systems, acceptance support systems decisions, prediction of stock prices, security systems, text analysis [5]. Also, deep learning approaches have shown promising results in the field of computer vision, namely in object tracking [6], object detection [7], and image classification [8].

Classifiers based on convolutional neural networks (CNN) are optimal and accurate in recognizing pigmented skin lesions [9]. This feature is because CNNs are the most optimal neural network architecture for identifying visual images. Since the most common classification method of pigmented neoplasms is visual diagnostics, creating systems based on CNN will increase the efficiency and speed of diagnosis and start treatment at an earlier stage [10].

To date, intensive developments are underway in creating neural network classification systems for pigmented skin lesions. In [11], the authors perform a comparative analysis of the efficiency of classifying dermatoscopic images in dermatologists with different levels of experience and computer programs based on CNN. Pigmented skin lesions were classified as benign or malignant. As a result, CNN's algorithm was able to outperform 136 out of 157 dermatologists. The accuracy of the system based on artificial neural networks was 76.00%. At the same time, the result of the expert diagnosis accuracy was 67.70%. The result of the study allows us to conclude that the use of systems based on artificial intelligence as an additional diagnostic tool will enable specialists to increase the efficiency in detecting skin cancer.

In [12], the authors presented an automated mobile system for classifying skin lesions based on deep neural networks, making it possible to achieve an accuracy of 66.70%. The use of mobile systems by patients with the supervision of specialists will allow monitoring the development of pigmented neoplasms in real time. Routine mobile diagnostics will make it possible to detect a malignant neoplasm at the initial stages and begin treatment on time, thereby reducing the lethality of skin cancer.

Despite the significant results in intelligent analysis of pigmented skin lesions, the problem of the low level of accuracy of recognition and classification systems remains. A possible solution to this problem is the development of various systems with stages of preliminary processing of diagnostically significant data and the use of existing high-precision pre-trained architectures.

Today, there is potential for creating artificial intelligence systems in dermatology that will quickly and accurately process huge amounts of data. Such systems allow medical specialists to improve the accuracy of visual diagnosis. The development of neural network systems will lay the foundation for highly effective medicine based on diagnostic and statistical data and reduce dependence on human resources.

## 2 Classification System Proposed for Dermatoscopic Images

The neural network includes neurons with activation function $\alpha$ and parameters $\theta = \{W, B\}$, $W$ is a vector of weights, and $B$ is a vector of displacements. Activation is a linear combination of input $x$ to the neuron and parameters followed by a transfer function $\sigma$.

$$\alpha = \sigma(WTx + b), \tag{1}$$

In artificial intelligence, the technology of convolutional networks is recognized as the most optimal for pattern recognition [13]. CNN is used today as the primary tool for the mining and classification of multidimensional visual data. CNN technology has also found applications in dermatology since dermatoscopic images are the most common type of data in diagnosing pigmented skin lesions. CNN is a deep learning technique in which trainable filters and merge operations are applied to raw input images, automatically extracting a set of complex high-level functions [14]. This makes it possible to achieve sufficiently high accuracy results in comparison with visual diagnostics by specialists. The difference between this type of neural network lies in the alternate use of convolutional layers and sampling layers.

The CNN architecture is shown in Fig. 1. The CNN's include input and output layers, several hidden layers represented by convolutional layers, sampling layers, and a fully connected classifier. Each convolutional layer creates a new image from the original image, known as a feature map. The convolutional layer contains small, square-shaped convolutional filters that are applied to the input image. The size of the filter used varies and depends on different types of applications. After the convolutional layer, there is a batch normalization layer that normalizes the output of the previous convolutional layer. An activation function such as a rectified linear unit (ReLU) or a sigmoid function is used at the end of the convolutional layer. The concatenation layer is used to reduce the size of a feature map created from a convolutional layer. The filter shift step is known as a stride. At the last stage of the CNN, one or more fully connected layers are present, and the classification level is soft-max.

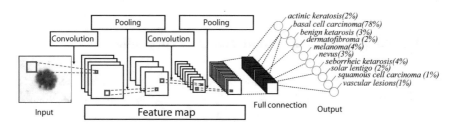

**Fig. 1** Neural network classification system for pigmented skin lesions based on dermatoscopic images

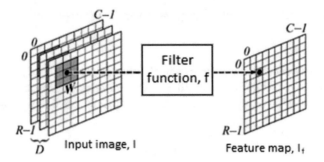

**Fig. 2** Procedure for obtaining a feature map in a convolutional layer

Suppose the CNN input receives a color image I. This image includes R rows, C columns, and D color components. In this case, for the RGB format $D = 3$, since the color components are represented by the image pixels' levels of red, green, and blue colors. The input is a three-dimensional function $I(x, y, z)$, where $0 \leq x < R$, $0 \leq y < C$ and $0 \leq z < D$ are spatial coordinates, and the amplitude I at any point with coordinates $(x, y, z)$ – is the intensity of the pixels at a given moment. The procedure for obtaining feature maps in a convolutional layer is

$$I_f(x, y) = b + \sum_{i=\frac{-w-1}{2}}^{\frac{w-1}{2}} \sum_{j=\frac{-w-1}{2}}^{\frac{w-1}{2}} \sum_{k=0}^{D-1} W_{i,j,k} I(x + i, y + j, K), \qquad (2)$$

where $I_f$ is a feature map; $W_{i,j,k}$ - coefficient of a 3D filter of size $w \times w$ for processing D two-dimensional arrays; $b$ - offset [15]. This procedure is shown in Fig. 2.

The activation of the last network layer is displayed through the *softmax* function with the distribution:

$$P(y|x; \theta) = softmax(x; \theta) = \frac{e^{(W_i^n)^T x + b_i^n}}{\sum_{k=1}^{K} e^{(W_k^n)^T x + b_k^n}}, \qquad (3)$$

where $W_i^n$ is the weight vector leading to the output node associated with the $i$ class.

## 3   Modeling of Dermatoscopic Images Processing

To simulate a neural network system for recognizing pigmented skin lesions, we used a database of clinical dermatoscopic images from the open archive of the International Cooperation in Skin Imaging (ISIC) project [16]. The ISIC archive is the largest confidential dermatological data set available with confirmed diagnoses. The ISIC archive contains 41,725 dermatoscopic photographs of various sizes, which represent a database of digital representative images of the 10 most important diagnostic categories in the field of pigmented skin lesions (Table 1).

In the course of the work, four neural network architectures were trained: the SqueezeNet neural network presented in [17], the AlexNet neural network [18], the GoogLeNet neural network [19], and the ResNet101 neural network [5]. All neural network architectures were pre-trained on the ImageNet natural image set. The selected neural network architectures are small in size and do not require extensive computing resources. This fact is associated with the potential for further implementation of the proposed system as a diagnostic tool for specialists. The simulation was carried out using the MATLAB R2020b software package for solving technical calculations. The calculations were performed on a PC with an Intel (R) Core (TM) i5-3230 M CPU @ 2.60 GHz with 8.0 GB of RAM and a 64-bit Windows 7 operating system.

The main part of the selected images has a size of $400 \times 600$ pixels. Most of the photographs are digitized transparencies of the Roffendal Skin Cancer Clinic in

**Table 1** Image categories for modeling neural network classification of pigmented skin neoplasms

| № | International name of the category in ISIC base (number of items in the base) | Sample image | № | International name of the category in ISIC base (number of items in the base) | Sample image |
|---|---|---|---|---|---|
| 1 | Actinic keratosis (869) | | 6 | Nevus (27878) | |
| 2 | Basal cell carcinoma (3393) | | 7 | Seborrheic keratosis (1464) | |
| 3 | Benign ketosis (1099) | | 8 | Solar lentigo (270) | |
| 4 | Dermatofibroma (246) | | 9 | Squamous cell carcinoma (656) | |
| 5 | Melanoma (5597) | | 10 | Vascular lesions (253) | |

**Table 2** Information about the neural network architectures used to recognize pigmented lesions in the skin

| Network | Depth | Size (MB) | Weights (mln) | Size of input images |
| --- | --- | --- | --- | --- |
| SqueezeNet | 18 | 5.20 | 1.24 | 227 × 227 |
| GoogLeNet | 22 | 27.00 | 7.00 | 224 × 224 |
| ResNet-101 | 101 | 167.00 | 44.60 | 224 × 224 |
| AlexNet | 8 | 227.00 | 61.00 | 227 × 227 |

Queensland, Australia, and the Department of Dermatology at the Medical University of Vienna, Austria [20]. For neural network architectures AlexNet and SqueezeNet, images were converted to 227 × 227 × 3. For CNN ResNet-101 and GoogLeNet, images were converted to 224 × 224 × 3. For further modeling, the dermatoscopic photo base was divided into images for training and images for checks in a percentage ratio of 80 to 20. For the training base, along with the pictures, labels with diagnoses were submitted. Testing was carried out on a base without labels. The predicted results obtained were compared with reliable diagnosis marks, and the system recognition accuracy and loss function were deduced.

Table 2 shows the depth of the neural network architectures used for training and the required sizes of the input images. The deepest of the selected architectures was ResNet-101, which contains a total of 101 layers. Learning this architecture was the most time-consuming. Despite its shallow depth, AlexNet has the largest number of options and also the largest architecture weight.

In the selected CNN architectures, the last layers were replaced to adjust to recognize pigmented skin lesions in 10 diagnostically significant categories. The original layers *loss3-classifier*, *prob*, and *output* in AlexNet and GoogLeNet architectures contained information on combining network-extracted features, class probabilities, loss values, and predicted labels across 1000 categories from the ImageNet natural image set. Instead, a *fully connected* layer, a *softmax* layer, and an *output* layer were added, adapted to the new dataset. In the SqueezeNet architecture, the *conv10* and *ClassificationLayer* layers were replaced, representing the final 1 × 1 × 512 convolution layer and the 1000 category classification layer. Instead, a *convolutional* layer with the number of filters equal to 10 classes and an *output* layer was added. During training, the parameters of the main layers of CNN architectures were not updated.

All parameters used in training the neural network are presented in Table 3. Each neural network architecture accepted a batch size of 128 dermatoscopic images as input. All neural network architectures were trained for 10 epochs. At the same time, each epoch was made mixing training and test data before validating the network. The CNN learning process is presented in the form of graphs in Figs. 3 and 4.

The SqueezeNet CNN was trained with a recognition accuracy of 78.22%, the AlexNet architecture achieved 80.15%, the GoogLeNet recognition accuracy was 77.32%, and ResNet101 allowed the accuracy of 74.66%.

**Table 3** Parameters and their values used in training a deep convolutional neural network

| Option | Value | Explanation |
|---|---|---|
| InitialLearnRate | 0.01 | Initial learning rate |
| MaxEpochs | 10 | Maximum number of epochs used for training |
| LearnRateDroperiod | 5 | The number of epochs to reduce the learning rate |
| LearnRateDropFactor | 0.3 | Factor for shedding learning rate |
| Shuffle | Every-epoch | Option for data shuffling. Shuffles training data before each training epoch and shuffles validation data before every network validation |
| MiniBatchSize | 128 | Mini-batch size to use in each training iteration |
| ValidationData | imdsValidation | Data to be used for validation during training |
| ValidationFrequency | 50 | Network check frequency in number of iterations |
| Plots | Training-progress | Graphs to display during network training |

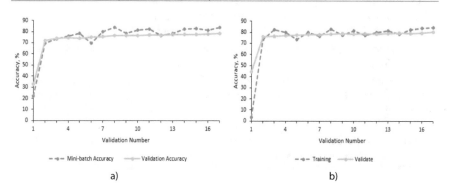

**Fig. 3** Learning outcomes graphs of neural network architectures for recognizing pigmented skin lesions based on CNN: **a** SqueezeNet; **b** AlexNet

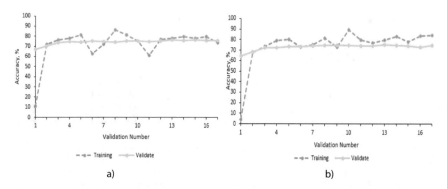

**Fig. 4** Learning outcomes graphs of neural network architectures for recognizing pigmented skin lesions based on CNN: **a** GoogLeNet; **b** ResNet101

# 4   Conclusion

Four neural network architectures were trained to carry out automated classification and diagnosis of pigmented skin neoplasms in work. The highest accuracy rate was achieved using AlexNet and amounted to 80.15%. This accuracy indicator is higher than the accuracy of the mobile neural network system from [12] by about 14%. Also, the accuracy rate of the AlexNet architecture surpassed the accuracy of the neural network system [11] by about 4.15%. Based on the data on the accuracy of the visual diagnosis of dermatologists in work [11], the use of the proposed system will significantly increase the efficiency of diagnosis. The proposed neural network system for the recognition and classification of dermatoscopic images of pigmented skin neoplasms will significantly reduce the effect of a specialist's experience and minimize the influence of the human factor. Timely detection of a malignant skin lesion will allow starting treatment at an earlier stage of the disease, which directly affects the percentage of survival and recovery of patients.

The main limitation in using the proposed system for detecting pigmented skin lesions is that the system is not a medical device and cannot independently diagnose patients. Specialists can only use the system as an additional diagnostic tool. A promising area of further research is constructing more complex systems for neural network classification of pigmented skin neoplasms with preliminary visual data processing.

**Acknowledgements** Section 2 was supported by Russian Science Foundation, project 21-71-00017. The rest of the paper was supported by North-Caucasus Center for Mathematical Research under agreement №. 075-02-2021-1749 with the Ministry of Science and Higher Education of the Russian Federation.

# References

1. Rogers HW (2010) Incidence estimate of nonmelanoma skin cancer in the United States. Arch Dermatol 146(3):283–287
2. Haenssle N, Fink C, Schneiderbauer R, Toberer F, Buhl T, Blum A (2018) Man against machine: diagnostic performance of a deep learning convolutional neural network for dermoscopic melanoma recognition in comparison to 58 dermatologists. Ann Oncol 29(8):1836–1842
3. Stern RS (2010) Prevalence of a history of skin cancer in 2007: results of an incidence-based model. Arch Dermatol 146(3):279–282
4. Siegel RL, Miller KD, Jemal A (2018) Cancer statistics 2018. CA Cancer J Clin 68(1):7–30
5. He K, Zhang X, Ren S, Sun J (2016) Deep residual learning for image recognition. In: Proceedings of the IEEE conference on computer vision and pattern recognition, pp 770–778
6. Wang N, Yeung DY (2013) Learning a deep compact image representation for visual tracking. In: Advances in neural information processing systems, pp 809–817
7. Erhan D, Szegedy C, Toshev A, Anguelov D (2014) Scalable object detection using deep neural networks. In: Proceedings of the IEEE conference on computer vision and pattern recognition, pp 2147–2154
8. Ciresan D, Meier U, Schmidhuber J (2012) Multi-column deep neural networks for image classification. IEEE Conf Comput Vis Pattern Recogn 2012:3642–3649

9. Esteva A, Kuprel B, Novoa RA, Ko J, Swetter SM, Blau HM (2017) Dermatologist-level classification of skin cancer with deep neural networks. Nature 542(7639):115–118

10. Korotkov K, Garcia R (2012) Computerized analysis of pigmented skin lesions: a review. Artif Intell Med 56(2):69–90

11. Brinker TJ, Hekler A, Enk AH, Klode J, Hauschild A, Berking C, Schrüfer P (2019) Deep learning outperformed 136 of 157 dermatologists in a head-to-head dermoscopic melanoma image classification task. Eur J Cancer 113:47–54

12. Ramlakhan K, Shang Y (2011) A mobile automated skin lesion classification. IEEE 23rd international conference on tools with artificial intelligence, pp 138–141

13. Zhang J, Shao K, Luo X (2018) Small sample image recognition using improved convolutional neural network. J Vis Commun Image Repres 55:640–647

14. Shuiwang J, Wei X, Ming Y, Kai Y (2013) 3D Convolutional neural networks for human action recognition. IEEE Trans Pattern Anal Mach Intell 35(1):221–231

15. Chervyakov NI, Lyakhov PA, Valueva MV (2017) Increasing of convolutional neural network performance using residue number system. International multi-conference on engineering, computer and information sciences (SIBIRCON), pp 135–140

16. ISIC Melanoma Project (2020) https://www.isic-archive.com/

17. Iandola FN (2016) SqueezeNet: AlexNet-level accuracy with 50x fewer parameters and <0.5 MB model size. https://arxiv.org/abs/1602.07360

18. Krizhevsky A, Sutskever I, Hinton GE (2012) Imagenet classification with deep convolutional neural networks. In: Advances in neural information processing systems, pp 1097–1105

19. Szegedy C (2015) Going deeper with convolutions. In: Proceedings of the IEEE conference on computer vision and pattern recognition, pp 1–9

20. Tschandl P, Rosendahl C, Kittler H (2018) The HAM10000 dataset, a large collection of multi-source dermatoscopic images of common pigmented skin lesions. Sci Data 5(1):1–9

# Decentralized Role-Based Secure Management in Wireless Sensor Networks

**Vasily Desnitsky**◉

**Abstract** The paper encompasses an approach to construction of a decentralized secure management mechanism in wireless sensor networks. The decentralization relies on role-based management models. The particular roles are specified on a management layer of the wireless sensor network. The core features of such network operation are exposed. The approach is targeted on a secure distributed modular computing and allows the use of data analysis methods to intellectualize and improve the security of the services provided. Moreover, each role used by one or more nodes defines a specific set of application functions and functions for data collection, their processing and control of the wireless sensor network. The approach is realized in the shape of a multiparty decentralized protocol. The most crucial kinds of malicious influences towards the defeat of the protocol as well as some protection techniques against them are disclosed in the paper. The novelty of the work includes introduction of role-based application level decentralization mechanisms against attacks exploiting network decentralization vulnerabilities.

**Keywords** Wireless sensor network · Security · Role-based management · Decentralized modular computing

## 1 Introduction

At present wireless sensor networks (WSN) are becoming more and more widespread, aimed at performing functions of monitoring critical characteristics of the environment, technical objects, constructions and human. The distributed dynamic nature of such systems will determine the need for decentralized interaction of the WSN nodes with the implementation of specific application functions by each of the nodes. The need for secure decentralized management in such networks

V. Desnitsky (✉)
St. Petersburg Federal Research Center of the Russian Academy
of Sciences (SPC RAS), St. Petersburg, Russia
e-mail: desnitsky@comsec.spb.ru

© The Author(s), under exclusive license to Springer Nature Switzerland AG 2022          51
A. Tchernykh et al. (eds.), *Mathematics and its Applications in New Computer Systems*,
Lecture Notes in Networks and Systems 424,
https://doi.org/10.1007/978-3-030-97020-8_6

causes the construction of network protocols that can ensure, first, the correct and consistent interaction of heterogeneous devices, i.e. network nodes, and, second, the proper level of protection against attacks aimed at disrupting the operation of the protocol.

Currently the presentation of wireless networks in most cases is reduced to two-tier structures. According to such models at the lower level primary data is collected, and sent directly or indirectly, through other nodes, to base stations located at the upper level. The data are used for its further processing and analysis within the framework of centralized approaches to the organization of computing systems [2].

The absence and complexity of deploying security tools, such as authentication means and trust mechanisms [6], which are widely used in traditional software and hardware systems, networks and clusters consisting of general-purpose computers, are also a deterrent to the formation of a WSN with more complex structures and network topologies. Nevertheless, there is a tendency that modern network WSN protocols implement a certain degree of protection at the level of routing protocols [11], by using security policies [7] or at the level of transforming data generated by the nodes [8]. At the same time, the vulnerability of the protocol at the application level remains a task specific to a particular scenario and firmware. And, as a rule, such issues are resolved ad-hoc and centrally, taking into account the available initial data and network limitations.

With regard to role-based security control mechanisms in WSN, the issues related to the adaptation and implementation of the RBAC (role-based access control) model [3] for organizing secure access to network nodes and its services by using role mechanisms are distinguished [4, 12]. The formation of a system of roles is possible both on the basis of application features and the purpose of certain types of nodes, and taking into account reputation mechanisms [5]. In addition to organizing secure access in WSN, secure routing tools are also implemented on the base of role mechanisms [9]. Particular role-based and hierarchical routing protocols are proposed, and the increase in security occurs due to the introduction of a few clusters of devices, i.e. roles, communicating with different types of cryptographic keys [10].

In this paper we propose an approach to the development of a protocol for decentralized interaction of WSN nodes for solving problems of monitoring critical parameters of the environment. The new, peculiar features of the approach that distinguish it from alternative solutions in the field include the application of the principles of decentralization at the level of the application protocol and calculations based on the role-based mechanism, taking into account the current types of attacks that exploit the properties of network decentralization.

## 2   Materials and Methods

The developed protocol for decentralized interaction of WSN nodes corresponds to the application level of the network presentation model. The protocol provides the functions of forming a list of free network nodes, linking them on the base of roles

**Table 1** WSN node roles—data collector and data hoarder

| Role | Characteristic |
|------|----------------|
| Data collector | It collects data from sensors connected to the node, various peripheral equipment associated with the node, and user interface elements. Also the following functions are imposed on the data collection nodes, namely nodal data preprocessing, filtering and normalization, which make it possible to reduce the amount of initial data, remove unnecessary and duplicated ones and transform them to some unified form |
| Data hoarder | A node including a network data storage module. Data is sent here from all collector nodes and stored. Besides nodes with the roles of a data processor and analyzer also gain access to this storage. Those nodes are able to request data from this storage and add new data to it, i.e. the results of their own work. In addition to providing an interface for storing data from the nodes, a hoarder node can also store its own data in a single repository. Besides that, the collector continuously monitors the integrity of data structures, as well as controls the filling of the storage volumes |

and assigning tasks for them, depending on the application goals of such a network. It is assumed that the functions of self-organization of the network in the context of addressing and establishment of physical wireless communication channels rely entirely upon the lower-level protocols [1]. In addition, although the control functions and low power consumption in a WSN are mainly provided within the underlying layers of the network model, the control functions located at the application protocol layer can also affect the power consumption directly.

Note that the considered application protocol is focused on a specific network application and some expected types of scenarios for the operation of network nodes. Changes in the planned nature of the operation of the network should be taken into account at the level of this protocol, and if necessary, such a protocol should be subjected to an adjustment stage.

The proposed and underlying system of roles for network nodes in the protocol includes the following main roles: data collector, data hoarder, data processor, data analyzer and network controller. Each of the network nodes can assume one or a few roles at the same time, performing the functions of each role in parallel or sequentially. If a node does not have any role, nevertheless it does not leave the network, but it stays in a reserve. Tables 1 and 2 shows the main roles of WSN nodes and their characteristics.

In fact, the collector node is an element of centralization, although not a completely, but partially decentralized network. It is necessary for the smooth and consistent operation of network nodes. At the same time, it is possible to dynamically replace it with some other node, either planned or situational if the current controller is unable to continue the correct execution of its functions. The specified list of roles is not exhaustive and can be expanded. Individual roles can be specified depending on the specificity of a particular network and the tasks assigned to it.

**Table 2** WSN node roles—data processor, data analyzer and network controller

| Role | Characteristic |
|------|----------------|
| Data processor | The node's functions include data processing and, in particular, aggregation of data, performed by taking into account the availability of data from sets of network nodes, as well as historical data for certain periods of time. The data processor receives data from the hoarder and, after the completion of the execution process or during its process, initiates sending the processed data to the hoarder |
| Data analyzer | A node that analyzes the data received from the hoarder and conducts data mining based on rules, statistical analysis, analysis based on heuristics, analysis using machine learning methods and artificial neural networks. In particular, analysis of security incidents and monitoring of network security are performed. The results of this analysis are sent to the hoarder node |
| Network controller | The controlling node is responsible for making decisions about reorganizing the network, changing the distribution of roles based on its own network management model and service messages from other nodes. In particular, the decision to replace a node with a certain role or allocate additional nodes may be made as a result of an overflow of the hoarder's permanent memory or the need to use additional nodes for cooperative analysis of incidents. In addition, the hoarder is responsible for assessing and tracking the current resource consumption indicators of nodes of each role (including the computational load of the node and the used amount of its outgoing and incoming network bandwidth). Unlike other roles, always there is only one network controller node in a network. Its initial selection is made by multiple handshake of all nodes located within the existing network context, having the same unique network identifier, i.e. PAN ID, and configured to form an application network |

Note that a seamless (although probably with some time delays) network reconfiguration is possible with the replacement of the roles of one or more nodes. Seamless is understood as such a WSN reconfiguration, which there is no loss of current or previously accumulated data in, and there are no interruptions to the application or service functions of the network.

In the simplest case, a WSN can include several data collector nodes and one node each with the roles of collector, processor, analyzer and network controller. The impossibility of one of the nodes with a certain role to fully meet all the needs of this role, including the limited software and hardware functions of one network node, makes it possible to increase the number of nodes with a certain role to increase the achievement of targets related to the given role. Table 3 exposes the results of an analysis of the possibilities for the distribution of functions of each role among two or more nodes.

**Table 3** Possibilities for the functions distribution

| Role | Possibilities for the functions distribution |
|---|---|
| Data collector | The most natural scenario seems to be the delegation of the role of the collector to all or almost all WSN nodes that have any sensors. In general, the collection of data from the maximum possible number of network nodes contributes to an increase in the completeness of the collected data about the physical and software/hardware environment of the node and the objects in it |
| Data hoarder | The distribution of the data storage allows, first, to increase the volume of stored data and, second, to raise the speed of access to data from various network nodes, as well as to increase the reliability and availability of data storage and ensure their backup |
| Data processor | It is possible to increase the number of devices with the processor role to organize decentralized and modular computing. Parallelization of the computation process will increase the speed of execution. In particular, it turns out to be possible to increase the processing speed due to group processing of data and various optimizations of the computation processes due to multiprocessor computational models |
| Data analyzer | Collaborative computing is possible, including the use of federated learning methods to operate complex composite data analysis models at network nodes, as well as information security monitoring and security analysis functions |
| Network controller | It is assumed that there is a single node with this role and it performs the assignment, distribution of roles in the network and provides the functions of configuring and monitoring the network. At the same time, the network controller is indirectly responsible for ensuring the decentralization of the entire network in the form of assignment and redistribution of roles, as well as ensuring synchronization in the operation of a set of nodes with a certain role. If a set of devices with a certain role reaches a definite threshold value of resource consumption, the network controller can announce the need to expand the number of devices with this role |

## 3 Results

The proposed protocol for decentralized interaction of WSN nodes includes the following stages.

1. Stage of formation of a decentralized network. At this stage, on the basis of a physically formed wireless network, a decentralized network is constructed at a logical level with the choice of a controller node and assigning roles to other network nodes.

2. Stage of network functioning. This stage includes the implementation of application functions of the network, as well as its rebuilding, including an increase or decrease in the number of network nodes.

3. Network reconfiguration. This stage initiates the process of redistributing roles among network nodes, depending on the current program/network context and the availability of network nodes.

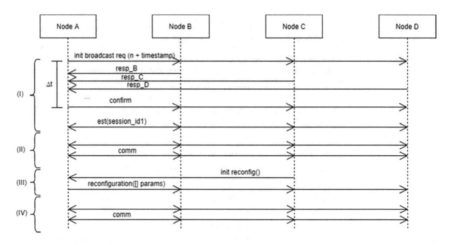

**Fig. 1** Decentralized communication protocol scheme

The Fig. 1 discloses a generalized diagram of a simple version of the protocol for decentralized interaction of WSN nodes in the form of a UML sequence diagram. Participants of information exchange (nodes A–D), which send messages to each other are marked along the vertical axis. The main communications are summarized by arrows between the participants in the information exchange. In phase (I), node A initiates the message exchange by generating and sending a broadcast request containing an arbitrary natural number n and a timestamp. By default, host A takes the role of the network controller. During a set time interval $\Delta t$, node A waits for responses from nodes that are ready to connect to the formed decentralized network. If, within the interval $\Delta t$, some other node also initiates the creation of a network, all other participants will identify a later initiation time by its time stamp.

As a result, such requests will be ignored until a confirm message is received from node A or a control time period has passed, after which the request for initial initiation from node A is considered outdated. The establishment of a session ends with the distribution of roles based on their direct assignment by the controller node, i.e. by sending a series of commands of the form *est(session_id1)*.

In essence, the session being formed is multilateral, i.e. multicast communication is performed between all participants in the information exchange. Being a key element of a decentralized network, and providing a non pre-determined distribution of application roles among network nodes, the network controller node, the initiator of the network, nevertheless, represents a certain element of centralization of this network architecture. Within the framework of a simple version of the protocol, the controller's leaving the network destroys the functioning of the decentralized network, transferring its nodes into a state of physical connectivity by wireless data channels, but without any logical role-based superstructure. In this case, in the event of a controller node disconnecting or its transition to an inoperative state, both as a result of deliberate attacks and any software/hardware failures, the set of WSN nodes, in fact, returns to the state preceding the establishment of a multilateral session.

In normal operation, after the completion of the session establishment, in phase (II), the exchange of application data between the network nodes is conducted. Packages of the current session are characterized by the presence of a common identifier shared by all participants. Phase (III) is optional and occurs when it becomes necessary to change the distribution of roles between network nodes. Such a change may be related to the notification of the controller node that a certain node with a specified role signals its intention to terminate its functioning within the cscope of the decentralized network. With a regular redistribution of the role, the existing program context and the necessary data are sent from the node that terminates its work to some new node.

Note that the reconfiguration of the network can also be caused by a spontaneous termination of the functioning of some node of a given role. In this case, in the absence of an up-to-date copy of the program context of the left node in the network, such a context has to be formed on a new node. As consequences it can affect the application services provided by the network as well as lead to the loss of part or all of the historical data that could later be used in the analysis of network security. In addition, it can also lead to a temporary decrease in the precision or recall of monitoring data for the environment of the WSN. Similarly to phase (II), phase (IV) includes the exchange of application and service data as part of the normal operation of the network.

## 4 Discussion and Conclusion

As elements of security analysis of the proposed protocol, consider some possible actions of an intruder aimed at disrupting the correct functioning and compromising this protocol. Type 1 attack is a phase (I) DoS (Denial-of-Service) attack that prevents the controller node from completing the initiation of a multiparty session. By changing the timestamp to an earlier one, the attacking node is able to generate its own request to create a session on each initiation, but without the further confirmation (Fig. 1). An alternative option of the intruder would be to perform a replay attack that reproduces a record of some previous session request from some legitimate node. As means of increasing protection against this type of attack, one can apply time control using hardware certification tools such as TPM and implement node reputation mechanisms to identify participants with a large number of unsuccessfully established sessions.

To increase protection against attacks of type 1, it is also possible to introduce forced delays before re-attempting to initiate a multilateral session from a certain node. This circumstance can be taken into account by the attacker for sequential imitation of incorrect session initiations from all legitimate nodes. This is possible by substituting the outgoing addresses of network nodes in the traffic, which in turn will cause delays for all participants in the information exchange. To prevent such situations, it will be necessary to strengthen the means of nodes authentication on the base of the analysis of their structural and behavioral characteristics.

As an attack of type 2, consider an attack of unauthorized change of the node's role. The attack can be fulfilled in phases (II) or (IV). The attack includes, first, intercepting the role identifier of another node, second, redefining the role on oneself and setting the context, third, blocking the victim's node, for example, by remotely setting incorrect addressing parameters, and, fourth, functioning under a new role. The purpose of such an attack can be the modification of the application functions of this role, for example, the distortion of data stored within the hoarder node or the substitution of the results of data processing and analysis. To increase the security against this type of attacks, it is advisable to use means of control of the immutability of the node carrying this role within the framework of the reconfiguration phase and the subsequent phase of operation. Particularly methods of remote attestation could be applied. As the further research, it is planned to develop means of formal verification of the developed protocol as a tool for checking its correctness and security.

# References

1. Desnitsky VA, Kotenko IV (2018) Security event analysis in XBee-based wireless mesh networks. In: Proceedings of ElConRus' 2018: The 2018 IEEE conference of Russian young researchers in electrical and electronic engineering, pp 42–44, St. Petersburg, Russia
2. Kodali RK, Narasimha Sarma NVS (2013) Experimental WSN setup using XMesh networking protocol. In: Proceedings of ICAES' 2013: international conference on advanced electronic systems, pp 267–271
3. Maerien, J, Michiels S, Huygens C, Hughes D, Joosen W (2013) Access control in multi-party wireless sensor networks. In: Proceedings of: EWSN' 2013: 10th European conference wireless sensor networks. Lecture notes in computer science, vol 7772. Springer, Heidelberg
4. Maw HA, Xiao H, Christianson B, Malcolm JA (2014) A survey of access control models in wireless sensor networks. J Sens Actuator Netw 3:150–180
5. Misra S, Vaish A (2011) Reputation-based role assignment for role-based access control in wireless sensor networks. Comput Commun 34(3):281–294
6. Moinet A, Darties B, Baril J-L (2017) Blockchain based trust & authentication for decentralized sensor networks. arXiv: 1706.01730
7. Ndia JA (2017) Survey of WSN security protocols. Int J Appl Comput Sci (IJACS), 1(2), 1–11 (2017)
8. Olakanmi O, Dada A (2020) Wireless Sensor Networks (WSNs): security and privacy issues and solutions. Wireless mesh networks - security, architectures and protocols, pp 1–17. IntechOpen, UK
9. Sarma HKD, Kar A, Mall R (2013) Role based secure routing in large wireless sensor networks. Int J Innov Manage Technol 4(1):51–55
10. Sarma HKD, Kar A, Mall R (2016) A hierarchical and role based secure routing protocol for mobile wireless sensor networks. Int J Wirel Pers Commun 90(3):1067–1103
11. Senthilkumar A, Chandrasekar C (2010) Secure routing in wireless sensor networks: routing protocols. Int J Comput Sci Eng 2(4):1266–1270
12. Singh AK, Alshehri M, Bhushan S, Kumar M, Alfarraj O, Pardarshani KR (2021) Machine learning in detecting Schizophrenia: an overview. Intell Autom Soft Comput 27(3):761–769

# Numerical Study of Longitudinal-Transverse Vibrations of Objects with a Moving Boundary

Valeriy N. Anisimov⬛, Inna V. Korpen, and Vladislav L. Litvinov⬛

**Abstract** Difference numerical schemes for solutions of problems described the longitudinal-cross oscillations of object with moving boundaries is noted. The scheme allows to solve the Cauchy problem for a system of nonlinear differential equations with nonlinear boundary conditions and to take into account the energy exchange between the parts of the vibrating object on the left and right of the moving boundary. Grid is divided into equally spaced time layers by the time variable. Grid is divided in to a fixed number of parts equidistant nodes by the space variable in each time step to the left and right of the moving boundary. Partitioning step in temporary layers are different in connection with the movement of the boundary. Such a partition avoids the transition moving boundary through the nodes of the grid. To find the functions and their derivatives are used finite difference approximation. Approximation error is of second order of smallness relative to the grid spacing on the space and time variables. The solution obtained by successive transition from one time to another layer. The accuracy of the numerical solution is confirmed by the coincidence of the solutions of the linear and nonlinear models at low vibration amplitudes.

**Keywords** Longitudinal-cross oscillations of object with moving boundaries · Nonlinear system of partial derivative · Numerical methods for solving mathematical physics

**MSC** MSC2010 · 35R37 · 35G30 · 35Q70

V. N. Anisimov · I. V. Korpen · V. L. Litvinov (✉)
Department of General-Theoretical Disciplines, Syzran' Branch of Samara State Technical University, 45, Sovetskaya Street, Syzran', Samara Region 446001, Russian Federation
e-mail: vladlitvinov@rambler.ru

V. L. Litvinov
Moscow State University, GSP-1, Leninskie Gory, Moscow 119991, Russian Federation

59

# 1 Introduction

Among all the many problems of the dynamics of elastic systems from the point of view of technical applications, the problems of oscillations in systems with time-varying geometric dimensions are highly relevant. Studies of many authors on the dynamics of hoisting ropes have led to the need to formulate new problems in mechanics concerning the dynamics of one-dimensional objects of variable length [1–12]. In a mathematical setting, this is reduced to new problems in mathematical physics - to the study of the corresponding equations of hyperbolic type in variable ranges of variation of both arguments [25].

Until now, there is no general approach to the formulation of such problems, and the authors in each specific case adapt the existing methods to solve the problem under consideration [1–9]. Here we note that the methods for solving these equations in variable geometric domains are qualitatively different from the classical methods of mathematical physics. For example, for vibrations of strings of variable length, the concepts of eigenfrequencies and phases, that is, eigenvalues and eigenfunctions, lose their usual meaning, since the vibrational frequencies of a string of variable length will be some functions of time. The independence of individual vibration tones is lost. In other words, the studied dynamic process develops over time.

The problems of oscillation of systems with moving boundaries have been solved mainly with a linear setting and rigid fixation of boundaries, when there is no energy exchange across the boundary [1–15, 19–24]. In rare cases, the effect of damping forces was taken into account. Real technical objects are much more complicated.

In connection with the intensive development of numerical methods, it became possible to describe such objects more accurately, taking into account a large number of factors. In [17], a nonlinear formulation of problems describing the longitudinal-transverse vibrations of objects with moving boundaries was made. This article describes a difference scheme for the numerical solution of such problems.

# 2 Formulation of the problem

The formulation of the problem of longitudinal-transverse vibrations of objects with moving boundaries was made in [17]. The oscillation region is shown in Fig. 1. The figure indicates: $x$ - spatial coordinate; $t$—is time; $L(t)$—the law of movement of the border; $i$, $j$—numbers of grid nodes; $H_t$—grid step in $t$ variable.

Let's write the problem in a generalized form.

A system of two nonlinear partial differential equations:

$$\begin{cases} u_{1,tt} = F_2\big(x, t, u_k, u_{k,x}, u_{k,xx}, u_{k,t}, u_{k,xt}, u_{k,xxt}\big); \\ u_{2,tt} = F_2\big(x, t, u_k, u_{k,x}, u_{k,xx}, u_{k,xxxx}, u_{k,t}, u_{k,xt}, u_{k,xxt}, u_{k,xxxxt}\big); \\ \qquad\qquad k = \overline{1, 2}. \end{cases} \quad (1)$$

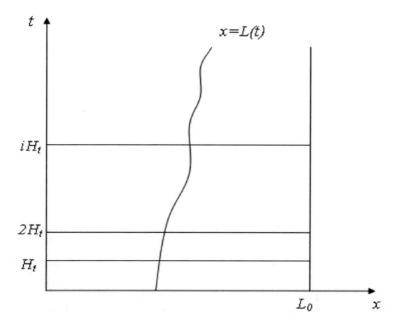

**Fig. 1** Longitudinal-transverse oscillation region

Here and below, $k = \overline{1, 2}$.

At the fixed ends ($x = 0, x = L_0$) we take the boundary conditions in the form:

$$u_1(0, t) = 0; \ u_1(L_0, t) = 0; \ u_2(0, t) = 0; \ u_{2,xx}(0, t) = 0; \ u_2(L_0, t) = 0; \ u_{2,x}(L_0, t) = 0. \tag{2}$$

Let us write down the conditions on the moving boundary in a generalized form:

$$\frac{d^2 u_1(L(t), t)}{dt^2} = f_1\big(x, t, u_k(L(t), t), u_{k,x}(L(t) \mp 0, t), u_{k,t}(L(t), t), u_{k,xt}(L(t) \mp 0, t)\big); \tag{3}$$

$$\frac{d^2 u_2(L(t), t)}{dt^2} = f_2\big(x, t, u_2(L(t), t), u_{2,xxx}(L(t) \mp 0, t), u_{2,xxxt}(L(t) \mp 0, t)\big); \tag{4}$$

$$\frac{d^2 u_{2,x}(L(t), t)}{dt^2} = f_3\big(x, t, u_{2,x}(L(t), t), u_{2,xx}(L(t) \mp 0, t), u_{2,xxt}(L(t) \mp 0, t)\big). \tag{5}$$

Moving border ratios:

$$u_k(L(t) - 0, t) = u_k(L(t) + 0, t); \ u_{2,x}(L(t) - 0, t) = u_{2,x}(L(t) + 0, t). \tag{6}$$

Initial conditions:

$$u_1(x, 0) = \varphi_1(x); \ u_2(x, 0) = \varphi_2(x); \ u_{1,t}(x, 0) = \varphi_3(x); \ u_{2,t}(x, 0) = \varphi_4(x). \quad (7)$$

In the posed problem (1)–(7), the following designations are used: $u_1(x, t)$ and $u_2(x, t)$ - longitudinal and lateral displacement of a point of an object with coordinate $x$ at time $t$;

$u_k(L(t) - 0, t)$ and $u_k(L(t) + 0, t)$ - are the values of functions to the left and right of the moving boundary; $F_1$, $F_2$, $f_m(m = \overline{1, 3})$, $\varphi_n(x)(n = \overline{1, 4})$ - are given class $C^2$ functions.

## 3   Description of the Numerical Method

Along the $t$-axis, the region is divided into layers with a step $H_t$ number $i$ $(i = 0, 1, 2, \ldots \ldots)$ and time values $t_i = i H_t$. Along the $x$-axis, time layers to the left of the moving boundary are divided into $Nl$ parts with a step $Hl_i = L(t_i)/Nl$, and on the right into $Np$ parts with a step $Hp_i = (L_0 - L(t_i))/Np$. The node number along the $x$-axis is denoted by the subscript $j$. Index $j$ to the left of the moving border changes from $0$ to $Nl$, and to the right from $Nl$ to $N$ $(N = Nl + Np)$.

Let us call the numerical scheme a scheme with a variable step in a spatial variable, since in time layers, the steps $Hl_i$ and $Hp_i$ are different. The $x$-values at grid points are determined by the following equalities:

$$\begin{cases} x_{i,j} = Hl_i * j, 0 \le j \le Nl; \\ x_{i,j} = Hl_i * Nl + Hp_i * (j - Nl), Nl \le j \le N. \end{cases}$$

We denote by $u_k(x_{i,j}, t_i)$ the values of the functions at the grid nodes. A fragment of the mesh is shown in Fig. 2.

The values of the sought functions in the layers $t_0$ and $t_1$ are found from the initial conditions (7):

$$u_1(x_{0,j}, t_0) = \varphi_1(x_{0,j}); \ u_2(x_{0,j}, t_0) = \varphi_2(x_{0,j});$$

$$u_1(x_{1,j}, t_1) = \varphi_1(x_{1,j}) + \varphi_3(x_{1,j})H_t; \ u_2(x_{1,j}, t_1) = \varphi_2(x_{1,j}) + \varphi_4(x_{1,j})H_t.$$

The values of the derivatives with respect to $x$ at intermediate points of the layer $t_1$ are found by the formulas:

$$u_{1,x}(x_{1,j}, t_1) = \varphi_1'(x_{1,j}) + \varphi_3'(x_{1,j})H_t; \ u_{1,xx}(x_{1,j}, t_1) = \varphi_1''(x_{1,j}) + \varphi_3''(x_{1,j})H_t;$$

$$u_{2,x}(x_{1,j}, t_1) = \varphi_2'(x_{1,j}) + \varphi_4'(x_{1,j})H_t; \ u_{2,xx}(x_{1,j}, t_1) = \varphi_2''(x_{1,j}) + \varphi_4''(x_{1,j})H_t;$$

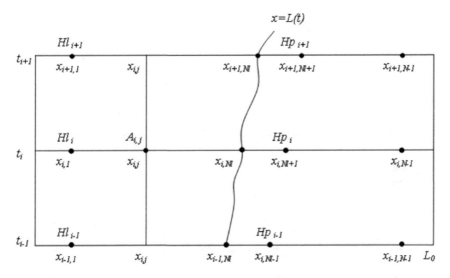

**Fig. 2** Variable pitch scheme

$$u_{2,xxxx}(x_{1,j}, t_1) = \varphi_2''''(x_{1,j}) + \varphi_4''''(x_{1,j})H_t.$$

Here and below, at intermediate points to the left of the moving boundary $j = \overline{1, NL - 1}$, and on the right $j = \overline{NP + 1, N - 1}$.

The values of the derivatives with respect to $t$ at intermediate points of the $t_1$ layer, up to values of the order of $H_t$ are assumed to be equal to the values in the $t_0$ layer:

$$u_{1,t}(x_{1,j}, t_1) = \varphi_3(x_{1,j}); \quad u_{2,t}(x_{1,j}, t_1) = \varphi_4(x_{1,j});$$

$$u_{1,xt}(x_{1,j}, t_1) = \varphi_3'(x_{1,j}); \quad u_{2,xt}(x_{1,j}, t_1) = \varphi_4'(x_{1,j});$$

$$u_{1,xxt}(x_{1,j}, t_1) = \varphi_3''(x_{1,j}); \quad u_{2,xxt}(x_{1,j}, t_1) = \varphi_4''(x_{1,j});$$

$$u_{2,xxxxt}(x_{1,j}, t_1) = \varphi_4''''(x_{1,j}).$$

The values of functions in time layers $t_i$ are found by sequential transition from one layer to another. When finding functions and their derivatives, approximations are used up to terms of the second order of smallness with respect to $Hl_i$, $Hp_i$, $H_t$. In accordance with the boundary conditions (2), the following equalities hold on the fixed boundaries:

$$u_k(x_{i,0}, t_i) = 0; \quad u_k(x_{i,N}, t_i) = 0; \quad u_{k,t}(x_{i,0}, t_i) = 0; \quad u_{k,t}(x_{i,N}, t_i) = 0.$$

The values of functions at interior points are found using the system of differential Eq. (1). In time layers $t_{i-1}$ and $t_i$ the values of functions and their derivatives are known. Using the approximation

$$u_{k,tt}(x_{i,j}, t_i) = \frac{u_k(x_{i,j}, t_{i-1}) - 2u_k(x_{i,j}, t_i) + u_k(x_{i,j}, t_{i+1})}{H_t^2}$$

we get the values of the functions at intermediate points of the layer $t_{i+1}$

$u_1(x_{i,j}, t_{i+1})$
$= F_1(x_{i,j}, t_i, u_k(x_{i,j}, t_i), u_{k,x}(x_{i,j}, t_i), u_{k,xx}(x_{i,j}, t_i), u_{k,t}(x_{i,j}, t_i), u_{k,xt}(x_{i,j}, t_i), u_{k,xxt}(x_{i,j}, t_i))$
$-u_1(x_{i,j}, t_{i-1}) + 2u_1(x_{i,j}, t_i);$
$u_2(x_{i,j}, t_{i+1})$
$= F_2(x_{i,j}, t_i, u_k(x_{i,j}, t_i), u_{k,x}(x_{i,j}, t_i), u_{k,xx}(x_{i,j}, t_i), u_{k,xxxx}(x_{i,j}, t_i), u_{k,t}(x_{i,j}, t_i), u_{k,xt}(x_{i,j}, t_i),$
$u_{k,xxt}(x_{i,j}, t_i), u_{k,xxxxt}(x_{i,j}, t_i)) - u_2(x_{i,j}, t_{i-1}) + 2u_2(x_{i,j}, t_i).$

The values of the functions $u_k(x_{i,j}, t_{i-1})$ are unknown, since nodes in layer $t_{i-1}$ are located at points $x_{i-1,j}$.

To find the functions, we use the approximations:

$$u_k(x_{i,j}, t_{i-1}) = \frac{1}{2H_{i-1}^2}(x_{i,j} - x_{i-1,j})^2(u_k(x_{i-1,j-1}, t_{i-1}) - 2u_k(x_{i-1,j}, t_{i-1}) + u_k(x_{i-1,j+1}, t_{i-1}))$$
$$+\frac{1}{2H_{i-1}}(x_{i,j} - x_{i-1,j})(u_k(x_{i-1,j+1}, t_{i-1}) - u_k(x_{i-1,j-1}, t_{i-1})) + u_k(x_{i-1,j}, t_{i-1}).$$
(8)

Formula (8) was obtained by approximating functions at three nearby nodes using the Lagrange formula. Here and below, for $j = \overline{1, NL-1}$ instead of $H_{i-1}$ it is necessary to take $Hl_{i-1}$, and for $j = \overline{NP+1, NHp_{i-1}}$.

The total derivatives with respect to $t$ at the moving boundary in the $t_i$ layer are found by the following formulas:

$$\frac{d^2u_k(L(t), t)}{dt^2}|_{t=t_i} = \frac{1}{H_t^2}(u_k(x_{i-1,Nl}, t_{i-1}) - 2u_k(x_{i,Nl}, t_i) + u_k(x_{i+1,Nl}, t_{i+1}));$$
$$\frac{d^2u_{2,x}(L(t), t)}{dt^2}|_{t=t_i} = \frac{1}{H_t^2}(u_{2,x}(x_{i-1,Nl}, t_{i-1}) - 2u_{2,x}(x_{i,Nl}, t_i) + u_{2,x}(x_{i+1,Nl}, t_{i+1})).$$

From the boundary conditions (3)–(5) we obtain:

$u_1(x_{i+1,Nl}, t_{i+1}) = f_1(x_{i,Nl}, t_i, u_k(x_{i,Nl}, t_i), u_{k,x}(x_{i,Nl} \pm 0, t_i), u_{k,t}(x_{i,Nl}, t_i), u_{k,xt}(x_{i,Nl} \pm 0, t_i))$
$H_t^2 + 2u_1(x_{i,Nl}, t_i) - u_1(x_{i-1,Nl}, t_{i-1});$

$u_2(x_{i+1,Nl}, t_{i+1})$
$= f_2(x_{i,Nl}, t_i, u_2(x_{i,Nl}, t_i), u_{2,t}(x_{i,Nl}, t_i), u_{2,xxx}(x_{i,Nl} \pm 0, t_i), u_{2,xxxt}(x_{i,Nl} \pm 0, t_i))$
$H_t^2 + 2u_2(x_{i,Nl}, t_i) - u_2(x_{i-1,Nl}, t_{i-1});$

$$u_{2,x}(x_{i+1,NI}, t_{i+1})$$
$$= f_3(x_{i,NI}, t_i, u_{2,x}(x_{i,NI}, t_i)u_{2,xt}(x_{i,NI}, t_i)u_{2,xx}(x_{i,NI} \pm 0, t_i), u_{2,xxt}(x_{i,NI} \pm 0, t_i))$$
$$H_t^2 + 2u_{2,x}(x_{i,NI}, t_i) - u_{2,x}(x_{i-1,NI}, t_{i-1}).$$

To find the functions at the internal nodes of the layer $t_{i+1}$ we use an approximation similar to (8):

$$u_k(x_{i+1,j}, t_{i+1}) = \frac{1}{2H_{i+1}^2}(x_{i+1,j} - x_{i,j})^2(u_k(x_{i,j-1}, t_{i+1}) - 2u_k(x_{i,j}, t_{i+1}) + u_k(x_{i,j+1}, t_{i+1}))$$

$$+\frac{1}{2H_{i+1}}(x_{i+1,j} - x_{i,j})(u_k(x_{i,j+1}, t_{i+1}) - u_k(x_{i,j-1}, t_{i+1})) + u_k(x_{i,j}, t_{i+1}).$$

To find $u_{k,x}$ at the inner nodes of the layer $t_{i+1}$ we use the formula:

$$u_{k,x}(x_{i+1,j}, t_{i+1}) = \frac{1}{2H_{i+1}}(u_k(x_{i+1,j+1}, t_{i+1}) - u_k(x_{i+1,j-1}, t_{i+1})).$$

To find $u_{1,x}$ at the boundaries and $u_{2,x}(x_{i+1,0}, t_{i+1})$ in the $t_{i+1}$ layer, forward and backward approximations are used:

$$u_{1,x}(x_{i+1,0}, t_{i+1}) = \frac{1}{2Hl_{i+1}}(-3u_1(x_{i+1,0}, t_{i+1}) + 4u_1(x_{i+1,1}, t_{i+1}) - u_1(x_{i+1,2}, t_{i+1}));$$

$$u_{1,x}(x_{i+1,N}, t_{i+1}) = \frac{1}{2Hp_{i+1}}(u_1(x_{i+1,N-2}, t_{i+1}) - 4u_1(x_{i+1,N-1}, t_{i+1}) + 3u_1(x_{i+1,N}, t_{i+1}));$$

$$u_{1,x}(x_{i+1,NI} - 0, t_{i+1}) = \frac{1}{2Hl_{i+1}}(u_1(x_{i+1,NI-2}, t_{i+1}) - 4u_1(x_{i+1,NI-1}, t_{i+1}) + 3u_1(x_{i+1,NI}, t_{i+1}));$$

$$u_{1,x}(x_{i+1,NI} + 0, t_{i+1}) = \frac{1}{2Hp_{i+1}}(-u_1(x_{i+1,NI-2}, t_{i+1}) + 4u_1(x_{i+1,NI-1}, t_{i+1}) - 3u_1(x_{i+1,NI}, t_{i+1}));$$

$$u_{2,x}(x_{i+1,0}, t_{i+1}) = \frac{1}{2Hl_{i+1}}(-3u_2(x_{i+1,0}, t_{i+1}) + 4u_2(x_{i+1,1}, t_{i+1}) - u_2(x_{i+1,2}, t_{i+1})).$$

The derivatives $u_{2,x}(x_{i+1,NI}, t_{i+1})$ and $u_{2,x}(x_{i+1,N}, t_{i+1})$ are known from boundary conditions.

To find $u_{k,xx}$ at the inner nodes of the layer $t_{i+1}$ we use the formula:

$$u_{k,xx}(x_{i+1,j}, t_{i+1}) = \frac{1}{2H_{i+1}}(u_{k,x}(x_{i+1,j+1}, t_{i+1}) - u_{k,x}(x_{i+1,j-1}, t_{i+1})).$$

The derivative $u_{2,xx}(x_{i+1,0}, t_{i+1})$ equal to zero. The rest of the derivatives $u_{2,xx}$ at the boundaries in the layer $t_{i+1}$ are found using forward and backward approximations:

$$u_{2,xx}(x_{i+1,N}, t_{i+1}) = \frac{1}{2Hp_{i+1}}(u_{2,x}(x_{i+1,N-2}, t_{i+1}) - 4u_{2,x}(x_{i+1,N-1}, t_{i+1}) + 3u_{2,x}(x_{i+1,N}, t_{i+1}));$$

$$u_{2,xx}(x_{i+1,Nl} - 0, t_{i+1})$$

$$= \frac{1}{2Hl_{i+1}}(u_{2,x}(x_{i+1,Nl-2}, t_{i+1}) - 4u_{2,x}(x_{i+1,Nl-1}, t_{i+1}) + 3u_{2,x}(x_{i+1,Nl}, t_{i+1}));$$

$$u_{2,xx}(x_{i+1,Nl} + 0, t_{i+1})$$

$$= \frac{1}{2Hp_{i+1}}(-u_{2,x}(x_{i+1,Nl-2}, t_{i+1}) + 4u_{2,x}(x_{i+1,Nl-1}, t_{i+1}) - 3u_{2,x}(x_{i+1,Nl}, t_{i+1})).$$

The partial derivatives $u_{2,xxx}$ are found by the formulas for $u_{2,xx}$ only instead of the function $u_{2,x}$ you must use $u_{2,xx}$.

To find $u_{2,xxxx}$ at the inner nodes of the layer $t_{i+1}$ we use the well-known approximation:

$$u_{2,xxxx}(x_{i+1,j}, t_{i+1}) = \frac{1}{2H_{i+1}^2}(u_{k,xx}(x_{i,j-1}, t_{i+1}) - 2u_{k,xx}(x_{i,j}, t_{i+1}) + u_{k,xx}(x_{i,j+1}, t_{i+1})).$$

To find the derivatives with respect to $t$ at intermediate points of the layer $t_{i+1}$ the forward approximation is used:

$$u_{k,t}(x_{i,j}, t_{i+1}) = \frac{1}{2H_t}(u_k(x_{i,j}, t_{i-1}) - 4u_k(x_{i,j}, t_i) + 3u_k(x_{i,j}, t_{i+1})).$$

The derivatives $u_{k,t}$ at the fixed ends are equal to zero. The total derivative on the moving boundary is equal to:

$$\frac{du_k(L(t), t)}{dt}\Big|_{t=t_{i+1}} = \frac{1}{2H_t}(u_k(x_{i-1,Nl}, t_{i-1}) - 4u_k(x_{i,Nl}, t_i) + 3u_k(x_{i+1,Nl}, t_{i+1}).$$

Considering that

$$\frac{du_k(L(t), t)}{dt} = u_{k,x}(L(t) \mp 0, t)L'(t) + u_{k,t}(L(t) \mp 0, t)$$

get

$$u_{k,t}(x_{i+1,Nl} \mp 0, t_{i+1}) = \frac{du_k(L(t), t)}{dt}\Big|_{t=t_{i+1}} - u_{k,x}(x_{i+1,Nl} \mp 0, t_{i+1})L'(t_{i+1}).$$

To find the functions and their derivatives with respect to $t$ at the nodes of the layer $t_{i+1}$ we use approximations similar to (8):

$$u_{k,t}(x_{i+1,j}, t_{i+1})$$

$$= \frac{1}{2H_{i+1}^2}(x_{i+1,j} - x_{i,j})^2(u_{k,t}(x_{i,j-1}, t_{i+1}) - 2u_{k,t}(x_{i,j}, t_{i+1}) + u_{k,t}(x_{i,j+1}, t_{i+1}))$$

$$+ \frac{1}{2H_{i+1}}(x_{i+1,j} - x_{i,j})(u_{k,t}(x_{i,j+1}, t_{i+1}) - u_{k,t}(x_{i,j-1}, t_{i+1})) + u_{k,t}(x_{i,j}, t_{i+1}).$$

To find the derivatives $u_{k,xt}$, $u_{k,xxt}u_{2,xxxt}$, $u_{2,xxxxt}$ in the layer $t_{i+1}$ it is necessary to apply the formulas for $u_{k,x}$, $u_{k,xx}u_{2,xxx}$, $u_{2,xxxxt}$ only use $u_k$ instead of the $u_{k,t}$ functions.

## 4 Longitudinal-Transverse Vibrations of a String Taking into Account Geometric Nonlinearity

In [17], using Hamilton's variational principle, the problem of longitudinal-transverse vibrations of a string was formulated taking into account geometric nonlinearity. The obtained mathematical model makes it possible to describe oscillations of systems with high-intensity moving boundaries. The problem posed was solved numerically according to the method described in Sect. 3.

By comparing the exact solution of the wave equation [25] and the numerical solution of the nonlinear problem, the correctness of the description of high-intensity oscillations by the wave equation has been investigated. A comparative analysis of the linear and nonlinear models showed that the incorrectness of the linear model is associated with an increase in the string tension with an increase in the vibration intensity, which is not taken into account by the linear model. The accuracy of the numerical solution is confirmed by the coincidence of the solutions of the linear and nonlinear models at small vibration amplitudes [17].

## 5 Conclusion

The developed scheme allows solving the Cauchy problem for a system of nonlinear partial differential equations with nonlinear boundary conditions, as well as taking into account the energy exchange between parts of an oscillating object to the left and right of the moving boundary. In the work, all functions and derivatives are defined that allow you to go to the next time layer. Thus, passing from one layer to another, one can find a solution to the problem for any value of t.

The solutions presented can be used in the study of the longitudinal-transverse vibrations of the ropes of hoisting installations [9, 16, 18–20, 23, 24, 26, 27], flexible transmission links [1, 5, 6, 15, 20], rods of solid fuel and beams of variable length [2, 4, 10, 11], drill strings [8], railway contact network [3, 7, 12, 14, 22], belt conveyors [1], etc.

## References

1. Boyle JM Jr, Bhatat B (2006) Vibration modeling of magnetic tape with vibro–impact of tape–guide contact. J Sound Vibr 3:632–655

2. Brake MR, Wickert JA (2008) Frictional vibration transmission from a laterally moving surface to a traveling beam. J Sound Vibr 3:663–675
3. Cho YH (2008) Numerical simulation of the dynamic responses of railway overhead contact lines to a moving pantograph, considering a nonlinear dropper. J Sound Vibr 3:433–454
4. Ding H, Chen L-Q (2010) Galerkin methods for natural frequencies of high–speed axially moving beams. J Sound Vibr 17:3484–3494
5. Inacio O, Antunes J, Wright MCM (2008) Computational modelling of string–body interaction for the violin family and simulation of wolf notes. J Sound Vibr 1–2:260–286
6. Nakagawa C, Shimamune R, Watanabe K, Erimitsu S (2010) Fundamental study on the effect of high–frequency vibration in the vertical and lateral directions on ride comfort. Q Rep Railway Tech Res Inst 2:101–105
7. Ryue J, Thompson DJ, White PR, Thompson DR (2009) Decay rates of propagating waves in railway tracks at high frequencies. J Sound Vibr 4–5:955–976
8. Sahebkar SM, Ghazavi MR, Khadem SE, Ghayesh MH (2011) Nonlinear vibration analysis of an axially moving drillstring system with time dependent axial load and axial velocity in inclined well. Mech Mach Theory 5:743–760
9. Shi Y, Wu L, Wang Y (2006) Нелинейный анализ собственных частот тросовой системы [Текст]. J Vibr Eng 2:173–178
10. Sun L, Luo F (2008). Steady–state dynamic response of a Bernoulli–Euler beam on a viscoelastic foundation subject to a platoon of moving dynamic loads. Trans ASME J Vibr Acoust 5:051002/1–051002/19
11. Teng, Y-Feng, Teng N-G, Kou X-Y (2008) Vibration analysis of maglev three–span continuous guideway considering control system. J Zhejiang Univ Sci A Int Appl Phys Eng J 1:8–14
12. Wang L, Zhao Y (2008) Multiple internal resonances and non–planar dynamics of shallow suspended cables to the harmonic excitations. J Sound Vibr 1–2:1–14
13. Yagci B, Sinan F, Louis R, Ozdoganlar L, Burak O (2009) A spectral–Tchebychev technique for solving linear and nonlinear beam equations. J Sound Vibr 1–2:375–404
14. Zhao Y, Wang L (2006) On the symmetric modal interaction of the suspended cable: three–to–one internal resonance. J Sound Vibr 4–5, 1073–1093
15. Zhu WD, Zheng NA (2008) Exact response of a translating string with arbitrarily varying length under general excitation. Trans. ASME. J. Appl. Mech. 3:031003/4–031003/14
16. Anisimov VN, Litvinov VL (2009) Investigation of resonant properties of mechanical objects with moving boundaries using the Kantorovich-Galerkin method. Bull Samara State Tech Univ Ser "Phys Math Sci" 1(18):149–158
17. Anisimov VN, Litvinov VL (2015) Mathematical models of longitudinal-transverse vibrations of objects with moving boundaries. Bull Samara State Tech Univ Ser "Phys Math Sci" 2(19):382–397
18. Anisimov VN, Litvinov VL (2018) Application of the Kantorovich–Galerkin method for solving boundary value problems with conditions on moving boundaries. Bull Russ Acad Sci Rigid Body Mech 2:70–77
19. Berlioz A, Lamarque C-H (2005) A non-linear model for the dynamics of an inclined cable. J Sound Vibr 279:619–639
20. Sandilo SH, van Horssen WT (2014) On variable length induced vibrations of a vertical string. J Sound Vibr 333:2432–2449
21. Liu Z, Chen G (2007) Analysis of plane nonlinear free vibrations of a carrying rope taking into account the influence of flexural rigidity. J Vibr Eng 1:57–60
22. Zhang W, Tang Y (2002) Global dynamics of the cable under combined parametrical and external excitations. Int J Non-linear Mech 37:505–526
23. Palm J et al (2013) Simulation of mooring cable dynamics using a discontinuous Galerkin method. In: International conference on computational methods in marine engineering
24. Faravelli L, Fuggini C, Ubertini F (2010) Toward a hybrid control solution for cable dynamics: theoretical prediction and experimental validation. Struct Control Health Monitoring 17:386–403

25. Anisimov VN, Litvinov VL, Korpen IV (2012) On a method for obtaining an analytical solution of a wave equation describing vibrations of systems with moving faces. Bulletin of the Samara state technical university. Ser "Physical and mathematical sciences", vol 3(28), pp 145–151
26. Litvinov VL (2020) Solution of boundary value problems with moving boundaries using an approximate method for constructing solutions of integro-differential equations. Tr Inst Math Mech Ural Branch Russ Acad Sci 26(2):188–199
27. Litvinov VL, Anisimov VN (2017) Transverse vibrations of a rope moving in the longitudinal direction. In: Proceedings of the Samara scientific center of the Russian Academy of Sciences, vol 19(4), 161–165

# On the Algorithmic Complexity of Digital Image Processing Filters with Winograd Calculations

**Pavel Lyakhov**⊙ **and Albina Abdulsalyamova**⊙

**Abstract** This paper analyzes the computational complexity of methods of digital filtering of images by linear spatial filters. A comparison of the direct method of realization of two-dimensional filtering and the method of realization of two-dimensional filters by Winograd is carried out. It is shown that the best result using the Winograd method is obtained for averaging Gaussian filters and for symmetrical filters of a general form with different coefficients. The Winograd method application for Gauss filter reduces the number of multiplication operations 3 times at an increase of several addition operations 1.84 times. For symmetric general filters with different coefficients, multiplication operations are reduced to 2.12 at a rise in the number of addition operations 2 times.

**Keywords** Winograd method · Digital image processing · Linear filters · Symmetric filters · Digital filtering

## 1 Introduction

In recent years, digital image processing has been put into practice in various fields. Among them, the development of methods of intellectual analysis of visual information has had a particularly great influence on our life. Active work on research and the introduction of intelligent methods of image processing is currently carried out in astronomy [1], medicine [2], earth sciences [3], and many others.

Usually, images generated by different information systems are distorted by interference, complicating their visual analysis by a human operator and automatic computer processing. When solving image processing problems, some or other components of the image itself may also interfere. For example, when analyzing a space image of the Earth's surface [4], the task may be to determine the boundaries between its separate areas - forest and field, water and land, etc. From the point

P. Lyakhov · A. Abdulsalyamova (✉)
North-Caucasus Center for Mathematical Research, North-Caucasus Federal University, Stavropol, Russia
e-mail: a.abdulsalyamova@mail.ru

© The Author(s), under exclusive license to Springer Nature Switzerland AG 2022   71
A. Tchernykh et al. (eds.), *Mathematics and its Applications in New Computer Systems*,
Lecture Notes in Networks and Systems 424,
https://doi.org/10.1007/978-3-030-97020-8_8

of view of this task, separate image details inside the separated areas are interference. The interference effect is attenuated by filtering [5]. In filtering, the brightness (signal) of each point of the original image, distorted by noise, is replaced by another brightness value, which is recognized as the least distorted by the interference.

There are various methods of accelerating digital image processing, but each has its significant drawbacks. Thus, although adaptive two-sided filtering removes noise but cannot improve the sharpness of edges [6], accelerated two-sided filtering with skipping blocks worsens the accuracy of filtering [7].

This paper will propose a way to accelerate digital image processing by using the Winograd method. The number of addition and multiplication operations for direct and Winograd filtering of images is calculated and compared.

## 2  Organization of Computations in Digital Image Processing Filters by the Winograd Method

A spatial filter consists of a neighborhood (usually a small rectangle) and a given operation performed on image pixels falling into the neighborhood. The filtering creates a new pixel whose value depends on the filter operator, and the coordinates coincide with the coordinates of the neighborhood center [8]. The processed (filtered) image arises in the process of scanning the original image by the filter. Linear spatial filtering coefficients in the general format filter mask can be written in the form of a matrix.

$$W = \begin{bmatrix} w(-1,-1) & w(-1,0) & w(-1,1) \\ w(0,-1) & w(0,0) & w(0,1) \\ w(1,-1) & w(1,0) & w(1,1) \end{bmatrix}. \tag{1}$$

Suppose we have an image fragment of size $4 \times 4$

$$P = \begin{bmatrix} p_{11} & p_{12} & p_{13} & p_{14} \\ p_{21} & p_{22} & p_{23} & p_{24} \\ p_{31} & p_{32} & p_{33} & p_{34} \\ p_{41} & p_{42} & p_{43} & p_{44} \end{bmatrix} \tag{2}$$

and a $3 \times 3$ filter.

$$F = \begin{bmatrix} f_{11} & f_{12} & f_{13} \\ f_{21} & f_{22} & f_{23} \\ f_{31} & f_{32} & f_{33} \end{bmatrix}. \tag{3}$$

Then the result of the convolution will be calculated by the formula:

$$P' = \begin{bmatrix} p_{11} & p_{12} & p_{13} & p_{14} \\ p_{21} & p_{22} & p_{23} & p_{24} \\ p_{31} & p_{32} & p_{33} & p_{34} \\ p_{41} & p_{42} & p_{43} & p_{44} \end{bmatrix} * \begin{bmatrix} f_{11} & f_{12} & f_{13} \\ f_{21} & f_{22} & f_{23} \\ f_{31} & f_{32} & f_{33} \end{bmatrix} = \begin{bmatrix} p'_{11} & p'_{21} \\ p'_{12} & p'_{22} \end{bmatrix}, \tag{4}$$

where

$$p'_{11} = \sum_{i=1}^{3}\sum_{j=1}^{3} p_{ij}f_{ij},$$

$$p'_{21} = \sum_{i=1}^{3}\sum_{j=1}^{3} p_{i,j+1}f_{ij},$$

$$p'_{12} = \sum_{i=1}^{3}\sum_{j=1}^{3} p_{i+1,j}f_{ij},$$

$$p'_{22} = \sum_{i=1}^{3}\sum_{j=1}^{3} p_{i+1,j+1}f_{ij}. \tag{5}$$

When filtering the image by the direct method using the formulas (5), you need to perform 36 multiplication operations and 32 addition operations.

Filtration by Winograd's method in matrix form has the form [9, 10]:

$$z = A^{T}\big((GFG^{T}) \odot (B^{T}PB)\big)A, \tag{6}$$

where $\odot$ denotes element-by-element multiplication, and

$$A = \begin{bmatrix} 1 & 0 \\ 1 & 1 \\ 1 & -1 \\ 0 & -1 \end{bmatrix}, B = \begin{bmatrix} 1 & 0 & 0 & 0 \\ 0 & 1 & -1 & 1 \\ -1 & 1 & 1 & 0 \\ 0 & 0 & 0 & -1 \end{bmatrix}, G = \begin{bmatrix} 1 & 0 & 0 \\ \frac{1}{2} & \frac{1}{2} & \frac{1}{2} \\ \frac{1}{2} & -\frac{1}{2} & \frac{1}{2} \\ 0 & 0 & 1 \end{bmatrix}. \tag{7}$$

Let us rewrite (6) as:

$$z = A^{T}(\Gamma \odot D)A, \tag{8}$$

where $\Gamma = GFG^{T}$, $D = B^{T}PB$. Further, let us represent (8) as

$$z = A^{T}MA, \tag{9}$$

where the matrix $M = \Gamma \odot D$ and consists of elements

$$M = \begin{bmatrix} m_{11} & m_{12} & m_{13} & m_{14} \\ m_{21} & m_{22} & m_{23} & m_{24} \\ m_{31} & m_{32} & m_{33} & m_{34} \\ m_{41} & m_{42} & m_{43} & m_{44} \end{bmatrix}. \tag{10}$$

Let P be the image from (2). Then the general view of D will be written in the form:

$$\begin{bmatrix} 1 & 0 & 0 & 0 \\ 0 & 1 & -1 & 1 \\ -1 & 1 & 1 & 0 \\ 0 & 0 & 0 & -1 \end{bmatrix} \cdot \begin{bmatrix} p_{11} & p_{12} & p_{13} & p_{14} \\ p_{21} & p_{22} & p_{23} & p_{24} \\ p_{31} & p_{32} & p_{33} & p_{34} \\ p_{41} & p_{42} & p_{43} & p_{44} \end{bmatrix} \cdot \begin{bmatrix} 1 & 0 & 0 & 0 \\ 0 & 1 & -1 & 1 \\ -1 & 1 & 1 & 0 \\ 0 & 0 & 0 & -1 \end{bmatrix}^T$$
$$= \begin{bmatrix} d_1 & d_2 & d_3 & d_4 \end{bmatrix}, \tag{11}$$

where

$$d_1 = \begin{bmatrix} p_{11} - p_{31} - p_{13} + p_{33} \\ p_{21} + p_{31} - p_{23} - p_{33} \\ -p_{21} + p_{31} + p_{23} - p_{33} \\ p_{21} - p_{41} - p_{23} + p_{43} \end{bmatrix}, d_2 = \begin{bmatrix} p_{12} - p_{32} + p_{13} - p_{33} \\ p_{22} + p_{32} + p_{23} + p_{33} \\ -p_{22} + p_{32} - p_{23} + p_{33} \\ p_{22} - p_{42} + p_{23} - p_{43} \end{bmatrix},$$

$$d_3 = \begin{bmatrix} -p_{12} + p_{32} + p_{13} - p_{33} \\ -p_{22} - p_{32} + p_{23} + p_{33} \\ -p_{22} + p_{32} - p_{23} + p_{33} \\ p_{22} - p_{42} + p_{23} - p_{43} \end{bmatrix}, d_4 = \begin{bmatrix} p_{12} - p_{32} - p_{14} + p_{34} \\ p_{22} + p_{32} - p_{24} - p_{34} \\ -p_{22} + p_{32} + p_{24} - p_{34} \\ p_{22} - p_{42} - p_{24} + p_{44} \end{bmatrix}. \tag{12}$$

Let

$$d_1 = \begin{bmatrix} d_{11} \\ d_{12} \\ d_{13} \\ d_{14} \end{bmatrix}, d_2 = \begin{bmatrix} d_{21} \\ d_{22} \\ d_{23} \\ d_{24} \end{bmatrix}, d_3 = \begin{bmatrix} d_{31} \\ d_{32} \\ d_{33} \\ d_{34} \end{bmatrix}, d_4 = \begin{bmatrix} d_{41} \\ d_{42} \\ d_{43} \\ d_{44} \end{bmatrix}, \tag{13}$$

where

$$d_{11} = p_{11} - p_{31} - p_{13} + p_{33}, d_{12} = p_{21} + p_{31} - p_{23} - p_{33},$$
$$d_{13} = -p_{21} + p_{31} + p_{23} - p_{33}, d_{14} = p_{21} - p_{41} - p_{23} + p_{43},$$
$$d_{21} = p_{12} - p_{32} + p_{13} - p_{33}, d_{22} = p_{22} + p_{32} + p_{23} + p_{33},$$
$$d_{23} = -p_{22} + p_{32} - p_{23} + p_{33}, d_{24} = p_{22} - p_{42} + p_{23} - p_{43},$$
$$d_{31} = -p_{12} + p_{32} + p_{13} - p_{33}, d_{32} = -p_{22} - p_{32} + p_{23} + p_{33},$$
$$d_{33} = -p_{22} + p_{32} - p_{23} + p_{33}, d_{34} = p_{22} - p_{42} + p_{23} - p_{43},$$
$$d_{41} = p_{12} - p_{32} - p_{14} + p_{34}, d_{42} = p_{22} + p_{32} - p_{24} - p_{34},$$

$$d_{43} = -p_{22} + p_{32} + p_{24} - p_{34}, \quad d_{44} = p_{22} - p_{42} - p_{24} + p_{44}, \quad (14)$$

Then, the general form of (9) can be written in the form:

$$Z = \begin{bmatrix} 1 & 0 \\ 1 & 1 \\ 1 & -1 \\ 0 & -1 \end{bmatrix}^{T} \cdot \begin{bmatrix} m_{11} & m_{12} & m_{13} & m_{14} \\ m_{21} & m_{22} & m_{23} & m_{24} \\ m_{31} & m_{32} & m_{33} & m_{34} \\ m_{41} & m_{42} & m_{43} & m_{44} \end{bmatrix} \cdot \begin{bmatrix} 1 & 0 \\ 1 & 1 \\ 1 & -1 \\ 0 & -1 \end{bmatrix} = \begin{bmatrix} q_{11} & q_{21} \\ q_{12} & q_{22} \end{bmatrix}, \quad (15)$$

where

$$
\begin{aligned}
q_{11} &= (m_{11} + m_{21} + m_{31}) + (m_{12} + m_{22} + m_{32}) + (m_{13} + m_{23} + m_{33}), \\
q_{21} &= (m_{21} - m_{31} - m_{41}) + (m_{22} - m_{32} - m_{42}) + (m_{23} - m_{33} - m_{43}), \\
q_{12} &= (m_{12} + m_{22} + m_{32}) - (m_{13} + m_{23} + m_{33}) - (m_{14} + m_{24} + m_{34}), \\
q_{22} &= (m_{22} - m_{32} - m_{42}) - (m_{23} - m_{33} - m_{43}) - (m_{24} - m_{34} - m_{44}).
\end{aligned}
\quad (16)
$$

When filtering by the Winográd method, computing the general case of the matrix $Z$ requires performing 32 multiplication operations and 16 addition operations.

Let us consider the application of the Winograd method for different linear filters. We will check its efficiency on Laplace, Sobel, averaging, and Gauss filters.

The simplest isotropic operator based on derivatives is the Laplacian (Laplace operator). The discrete formulation of the two-dimensional Laplacian of two variables is written as

$$\nabla^2 f(x, y) = f(x+1, y) + f(x-1, y) + f(x, y+1) + f(x, y-1) - 4f(x, y).$$

This equation can be implemented using a filter mask

$$L = \begin{bmatrix} 0 & 1 & 0 \\ 1 & -4 & 1 \\ 0 & 1 & 0 \end{bmatrix}, \quad (17)$$

which gives isotropic results for rotations at angles divisible by $90°$.

The number of products and additives when substituting a mask from (17) to (4) is calculated using a direct method of image filtration. Laplace filter realization needs to perform 4 multiplication operations and 16 addition operations. Calculate their number for filtering the image with the Laplace filter by the Winograd method. Substituting (17) into formula $\Gamma = GFG^T$ instead of $F$, we obtain the matrix

$$L_\Gamma = \begin{bmatrix} 0 & \frac{1}{2} & -\frac{1}{2} & 0 \\ \frac{1}{2} & 0 & 1 & \frac{1}{2} \\ -\frac{1}{2} & 1 & -2 & -\frac{1}{2} \\ 0 & \frac{1}{2} & -\frac{1}{2} & 0 \end{bmatrix}. \quad (18)$$

Using (18) and (11)–(13), obtain a special case of the matrix M for the Laplace filter

$$
L_M = \begin{bmatrix} 0 & \frac{d_{21}}{2} & -\frac{d_{31}}{2} & 0 \\ \frac{d_{12}}{2} & 0 & d_{32} & -\frac{d_{42}}{2} \\ -\frac{d_{13}}{2} & d_{23} & -2d_{33} & -\frac{d_{43}}{2} \\ 0 & \frac{d_{24}}{2} & -\frac{d_{34}}{2} & 0 \end{bmatrix}. \tag{19}
$$

Let us calculate the finite matrix Z by substituting (19) in (9):

$$
Z_L = \begin{bmatrix} z_{l1} & z_{l2} \\ z_{l3} & z_{l4} \end{bmatrix},
$$

where

$$
z_{l1} = \frac{(d_{12} - d_{13})}{2} + \left( \frac{d_{21}}{2} + d_{23} \right) + \left( -\frac{d_{13}}{2} + d_{32} - 2d_{33} \right),
$$

$$
z_{l2} = \frac{(d_{21} + d_{13})}{2} + \left( -d_{23} - \frac{d_{24}}{2} \right) + \left( d_{32} + 2d_{33} + \frac{d_{34}}{2} \right),
$$

$$
z_{l3} = \left( \frac{d_{21}}{2} + d_{23} \right) - \left( -\frac{d_{13}}{2} + d_{32} - 2d_{33} \right) - \frac{(-d_{42} - d_{43})}{2},
$$

$$
z_{l4} = \left( -d_{23} - \frac{d_{24}}{2} \right) - \left( d_{32} + 2d_{33} + \frac{d_{34}}{2} \right) + \frac{(-d_{42} - d_{43})}{2}.
$$

When calculating the matrix $Z_L$ the number of additions and multiplications used in Winograd filtering with Laplace filter can be calculated. Finally, the Laplace filter by the Winograd method requires 9 multiplications and 58 additions.

Maskin

$$
S_1 = \begin{bmatrix} -1 & -2 & -1 \\ 0 & 0 & 0 \\ 1 & 2 & 1 \end{bmatrix}, \; S_2 = \begin{bmatrix} -1 & 0 & 1 \\ -2 & 0 & 2 \\ -1 & 0 & 1 \end{bmatrix}, \tag{20}
$$

is called the Sobel operator. The value 2 in the central weight coefficients (in rows and columns) is based on the desire to achieve greater smoothness by assigning greater significance to the central points. To analyze the computational cost of implementing the Sobel filters by the Winograd method, we use the $S_1$ mask. Let us substitute it into $\Gamma = GFG^T$ and obtain the matrix.

$$S_\Gamma = \begin{bmatrix} -1 & -2 & 0 & -1 \\ 0 & 0 & 0 & 0 \\ 0 & 0 & 0 & 0 \\ 1 & 2 & 0 & 1 \end{bmatrix}. \tag{21}$$

Using (21) and (11)–(13) obtain a special case of the matrix M for the Sobel filter

$$S_M = \begin{bmatrix} -d_{11} & -2d_{21} & 0 & -d_{41} \\ 0 & 0 & 0 & 0 \\ 0 & 0 & 0 & 0 \\ d_{14} & 2d_{24} & 0 & d_{44} \end{bmatrix}. \tag{22}$$

Let's find a finite matrix Z

$$Z_S = \begin{bmatrix} z_{s1} & z_{s2} \\ z_{s3} & z_{s4} \end{bmatrix},$$

where

$$z_{s1} = -d_{11} - 2d_{21},$$

$$z_{s2} = d_{14} + 2d_{24},$$

$$z_{s3} = -d_{21} + d_{41},$$

$$z_{s4} = -2d_{24} + d_{44}).$$

Calculating the matrix $Z_S$ requires 6 multiplication operations and 39 addition operations.

The output (response) of the simplest linear smoothing spatial filter is the average of the elements in the neighborhood covered by the filter mask. Such filters are sometimes called averaging or smoothing filters. By replacing the initial values of the image elements with the average values of the filter mask, the "abrupt" transitions of brightness levels are reduced. Since random noise is precisely characterized by sharp jumps in brightness levels, the most obvious application of smoothing is noise suppression. The mask of the averaging filter in the vicinity of $3 \times 3$

$$Q = \frac{1}{9} \begin{bmatrix} 1 & 1 & 1 \\ 1 & 1 & 1 \\ 1 & 1 & 1 \end{bmatrix}. \tag{23}$$

Substituting (23) into formula $\Gamma = GFG^T$ and substituting the resulting matrix into (11), we obtain a special case of matrix M for the averaging filter

$$Q_M = \begin{bmatrix} \frac{d_{11}}{9} & \frac{d_{21}}{6} & \frac{d_{31}}{18} & \frac{d_{41}}{9} \\ \frac{d_{12}}{6} & \frac{d_{22}}{4} & \frac{d_{32}}{12} & \frac{d_{42}}{6} \\ \frac{d_{13}}{18} & \frac{d_{23}}{12} & \frac{d_{33}}{36} & \frac{d_{43}}{18} \\ \frac{d_{14}}{9} & \frac{d_{24}}{6} & \frac{d_{34}}{18} & \frac{d_{44}}{9} \end{bmatrix}. \tag{24}$$

Let us calculate the finite matrix $Z_Q$ by substituting (24) in (9):

$$Z_Q = \begin{bmatrix} z_{q1} & z_{q2} \\ z_{q3} & z_{q4} \end{bmatrix},$$

where

$$z_{q1} = \left( \frac{d_{11}}{9} + \frac{d_{12}}{6} + \frac{d_{13}}{18} \right) + \left( \frac{d_{21}}{6} + \frac{d_{22}}{4} + \frac{d_{23}}{12} \right) + \left( \frac{d_{31}}{18} + \frac{d_{32}}{12} + \frac{d_{33}}{36} \right),$$

$$z_{q2} = \left( \frac{d_{12}}{6} - \frac{d_{13}}{18} - \frac{d_{14}}{9} \right) + \left( \frac{d_{22}}{4} - \frac{d_{23}}{12} - \frac{d_{24}}{6} \right) + \left( \frac{d_{32}}{12} - \frac{d_{33}}{36} - \frac{d_{34}}{18} \right),$$

$$z_{q3} = \left( \frac{d_{21}}{6} + \frac{d_{22}}{4} + \frac{d_{23}}{12} \right) - \left( \frac{d_{31}}{18} + \frac{d_{32}}{12} + \frac{d_{33}}{36} \right) - \left( \frac{d_{41}}{9} + \frac{d_{42}}{6} + \frac{d_{43}}{18} \right),$$

$$z_{q4} = \left( \frac{d_{22}}{4} - \frac{d_{23}}{12} - \frac{d_{24}}{6} \right) - \left( \frac{d_{32}}{12} - \frac{d_{33}}{36} - \frac{d_{34}}{18} \right) - \left( \frac{d_{42}}{6} - \frac{d_{43}}{18} - \frac{d_{44}}{9} \right).$$

When calculating the matrix $Z_Q$, we can count the number of multiplication and addition operations used for the Winograd method. Thus, 16 multiplication operations and 64 addition operations are required to calculate the averaging filter by the Winograd method.

The Gaussian operator is a filter whose application is visually manifested by blurring the image, and its kernel is calculated using the Gaussian distribution function for the two-dimensional case $G(x) = \frac{1}{\sqrt{2\pi\sigma^2}} e^{\frac{x^2+y^2}{2\sigma^2}}$ where x is the X-axis distance from the origin, y is the Y-axis distance from the origin, $\sigma$ is the standard deviation of the Gaussian distribution.

A filter with such a kernel implements the idea of calculating a weighted average of each pixel. The effect of each pixel on the result depends on its distance to the pixel in question. Consider a Gaussian filter with coefficients:

$$N = \frac{1}{15} \begin{bmatrix} 1 & 2 & 1 \\ 2 & 3 & 2 \\ 1 & 2 & 1 \end{bmatrix}. \tag{25}$$

The matrix obtained by substituting (25) into $\Gamma = \text{GFG}^{\text{T}}$ is replaced in (11). We receive a special case of the matrix $M$ for the Gaussian filter

$$N_M = \begin{bmatrix} \frac{d_{11}}{15} & \frac{2d_{21}}{15} & 0 & \frac{d_{41}}{15} \\ \frac{2d_{12}}{15} & \frac{d_{22}}{4} & \frac{d_{32}}{60} & \frac{2d_{42}}{15} \\ 0 & \frac{d_{23}}{60} & -\frac{d_{33}}{60} & 0 \\ \frac{d_{14}}{15} & \frac{2d_{24}}{15} & 0 & \frac{d_{44}}{15} \end{bmatrix}. \tag{26}$$

Calculate the matrix $Z_N$ at filtering by Winograd method with Gauss filter by substituting (26) in (9):

$$Z_N = \begin{bmatrix} z_{n1} & z_{n2} \\ z_{n3} & z_{n4} \end{bmatrix},$$

where

$$z_{n1} = \left( \frac{d_{11}}{15} + \frac{2d_{12}}{15} \right) + \left( \frac{2d_{21}}{15} + \frac{d_{22}}{4} + \frac{d_{23}}{60} \right) + \left( \frac{d_{32}}{60} - \frac{d_{33}}{60} \right),$$

$$z_{n2} = \left( \frac{2d_{12}}{15} - \frac{d_{14}}{15} \right) + \left( \frac{d_{22}}{4} - \frac{d_{23}}{60} - \frac{2d_{24}}{15} \right) + \left( \frac{d_{32}}{60} + \frac{d_{33}}{60} \right),$$

$$z_{n3} = \left( \frac{2d_{21}}{15} + \frac{d_{22}}{4} + \frac{d_{23}}{60} \right) - \left( \frac{d_{32}}{60} - \frac{d_{33}}{60} \right) - \left( \frac{d_{41}}{15} + \frac{2d_{42}}{15} \right),$$

$$z_{n4} = \left( \frac{d_{22}}{4} - \frac{d_{23}}{60} - \frac{2d_{24}}{15} \right) - \left( \frac{d_{32}}{60} + \frac{d_{33}}{60} \right) - \left( \frac{2d_{42}}{15} - \frac{d_{44}}{15} \right).$$

Thus, when calculating the matrix $Z_N$ by Winograd's method, you need to perform 16 multiplication operations and 64 addition operations.

Table 1 shows the results of calculating the number of addition and multiplication operations for different linear filters. Based on these data, it is possible to conclude the efficiency of the Winograd method application. For convolution of the image

**Table 1** The number of addition and multiplication operations for partial filters

| Filter | Implementation method | | | |
|---|---|---|---|---|
| | Direct | | Winograd | |
| | × | + | × | + |
| General view | 36 | 32 | 32 | 16 |
| Laplace | 4 | 16 | 9 | 58 |
| Sobel | 8 | 20 | 6 | 39 |
| Averaging | 16 | 32 | 16 | 64 |
| Gauss | 36 | 32 | 12 | 59 |

with the Laplace filter, it is preferable to use the direct method. Using the Winograd method will increase the number of multiplications by 2.25 times and additions by 3.62. For averaging filter, it is also better to use the direct convolution method, as the number of additions is 2 times less than in the Winograd method, with the same number of multiplications. It is better to use Winograd's method for convolution with a Gaussian filter. The number of multiplications is 3 times less than with the direct method, and the number of additions is 1.84 more. It is impossible to say exactly which method is preferable without a practical test when convolution with the Sobel filter.

## 3   Analysis of Algorithmic Complexity of Digital Image Processing Filters

In this section, we analyze the efficiency of the Winograd method for filters with arbitrary coefficients. Consider filters with different coefficient symmetry $\{F_1, \ldots, F_5\}$

$$
F_1 = \begin{bmatrix} 0 & B & 0 \\ B & A & B \\ 0 & B & 0 \end{bmatrix}, F_2 = \begin{bmatrix} C & B & C \\ B & A & B \\ C & B & C \end{bmatrix}, F_3 = \begin{bmatrix} 0 & A & 0 \\ A & A & A \\ 0 & A & 0 \end{bmatrix}, F_4 = \begin{bmatrix} B & B & B \\ B & A & B \\ B & B & B \end{bmatrix},
$$

$$
F_5 = \begin{bmatrix} A & A & A \\ A & A & A \\ A & A & A \end{bmatrix}, \tag{27}
$$

where A, B, C $\neq \{\pm 1, 0\}$, because otherwise, all the calculations performed would be greatly simplified.

Consider the convolution of the image $D$ with the filter $F_1$

$$
\begin{bmatrix} p_{11} & p_{12} & p_{13} & p_{14} \\ p_{21} & p_{22} & p_{23} & p_{24} \\ p_{31} & p_{32} & p_{33} & p_{34} \\ p_{41} & p_{42} & p_{43} & p_{44} \end{bmatrix} * \begin{bmatrix} 0 & B & 0 \\ B & A & B \\ 0 & B & 0 \end{bmatrix} = \begin{bmatrix} z_1 & z_2 \\ z_3 & z_4 \end{bmatrix}, \tag{28}
$$

where

$$
z_1 = B(p_{12} + p_{21} + p_{23} + p_{32}) + Ap_{22},
$$

$$
z_2 = B(p_{13} + p_{22} + p_{24} + p_{33}) + Ap_{23},
$$

$$
z_3 = B(p_{22} + p_{31} + p_{33} + p_{42}) + Ap_{32},
$$

$$z_4 = B(p_{23} + p_{32} + p_{34} + p_{43}) + Ap_{33}.$$

When filtering an image fragment of size $4 \times 4$ by the $F_1$ filter mask by the direct method, we need to perform 15 multiplication operations and 12 addition operations. Calculate their number when filtering the image by the Winograd method. Find the matrix $M$:

$$D \odot \left( \begin{bmatrix} 1 & 0 & 0 \\ \frac{1}{2} & \frac{1}{2} & \frac{1}{2} \\ \frac{1}{2} & -\frac{1}{2} & \frac{1}{2} \\ 0 & 0 & 1 \end{bmatrix} \cdot \begin{bmatrix} 0 & B & 0 \\ B & A & B \\ 0 & B & 0 \end{bmatrix} \cdot \begin{bmatrix} 1 & 0 & 0 \\ \frac{1}{2} & \frac{1}{2} & \frac{1}{2} \\ \frac{1}{2} & -\frac{1}{2} & \frac{1}{2} \\ 0 & 0 & 1 \end{bmatrix}^T \right)$$

$$= \begin{bmatrix} 0 & \frac{1}{2}Bd_{21} & -\frac{1}{2}Bd_{31} & 0 \\ \frac{1}{2}Bd_{12} & (B + \frac{1}{4}A)d_{22} & -\frac{1}{4}Ad_{32} & \frac{1}{2}Bd_{42} \\ -\frac{1}{2}Bd_{13} & -\frac{1}{4}Ad_{23} & (-B + \frac{1}{4}A)d_{33} & -\frac{1}{2}Bd_{43} \\ 0 & \frac{1}{2}Bd_{24} & -\frac{1}{2}Bd_{34} & 0 \end{bmatrix} \tag{29}$$

Calculate the finite matrix $Z$:

$$Z_{F1} = \begin{bmatrix} z_{f11} & z_{f21} \\ z_{f12} & z_{f22} \end{bmatrix},$$

where

$$z_{f11} = \left( \frac{1}{2}Bd_{12} - \frac{1}{2}Bd_{13} \right) + \left( \frac{1}{2}Bd_{21} + \left( B + \frac{1}{4}A \right)d_{22} - \frac{1}{4}Ad_{23} \right)$$
$$+ + \left( -\frac{1}{2}Bd_{31} - \frac{1}{4}Ad_{32} + \left( -B + \frac{1}{4}A \right)d_{33} \right),$$

$$z_{f21} = \left( \frac{1}{2}Bd_{12} + \frac{1}{2}Bd_{13} \right) + \left( \left( B + \frac{1}{4}A \right)d_{22} + \frac{1}{4}Ad_{23} - \frac{1}{2}Bd_{24} \right)$$
$$+ + \left( -\frac{1}{4}Ad_{32} - \left( -B + \frac{1}{4}A \right)d_{33} + \frac{1}{2}Bd_{34} \right),$$

$$z_{f12} = \left( \frac{1}{2}Bd_{21} + \left( B + \frac{1}{4}A \right)d_{22} - \frac{1}{4}Ad_{23} \right)$$
$$- \left( -\frac{1}{2}Bd_{31} - \frac{1}{4}Ad_{32} + + \left( -B + \frac{1}{4}A \right)d_{33} \right) - \left( \frac{1}{2}Bd_{42} - \frac{1}{2}Bd_{43} \right),$$

$$z_{f22} = \left( \left( B + \frac{1}{4}A \right)d_{22} + \frac{1}{4}Ad_{23} - \frac{1}{2}Bd_{24} \right) - \left( -\frac{1}{4}Ad_{32} - \left( -B + \frac{1}{4}A \right)d_{33} + + \frac{1}{2}Bd_{34} \right)$$
$$- \left( \frac{1}{2}Bd_{42} + \frac{1}{2}Bd_{43} \right).$$

Thus, to filter a $4 \times 4$ fragment of the image by the filter mask $F_1$ by the method of Winograd, it is required to perform 12 multiplication operations and 28 addition operations. Calculate the number of operations when filtering the image by $F_2$ mask

$$
\begin{bmatrix} p_{11} & p_{12} & p_{13} & p_{14} \\ p_{21} & p_{22} & p_{23} & p_{24} \\ p_{31} & p_{32} & p_{33} & p_{34} \\ p_{41} & p_{42} & p_{43} & p_{44} \end{bmatrix} * \begin{bmatrix} C & B & C \\ B & A & B \\ C & B & C \end{bmatrix} = \begin{bmatrix} z_1 & z_2 \\ z_3 & z_4 \end{bmatrix}, \tag{30}
$$

where

$$
\begin{aligned}
z_1 &= Ap_{22} + B(p_{12} + p_{21} + p_{23} + p_{32}) \\
&\quad + C(p_{11} + p_{13} + p_{31} + p_{33}),
\end{aligned}
$$

$$
\begin{aligned}
z_2 &= Ap_{23} + B(p_{13} + p_{22} + p_{24} + p_{33}) \\
&\quad + C(p_{12} + p_{14} + p_{32} + p_{34}),
\end{aligned}
$$

$$
\begin{aligned}
z_3 &= Ap_{32} + B(p_{22} + p_{31} + p_{33} + p_{42}) \\
&\quad + C(p_{21} + p_{23} + p_{41} + p_{43}),
\end{aligned}
$$

$$
\begin{aligned}
z_4 &= Ap_{33} + B(p_{23} + p_{32} + p_{34} + p_{43}) \\
&\quad + C(p_{22} + p_{24} + p_{42} + p_{44}).
\end{aligned}
$$

Thus, when filtering an image fragment of size $4 \times 4$ with a filter mask $F_2$ by the direct method, it is required to perform 34 multiplication operations and 32 addition operations. At filtering by Winograd method, we get:

$$
\begin{aligned}
D \odot &\left( \begin{bmatrix} 1 & 0 & 0 \\ \frac{1}{2} & \frac{1}{2} & \frac{1}{2} \\ \frac{1}{2} & -\frac{1}{2} & \frac{1}{2} \\ 0 & 0 & 1 \end{bmatrix} \cdot \begin{bmatrix} C & B & C \\ B & A & B \\ C & B & C \end{bmatrix} \cdot \begin{bmatrix} 1 & 0 & 0 \\ \frac{1}{2} & \frac{1}{2} & \frac{1}{2} \\ \frac{1}{2} & -\frac{1}{2} & \frac{1}{2} \\ 0 & 0 & 1 \end{bmatrix}^T \right) \\
&= \begin{bmatrix} Cd_{11} & (C + \frac{1}{2}B)d_{21} & (C - \frac{1}{2}B)d_{31} & Cd_{41} \\ (C + \frac{1}{2}B)d_{12} & (C + B + \frac{1}{4}A)d_{22} & (C - \frac{1}{4}A)d_{32} & (C + \frac{1}{2}B)d_{42} \\ (C - \frac{1}{2}B)d_{13} & (C - \frac{1}{4}A)d_{23} & (C - B + \frac{1}{4}A)d_{33} & (C - \frac{1}{2}B)d_{43} \\ Cd_{14} & (C + \frac{1}{2}B)d_{24} & (C - \frac{1}{2}B)d_{34} & Cd_{44} \end{bmatrix}.
\end{aligned} \tag{31}
$$

Calculate the finite matrix $Z$:

$$
Z_{F2} = \begin{bmatrix} z_{f11} & z_{f21} \\ z_{f12} & z_{f22} \end{bmatrix},
$$

where

$$z_{f11} = \left(Cd_{11} + \left(C + \frac{1}{2}B\right)d_{12} + \left(C - \frac{1}{2}B\right)d_{13}\right)$$
$$+\left(\left(C + \frac{1}{2}B\right)d_{21} + +\left(C + B + \frac{1}{4}A\right)d_{22} + \left(C - \frac{1}{4}A\right)d_{23}\right)$$
$$+\left(\left(C - \frac{1}{2}B\right)d_{31} + \left(C - \frac{1}{4}A\right)d_{32} + +\left(C - B + \frac{1}{4}A\right)d_{33}\right),$$

$$z_{f21} = \left(\left(C + \frac{1}{2}B\right)d_{12} - \left(C - \frac{1}{2}B\right)d_{13} - Cd_{14}\right)$$
$$+\left(\left(C + B + \frac{1}{4}A\right)d_{22} - -\left(C - \frac{1}{4}A\right)d_{23} - \left(C + \frac{1}{2}B\right)d_{24}\right)$$
$$+\left(\left(C - \frac{1}{4}A\right)d_{32} - \left(C - B + \frac{1}{4}A\right)d_{33} - \left(C - \frac{1}{2}B\right)d_{34}\right),$$

$$z_{f12} = \left(Cd_{11} + \left(C + \frac{1}{2}B\right)d_{12} + \left(C - \frac{1}{2}B\right)d_{13}\right)$$
$$-\left(\left(C + \frac{1}{2}B\right)d_{21} + +\left(C + B + \frac{1}{4}A\right)d_{22} + \left(C - \frac{1}{4}A\right)d_{23}\right)$$
$$-\left(Cd_{41} + \left(C + \frac{1}{2}B\right)d_{42} + +\left(C - \frac{1}{2}B\right)d_{43}\right),$$

$$z_{f22} = \left(\left(C + \frac{1}{2}B\right)d_{12} - \left(C - \frac{1}{2}B\right)d_{13} - Cd_{14}\right)$$
$$-\left(\left(C + B + \frac{1}{4}A\right)d_{22} - -\left(C - \frac{1}{4}A\right)d_{23} - \left(C + \frac{1}{2}B\right)d_{24}\right)$$
$$-\left(\left(C + \frac{1}{2}B\right)d_{42} - \left(C - \frac{1}{2}B\right)d_{43} - Cd_{44}\right).$$

Thus, to filter a $4 \times 4$ fragment of the image with a filter mask $F_2$ by the Winograd method, it is required to perform 16 multiplication operations and 64 addition operations. Calculate the number of operations needed for the filter $F_3$

$$\begin{bmatrix} p_{11} & p_{12} & p_{13} & p_{14} \\ p_{21} & p_{22} & p_{23} & p_{24} \\ p_{31} & p_{32} & p_{33} & p_{34} \\ p_{41} & p_{42} & p_{43} & p_{44} \end{bmatrix} * \begin{bmatrix} 0 & A & 0 \\ A & A & A \\ 0 & A & 0 \end{bmatrix} = \begin{bmatrix} z_1 & z_2 \\ z_3 & z_4 \end{bmatrix}, \tag{32}$$

where

$$z_1 = A(p_{12} + p_{21} + p_{22} + p_{23} + p_{32}),$$

$$z_2 = A(p_{13} + p_{22} + p_{23} + p_{24} + p_{33}),$$

$$z_3 = A(p_{22} + p_{31} + p_{32} + p_{33} + p_{42}),$$

$$z_4 = A(p_{23} + p_{32} + p_{33} + p_{34} + p_{43}).$$

Thus, when filtering a $4 \times 4$ image fragment with $F_3$ filter mask by the direct method, you need to perform 13 multiplication operations and 16 addition operations. When filtering by Winograd method, we get:

$$
D \odot \left( \begin{bmatrix} 1 & 0 & 0 \\ \frac{1}{2} & \frac{1}{2} & \frac{1}{2} \\ \frac{1}{2} & -\frac{1}{2} & \frac{1}{2} \\ 0 & 0 & 1 \end{bmatrix} \cdot \begin{bmatrix} 0 & A & 0 \\ A & A & A \\ 0 & A & 0 \end{bmatrix} \cdot \begin{bmatrix} 1 & 0 & 0 \\ \frac{1}{2} & \frac{1}{2} & \frac{1}{2} \\ \frac{1}{2} & -\frac{1}{2} & \frac{1}{2} \\ 0 & 0 & 1 \end{bmatrix}^{\mathrm{T}} \right)
$$

$$
= \begin{bmatrix} 0 & \frac{1}{2}Ad_{21} & -\frac{1}{2}Ad_{31} & 0 \\ \frac{1}{2}Ad_{12} & \frac{5}{4}Ad_{22} & -\frac{1}{4}Ad_{32} & \frac{1}{2}Ad_{42} \\ -\frac{1}{2}Ad_{13} & -\frac{1}{4}Ad_{23} & -\frac{3}{4}Ad_{33} & -\frac{1}{2}Ad_{43} \\ 0 & \frac{1}{2}Ad_{24} & -\frac{1}{2}Ad_{34} & 0 \end{bmatrix}
$$

(33)

Let's calculate the finite matrix $Z$:

$$Z_{F3} = \begin{bmatrix} z_{f11} & z_{f21} \\ z_{f12} & z_{f22} \end{bmatrix},$$

where

$$z_{f11} = \left( \frac{1}{2}Ad_{12} - \frac{1}{2}Ad_{13} \right) + \left( \frac{1}{2}Ad_{21} + \frac{5}{4}Ad_{22} - \frac{1}{4}Ad_{23} \right) + \left( -\frac{1}{2}Ad_{31} - \frac{1}{4}Ad_{32} - \frac{3}{4}Ad_{33} \right),$$

$$z_{f21} = \left( \frac{1}{2}Ad_{12} + \frac{1}{2}Ad_{13} \right) + \left( \frac{5}{4}Ad_{22} + \frac{1}{4}Ad_{23} - \frac{1}{2}Ad_{24} \right) + \left( -\frac{1}{4}Ad_{32} + +\frac{3}{4}Ad_{33} - \frac{1}{2}Ad_{34} \right),$$

$$z_{f12} = \left( \frac{1}{2}Ad_{21} + \frac{5}{4}Ad_{22} - \frac{1}{4}Ad_{23} \right) - \left( -\frac{1}{2}Ad_{31} - \frac{1}{4}Ad_{32} - \frac{3}{4}Ad_{33} \right) - - \left( -\frac{1}{2}Ad_{42} - \frac{1}{2}Ad_{43} \right),$$

$$z_{f22} = \left( \frac{5}{4}Ad_{22} + \frac{1}{4}Ad_{23} - \frac{1}{2}Ad_{24} \right) - \left( -\frac{1}{4}Ad_{32} + \frac{3}{4}Ad_{33} - \frac{1}{2}Ad_{34} \right) - - \left( \frac{1}{2}Ad_{42} - \frac{1}{2}Ad_{43} \right).$$

Thus, when filtering a $4 \times 4$ image fragment with $F_3$ filter mask by Winograd method, it is necessary to perform 13 multiplication operations and 16 addition operations. The direct method for $F_4$ the filter looks like this:

$$
\begin{bmatrix} p_{11} & p_{12} & p_{13} & p_{14} \\ p_{21} & p_{22} & p_{23} & p_{24} \\ p_{31} & p_{32} & p_{33} & p_{34} \\ p_{41} & p_{42} & p_{43} & p_{44} \end{bmatrix} * \begin{bmatrix} B & B & B \\ B & A & B \\ B & B & B \end{bmatrix} = \begin{bmatrix} z_1 & z_2 \\ z_3 & z_4 \end{bmatrix},
$$

(34)

where

$$z_1 = B(p_{11} + p_{12} + p_{13} + p_{21} + p_{23} + p_{31} + p_{32} + p_{33}) + Ap_{22},$$

$$z_2 = B(p_{12} + p_{13} + p_{14} + p_{22} + p_{24} + p_{32} + p_{33} + p_{34}) + Ap_{23},$$

$$z_3 = B(p_{21} + p_{22} + p_{23} + p_{31} + p_{33} + p_{41} + p_{42} + p_{43}) + Ap_{32},$$

$$z_4 = B(p_{22} + p_{23} + p_{43} + p_{32} + p_{34} + p_{42} + p_{43} + p_{44}) + Ap_{33}.$$

We obtain that filtering an image fragment of size $4 \times 4$ with the mask of filter $F_4$ by the direct method requires 20 multiplication operations and 32 addition operations. With the Winograd method for the $F_4$ the filter we get:

$$
D \odot \left( \begin{bmatrix} 1 & 0 & 0 \\ \frac{1}{2} & \frac{1}{2} & \frac{1}{2} \\ \frac{1}{2} & -\frac{1}{2} & \frac{1}{2} \\ 0 & 0 & 1 \end{bmatrix} \cdot \begin{bmatrix} B & B & B \\ B & A & B \\ B & B & B \end{bmatrix} \cdot \begin{bmatrix} 1 & 0 & 0 \\ \frac{1}{2} & \frac{1}{2} & \frac{1}{2} \\ \frac{1}{2} & -\frac{1}{2} & \frac{1}{2} \\ 0 & 0 & 1 \end{bmatrix}^{\mathrm{T}} \right)
$$

$$
= \begin{bmatrix} Bd_{11} & \frac{3}{2}Bd_{21} & \frac{1}{2}Bd_{31} & Bd_{41} \\ \frac{3}{2}Bd_{12} & (2B + \frac{1}{4})Ad_{22} & (B - \frac{1}{4})Ad_{32} & \frac{3}{2}Bd_{42} \\ \frac{1}{2}Bd_{13} & (B - \frac{1}{4}A)d_{23} & \frac{1}{4}Ad_{33} & \frac{1}{2}Bd_{43} \\ Bd_{14} & \frac{3}{2}Bd_{24} & \frac{1}{2}Bd_{34} & Bd_{44} \end{bmatrix}.
$$

(35)

Calculate the finite matrix $Z$:

$$
Z_{F4} = \begin{bmatrix} z_{f11} & z_{f21} \\ z_{f12} & z_{f22} \end{bmatrix},
$$

where

$$
z_{f11} = \left( Bd_{11} + \frac{3}{2}Bd_{12} + \frac{1}{2}Bd_{13} \right) + \left( \frac{3}{2}Bd_{21} + \left( 2B + \frac{1}{4} \right)Ad_{22} + (B - -\frac{1}{4}A)d_{23} \right)
$$
$$
+ \left( \frac{1}{2}Bd_{31} + \left( B - \frac{1}{4} \right)Ad_{32} + \frac{1}{4}Ad_{33} \right),
$$
$$
z_{f21} = \left( \frac{3}{2}Bd_{12} - \frac{1}{2}Bd_{13} - Bd_{14} \right) + \left( \left( 2B + \frac{1}{4} \right)Ad_{22} - (B - \frac{1}{4}A)d_{23} - -\frac{3}{2}Bd_{24} \right)
$$
$$
+ \left( \left( B - \frac{1}{4} \right)Ad_{32} - \frac{1}{4}Ad_{33} - \frac{1}{2}Bd_{34} \right),
$$
$$
z_{f12} = \left( \frac{3}{2}Bd_{21} + \left( 2B + \frac{1}{4} \right)Ad_{22} + (B - \frac{1}{4}A)d_{23} \right) - \left( \frac{1}{2}Bd_{31} + \left( B - \frac{1}{4} \right)Ad_{32} + +\frac{1}{4}Ad_{33} \right)
$$
$$
- \left( Bd_{41} + \frac{3}{2}Bd_{42} + \frac{1}{2}Bd_{43} \right),
$$

$$z_{f22} = \left(\left(2B + \frac{1}{4}\right)Ad_{22} - \left(B - \frac{1}{4}A\right)d_{23} - \frac{3}{2}Bd_{24}\right) - \left(\left(B - \frac{1}{4}\right)Ad_{32} - \frac{1}{4}Ad_{33} - -\frac{1}{2}Bd_{34}\right)$$
$$- \left(\frac{3}{2}Bd_{42} - \frac{1}{2}Bd_{43} - Bd_{44}\right).$$

We obtain that the filtering of an image fragment of size $4 \times 4$ with a filter mask $F_4$ by Winograd method requires 16 multiplication operations and 64 addition operations. Calculate the direct method for $F_5$:

$$\begin{bmatrix} p_{11} & p_{12} & p_{13} & p_{14} \\ p_{21} & p_{22} & p_{23} & p_{24} \\ p_{31} & p_{32} & p_{33} & p_{34} \\ p_{41} & p_{42} & p_{43} & p_{44} \end{bmatrix} * \begin{bmatrix} A & A & A \\ A & A & A \\ A & A & A \end{bmatrix} = \begin{bmatrix} z_1 & z_2 \\ z_3 & z_4 \end{bmatrix}, \tag{36}$$

where

$$z_1 = A(p_{11} + p_{12} + p_{13} + p_{21} + p_{22} + p_{23} + p_{31} + p_{32} + p_{33}),$$

$$z_2 = A(p_{12} + p_{13} + p_{14} + p_{22} + p_{23} + p_{24} + p_{32} + p_{33} + p_{34}),$$

$$z_3 = A(p_{21} + p_{22} + p_{23} + p_{31} + p_{32} + p_{33} + p_{41} + p_{42} + p_{43}),$$

$$z_4 = A(p_{22} + p_{23} + p_{24} + p_{32} + p_{33} + p_{34} + p_{42} + p_{43} + p_{44}).$$

Thus, filtering a $4 \times 4$ image fragment with $F_5$ filter mask by the direct method requires 16 multiplication operations and 32 addition operations. With Winograd method:

$$D \odot \left(\begin{bmatrix} 1 & 0 & 0 \\ \frac{1}{2} & \frac{1}{2} & \frac{1}{2} \\ \frac{1}{2} & -\frac{1}{2} & \frac{1}{2} \\ 0 & 0 & 1 \end{bmatrix} \cdot \begin{bmatrix} A & A & A \\ A & A & A \\ A & A & A \end{bmatrix} \cdot \begin{bmatrix} 1 & 0 & 0 \\ \frac{1}{2} & \frac{1}{2} & \frac{1}{2} \\ \frac{1}{2} & -\frac{1}{2} & \frac{1}{2} \\ 0 & 0 & 1 \end{bmatrix}^T \right)$$
$$= \begin{bmatrix} Ad_{11} & \frac{3}{2}Ad_{21} & \frac{1}{2}Ad_{31} & Ad_{41} \\ \frac{3}{2}Ad_{12} & \frac{9}{4}Ad_{22} & \frac{3}{4}Ad_{32} & \frac{3}{2}Ad_{42} \\ \frac{1}{2}Ad_{13} & \frac{3}{4}Ad_{23} & \frac{1}{4}Ad_{33} & \frac{1}{2}Ad_{43} \\ Ad_{14} & \frac{3}{2}Ad_{24} & \frac{1}{2}Ad_{34} & Ad_{44} \end{bmatrix}. \tag{37}$$

Calculate the finite matrix $Z$:

$$Z_{F5} = \begin{bmatrix} z_{f11} & z_{f21} \\ z_{f12} & z_{f22} \end{bmatrix},$$

где

$$z_{f11} = \left( Ad_{11} + \frac{3}{2}Ad_{12} + \frac{1}{2}Ad_{13} \right) + \left( \frac{3}{2}Ad_{21} + \frac{9}{4}Ad_{22} + \frac{3}{4}Ad_{23} \right)$$
$$+ + \left( \frac{1}{2}Ad_{31} + \frac{3}{4}Ad_{32} + \frac{1}{4}Ad_{33} \right),$$

$$z_{f21} = \left( \frac{3}{2}Ad_{12} - \frac{1}{2}Ad_{13} - Ad_{14} \right) + \left( \frac{9}{4}Ad_{22} - \frac{3}{4}Ad_{23} - \frac{3}{2}Ad_{24} \right)$$
$$+ + \left( \frac{3}{4}Ad_{32} - \frac{1}{4}Ad_{33} - \frac{1}{2}Ad_{34} \right),$$

$$z_{f12} = \left( \frac{3}{2}Ad_{21} + \frac{9}{4}Ad_{22} + \frac{3}{4}Ad_{23} \right) - \left( \frac{1}{2}Ad_{31} + \frac{3}{4}Ad_{32} + \frac{1}{4}Ad_{33} \right)$$
$$- - \left( Ad_{41} + \frac{3}{2}Ad_{42} + \frac{1}{2}Ad_{43} \right),$$

$$z_{f22} = \left( \frac{9}{4}Ad_{22} - \frac{3}{4}Ad_{23} - \frac{3}{2}Ad_{24} \right) - \left( \frac{3}{4}Ad_{32} - \frac{1}{4}Ad_{33} - \frac{1}{2}Ad_{34} \right)$$
$$- - \left( \frac{3}{2}Ad_{42} - \frac{1}{2}Ad_{43} - Ad_{44} \right).$$

Thus, filtering a $4 \times 4$ image fragment with $F_5$ filter mask by Winograd method requires 16 multiplication operations and 64 addition operations.

Table 2 shows the calculation results of multiplication and addition operations using direct and Winograd methods for filters of different symmetry. At convolution by filter $F_2$ it is preferable to use the Winograd method since the number of multiplications decreases 2.12 times than the direct method. For filters $F_1$, $F_3$, $F_4$ it is possible to determine which method is better only after a practical test. For filter $F_5$ it is much better to apply filtering by the direct method, as the number of products

**Table 2** The number of addition and multiplication operations for filters with arbitrary coefficients and different symmetry

| Filter | Implementation method | | | |
|--------|-----------------------|-----|----------|-----|
|        | Direct | | Winograd | |
|        | × | + | × | + |
| $F_1$ | 15 | 12 | 12 | 28 |
| $F_2$ | 34 | 32 | 16 | 64 |
| $F_3$ | 13 | 16 | 12 | 28 |
| $F_4$ | 20 | 32 | 16 | 64 |
| $F_5$ | 16 | 32 | 16 | 64 |

in the direct method and the Winograd method is the same. Still, the additions are 2 times less in the direct method of realization.


# 4 Conclusions

In this paper, the computational complexity of digital image filtering methods by linear spatial filters was analyzed. Comparison of the direct realization method and Winograd method of two-dimensional filtering showed that the Winograd method is not preferable for all spatial filters. The best result when using the Winograd method is achieved for averaging Gaussian filter and for symmetric filters of general form with F_2 mask. The Winograd method application for Gauss filter allows reducing 3 times number of multiplication operations with 1.84 times increase of addition operations. Winograd method for filters with the mask of F_2 form allows decreasing multiplication operations by 2.12 times while the number of addition operations increases by 2 times. For all other considered filters, the benefits from the Winograd method application are not so obvious and require practical testing in conditions of real image processing by software or hardware.

An interesting direction of further research is the development of hardware architecture of Gaussian filter and filter with F_2 mask, using the calculation by Winograd method, with further implementation on FPGA. The resulting devices can be used in various areas of image processing in practice, which require real-time processing.

**Acknowledgements** The authors would like to thank the North-Caucasus Federal University for supporting in the contest of projects competition of scientific groups and individual scientists of the North-Caucasus Federal University. The work is supported by North-Caucasus Center for Mathematical Research under agreement № 075-02-2021-1749 with the Ministry of Science and Higher Education of the Russian Federation.


# References

1. Brill A, Feng Q, Humensky TB, Kim B, Nieto D, Miener T (2019) Investigating a deep learning method to analyze images from multiple gamma-ray telescopes. In: Proceedings of NYSDS: 2019 New York scientific data summit. IEEE, New York
2. Litjens G et al (2017) A survey on deep learning in medical image analysis. Med Image Anal 42:60–88
3. Fang W et al (2019) Recognizing global reservoirs from landsat 8 images: a deep learning approach. IEEE J Sel Top Appl Earth Observ Remote Sens 12(9):3168–3177
4. Xuan S, Li S, Han M, Wan X, Xia G-S (2020) Object tracking in satellite videos by improved correlation filters with motion estimations. IEEE Trans Geosci Remote Sens 58:1074–1086
5. Kim JH, Ahn IJ, Nam WH, Ra JB (2015) An effective post-filtering framework for 3-DPET image denoising based on noise and sensitivity characteristics. IEEE Trans Nucl Sci 62:137–147
6. Gavaskar RG, Chaudhury KN (2019) Fast adaptive bilateral filtering. IEEE Trans Image Process 28:779–790

7. Li S, Zhao S, Cheng B, Chen J (2018) Accelerated particle filter for real-time visual tracking with decision fusion. IEEE Signal Process Lett 25:1094–1098
8. Gonzalez RC, Woods RE (2018) Digital image processing, 4th edn. Pearson Education Limited, Harlow
9. Winograd S (1980) Arithmetic Complexity of Computations. Society for Industrial and Applied Mathematics, Pennsylvania
10. Liu ZG, Mattina M (2020) Efficient residue number system based winograd convolution. Springer, pp 53–68

# Non-classical Methods of Determination Critical Forces of Compressed Rods

**Kh. P. Kulterbaev** and **M. M. Lafisheva**

**Abstract** Determination of the critical forces of compressed rods both in their study in universities and in the design of structures of mechanical systems is one of the difficult tasks. The analytical methods of solution used so far are of a particular nature and are based on the use of the basic homogeneous differential equation describing the buckling of a rod with a constant cross-section along the length, supplemented by simple boundary conditions at the ends, etc. It is proposed using computer technologies to significantly simplify the solution of complex problems of the stability of a compressed rod and at the same time achieve greater clarity and versatility. The application of the proposed analytical-graphic and numerical-graphic methods is shown on specific examples. The solution of test and real problems of determining the critical forces for rods of constant and variable cross-sections is shown. The proposed mathematical models and decision algorithms are verified using computational experiments. On the basis of the results obtained, practical conclusions are drawn.

**Keywords** Ordinary homogeneous differential equation · Boundary conditions · Mathematical model · Critical force · Loss of stability · Analytical-graphical and numerical-graphical methods · Characteristic equation

## 1 Introduction

Determination of the critical forces of compressed rods both in their study in universities and in the design of structures of mechanical systems is one of the most difficult, but always urgent problems. Therefore, the solution of Euler's simple problem in 1744 continued for numerous schemes of rods, for structures in the form of trusses, beams, arches, frames, thin-walled rods, plates, shells, etc. Achievements in these

Kh. P. Kulterbaev (✉) · M. M. Lafisheva
NCFU, North Caucasian Center for Mathematical Research, Stavropol, Russia
e-mail: kulthp@mail.ru

© The Author(s), under exclusive license to Springer Nature Switzerland AG 2022
A. Tchernykh et al. (eds.), *Mathematics and its Applications in New Computer Systems*,
Lecture Notes in Networks and Systems 424,
https://doi.org/10.1007/978-3-030-97020-8_9

areas have an extensive bibliography [1–7]. The stability of multi-span rods is considered in [8, 9]. In article [10], the problem of stability of a rod of variable cross-section
was solved for the first time by a numerical-graphical method.

Analytical methods are quite effective for simple design schemes, but they become
too inconvenient for more complex rods. The boundary conditions used by these
methods take into account the kinematic and static actions in the end sections, but
do not take into account the properties of the bar between them. Another drawback
of analytical methods is associated with the inevitable appearance in algorithms
of solutions of systems of transcendental equations containing the required critical
forces.

In such cases, analytical-graphic and numerical-graphic methods are more effective, which are devoid of the listed disadvantages. Solutions with their help are
carried out easily with the help of computer standard subroutines available in almost
all computing systems. It is even more convenient to use the characteristic equation
of an algebraic system containing critical forces. The roots of such a single equation
are easily calculated graphically.

In this paper, the main attention will be focused on the application of numerical
methods that have standard non-classical methods for determining the critical forces.
For this purpose, let us turn to one of the calculation schemes (Fig. 1), proposed
in [3], and we will work out on it calculations based on analytical-graphic and
numerical-graphic methods.

In the latter case, we use the finite difference method (FCD). In this case, the
design scheme will be used twice: the axial moment of inertia is a constant value
$J = const$, the axial moment of inertia is variable along the axis $J = J(x)$. This
approach has a double chain: in the first case, the reliability of the results obtained

**Fig. 1** Design scheme of the
rod

by the analytical-graphic method is checked, in the second case, the effectiveness of MCS in solving a complex problem that has no solution by classical methods is proved.

In the Russian Federation, MCS was deeply developed and introduced in the works of A.A. Samarsky and his disciples. The work is devoted to theoretical questions of the application of numerical methods for solving problems of mathematical physics. In the early stages and further, the development of numerical methods was facilitated by textbooks [11–15]. The authors of works considered general and special issues of using numerical methods in calculations and design of building and machine-building structures.

## 2   Constant Cross-section Bar

We begin the determination of the critical force by the analytical-graphic method by considering the classical Euler problem (Fig. 2). A rectilinear rod with hinged ends with a constant cross-section under the action of an axial longitudinal force equal to the critical $F_n$, goes over from a rectilinear.

shape to the curvilinear form $v(x)$, depicted by dotted lines. At the same time, the critical force is easily determined analytically using a mathematical model consisting of a basic ordinary differential equation compiled for a curved axis [6]

$$v^{IV}(x) + k^2 v'''(x) = 0, k^2 = F/b, b = EJ, \forall x \in (0, l). \tag{1}$$

**Fig. 2** Design scheme

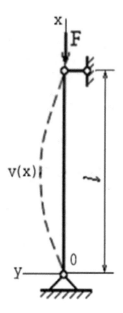

and boundary conditions.

$$\text{At the lower end: } v(0) = 0, \qquad v''(0) = 0. \tag{2}$$

$$\text{At the upper end:} v(l) = 0, \qquad v''(l) = 0. \tag{3}$$

Here v (x) is the function of the curved axis of the bar at loss of stability, the dashes in the superscripts denote derivatives with respect to the argument. The rest of the designations are generally accepted.

It is easy to show that in this case the critical force has the form of the set.

$$F_n = \frac{n^2 \pi^2 E J}{l^2}, n = 1, 2, 3 \ldots \tag{4}$$

At the same time, it should be pointed out that this case is the simplest. In most cases of design schemes, the use of the analytical method is associated with great difficulties and possible losses of critical forces from sets similar to (4). Let us solve problem (1)–(3) as a test problem for testing the analytical-graphic method. The general solution to Eq. (1) has the form

$$v(x) = a_1 \, sin \, kx + a_2 \, cos \, kx + a_3 x + a_4. \tag{5}$$

Then the second derivative of the function will be as follows

$$v''(x) = -a_1 sin \, k^2 sinkx - a_2 k^2 cos \, cos \, kx \tag{6}$$

Equations (2), (3), (5), (6) imply the equalities $a_2, a_3, a_4 = 0, a_1 \, \sin \, kl = 0$. For the function v (x) to be identically non-zero, the condition $\sin \, kl = 0$. Then the argument $kl$ must have the values $kl = n\pi$. From here and from (1) we obtain the critical force formula (4), which completes the analytical solution.

Now let us consider an example of solving the same problem by the analytical-graphical method for testing it.

*Example 1.* For the design scheme in Fig. 2 are given: $l = \pi$, $EJ = 1$. It is required to determine the first three values $F_n$ by the analytical-graphical method with the initial dimensionless data.

Equations (2), (3), (5), (6) give:

$$a_2 + a_4 = 0, -a_2 k^2 = 0,$$

$$a_1 sin \, kl + a_2 cos \, kl + a_3 l + a_4 = 0, -a_1 sin \, k^2 sin \, kl - a_2 k^2 \, cos \, kl = 0$$

We represent the system of equations in matrix-vector form

$$BA^T = 0, \tag{7}$$

where A - column vector $(a_1, a_2, a_3, a_4)^T$, T - sign of the vector's transposition, B - matrix of coefficients. Let us write (7) in expanded form

$$BA^T = \begin{bmatrix} 0 & 1 & 0 & 1 \\ 0 & -k^2 & 0 & 0 \\ \sin kl & -\cos kl & l & 1 \\ k^2\sin kl & -k^2\cos kl & 0 & 0 \end{bmatrix} \cdot \begin{bmatrix} a_1 \\ a_2 \\ a_3 \\ a_4 \end{bmatrix} = \begin{bmatrix} 0 \\ 0 \\ 0 \\ 0 \end{bmatrix}. \qquad (8)$$

For vector A to be nonzero, it is required that the determinant of the matrix is zero

$$det\, B(F) = 0. \qquad (9)$$

Equation (9) is transcendental and its solution presents significant difficulties. Therefore, we will use a non-classical graphical method.

Computing complex Matlab has very convenient software tools for calculating determinants of matrices, plotting curves F - det. along two vectors with given elements and multiple enlargement of the fragments of the picture. Using them and following the calculation algorithm outlined above, a graph of the function is obtained on the monitor screen (Fig. 3). Intersection ordinates of the curve with axis F and are equal to the critical forces $F_n$. The ordinates of the intersection of the curve with the F-axis are marked with dots. They are exactly equal to the critical forces $F_n$, for n = 1, 2, 3, 4, ie, F = {1, 4, 9, 16}. The results of this example confirm that the analytical-graphical method gives highly accurate results for the critical forces of compressed bars.

Next, we will show how the method is implemented for a real example. For this purpose, we will use the calculation scheme in Fig. 4, which shows the post-critical

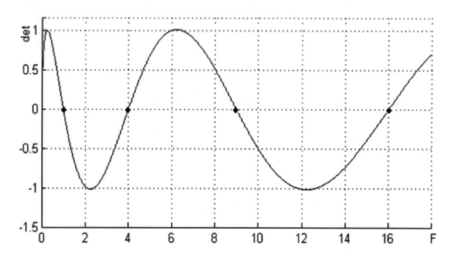

**Fig. 3** Values of critical forces

**Fig. 4** Design scheme

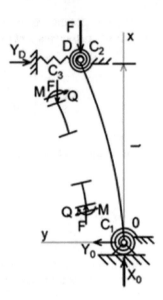

state of the rod. For it, the basic Eq. (1) is preserved, but the boundary conditions will be different

$$\text{At the lower end: } v(0) = 0, \ v''(0) - c_1 v'(0) = 0. \tag{10}$$

$$\text{At the upper end: } bv''(l) - c_2 v'(l) = 0, \ bv'''(l) - c_3 v(l) + F v'(l) = 0. \tag{11}$$

Together, (1), (10), (11) represent a system of five homogeneous differential equations, for which it is required to determine the set of eigenvalues $F_n$, $n = 1, 2, \dots$
As in the previous case, we take the general solution in the form

$$v(x) = a_1 \sin kx + a_2 \cos kx + a_3 x + a_4 x^2. \tag{12}$$

Then the derivatives of the function will be as follows

$$v'(x) = a_1 kx - a_2 \sin kx + a_3 + 2a_4 x,$$

$$v''(x) = -a_1 \sin k^2 \sin kx - a_2 k^2 \cos \cos kx - 2a_4, \tag{13}$$

$$v'''(x) = -a_1 \cos k^3 \cos kx + a_2 k^3 \cos kx.$$

We use (10)–(13) and write down four equations containing the vector A and the matrix B of the fourth order.

$$a_2 = 0, \theta a_1 + \kappa a_2 + \rho a_3 + 2ba_4 = 0, \alpha a_1 + \beta a_2 - c_2 a_3 + \lambda a_4 = 0,$$
$$\varepsilon a_1 + \delta a_2 + \eta a_3 + \mu a_4 = 0. \tag{14}$$

Здесь

$$\theta = -c_1 k, \kappa = -bk^2, \rho = -c_1, \alpha = -bk^2 \sin kl - c_2 k \cos kl, \ \beta = -bk^2 \cos kl - c_2 k \sin \kappa l,$$

$$\lambda = 2(b - c_2 l), \ \varepsilon = -bk^3 \cos kl - c_3 k \sin \kappa l + F \cos kl,$$
$$\delta = -c_3 \cos kl - F k \sin kl, \eta = -c_3 l, \ \mu = -c_3 l^2 + 2Fl.$$

Further, the usually used classical method for determining the critical forces is to represent Eqs. (14) in matrix-vector forms (7) and (8).

$$BA^T = 0, \tag{15}$$

where A—column vector $(a_1, a_2, a_3, a_4)^T$, т—sign of the vector's transposition, B—matrix of coefficients determined by boundary conditions (10) and (11). In this case, the components of the vector A are unknown numbers, the elements of the matrix B contain the required critical force F.

Let us write (15) in expanded form

$$BA^T = \begin{bmatrix} 0 & 1 & 0 & 1 \\ \theta & \kappa & \rho & 2b \\ \alpha & \beta & -c_2 & \lambda \\ \varepsilon & \delta & \eta & \mu \end{bmatrix} \cdot \begin{bmatrix} a_1 \\ a_2 \\ a_3 \\ a_4 \end{bmatrix} = \begin{bmatrix} 0 \\ 0 \\ 0 \\ 0 \end{bmatrix} \tag{16}$$

Obviously, the system of Eqs. (16) has a trivial solution $A = 0$ and it corresponds to a non-bent vertical state of the rod at small values of the force F. A nontrivial solution $A \neq 0$ exists only if the determinant of the matrix B is equal to zero, i.e.

$$\det B(F) = 0. \tag{17}$$

Equation (17) is transcendental and its composition and solution for the calculation scheme, for example, according to Fig. 4 presents significant difficulties. Therefore, we will use a non-classical graphical method using a curve $F - \det B(F)$ displayed on a computer monitor. The ordinates of the intersection of the curve with the F axis and are equal to the critical forces $F\kappa$.

A computer program in the environment of the Matlab computing complex produced a curve in Fig. 3. The ordinates of the marked points, read from the monitor screen (Fig. 3), coincide with the exact ones, namely, differ from them by an almost indistinguishable value of the order of 0.00001. Hence the conclusion: the analytical-graphical method used here is a simple and effective means of determining the critical

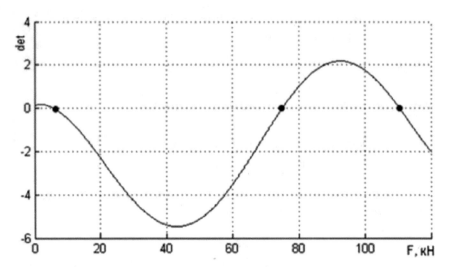

**Fig. 5** Critical force values

forces of a compressed bar. Let's move on to solving a real problem. We choose the design scheme according to Fig. 4.

*Example 2.* A rod from a steel pipe, length $l = 6$ m, $E = 2^{10}$ GPa, and an axial moment of inertia of $J = 66, 61$ cm$^4$. Spring stiffness coefficients: $c_1 = 2000$ Nm/rad, $c_2 = 1000$ Nm/rad, $c_3 = 1500$ N/m.

Using the Matlab computer complex and following the calculation algorithm described above, a graph of the function F—det was obtained on the monitor screen (Fig. 5). By increasing the fragments of the point of intersection of the curve of the F axis, the first three values of the critical force are read with a high degree of accuracy.

$$F = \{5, 914; 75, 086; 110, 472\}kN.$$

At the same time, it is obvious that when solving Eq. (17) by analytical methods, difficult to overcome difficulties would arise.

Noting the comparative ease of solving the problem by the analytical-graphic method, it should be noted that a rod of constant cross-section was considered. It is necessary to deal with more complex design schemes of rods: with continuously variable sections along the length, with stepwise variable sections along the axis, with forces applied not only to the ends, but also in the intervals between them, etc.

## 2.1 Bar of Variable Cross-section

Now let the geometric characteristics of the section of the bar be continuously variable along the length, in particular, the moment of inertia is represented by some given function $J = J(x)$. We select the design scheme according to Fig. 4, where the rod is shown in a tilted position. In this case, in the mathematical model of the problem, the basic equation has the form.

$$[b(x)v''(x)]'' + Fv''(x) = 0, b(x) = EJ(x), \forall x \in (0, l). \tag{18}$$

Boundary conditions (10, (11), given above) are added to it. It is obvious that the solution of such a problem by classical methods will be fraught with great difficulties. Therefore, a numerical-graphic method is proposed.

Let us pass from differential operators to finite difference ones, to the method of finite differences (FDM). The region of continuous variation of the argument of the function $x \in [0, l]$ is replaced by a uniform grid with step h

$$Q_h = \{x_i = (i - 1)h, h = l/(n - 1), i = 1, 2, \ldots, n, i = 1, 2, \ldots n\}.$$

In this case, the continuous function $v(x)$ is associated with the grid function $v_i(x_i)$ at the nodes, i.e. $v_i(x_i) \approx v(x_i)$. In what follows, we will use finite-difference operators that ensure accuracy $O(h^2)$ [16]. Their use and elementary transformations give an algebraic system instead of the basic Eq. (18).

To them are added the boundary conditions at the lower and upper ends, given above (2), (3)

$$b_{i-1}v_{i-2} + p_i v_{i-1} + q_i v + r_i v_{i+1} + b_{i+1}v_{i+2} = 0, i = 3, 4, \ldots, n - 2. \tag{19}$$

$$\alpha_i = -2b_{i-1} - 2b_i, \beta_i = b_{i-1} + 4b_i + b_{i+1}, \gamma_i = -2b_i - 2b_{i+1},$$
$$p_i = \alpha_i + Fh^2, q_i = \beta_i - 2Fh^2, r_i = \gamma_i + Fh^2,$$

The boundary conditions at the lower and upper ends given above (2), (3) are added to them

$$v_1 = 0, \kappa v_1 + \lambda v_2 + \mu v_3 - v_4 = 0, \tag{20}$$

$$3v_{n-4} - 14v_{n-3} + \varepsilon v_{n-2} + \eta v_{n-1} + \theta v_n = 0, \tag{21}$$

$$-v_{n-3} + \sigma v_{n-2} + \xi v_{n-1} + \eta v_{n-1} + \delta v_n = 0, \tag{22}$$

$$\kappa = 2 + \frac{3c_1 h}{2b_1}, \lambda = -5 - \frac{2c_1 h}{b_1}, \mu = 4 + \frac{c_1 h}{2b_1}, \varepsilon = 24 + \frac{Fh^2}{b_{n-1}}, \eta = -18 - \frac{4Fh^2}{b_{n-1}},$$

$$\theta = 5 - \frac{2h^3 c_3}{b_{n-1}}, \sigma = 4 - \frac{hc_2}{2b_n}, \xi = -5 - \frac{hc_2}{2b_n}, \delta = 2 - \frac{3hc_2}{2b_n}.$$

Equations (19)–(22) form an algebraic system in matrix-vector form

$$BV^T = 0. \tag{23}$$

Here B is the matrix of coefficients in Eqs. (19)–(22), V is the column vector of displacements of the nodes of the grid region, $V = (v_1, v_2, \ldots v_n)^T$, T—is the sign of the vector transposition.

$$
B = \begin{pmatrix}
1 & & & & & \vdots & & & & & \\
\kappa & \lambda & \mu & -1 & & \vdots & & & & & \\
b_{i\text{-}1} & p_i & q_i & r_i & b_{i+1} & \vdots & & & & & \\
& b_{i\text{-}1} & p_i & q_i & r_i & b_{i+1} & \vdots & & & & \\
\cdots & \cdots & \cdots & \cdots & \cdots & \cdots & \cdots & \cdots & \cdots & \cdots & \cdots \\
& & & & b_{i\text{-}1} & p_i & q_i & r_i & b_{i+1} & & \\
\cdots & \cdots & \cdots & \cdots & \cdots & \vdots & \cdots & \cdots & \cdots & \cdots & \cdots \\
& & & & \vdots & & b_{i\text{-}1} & p_i & q_i & r_i & b_{i+1} \\
& & & & \vdots & & & b_{i\text{-}1} & p_i & q_i & r_i & b_{i+1} \\
& & & & \vdots & & & & 3 & -14 & \varepsilon & \eta & \theta \\
& & & & \vdots & & & & & -1 & \sigma & \xi & \delta
\end{pmatrix},
$$

$$V = (v_1, v_2, \ldots, v_n)^T.$$

Zero elements are not written out. The characteristic equation has the form (17). Its roots are found graphically using computer calculations.

Example 3. Vertical bar with a design scheme according to Fig. 1 from a steel pipe, tapered upward, has the initial data $l = 7$ m, $E = 2, 1 \cdot 10^{11}$ Pa, $D_1 = 12$ cm, $D_2 = 9$ cm, $t = 1,4$ mm, $c_1 = 2000$ Nm/rad, $c_2 = 1000$ Nm/rad, $c_3 = 1500$ N/m, $m = 10001$, $m$—number of points along the $F$ axis. $D_1$, $D_2$—diameters of the lower and upper sections of the pipe, t - wall thickness.

It is required to determine the first three eigenvalues by the numerical-graphical method with the buckling of the bar.

To determine the eigenvalues, a small computer program was developed in the Matlab environment. She displayed the graph of the function F—det on the monitor screen (Fig. 6). In the figure, bold dots mark the values of the critical forces $F$. Their abscissas correspond to the three smallest values of the critical force.

$$F = \{25, 230; 100, 764; 226, 649\}kN.$$

At the same time, note that Matlab has a "magnifying glass" that allows you to view the neighborhood of these points in a several thousand times enlarged form. Therefore, the eigenvalues read with its help should be recognized as highly accurate, despite their graphic origin.

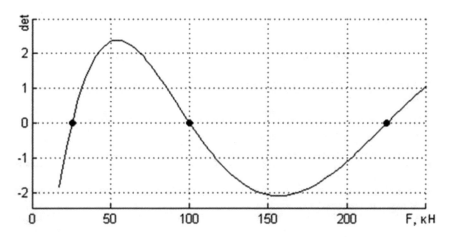

**Fig. 6** Critical forces

This allows us to assert that the finite difference method can be used to solve complex problems without resorting to analytical methods.


# 3  Conclusions

1.  The proposed non-classical analytical-graphic method easily solves complex problems of determining the critical forces of compressed rods, which is confirmed by test and real examples.
2.  According to the three methods discussed above, you can make recommendations for their application:
3.  – classical analytical for bars of constant cross-section with simple boundary conditions;

    – analytical-graphical for bars of constant cross-section with complex boundary conditions;
    – numeric-graphical for bars of variable cross-section.

4.  The methods proposed here can be used to calculate the eigenvalues in the problems of the period of oscillations to determine the frequencies of free oscillations and to solve elliptic differential equations of the Sturm-Liouville type.
5.  It is advisable to include analytical-graphic and numerical-graphic methods in the curricula of courses on structural mechanics, strength of materials, technical mechanics, mechanics of materials and structures.

**Acknowledgements**  The work is supported by North-Caucasus Center for Mathematical Research under agreement №. 075-02-2021-1749 with the Ministry of Science and Higher Education of the Russian Federation and part study was funded by RFBR, project number 20-37-70023.

# References

1. Feodosiev VI (2003) Strength of Materials, 10th edn. Publishing House of MSTU im. N.E.Bauman, Moscow
2. Korobko VI, Korobko AV (2008) Dynamics and stability of rod systems. ASV Publishing House, Moscow
3. Maslennikov AM (2016) Dynamics and stability of structures. Yurayt Publishing House, Moscow
4. Makarov EG (2009) Resistance of materials using computing systems: In 2 books. The main course. Higher School, Moscow
5. Vardanyan GS, Nikolai MA, Gorshkov AA (2011) Resistance of materials with the basics of structural mechanics
6. Gorodetsky AA, Barabash MS, Sidorov VN (2016) Computer modeling in problems of structural mechanics. ASV Publishing House
7. Gorshkov AG, Troshin VN, Shalashilin VI (2002) Strength of materials. FIZMATLIT, Moscow
8. Kulterbaev KhP, Karmokov KA (2013) On the stability of a multi-span rod of variable stiffness on flexible supports. Bull VolgGASU Ser Build Archit 34(53):90–98
9. Chechenov TYu (2011) On the stability of a multi-span rod on flexible supports. Vestnik Nizhegorodskogo gosudarstvennogo universiteta im. N.I. Lobachevsky. In: Proceedings of the X All-Russian Congress on fundamental problems of theoretical and applied mechanics, no 4, part 2. Publishing House of NNSU named after N.I. Lobachevsky, pp 1850–1851
10. Kulterbaev KhP, Baragunova LA, Shogenova MM, Senov KhM (2018) About a high-precision graphoanalytical method of determination of critical forces of an oblate rod. In: Proceedings of IEEE International Conference "Quality Management, Transport and Information Security, Information Technologies" (IT & QM & IS), St. Petersburg. Russia, 24–28 September 2018, pp 794–796
11. Formalev VF, Reviznikov DL (2006) Numerical methods. Fizmatlit, Moscow
12. Zolotov AB, Akimov PA, Sidorov VN, Mozgaleva ML (2008) Mathematical methods in structural mechanics (with the foundations of the theory of generalized functions). Publishing House ASV
13. Zolotov AB, Akimov PA, Sidorov VN, Mozgaleva ML (2009) Numerical and analytical methods for calculating building structures. ASV, Moscow
14. Verzhbitsky VM (2002) Foundations of numerical methods. Russia, Moscow
15. Ilyin VP, Karpov VV, Maslennikov AM (2005) Numerical methods for solving problems of structural mechanics. Publishing House ASV, Moscow

# Numerical Simulation of the Control System of the Propulsion-Steering System of Tele-Controlled Uninhabited Vehicles

**M. Dantsevich⬤, M. N. Lyutikova⬤, and I. V. Samoylenko⬤**

**Abstract** The article "Numerical simulation of the control system of the propulsion-steering system of tele-controlled uninhabited vehicles" describes a method simulation of the control system of the propulsion-steering system of tele-controlled uninhabited vehicles, in which movement, change of direction of movement and positioning are ensured by four thrusters located in the plan rectangular bluff platform. Vertical movement together with the propulsion and steering complex is ensured by a separate vertical movement system.

The control system is fundamentally non-linear, arising from the characteristics of the propulsive force of the thrusters. The method of modeling the control system of the propulsion and steering complex considers static errors of the control system and refers to numerical methods.

**Keywords** Remotely Operated Vehicle (ROV) · Engine-steering system (ESS) · Control system · Numerical methods · Positioning · Force of propellers · Mathematical model · Transformation matrix · Command signal · Transfer coefficients

## 1 Introduction

The remotely operated underwater vehicle is a rectangular platform, along the edges of which there are propellers. The propellers do not turn and are controlled by switching the thrust of the propellers, due to the resulting skew, the resulting vector of movement of the tele-controlled uninhabited vehicle (ROV) changes [1]. Four propellers are installed according to the $4 \times 45°$ scheme in the corners of the square to control the maneuvers of the ROV in the horizontal plane (propulsion thrusters) (Fig. 1) and three thrusters to control the diving and surfacing maneuvers (vertical

M. Dantsevich · M. N. Lyutikova (✉)
Admiral Ushakov Maritime State University, Novorossiysk, Russian Federation
e-mail: mnlyutikova@mail.ru

I. V. Samoylenko
Stavropol State Agrarian University, Stavropol, Russian Federation

© The Author(s), under exclusive license to Springer Nature Switzerland AG 2022          103
A. Tchernykh et al. (eds.), *Mathematics and its Applications in New Computer Systems*,
Lecture Notes in Networks and Systems 424,
https://doi.org/10.1007/978-3-030-97020-8_10

thrusters) [2, 3]. The pilot of the ROV controls the propulsive force of the thrusters using two joystick manipulators, the handle of each of which has two degrees of freedom in mutually perpendicular directions $x_i$, $y_i$, i = 1,2 (Fig. 2).

The propulsion devices are controlled by deviating from the average position of the manipulator handle M1 along the $x_1$, $y_1$ axes and the handle of the M2 manipulator

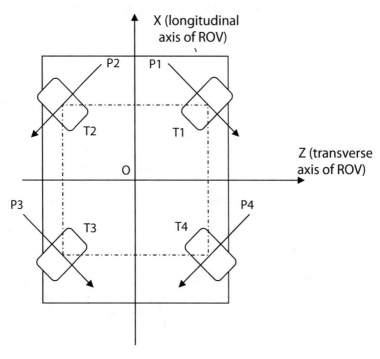

**Fig. 1** The layout of the propulsion thrusters T1, T2, T3, T4 on ROV

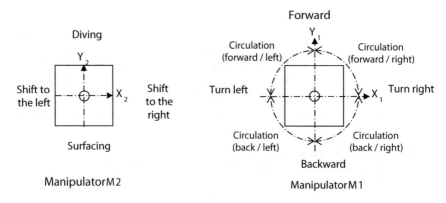

**Fig. 2** Manipulator working areas

along the $x_2$ axis. The vertical thrusters are controlled by deviation from the average position of the M2 manipulator handle along the $y_2$ axis.

The output signals of manipulators M1 and M2 usually change discretely at 16 levels from zero when the joystick handles deviate from the middle position $x_1 = y_1 = x_2 = y_2 = 0$ (zero signal values) to the maximum values $\pm x_m$, $\pm y_m$ in two perpendicular directions ($-x_m \leq x_i \leq x_m$, $-y_m \leq y_i \leq y_m$, $i = 1,2$) [4].

## 2   Materials and Methods

The formation of the movement of the ROV by means of manipulators M1 and M2 requires the development of a model that implements the propulsion control algorithm. While the expected movement of the ROV and the polarity of the deviation of the handles that control the propulsion devices should be coordinated as follows [4]:

– direction (+) along the $y_1$ axis sets the forward movement of the ROV;
– direction (−) along the $y_1$ axis sets the ROV movement backward;
– direction (+) along the $x_1$ axis sets the turn of the ROV to the right relative to the center of the ROV ($y1 = 0$);
– direction (−) along the $x_1$ axis sets the turn of the ROV to the left relative to the center of the ROV ($y1 = 0$);
– simultaneous deflection of the M1 handle along the $y_1$ and $x_1$ axes sets the circulation of the ROV (translational movement of the ROV forward ($y_1 > 0$) or backward ($y_1 < 0$) with a simultaneous turn of the ROV to the right ($x_1 > 0$) or to the left ($x_1 < 0$)
– direction (+) along the $x_2$ axis sets the translational movement of the ROV to the right perpendicular to its longitudinal axis;
– direction (−) along the $x_2$ axis sets the translational movement of the ROV to the left perpendicular to its longitudinal axis;
– backward ($y_1 < 0$) with a simultaneous turn of the ROV to the right ($\times 1 > 0$) or to the left ($\times 1 < 0$));
– direction (+) along the $x_2$ axis sets the translational movement of the ROV to the right perpendicular to its longitudinal axis
– direction (−) along the $x_2$ axis sets the translational movement of the ROV to the left perpendicular to its longitudinal axis.

We define the OXYZ system, the right triplet (Fig. 1), by the associated ROV coordinate system.

The center of gravity of the ROV is usually placed in the center and the center of mass of the platform is reached by the center of mass, the X axis is forward direction, the Y axis is considered perpendicular to the plane of the ROV, the Z axis is the lateral coordinate corresponding to the transverse plane [4, 5].

The numbering of the thrusters corresponds to the movement of the vector counterclockwise from T1 in the first quadrant to T4 in the fourth. An important indicator is the installation angles of the propellers $\varphi_1$, $\varphi_2$, $\varphi_3$, $\varphi_4$, also a positive direction corresponds to counterclockwise movement [6, 7].

The points of application of the propulsive forces of the thrusters are determined by the system of radius vectors:

$$\vec{r}_i = \vec{i} x_{Pi} + \vec{j} 0 + \vec{k} z_{Pi} , \tag{1}$$

where $\vec{i}, \vec{j}, \vec{k}$ is orthogonal basis triplet OXYZ; $x_{Pi}$, $y_{Pi} = 0$, $z_{Pi}$ are the current coordinates of the points of application of the propulsive force Pi. For a bound coordinate system, the values of the stop forces (thrust) are:

$$P_x = \sum_{i=1}^{4} P_i sin(\phi_i) , \quad P_z = \sum_{i=1}^{4} P_i cos(\phi_i) . \tag{2}$$

Consequently, for the moment created by the propulsive force relative to the center of mass:

$$\vec{M} = \sum_{i=1}^{4} \vec{r}_i \times \vec{P}_i = \sum_{i=1}^{4} \begin{vmatrix} \vec{i} & \vec{j} & \vec{k} \\ x_{Pi} & 0 & z_{Pi} \\ P_i sin(\phi_i) & 0 & P_i cos(\phi_i) \end{vmatrix} = \vec{j} M_y$$

$$= \vec{j} \sum_{i=1}^{4} P_i(z_{Pi} sin(\phi_i) - x_{Pi} cos(\phi_i)). \tag{3}$$

The forces $P_x$, $P_z$ and the moment $M_y$ are control actions that cause the maneuvers of the ROV set by the pilot.

Formulas (2) and (3) written in matrix form:

$$F = BP, \tag{4}$$

where $F = (P_x \ M_y \ P_z)^T$ - vector of control actions;
$P = (P_1 \ P_2 \ P_3 \ P_4)^T$ - propulsive force vector;
$B = (b_{ij})_{3 \times 4}$ - coefficient matrix.
The elements of matrix B have the form:

$$b_{1j} = sin(\phi_j); \ b_{2j} = z_{Pj} sin(\phi_j) - x_{Pj} cos(\phi_j); \ b_{3j} = cos(\phi_j); \ j \in 1, 4. \tag{5}$$

With a symmetrical arrangement of the thrusters relative to the origin of the OXYZ coordinate system, we denote:

$$|x_{Pi}| = a, \ |z_{Pi}| = b, \ \rightleftarrows \ (i \in 1,4). \tag{6}$$

In Table 1, we present the values of the mathematical expectations of the coordinates $x_{Pi}$, $z_{Pi}$, the angles $\phi_i$ и and the values of the functions $sin(\phi_i)$, $cos(\phi_i)$.

**Table 1** Values of mathematical expectations of coordinates and function values

| i | $X_{Pi}$ | $Z_{Pi}$ | $\varphi_i$ (degree) | $Sin(\varphi_i)$ | $Cos(\varphi_i)$ |
|---|---|---|---|---|---|
| 1 | a | b | 135 | $\sqrt{2}/2$ | $-\sqrt{2}/2$ |
| 2 | a | $-b$ | 45 | $\sqrt{2}/2$ | $\sqrt{2}/2$ |
| 3 | $-a$ | $-b$ | 135 | $\sqrt{2}/2$ | $-\sqrt{2}/2$ |
| 4 | $-a$ | b | 45 | $\sqrt{2}/2$ | $\sqrt{2}/2$ |

Figure 3 shows the diagrams of the distribution of the directions of the propulsive forces of the main propellers P1, P2, P3, P4 and the control actions $P_x$, $M_y$, $P_z$, depending on the control signals set by the pilot of the ROV from the control panel [5, 8]. To implement a given maneuver, it is necessary to fulfill the following the conditions.

1. When controlling a "forward-backward" movement ($y_1 \neq 0$, $x_1 = x_2 = 0$):

$$\vec{F} = B\vec{P} = \vec{i}F_x;$$

2. When controlling a "left-right" turn ($x_1 \neq 0$, $y_1 = x_2 = 0$):

$$\vec{F} = B\vec{P} = \vec{j}M_y;$$

3. When controlling a "right-left" shift ($x_2 \neq 0$, $y_1 = x_1 = 0$):

$$\vec{F} = B\vec{P} = \vec{k}F_z.$$

We set the model of the control panel (CP) in the form of a structural diagram (Fig. 3).

Designations on the structural schemes of the CP: $\alpha_1$, $\alpha_2$, $\alpha_3$ are angles of rotation of the handles of manipulators M1 and M2 in directions $Y_1$, $X_1$, $X_2$ respectively; $u_1$, $u_2$, $u_3$ are output signals of manipulators, changing discretely within the limits – $u_m \leq u_i \leq u_m$ depending on the angle of rotation of the corresponding handle $\alpha_i$ from $-\alpha_m$ to $+\alpha_m$; $u_m$ is the limiting value of the output signals, which we will take equal to the number of levels of their change (in accordance with the initial data $u_m = 16$) [9, 10].

Signals $u_1$, $u_2$, $u_3$ form the controlling $f$ = vector:

$$U = (u_1, u_2, u_3)^T. \tag{7}$$

The computing device performs a linear transformation of the vector U into command signals arriving at each of the cruise engines:

$$N^* = AU, \tag{8}$$

where $\mathbf{N}^* = (n_i^*)_{4 \times 1}$ is a vector of command signals; $A = (a_{ij})_{4 \times 3}$ is transformation matrix.

Since each of the command signals depends on three manipulator outputs:

$$n_i^* = \sum_{j=1}^{3} a_{ij} u_j, \quad (i \in 1, 4), \tag{9}$$

then the value of the command $n_i^*$ with the combined control of ROV may exceed its operating range of variation. This means that the input circuits of the regulators that control the operating mode of the propellers must contain a matching block that limits the command signals.

In addition, the command signals that go directly to the inputs of the regulators of the operating modes of the propellers change in magnitude discretely. We denote the signals at the output of the matching unit by letters $n_i$, $(i \in 1, 4)$.

The limiting signal levels at the input and output of the matching unit are the same and equal [4]:

$$max(|n_i^*|) = max(|n_i|) = n_m. \tag{10}$$

By analogy with the manipulator model, let us assume that the value $n_m$ means the number of signal levels $n_i$.

The functional transformation of signals in the matching block can be written as an expression:

$$n_i = \begin{cases} round(n_i^*), \text{ если } |n_i^*| < n_m, \\ n_m sign(n_i^*), \text{ если } |n_i^*| \geq n_m, \end{cases} \tag{11}$$

where $round(\ldots)$ is a function that rounds its argument to an integer.

When constructing a model of propellers, we assume that the transient processes of the propeller reaching a given thrust occur instantly.

The dependence of the propulsion $P_i$ on the command signal $n_i$ is determined by a piecewise linear function:

$$P_i = \begin{cases} k_p n_i, & if \ n_i \geq 0, \\ k_m n_i, & if \ n_i < 0; \end{cases} \quad (i \in 1, 4). \tag{12}$$

Function (12) can be written as the equation:

$$P_i = \frac{1}{2}[k_p + k_m + (k_p - k_m)sign(n_i)]n_i, \ (i \in 1, 4), \tag{13}$$

Where

$$k_p = \frac{P_{max}}{n_m}, \ k_m = \frac{P_{min}}{n_m}, \ sign(n_i) = \begin{cases} 1, if \ n_i \geq 0, \\ -1, if \ n_i < 0. \end{cases} \tag{14}$$

Since $P_{max} > P_{min}$, then for the steepness coefficients of the positive and negative sections of the characteristic $P_i(n_i)$ we can write the inequality $k_P > k_m$.

The ROV maneuver control system includes a computing device CD that generates continuous values of the **N\*** command signals at least in accordance with the formula (8).

The block with the transfer function **B** performs a linear transformation of the vector **P** in accordance with the matrix Eq. (4). In order for the ROV maneuver control system [11] to function, it is necessary to determine the elements of the matrix **A**.

We will consider controlling the ROV by exposing each of thrusters with one of three signals separately ui, $i = 1, 2, 3$ and in turn (Fig. 3).

## 3   Results

Forward-backward movement control (Fig. 3 a, b). In this case, $u_1 \neq 0$, $u_2 = u_3 = 0$. With a positive signal $u_1$, the propulsive forces of all propellers are directed forward:

$$n_i = a_{i1}u_1 > 0, \quad P_i = k_p a_{i1}u_1 > 0. \tag{15}$$

Expressions for the projections of the vector of control actions F on the axes of the associated coordinate system can be written as:

$$P_x = 2\sqrt{2}k_p a_{i1}u_1,$$

$$M_y = k_p a_{i1}u_1 \frac{\sqrt{2}}{2}[(b + a) + (-b - a) + (-b - a) + (b + a))] = 0,$$

$$P_z = \frac{\sqrt{2}}{2}\left[-k_p a_{i1}u_1 + k_p a_{i1}u_1 - k_p a_{i1}u_1 + k_p a_{i1}u_1\right] = 0. \tag{16}$$

As can be seen from expressions (16), for any values of $a_{i1}$, $(i = 1, 2, 3, 4)$, the cross-control actions $M_y$ and $P_z$ are equal to zero.

With a negative value of the signal $u_1$, the thrust of all propellers are directed backwards:

$$n_i = a_{i1}u_1 < 0, \quad P_i = k_m a_{i1}u_1 < 0. \tag{17}$$

Obviously, in this case, the cross control actions $M_y$ and $P_z$ will also be equal to zero.

Conclusion: the coefficients $a_{i1}$ do not require correction when the polarity of the signal $u_1$ changes and their nominal values can be determined by the formula:

$$a_{i1} = q_{ix} = q_x = \frac{n_m}{u_m}, \ (i \in 1, 4). \tag{18}$$

Left-right turn control (Fig. 3 c, d).

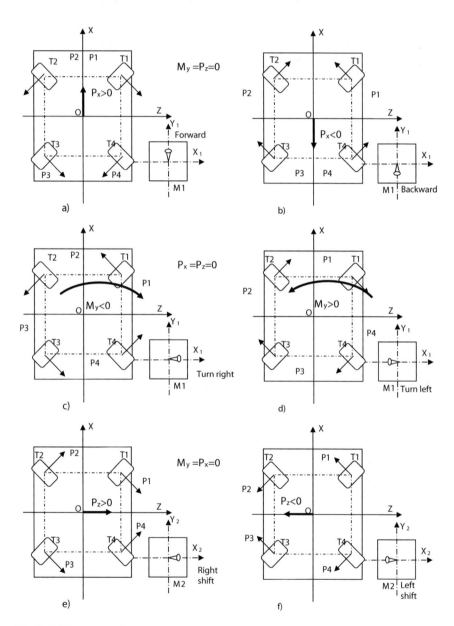

**Fig. 3** ROV control options

In this case, $u_2 \neq 0$, $u_1 = u_3 = 0$. With a positive signal $u_2$, the second and third movers create a thrust "forward", and the first and fourth - "backward" (Fig. 3 c):

$$
\begin{aligned}
n_1 &= a_{12}u_2 < 0, \; n_4 = a_{42}u_2 < 0, \\
n_2 &= a_{22}u_2 > 0, \; n_3 = a_{32}u_2 > 0, \\
P_1 &= k_m a_{12}u_2 < 0, \; P_4 = k_m a_{42}u_2 < 0, \\
P_2 &= k_p a_{22}u_2 > 0, \; P_3 = k_p a_{32}u_2 > 0.
\end{aligned}
\tag{19}
$$

From (19) we determine the polarities of the coefficients $a_{i2}$:

$$
a_{12} < 0, \; a_{42} < 0, \; a_{22} > 0, \; a_{32} > 0.
\tag{20}
$$

Since the thrust in the "forward" direction changes according to the signal $u_2$ with the proportionality coefficient $k_P > k_m$, then with equal values of the transfer coefficients $a_{i2}$, $(i \in 1, 4)$ a there will be a cross control action $P_x \neq 0$. Then $a_{12} = a_{42} = -q_y$, $a_{22} = a_{32} = q_y$.

The expressions for the projections of the vector of control actions $\mathbf{F}$ on the axes of the associated coordinate system can be written as:

$$
\begin{aligned}
P_x &= \frac{\sqrt{2}}{2}\left[2q_y(k_P - k_m)\right]u_2 > 0, \\
M_y &= -\frac{\sqrt{2}}{2}(a + b)q_y(2k_P + 2k_m)u_2, \\
P_z &= \frac{\sqrt{2}}{2}q_y\left[k_m + k_p - k_p - k_m\right] = 0.
\end{aligned}
\tag{21}
$$

With a negative polarity of the signal u2 (Fig. 3 d) for $P_x$ we get the expression:

$$
P_x = \frac{\sqrt{2}}{2}\left[2q_y(-k_P + k_m)\right]u_2 > 0,
\tag{22}
$$

only in this case, the thrust distortion will be created by the right longitudinal pair of propellers T1–T4. To align the thrust of longitudinal pairs of propellers working "forward" (for $u_2 > 0$ this is a pair of T2–T3, and for $u_2 < 0$ is T1–T4), the gear ratios $a_{i2}$, $(i \in 1, 4)$ must change when the signal polarity changes u2 so that all thrust is equal in magnitude.

We suppose that with negative thrust the gear ratio $q_y$ allows the maximum command level $n_i = -n_m$, $(i \in 1, 4)$.

In this case the value of $q_y$ is determined by the formula:

$$
q_y = \frac{n_m}{u_m}.
\tag{23}
$$

When a thrust value is positive, the gear ratio (we denote it $q_y^*$) should be less than $q_y$. Expressions for the gear ratios $a_{i2}$, $(i \in 1, 4)$ can be written as:

$$a_{12} = a_{42} = -\frac{1}{2}\left[q_y + q_y^* + \left(q_y - q_y^*\right)sign(u_2)\right];$$

$$a_{22} = a_{32} = \frac{1}{2}\left[q_y + q_y^* - \left(q_y - q_y^*\right)sign(u_2)\right]. \tag{24}$$

The value of the coefficient $q_y^*$ can be found from the condition $P_x = 0$. For example, for $u_2 > 0$ we get the expression:

$$P_x = \frac{\sqrt{2}}{2}\left(2k_P q_y^* - 2k_m q_y\right)u_2 = 0, \tag{25}$$

from which follows:

$$q_y^* = q_y \frac{k_m}{k_P} = \frac{n_m}{u_m} \frac{P_{min}}{P_{max}}. \tag{26}$$

The same solution (26) can be found if $u_2 < 0$.

Left-right shift control (Fig. 3 e, f). In this case $u_3 \neq 0$, $u_1 = u_2 = 0$. As can be seen from Fig. 3, when controlling the shift to the right-left, the diagonal pairs of propellers work synchronously. Thus, for $u_3 > 0$, positive thrust ("forward") is created by thrusters T2–T4, and negative ("backward") - by thrusters T1–T3. When the polarity of the signal $u_3$ (shift to the left, $u_3 < 0$) is reversed, the polarity of the thrust forces of the diagonal pairs of propellers will also change.

In the absence of equalization of positive propulsive forces of a diagonal pair of propellers, for any polarity of the signal $u_3$, a cross control action $P_x > 0$ will be created. To fulfill the condition $P_x = 0$, the transfer coefficients $a_{i3}$, $(i \in 1, 4)$ must be functions of the sign of the signal $u_3$. Expressions for the coefficients $a_{i2}$, $(i \in 1, 4)$, calculated similarly to expressions (24), have the form:

$$a_{13} = a_{33} = -\frac{1}{2}\left[q_z + q_z^* + \left(q_z - q_z^*\right)sign(u_3)\right];$$

$$a_{23} = a_{43} = \frac{1}{2}\left[q_z + q_z^* - \left(q_z - q_z^*\right)sign(u_3)\right] \tag{27}$$

where:

$$q_z = \frac{n_m}{u_m}$$

;

$$q_z^* = q_z \frac{k_m}{k_P} = \frac{n_m}{u_m} \frac{P_{min}}{P_{max}}. \tag{28}$$

Thus, to calculate the nominal values of the elements of the matrix A for the transformation of control signals, the following expressions can be formed:

$$a_{i1} = q_x = \frac{n_m}{u_m}, \ (i \in 1, 4);$$

$$a_{12} = a_{42} = -\frac{1}{2}\left[q_y + q_y^* + (q_y - q_y^*)sign(u_2)\right];$$

$$a_{22} = a_{32} = \frac{1}{2}\left[q_y + q_y^* - (q_y - q_y^*)sign(u_2)\right];$$

$$a_{13} = a_{33} = -\frac{1}{2}\left[q_z + q_z^* + (q_z - q_z^*)sign(u_3)\right];$$

$$a_{23} = a_{43} = \frac{1}{2}\left[q_z + q_z^* - (q_z - q_z^*)sign(u_3)\right]; \tag{29}$$

$$q_y = q_z = \frac{n_m}{u_m}; \ q_y^* = q_z^* = \frac{n_m}{u_m} \cdot \frac{P_{min}}{P_{max}};$$

$$sign(u_j) = \begin{cases} 1, if \ u_j \geq 0, \\ -1, if \ u_j < 0, \\ j = 2, 3. \end{cases}$$

## 4 Discussion

According to the synthesized algorithm, the thrusters of the DRC propellers were calculated with the following parameters:

- reversible thrust of Model 1020 propellers with a maximum forward thrust $P_{max} = 21.4$ kgf and a maximum backward thrust $P_{min} = 14.5$ kgf;
- output signals of manipulators M1 and M2, changing discretely at 16 levels from zero when the joystick handles deviate from the middle position $x_1 = y_1 = x_2 = y_2 = 0$ (zero signal values) to the maximum values $\pm x_m, \pm y_m$ in two perpendicular directions $(-x_m \leq x_i \leq x_m, -y_m \leq y_i \leq y_m, i = 1, 2)$.

When constructing control tables, it is necessary to calibrate the control system on the bench or in the experimental pool. As follows from Table 2, the rounded values of the control signals correspond to the calibration points of propulsion. According to the calibration points, we construct a control Table 2 for the variant left rotation with a given moment (when U1 = 0, U2 = −8, My = 120.19 H).

**Table 2**  ROV control table, by type of movement combination left rotation U2 = [−8] and forward-backward movement (U1)

| U1 | Px | My | P1 | P2 | P3 | P4 |
|---|---|---|---|---|---|---|
| −16 | −301.75 | 60.35 | −71.12 | −142.25 | −142.25 | −71.12 |
| −12 | −251.46 | 90.52 | −35.56 | −142.25 | −142.25 | −35.56 |
| −8 | −201.16 | 120.7 | 0 | −142.25 | −142.25 | 0 |
| −4 | −99.85 | 121.14 | 36.08 | −106.68 | −106.68 | 36.08 |
| 0 | −0.85 | 120.19 | 70.52 | −71.12 | −71.12 | 70.52 |
| 4 | 147.24 | 120.41 | 123.01 | −18.89 | −18.89 | 123.01 |
| 8 | 296.89 | 119.68 | 175.49 | 34.44 | 34.44 | 175.49 |
| 12 | 419.82 | 104.38 | 209.93 | 86.93 | 86.93 | 209.93 |
| 16 | 494.05 | 59.84 | 209.93 | 139.41 | 139.41 | 209.93 |

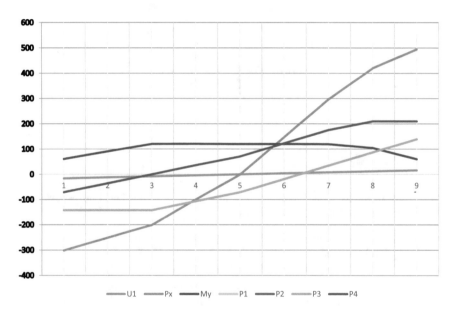

**Fig. 4**  Graphs of distribution of propulsion P1, P2, P3, P4

As follows from Table 2 and the graphs of the moments (Fig. 4), the power characteristics are equal in pairs (P1 = P4; P2 = P3).

## 5  Conclusion

In this work, an algorithm for controlling the propulsion-steering system of tele-controlled uninhabited vehicles was created, which matches the control signals $u_i$, i = 1,2,3, entered by the pilot of the ROV from the "joystick" manipulators, with the operating modes of the propellers.

The basis of the algorithm is the matrix $\mathbf{A}$ which transforms the control vector $\mathbf{U} = (u_1\ u_2\ u_3)^T$ into the vector of command signals $\mathbf{N}^* = (n_1{}^*\ n_2{}^*\ n_3{}^*\ n_4{}^*)^T$ coming from the regulators which control the propulsion modes.

The control algorithm of the motor-steering complex is implemented in the form of tabular control, according to the tables of the calculated values of the thrusts of the DRC. A methodology has been developed and formulas have been derived for calculating corrections of the transfer coefficients of the matrix $\mathbf{A}$ based on the results of bench tests of the ROV, an example of calculating corrections on the mathematical model of the control system with possible deviations of the system parameters (angles of installation of propellers, coordinates of points of application traction forces, traction characteristics of the propellers) from their nominal values was given.

# References

1. Yuh J, Ura T, Bekey GA (eds) (2012) Underwater robots. Springer, Cham
2. Lyutikova MN, Dantsevich IM Pankina SI (eds) (2021) The intelligent underwater laboratory. In: 1st International Conference on "Marine Geology and Engineering". IOP Conference Series: Earth and Environmental Science. IOP Publishing, Novorossiysk
3. Choi SK, Yuh J, Keevil N (1993). Design of omni-directional underwater robotic vehicle. In: Proceedings of OCEANS'93. IEEE, pp I192–I197
4. Dantsevich I, Zviagintsev N, Tarasenko A (2012) Control of unmanned underwater vehicles: monograph
5. Ma Y, Ma S, Wu Y, Pei X, Gorb SN, Wang Z, Zhou F (2018) Remote control over underwater dynamic attachment/detachment and locomotion. Adv Mater 30(30):1801595
6. Seto ML (ed) (2012) Marine robot autonomy. Springer, Canada
7. Wang YH, Wang SX, Xie CG (2007) Dynamic analysis and system design on an underwater glider propelled by temperature difference energy. J Tianjin Univ 40(2):133–138
8. Liao Y, Wang L, Li Y, Li Y, Jiang Q (2016) The intelligent control system and experiments for an unmanned wave glider. PLoS ONE 11(12):e0168792
9. Acristinius VA, Shiyanova TV (2020) Foreign methods and equipment used in the survey of soil dams of III and IV classes. Int J Appl Sci Technol "Integral", (2):220–226
10. Dantsevich IM, Lyutikova MN, Metreveli YuYu (2021) Formalization of the problem of movement in the longitudinal-transverse plane of remotely controlled underwater vehicles. Marine Intell Technol 4(2):168–177
11. Wang X, Yao X, Zhang L (2020) Path planning under constraints and path following control of autonomous underwater vehicle with dynamical uncertainties and wave disturbances. J Intell Rob Syst 99(3):891–908

# Numerical Method for Correcting Command Signals for Combined Control of a Multiengined Complex

Igor Dantsevich⬡, Marina Lyutikova⬡, and Vladimir Fedorenko⬡

**Abstract** The article describes a synthesis of a numerical method for compensating distortions in the control system of underwater tele-controlled uninhabited vehicles, consisting of a multiengine system, simultaneously performing the functions of a propulsion unit and a steering device. Combinations of control signals create the predominant type of movement and, at the same time, maneuvering of the vehicle. When two types of signals are generated on the same propellers, distortions occur, leading to errors in the semi-automatic (operator control) and programmed types of movement, while the actual type of propulsive movement and the propulsive force of the thrust devices of the vehicle propellers will differ from the specified (expected) ones. The proposed approach is universal and, under certain assumptions, can be used for any uninhabited robotic systems, including quadcopters.

**Keywords** Remotely Operated Vehicle (ROV) · Engine Steering System (ESS) · Control system · Numerical methods · Force of propellers · Mathematical model · Transformation matrix · Command signal · Transfer coefficients · Combined control of multi-engine complex

## 1 Introduction

The ROV controller performs linear transformations of the control vectors (or their combinations) $\mathbf{U_{ni}}$ and outputs to each of the 4 thrusters, as shown in the Fig. 1.

$$\mathbf{N^*} = \mathbf{AU}, \tag{1}$$

where $\mathbf{N^*} = \left(n_i^*\right)_{4 \times 1}$ is matrix of transformation of control signals; $A = \left(a_{ij}\right)_{4 \times 3}$ is transformation operator (matrix of coefficients) [1]. The setting of control signals

I. Dantsevich · M. Lyutikova (✉)
Admiral Ushakov Maritime State University, Novorossiysk, Russian Federation
e-mail: mnlyutikova@mail.ru

V. Fedorenko
North Caucasus Federal University, Stavropol, Russian Federation

during semi-automatic control (from the joystick) or the type of movement specified by the intelligent control program is provided by a combination of signals, $u_1$ is "forward-backward" propeller thrust, $u_2$ is switching the polarity of signals that provide "left-right" turns, $u_3$ is switching the polarity of signals implementing the "left-right" shift [2]:

$$n_i^* = \sum_{j=1}^{3} a_{ij} u_j, \ (i \in 1, 4). \tag{2}$$

Control of the movement of the apparatus in the vertical plane is implemented by another group of propellers (usually includes up to 4 propellers) and those are not considered in the articles [3–5].

Expressions (1–2) of the elements of the transformation matrix A are obtained under the condition that all the parameters characterizing the propulsion of the ROV have strictly nominal values. In real conditions, certain parameters of thrust characteristics (angles of installation of the propellers, coordinates of the application of vectors of thrust forces, the nature of the dependence of the thrust force of each propeller on control signals of different polarity, the position of the center of mass of the ROV in the coordinate system associated with the platform of the ROV, etc.) may differ from the nominal values [3, 14].

In this regard, the problem of correcting the nominal values of the coefficients calculated by the formulas (1–2) arises. We suppose that each element of matrix A can be corrected by an additional coefficient:

$$a_{ij} = \left(a_{ij}\right)_0 + \Delta a_{ij}, \ (i \in 1, 4, \ j \in 1, 3), \tag{3}$$

where $\left(a_{ij}\right)_0$ is the nominal value of the coefficient $a_{ij}$, $\Delta a_{ij}$ is the correction value of the coefficient $a_{ij}$.

## 2   Materials and Methods

To determine the expression of the correcting portions $\Delta a_{ij}$, we transform the formulas (1–2) to the form [3, 4, 7]:

$$a_{i1} = (q_x)_0 + \Delta q_{ix}, \ (i \in 1, 4);$$

$$a_{12} = -\frac{1}{2}\left((q_y)_0 + \Delta q_{1y}\right)\left[1 + \frac{(q_y^*)_0}{(q_y)_0} + \left(1 - \frac{(q_y^*)_0}{(q_y)_0}\right)sign(u_2)\right];$$

$$a_{42} = -\frac{1}{2}\left((q_y)_0 + \Delta q_{4y}\right)\left[1 + \frac{(q_y^*)_0}{(q_y)_0} + \left(1 - \frac{(q_y^*)_0}{(q_y)_0}\right)sign(u_2)\right];$$

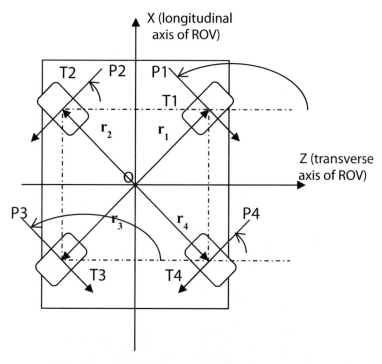

**Fig. 1** The layout of the propulsion thrusters on ROV [3]

$$a_{22} = \frac{1}{2}\big((q_y)_0 + \Delta q_{2y}\big)\left[1 + \frac{(q_y^*)_0}{(q_y)_0} - \left(1 - \frac{(q_y^*)_0}{(q_y)_0}\right)sign(u_2)\right];$$

$$a_{32} = \frac{1}{2}\big((q_y)_0 + \Delta q_{3y}\big)\left[1 + \frac{(q_y^*)_0}{(q_y)_0} - \left(1 - \frac{(q_y^*)_0}{(q_y)_0}\right)sign(u_2)\right];$$

$$a_{13} = -\frac{1}{2}\big((q_z)_0 + \Delta q_{1z}\big)\left[1 + \frac{(q_z^*)_0}{(q_z)_0} + \left(1 - \frac{(q_z^*)_0}{(q_z)_0}\right)sign(u_3)\right];$$

$$a_{33} = -\frac{1}{2}\big((q_z)_0 + \Delta q_{3z}\big)\left[1 + \frac{(q_z^*)_0}{(q_z)_0} + \left(1 - \frac{(q_z^*)_0}{(q_z)_0}\right)sign(u_3)\right];$$

$$a_{23} = \frac{1}{2}\big((q_z)_0 + \Delta q_{2z}\big)\left[1 + \frac{(q_z^*)_0}{(q_z)_0} - \left(1 - \frac{(q_z^*)_0}{(q_z)_0}\right)sign(u_3)\right];$$

$$a_{43} = \frac{1}{2}\big((q_z)_0 + \Delta q_{4z}\big)\left[1 + \frac{(q_z^*)_0}{(q_z)_0} - \left(1 - \frac{(q_z^*)_0}{(q_z)_0}\right)sign(u_3)\right], \tag{4}$$

where:

$$(q_x)_0 = (q_y)_0 = (q_z)_0 = \frac{n_m}{u_m};$$

$$(q_y^*) = (q_z^*)_0 = (q_z)_0 = \frac{n_m}{u_m} \cdot \frac{P_{min}}{P_{max}};$$

$$\frac{(q_y^*)_0}{(q_y)_0} = \frac{(q_z^*)_0}{(q_z)_0} = \frac{P_{min}}{P_{max}};\tag{5}$$

$$sign(u_j) = \begin{cases} 1, \text{если } u_j \geq 0, \\ -1, \text{если } u_j < 0, \\ \quad j = 2,3; \end{cases}$$

where $\Delta q_{ix}$, $\Delta q_{iy}$, $\Delta q_{iz}$, ($i \in 1,4$) are correction coefficients, $n_m$ is a number of discrete control levels, $u_m$ is a setting parameter, $P_{min}$ is the minimum value of the propulsive force of the thrusters, $P_{max}$ is the maximum value of the propulsive force of the thrusters.

From the expression (4) it follows [3, 6]:

$$\Delta a_{i1} = \Delta q_{ix}, \ (i \in 1,4);$$

$$\Delta a_{12} = -\frac{1}{2} \Delta q_{1y} \left[ 1 + \frac{(q_y^*)_0}{(q_y)_0} + \left(1 - \frac{(q_y^*)_0}{(q_y)_0}\right) sign(u_2) \right];$$

$$\Delta a_{42} = -\frac{1}{2} \Delta q_{4y} \left[ 1 + \frac{(q_y^*)_0}{(q_y)_0} + \left(1 - \frac{(q_y^*)_0}{(q_y)_0}\right) sign(u_2) \right];$$

$$\Delta a_{22} = \frac{1}{2} \Delta q_{2y} \left[ 1 + \frac{(q_y^*)_0}{(q_y)_0} - \left(1 - \frac{(q_y^*)_0}{(q_y)_0}\right) sign(u_2) \right];$$

$$\Delta a_{32} = \frac{1}{2} \Delta q_{3y} \left[ 1 + \frac{(q_y^*)_0}{(q_y)_0} - \left(1 - \frac{(q_y^*)_0}{(q_y)_0}\right) sign(u_2) \right];$$

$$\Delta a_{13} = -\frac{1}{2} \Delta q_{1z} \left[ 1 + \frac{(q_z^*)_0}{(q_z)_0} + \left(1 - \frac{(q_z^*)_0}{(q_z)_0}\right) sign(u_3) \right];$$

$$\Delta a_{33} = -\frac{1}{2} \Delta q_{3z} \left[ 1 + \frac{(q_z^*)_0}{(q_z)_0} + \left(1 - \frac{(q_z^*)_0}{(q_z)_0}\right) sign(u_3) \right];$$

$$\Delta a_{23} = \frac{1}{2} \Delta q_{2z} \left[ 1 + \frac{(q_z^*)_0}{(q_z)_0} - \left(1 - \frac{(q_z^*)_0}{(q_z)_0}\right) sign(u_3) \right];$$

$$\Delta a_{43} = \frac{1}{2} \Delta q_{4z} \left[ 1 + \frac{(q_z^*)_0}{(q_z)_0} - \left(1 - \frac{(q_z^*)_0}{(q_z)_0}\right) sign(u_3) \right].\tag{6}$$

From expressions (4) it follows that the correction coefficients $\Delta a_{i2}$, $\Delta a_{i3}$, $(i \in 1, 4)$ depend on the polarity of the signals $u_2$ and $u_3$ and thus have two values, but in fact, the sought quantities regardless of the polarity of the command signals will be 12 coefficients, and namely, combinations of the elements $\Delta q_{ix}$, $\Delta q_{iy}$, $\Delta q_{iz}$, $(i \in 1, 4)$.

Let the values of the parameters, on which the propulsion characteristics $F = \left(P_x M_y P_z\right)^T$ depend, differ from the nominal values by values having the first order of smallness [7–9]:

$$x_{Pi} = (x_{Pi})_0 + \Delta x_{Pi}, \; z_{Pi} = (z_{Pi})_0 + \Delta z_{Pi}, \; \phi_i = (\phi_i)_0 + \Delta\phi_i,$$
$$k_{Pi} = \left(k_p\right)_0 + \Delta k_{pi}, \; k_{mi} = (k_m)_0 + \Delta k_{mi}, \; (i \in 1, 4). \tag{7}$$

Let us find the dependences of the propulsion characteristics $F = \left(P_x M_y P_z\right)^T$ on the control signals $u_j$, $j \in 1, 3$ with separate control. In this case, the functions $n_i$, $i \in 1, 4$ will vary in the linear range from $-n_m$ to $+n_m$ [3, 10].

We will assume that in this case $n_i$, $i \in 1, 4$ will be continuous functions of their arguments $u_j$, $j \in 1, 3$ in accordance with the formula:

$$n_i = n_i^* = \sum_{j=1}^{3} a_{ij} u_j, \; (i \in 1, 4). \tag{8}$$

We will substitute conditions (7) into Eq. (8) [12].

$$n_i = \sum_{j=1}^{3} \left[(a_{ij})_0 + \Delta a_{ij}\right] u_j = (n_i)_0 + \Delta n_i, \; (i \in 1, 4),$$

where

$$(n_i)_0 = \sum_{j=1}^{3} \left(a_{ij}\right)_0 u_j, \; \Delta n_i = \sum_{j=1}^{3} \Delta a_{ij} u_j, \; (i \in 1, 4). \tag{9}$$

The propulsive forces can be determined by the formulas:

$$P_i = \frac{1}{2}\left[k_p + k_m + \left(k_p - k_m\right)sign((n_i)_0)\right]n_i = (P_i)_0 + \Delta P_i,$$
$$(P_i)_0 = \frac{1}{2}\left[(k_p)_0 + (k_m)_0 + \left((k_p)_0 - (k_m)_0\right)sign((n_i)_0)\right](n_i)_0, \; sought$$
$$\Delta P_i = \frac{1}{2}\left[\Delta k_{pi} + \Delta k_{mi} + \left(\Delta k_{pi} - \Delta k_{mi}\right)sign((n_i)_0)\right](n_i)_0$$
$$+ \frac{1}{2}\left[(k_p)_0 + (k_m)_0 + \left((k_p)_0 - (k_m)_0\right)sign((n_i)_0)\right]\Delta n_i, \; (i \in 1, 4). \tag{10}$$

## 3  Results

Let us consider forward-backward motion control, in this case $u_1 \neq 0$, $u_2 = u_3 = 0$.

When a signal $u_1$ is positive, then the thrust forces of all propellers are directed forward:

$$
\begin{aligned}
&(n_i)_0 = (q_x)_0 u_1 > 0, \quad \Delta n_i = \Delta q_{ix} u_1, \\
&P_i = \left[(k_p)_0 + \Delta k_p\right]\left[(q_x)_0 + \Delta q_{ix}\right]u_1 = (P_i)_0 + \Delta P_i, \\
&(P_i)_0 = (k_p)_0 (q_x)_0 u_1 > 0, \\
&\Delta P_i = \Delta k_p (q_x)_0 u_1 + (k_p)_0 \Delta q_{ix} u_1, \quad i \in 1, 4.
\end{aligned}
\tag{11}
$$

Let us write the expressions for the cross propulsive forces $M_y$, $P_z$:

$$
\begin{aligned}
M_y &= \sum_{i=1}^{4} P_i (z_{Pi} \sin(\phi_i) - x_{Pi} \cos(\phi_i)) = (M_y)_0 + \Delta M_y, \\
P_z &= \sum_{i=1}^{4} P_i \cos(\phi_i) = (P_z)_0 + \Delta P_z.
\end{aligned}
\tag{12}
$$

In formulas (12), the moment $(M_y)_0$ and the force $(P_z)_0$ are equal to zero, since they are determined at the nominal values of the parameters, which corresponds to movement in the longitudinal axis.

Variations $\Delta M_y$, $\Delta P_z$ can be written as:

$$
\Delta M_y = \Delta M_y^{\text{в}}(u_1^+) + \Delta M_y^{\text{к}}, \quad \Delta P_z = \Delta P_z^{\text{в}}(u_1^+) + \Delta P_z^{\text{к}},
\tag{13}
$$

where $\Delta M_y^{\text{в}}(u_1^+)$, $\Delta P_z^{\text{в}}(u_1^+)$ are variations of cross control actions caused by disturbing factors, if $u_1 > 0$ and $\Delta M_y^{\text{к}}$, $\Delta P_z^{\text{к}}$ are corrective variations of cross control actions required to compensate for disturbance[13].

For the sets of control signals $\Delta M_y^{\text{в}}(u_1^+)$, $\Delta P_z^{\text{в}}(u_1^+)$, $\Delta M_y^{\text{к}}$, $\Delta P_z^{\text{к}}$ we get the expressions:

$$
\begin{aligned}
\Delta M_y^{B}(u_1^+) &= \sum_{i=1}^{4}\{(P_i)_0[\Delta z_{Pi}\sin((\phi_i)_0) - \Delta x_{Pi}\cos((\phi_i)_0) \\
&+ ((z_{Pi})_0 \cos((\phi_i)_0) + (x_{Pi})_0 \sin((\phi_i)_0))\Delta\phi_i] \\
&+ \left[(z_{Pi})_0 \sin((\phi_i)_0) - (x_{Pi})_0 \cos((\phi_i)_0)\right](q_x)_0 \Delta k_p u_1\}; \\
\Delta P_z^{B}(u_1^+) &= \sum_{i=1}^{4}[\cos((\phi_i)_0)(q_x)_0 \Delta k_p u_1 - (P_i)_0 \sin((\phi_i)_0)\Delta\phi_i];
\end{aligned}
\tag{14}
$$

$$
\begin{aligned}
\Delta M_y^{\text{к}} &= \sum_{i=1}^{4}[(z_{Pi})_0 \sin((\phi_i)_0) - (x_{Pi})_0 \cos((\phi_i)_0)](k_p)_0 \Delta q_{ix} u_1; \\
\Delta P_z^{\text{к}} &= \sum_{i=1}^{4} \cos((\phi_i)_0)(k_p)_0 \Delta q_{ix} u_1.
\end{aligned}
\tag{15}
$$

Considering the nominal (calibrated in the experimental pool) values:

$$
\begin{aligned}
&(x_{P1})_0 = (x_{P2})_0 = a, \ (x_{P3})_0 = (x_{P4})_0 = -a; \\
&(z_{P1})_0 = (z_{P4})_0 = b, \ (z_{P2})_0 = (z_{P3})_0 = -b; \\
&\sin((\phi_i)_0) = \sqrt{2}/2, \ i \in 1,4; \\
&\cos((\phi_1)_0) = \cos((\phi_3)_0) = -\sqrt{2}/2, \ \cos((\phi_2)_0) = \cos((\phi_4)_0) = \sqrt{2}/2; \\
&(k_p)_0 = {}^{P_{max}}\!\Big/_{n_m (k_m)_0} \, {}^{P_{min}}\!\Big/_{n_m}
\end{aligned}
\tag{16}
$$

Expressions (16) can be written:

$$
\begin{aligned}
\Delta M_y^{\kappa} &= \frac{\sqrt{2}}{2} \frac{P_{max}}{n_m (a+b)(\Delta q_{1x} - \Delta q_{2x} - \Delta q_{3x} + \Delta q_{4x}) u_1}; \\
\Delta P_z^{\kappa} &= \frac{\sqrt{2}}{2} \frac{P_{max}}{n_m (-\Delta q_{1x} + \Delta q_{2x} - \Delta q_{3x} + \Delta q_{4x}) u_1}.
\end{aligned}
\tag{17}
$$

Let us introduce new variables to replace differences:

$$
\mu_{12} = \Delta q_{1x} - \Delta q_{2x}, \ \mu_{34} = \Delta q_{3x} - \Delta q_{4x}.
\tag{18}
$$

Substituting new variables (17) into (18), we obtain:

$$
\begin{aligned}
\Delta M_y^{\kappa} &= \frac{\sqrt{2}}{2} \frac{P_{max}}{n_m (a+b)(\mu_{12} - \mu_{34}) u_1}; \\
\Delta P_z^{\kappa} &= \frac{\sqrt{2}}{2} \frac{P_{max}}{n_m (-\mu_{12} - \mu_{34}) u_1}.
\end{aligned}
\tag{19}
$$

We will consider a balance mode in which expressions (19) are equal to zero:

$$
\Delta M_y^{B}(u_1^+) + \Delta M_y^{\kappa} = 0, \ \ \Delta P_z^{B}(u_1^+) + \Delta P_z^{\kappa} = 0.
\tag{20}
$$

Considering the assumptions made, Eqs. (18) will be written as follows:

$$
\begin{aligned}
\mu_{12} - \mu_{34} &= -\sqrt{2} \frac{n_m}{(a+b)} \frac{\Delta M_y^{B}(u_1^+)}{u_1 P_{max}}; \\
\mu_{12} + \mu_{34} &= \sqrt{2} \, n_m \frac{\Delta P_z^{B}(u_1^+)}{u_1 P_{max}}.
\end{aligned}
\tag{21}
$$

When a signal $u_1$ is negative, the propulsive forces of all propellers are directed backward:

$$
\begin{aligned}
&(n_i)_0 = (q_x)_0 u_1 < 0, \ \Delta n_i = \Delta q_{ix} u_1, \\
&P_i = [(k_m)_0 + \Delta k_m][(q_x)_0 + \Delta q_{ix}] u_1 = (P_i)_0 + \Delta P_i, \\
&(P_i)_0 = (k_m)_0 (q_x)_0 u_1 < 0,
\end{aligned}
$$

$$\Delta P_i = \Delta k_m (q_x)_0 u_1 + (k_m)_0 \Delta q_{ix} u_1, \qquad i \in 1, 4. \tag{22}$$

Variations of cross control actions caused by disturbing factors for $u_1 < 0$ will be denoted by $\Delta M_y^{\mathrm{B}}(u_1^-)$, $\Delta P_z^{\mathrm{B}}(u_1^-)$:

For variations $\Delta M_y^{\mathrm{B}}(u_1^-)$, $\Delta P_z^{\mathrm{B}}(u_1^-)$ $\Delta M_y^{\mathrm{K}}$, $\Delta P_z^{\mathrm{K}}$ we get expressions:

$$\Delta M_y^{\mathrm{B}}(u_1^-) = \sum_{i=1}^{4} \{ (P_i)_0 [\Delta z_{Pi} \sin((\phi_i)_0) - \Delta x_{Pi} \cos((\phi_i)_0)]$$

$$+ ((z_{Pi})_0 \cos((\phi_i)_0) + (x_{Pi})_0 \sin((\phi_i)_0)) \Delta \phi_i]$$

$$+ [(z_{Pi})_0 \sin((\phi_i)_0) - (x_{Pi})_0 \cos((\phi_i)_0)](q_x)_0 \Delta k_m u_1 \};$$

$$\Delta P_z^{\mathrm{B}}(u_1^-) = \sum_{i=1}^{4} [\cos((\phi_i)_0)(q_x)_0 \Delta k_m u_1 - (P_i)_0 \sin((\phi_i)_0) \Delta \phi_i] \tag{23}$$

$$\Delta M_y^{\mathrm{K}} = \sum_{i=1}^{4} [(z_{Pi})_0 \sin((\phi_i)_0) - (x_{Pi})_0 \cos((\phi_i)_0)](k_m)_0 \Delta q_{ix} u_1; \tag{24}$$

$$\Delta P_z^{\mathrm{K}} = \sum_{i=1}^{4} \cos((\phi_i)_0)(k_m)_0 \Delta q_{ix} u_1.$$

Considering the assumptions made (19) expressions (24) will get the form:

$$\Delta M_y^{\mathrm{K}} = \frac{\sqrt{2}}{2} \frac{P_{min}}{n_m (a+b)(\Delta q_{1x} - \Delta q_{2x} - \Delta q_{3x} + \Delta q_{4x})_1};$$

$$\Delta P_z^{\mathrm{K}} = \frac{\sqrt{2}}{2} \frac{P_{min}}{n_m (-\Delta q_{1x} + \Delta q_{2x} - \Delta q_{3x} + \Delta q_{4x})_1}. \tag{25}$$

We substitute in (25) the form (19):

$$\Delta M_y^{\mathrm{K}} = \frac{\sqrt{2}}{2} \frac{P_{min}}{n_m (a+b)(\mu_{12} - \mu_{34})_1};$$

$$\Delta P_z^{\mathrm{K}} = \frac{\sqrt{2}}{2} \frac{P_{min}}{n_m (-\mu_{12} - \mu_{34})_1}. \tag{26}$$

In the next step we will consider the cross components of forces and moments arising from the errors described by Eq. (12) equal to zero:

$$\Delta M_y^{\mathrm{B}}(u_1^-) + \Delta M_y^{\mathrm{K}} = 0, \quad \Delta P_z^{\mathrm{B}}(u_1^-) + \Delta P_z^{\mathrm{K}} = 0. \tag{27}$$

Let us substitute expressions (26) for cross-correcting control actions into Eqs. (27) and transform them to the form:

$$\mu_{12} - \mu_{34} = -\sqrt{2} \frac{n_m}{(a+b)} \frac{\Delta M_y^{\mathrm{B}}(u_1^-)}{u_1 P_{min}};$$

$$\mu_{12} + \mu_{34} = \sqrt{2}\, n_m \frac{\Delta P_z^{\mathrm{B}}(u_1^-)}{u_1 P_{min}}. \tag{28}$$

Let us compare Eqs. (20) and (28). The ratios $\frac{\Delta M_y^e(u_1^+)}{u_1}$, $\frac{\Delta P_z^e(u_1^+)}{u_1}$ for a series of measurements of the quantities $\Delta M_y^e(u_{1j}^+)$, $\Delta P_z^e(u_{1j}^+)$, $j \in 1, N$ for various values of $u_{1j}$, $j \in 1, N$ can be defined as the slope of the corresponding experimental characteristics $\Delta M_y^B(u_1^+) = f_{1p}(u_1)$, $\Delta P_z^B(u_1^+) = f_{2p}(u_1)$, approximated by the equations splines:

$$\Delta M_y^B(u_1^+) = \alpha_p + M_y^{u_1^+} u_1, \quad \Delta P_z^B(u_1^+) = \beta_p + P_z^{u_1^+} u_1. \tag{29}$$

The coefficients of Eqs. (29) can be determined from the results of bench tests using the least squares method (assuming that the coefficients $\alpha_p$ and $\beta_p$ should be close to zero). Then, when calculating the corrections $\mu_{12}$, $\mu_{34}$, instead of the ratios $\frac{\Delta M_y^e(u_1^+)}{u_1}$, $\frac{\Delta P_z^e(u_1^+)}{u_1}$, we substitute $\frac{\Delta M_y^e(u_1^+)}{u_1} \approx M_y^{u_1^+}$, $\frac{\Delta P_z^e(u_1^+)}{u_1} \approx P_z^{u_1^+}$ ..

As well as for $u_1 < 0$, we come to the conclusion $\frac{\Delta M_y^e(u_1^-)}{u_1} \approx M_y^{u_1^-}$, $\frac{\Delta P_z^e(u_1^-)}{u_1} \approx P_z^{u_1^-}$,,

where $M_y^{u_1^-}$, $P_z^{u_1^-}$ are the coefficients of linear equations:

$$\Delta M_y^B(u_1^-) = \alpha_m + M_y^{u_1^-} u_1, \quad \Delta P_z^B(u_1^-) = \beta_m + P_z^{u_1^-} u_1, \tag{30}$$

of approximating experimental characteristics $\Delta M_y^B(u_1^-) = f_{1m}(u_1)$, $\Delta P_z^B(u_1^-) = f_{2m}(u_1)$

We expect ratios $\frac{M_y^{u_1^+}}{P_{max}}$, $\frac{P_z^{u_1^+}}{P_{max}}$ to be close to ratios $\frac{M_y^{u_1^-}}{P_{min}}$, $\frac{P_z^{u_1^-}}{P_{min}}$, since when signal $u_1$ is positive, there is a greater change in the propulsive force of each propeller per unit of signal change, than when it is negative due to the nonlinearity of the thrust characteristics of the propellers, which should lead to the corresponding non-symmetry of the $M_y^B(u_1) = f_1(u_1)$, $\Delta P_z^B(u_1) = f_2(u_1)$ at different polarities of the signal $u_1$. Moreover, it can be expected that the slope coefficients of the approximating linear dependences (29) and (30) relate to each other in the same way as the slope coefficients of the propulsion characteristics:

$$k_p = \frac{P_{max}}{n_m}, \quad k_m = \frac{P_{min}}{n_m}, \quad sign(n_i) = \begin{cases} 1, & if\ n_i \geq 0, \\ -1, & if\ n_i < 0 \end{cases} \tag{31}$$

$$\frac{M_y^+}{M_y^-} \approx \frac{P_z^+}{P_z^-} \approx \frac{k_p}{k_m} = \frac{P_{max}}{P_{min}}, \tag{32}$$

Let us introduce the notation for the averaged estimates of the mathematical expectations of the ratios $\frac{M_y^{u_1^+}}{P_{max}}$, $\frac{P_z^{u_1^+}}{P_{max}}$, $\frac{M_y^{u_1^-}}{P_{min}}$, $\frac{P_z^{u_1^-}}{P_{min}}$:

$$\begin{aligned}
h_y^{u_1} &= \frac{1}{2} \cdot \left[ M\left\{ \frac{M_y^{u_1^+}}{P_{max}} \right\} + M\left\{ \frac{M_y^{u_1^-}}{P_{min}} \right\} \right]; \\
h_z^{u_1} &= \frac{1}{2} \cdot \left[ M\left\{ \frac{P_z^{u_1^+}}{P_{max}} \right\} + M\left\{ \frac{P_z^{u_1^-}}{P_{min}} \right\} \right].
\end{aligned} \tag{33}$$

**Table 1** The values of the control signals and the propellers' propulsion obtained in the pool (Model 1020 propellers) with the maximum forward thrust $P_{max} = 21.4$ kgf and the maximum backward thrust $P_{min} = 14.5$ kgf

| $U_1$ [B] | $P_x$ [H] | $M_y$ [HM] | $P_z$[H] | $P_1$[H] | $P_2$[H] | $P_3$[H] | $P_4$[H] |
|---|---|---|---|---|---|---|---|
| $-14$ | $-367.04$ | $-17.32$ | 15.23 | $-146.38$ | $-116.74$ | $-121.03$ | $-129.61$ |
| $-12$ | $-314.6$ | $-14.85$ | 13.06 | $-116.98$ | $-100.06$ | $-103.74$ | $-111.1$ |
| $-10$ | $-262.17$ | $-12.37$ | 10.88 | $-97.49$ | $-83.39$ | $-86.45$ | $-92.58$ |
| $-8$ | $-209.73$ | $-9.9$ | 8.7 | $-77.99$ | $-66.71$ | $-69.16$ | $-74.07$ |
| $-6$ | $-157.3$ | $-7.42$ | 6.53 | $-58.49$ | $-50.03$ | $-51.87$ | $-55.55$ |
| $-4$ | $-104.87$ | $-4.95$ | 4.35 | $-38.99$ | $-33.35$ | $-34.58$ | $-37.03$ |
| $-2$ | $-52.43$ | $-2.47$ | 2.18 | $-19.5$ | $-16.68$ | $-17.29$ | $-18.52$ |
| 0 | 0 | 0 | 0 | 0 | 0 | 0 | 0 |
| 2 | 77.27 | 4.17 | $-2.88$ | 28.82 | 24.28 | 25.14 | 27.84 |
| 4 | 154.53 | 8.35 | $-5.75$ | 57.63 | 48.56 | 50.28 | 55.67 |
| 6 | 231.8 | 12.52 | $-8.63$ | 86.45 | 72.84 | 75.41 | 83.51 |
| 8 | 309.07 | 16.7 | $-11.5$ | 115.27 | 97.12 | 100.55 | 111.34 |
| 10 | 386.33 | 20.87 | $-14.38$ | 144.08 | 121.4 | 125.69 | 139.18 |
| 12 | 463.6 | 25.04 | $-17.26$ | 172.9 | 145.68 | 150.83 | 167.02 |
| 14 | 540.86 | 28.22 | $-20.13$ | 201.72 | 169.96 | 175.97 | 194.85 |

Considering the notation (33), Eqs. (21) and (28) can be written as:

$$\mu_{12} - \mu_{34} = -\sqrt{2}\frac{n_m}{(a+b)}h_y^{u_1};$$

$$\mu_{12} + \mu_{34} = \sqrt{2}\, n_m h_z^{u_1}. \tag{34}$$

Then the solution of equation is (34):

$$\mu_{12} = \frac{\sqrt{2}}{2}n_m\left(h_z^{u_1} - \frac{h_y^{u_1}}{(a+b)}\right); \; \mu_{34} = \frac{\sqrt{2}}{2}n_m\left(h_z^{u_1} + \frac{h_y^{u_1}}{(a+b)}\right). \tag{35}$$

If we substitute (34) into expressions (18), we find the correction factors $\Delta q_{1x}, \Delta q_{2x}, \Delta q_{3x}, \Delta q_{4x}$ from the condition that the sum of the squares of their values will be minimal. As a result, we get:

$$\Delta q_{1x} = \frac{1}{2}\mu_{12} = \frac{\sqrt{2}}{4}n_m \left( h_z^{u_1} - \frac{h_y^{u_1}}{(a+b)} \right);$$

$$\Delta q_{2x} = -\frac{1}{2}\mu_{12} = -\frac{\sqrt{2}}{4}n_m \left( h_z^{u_1} - \frac{h_y^{u_1}}{(a+b)} \right); \qquad (36)$$

$$\Delta q_{3x} = \frac{1}{2}\mu_{34} = \frac{\sqrt{2}}{4}n_m \left( h_z^{u_1} + \frac{h_y^{u_1}}{(a+b)} \right);$$

$$\Delta q_{4x} = -\frac{1}{2}\mu_{34} = -\frac{\sqrt{2}}{4}n_m \left( h_z^{u_1} + \frac{h_y^{u_1}}{(a+b)} \right).$$

## 4 Discussion

Let us assess the results of the work by checking the calculation, and considering option 1 (control by the signal U1).

Based on Table 1 (the calibrated values of the control signals and thrust propellers obtained in the pool), we linearize the characteristics $M_y(U_1)$, $P_z(U_1)$ using the least squares method at different signal polarities U1 in accordance with formulas (32) and (33):

$$\Delta M_y^{\text{B}}(u_1^-) = \alpha_m + M_y^{u_1^-} u_1, \ \Delta P_z^{\text{B}}(u_1^-) = \beta_m + P_z^{u_1^-} u_1;$$

$$\Delta M_y^{\text{B}}(u_1^+) = \alpha_p + M_y^{u_1^+} u_1, \ \Delta P_z^{\text{B}}(u_1^+) = \beta_p + P_z^{u_1^+} u_1; \qquad (37)$$

where

$$\alpha_m = 0,0017; \ M_y^{u_1^-} = 1,2374; \ \beta_m = 0,00083; \ P_z^{u_1^-} = -1,0879;$$

$$\alpha_p = -0,00083; \ M_y^{u_1^+} = 2,0871; \ \beta_p = -0,00083; \ P_z^{u_1^+} = -1,4379.$$

Using formula (33), we find the coefficients $h_y^{u_1}$, $h_z^{u_1}$:

$$h_y^{u_1} = 0,00932; \ h_z^{u_1} = -0,00725.$$

Using formulas (36), we obtain the values of the correcting coefficients:

$$\Delta q_{1x} = -1,031; \ \Delta q_{2x} = 1,031; \ \Delta q_{3x} = 0,375; \ \Delta q_{4x} = -0,375.$$

**Table 2** Calculated corrections of the propulsion and steering complex

| Number of command signal levels $n_m$=12 and deviation $P_{max}$ nominal, kgf | | | | | | | |
|---|---|---|---|---|---|---|---|
| dpmax1 | | dpmax2 | | dpmax3 | | dpmax4 | |
| 2.1 | | -1.6 | | -0.9 | | 1.3 | |
| Deviation $P_{min}$ nominal, kgf | | | | | | | |
| dpmin1 | | dpmin2 | | dpmin3 | | dpmin4 | |
| 1.4 | | -0.9 | | -0.4 | | 0.6 | |
| Coefficients of model thrusters | | | | | | | |
| kp1 | kp2 | kp3 | kp4 | km1 | km2 | km3 | km4 |
| 1.801 | 1.517 | 1.571 | 1.74 | 1.219 | 1.042 | 1.081 | 1.157 |
| Coordinates of points of application of propulsion forces, м | | | | | | | |
| xp1 | xp2 | xp3 | xp4 | zp1 | zp2 | zp3 | zp4 |
| 0.32 | 0.285 | -0.27 | -0.322 | 0.286 | -0.321 | -0.27 | 0.282 |
| Angles of installation of propellers, radians | | | | | | | |
| $fi_1$ | | $fi_2$ | | $fi_3$ | | $fi_4$ | |
| 2.382 | | 0.829 | | 2.304 | | 0.846 | |
| Nominal values of gear ratios | | | | | | | |
| $k_{p0}$ | $k_{pm}$ | $q_x$ | $q_y$ | $q_z$ | $q_{ys}$ | $q_{zs}$ | |
| 1.64 | 1.111 | 8 | 8 | 8 | 5.421 | 5.421 | |
| Corrective values of gear ratios | | | | | | | |
| $dq_{x1}$ | | $dq_{x2}$ | | $dq_{x3}$ | | $dq_{x4}$ | |
| -1.031 | | 1.031 | | 0.375 | | -0.375 | |
| $dq_{y1}$ | | $dq_{y2}$ | | $dq_{y3}$ | | $dq_{y4}$ | |
| 0 | | 0 | | 0 | | 0 | |
| $dq_{z1}$ | | $dq_{z2}$ | | $dq_{z3}$ | | $dq_{z4}$ | |
| 0 | | 0 | | 0 | | 0 | |
| Matrix of corrections (coefficients) of propellers in the direction of movement | | | | | | | |

| Forces | thrusters | | | |
|---|---|---|---|---|
| | $P_1$ | $P_2$ | $P_3$ | $P_4$ |
| $F_x$ | 0.6884 | 0.7373 | 0.7431 | 0.749 |
| $M_y$ | 0.429 | -0.4292 | -0.3813 | 0.4246 |
| $F_z$ | -0.7254 | 0.6756 | -0.6691 | 0.6626 |

The numerical values of the functions $P_x(U_1)$, $M_y(U_1)$, $P_z(U_1)$ from Table 1 and Table 2 are placed in Table 3.

Figure 2 shows the graphs of the functions $P_x(U_1)$, $M_y(U_1)$, $P_z(U_1)$, built according to the data in Table 3.

As can be seen from the graphs, the corrected values of forces and moments are close to linear values. Good results were obtained by compensating the values of $P_x(U_1)$, $M_y(U_1)$, $P_z(U_1)$, especially relative to the moment and forces in the transverse plane of the ROV.

**Table 3** Corrected values of forces and moments for control type U1

| $U_1$ | Without corrections | | | With corrections | | |
|---|---|---|---|---|---|---|
| | $P_x$ H | $M_y$ H$_M$ | $P_z$ H | $P_{xk}$ H | $M_{yk}$ H$_M$ | $P_{zk}$ H |
| −14 | −367,04 | −17,32 | 15,23 | −365,73 | 0,78 | 0,45 |
| −12 | −314,6 | −14,85 | 13,06 | −314,31 | 0,82 | 0,68 |
| −10 | −262,17 | −12,37 | 10,88 | −261,21 | 0,94 | 0,96 |
| −8 | −209,73 | −9,9 | 8,7 | −208,98 | 0,57 | 0,47 |
| −6 | −157,3 | −7,42 | 6,53 | −156,75 | 0,2 | −0,02 |
| −4 | −104,87 | −4,95 | 4,35 | −105,33 | 0,25 | 0,21 |
| −2 | −52,43 | −2,47 | 2,18 | −52,23 | 0,37 | 0,49 |
| 0 | 0 | 0 | 0 | 0 | 0 | 0 |
| 2 | 77,27 | 4,17 | −2,88 | 76,89 | −0,01 | −0,42 |
| 4 | 154,53 | 8,35 | −5,75 | 155,08 | 0,71 | 0,32 |
| 6 | 231,8 | 12,52 | −8,63 | 230,8 | 1,3 | 0,95 |
| 8 | 309,07 | 16,7 | −11,5 | 307,69 | 1,29 | 0,54 |
| 10 | 386,33 | 20,87 | −14,38 | 384,58 | 1,28 | 0,12 |
| 12 | 463,6 | 25,04 | −17,26 | 462,77 | 2,01 | 0,86 |
| 14 | 540,86 | 29,22 | −20,13 | 538,49 | 2,59 | 1,49 |
| σ | 291,3615 | 14,9889 | 11,3241 | 290,4278 | 0,7335 | 0,4765 |

Due to the compensation of errors and parasitic forces, the proposed method allows to eliminate the arising moments and forces in the transverse plane, when ROV is moving in the longitudinal plane, to align the propulsion system and to provide proportional control of the DRC, both in the semi-automatic mode and in programmed control.

Compensated control in the implementation of controllers of the DRC thrusters facilitates the task of synthesizing a control system based on the observed dynamics; the problem of scaling the control system can be considered promising, i.e. proportional control with nonlinear characteristics of the thrusters of the propellers both in the modes of a small signal and a high slope of the control characteristic.

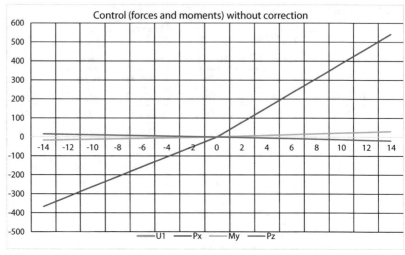

a)

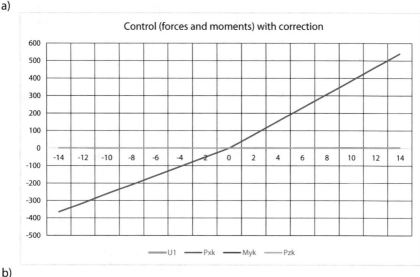

b)

**Fig. 2** Propulsion characteristics of the ROV when the U1 signal changes: a) uncompensated type of characteristics; b) compensated

# 5   Conclusion

The numerical method, which has been synthesized in work, for compensating control errors of the propulsion-steering complex propellers makes it possible to solve the problem of coordinating the thrusts of the propulsion devices of tele-controlled uninhabited vehicles in rectilinear motion. The influencing quantities are considered, and the scientific methodology for compensating for errors in the installation of the propellers, the irregularity of thrusts, deviations of the parameters of propeller characteristics and other parameters is determined.

For tele-controlled uninhabited vehicles, a single propulsion system is used both for marching and for maneuvering. The formed command system with the compensated type of control allows the implementation of the autopilot and scalable type of control, which also makes it possible to efficiently work with an on-board set of tools and manipulators.

# References

1. Pinciroli C, Trianni V, O'Grady R, Pini G, Brutschy A, Brambilla M, Dorigo M (2012) ARGoS: a modular, parallel, multi-engine simulator for multi-robot systems. Swarm Intell 6(4):271–295
2. Choi SK, Yuh J, Keevil N (1993) Design of omni-directional underwater robotic vehicle. In: Proceedings of OCEANS 1993. IEEE, pp I192–I197
3. Dantsevich I, Zviagintsev N, Tarasenko A (2012) Control of unmanned underwater vehicles: monograph
4. Antonelli G, Chiaverini S, Sarkar N (2001) External force control for underwater vehicle-manipulator systems. IEEE Trans Robot Autom 17(6):931–938
5. Seto ML (ed) (2012) Marine robot autonomy. Springer, Canada
6. Wang YH, Wang SX, Xie CG (2007) Dynamic analysis and system design on an underwater glider propelled by temperature difference energy. J Tianjin Univ 40(2):133–138
7. Liao Y, Wang L, Li Y, Li Y, Jiang Q (2016) The intelligent control system and experiments for an unmanned wave glider. PLoS One 11(12):e0168792
8. Wang X, Yao X, Zhang L (2020) Path planning under constraints and path following control of autonomous underwater vehicle with dynamical uncertainties and wave disturbances. J Intell Rob Syst 99(3):891–908
9. Dannigan MW, Russell GT (1998) Evaluation and reduction of the dynamic coupling between a manipulator and an underwater vehicle. IEEE J Oceanic Eng 23(3):260–273
10. Ura T, Obara T, Takagawa S, Gamo T (2001) Exploration of Teisi Knoll by autonomous underwater vehicle "R-One robot". In: MTS/IEEE oceans 2001. An ocean Odyssey. Conference proceedings (IEEE Cat. No. 01CH37295), vol 1. IEEE, pp 456–461
11. Joo MG, Qu Z (2015) An autonomous underwater vehicle as an underwater glider and its depth control. Int J Control Autom Syst 13(5):1212–1220
12. Wang W, Clark CM (2006) Modeling and simulation of the VideoRay Pro III underwater vehicle. In: OCEANS 2006-Asia Pacific. IEEE, pp 1–7
13. Dantsevich IM, Lyutikova MN (2021) Geoinformation laboratory for determining objects by unmanned aerial vehicles. In: IOP conference series: earth and environmental science, vol 745, no 1. IOP Publishing, p 012028
14. Lyutikova MN, Dantsevich IM, Tarasenko AA (October 2021) Management of towed geophysical systems when exploring a shelf. In: IOP conference series: earth and environmental science, vol 872, no 1. IOP Publishing, p 012017

# Hardware Implementation of the Kalman Filter for Video Signal Processing

Pavel Lyakhov⏺, Diana Kalita⏺, and Maxim Bergerman⏺

**Abstract** Probabilistic methods for detecting a moving object are widely used in solving problems of digital processing of a video signal in computer vision systems. At the same time, the primary tasks remain to increase the quantitative and qualitative characteristics of the digital processing of video information. This article discusses a probabilistic method for detecting a moving object in a video data stream using the Kalman filter as an example. A scheme for detecting a moving object based on the Kalman filter is proposed. The architectures of calculators for predicting the system state and covariance errors are built. Hardware simulations have shown the ability to reduce the computation time of the system state and covariance error by 5.4% when using high-speed parallel Carry-Save Adders (CSA) and Kogge-Stone prefix adders (KSA) compared to an architecture based on built-in addition and multiplication operations. Software modeling made it possible to implement the proposed algorithm for detecting a moving object in a video data stream under affine transformations.

**Keywords** Kalman filter · Digital video signal processing · Hardware implementation · Position prediction · System state prediction · Covariance error

## 1 Introduction

The widespread use of digital video surveillance systems entails working with digital information in a video data stream. The systems that analyze the video stream are faced with the solution of one of the important problems in this area, which boils down to selecting and detecting an object in the received video stream. Among such systems: security systems, navigation, observation, meteorology, as well as developing computer vision systems.

P. Lyakhov · M. Bergerman
North Caucasus Center for Mathematical Research, North-Caucasus Federal University, 355017 Stavropol, Russia

D. Kalita (✉)
Department of Mathematical Modeling, North-Caucasus Federal University, Stavropol, Russia
e-mail: diana.kalita@mail.ru

A. Tchernykh et al. (eds.), *Mathematics and its Applications in New Computer Systems*,
Lecture Notes in Networks and Systems 424,
https://doi.org/10.1007/978-3-030-97020-8_12

133

In turn, in systems that solve the problem of object detection, the search process is complicated by various kinds of distortions (affine and projective), receiver noise, and such cases as overlapping of the object under study by other objects. On the other hand, the designated range of systems must process the incoming video sequence at the real rate of receiving the data stream. For these reasons, the issue of developing new and high-quality computational algorithms is urgent.

Domestic and foreign researchers' wide range of publications confirms the relevance of this type of development. So, to solve the indicated range of problems, it is proposed to use a multi-interactive double decoder [4]. In [5], an approach is proposed for creating a distortion adaptation module to eliminate distortions caused by an equirectangular projection. A multiscale unit is also proposed for distinguishing objects in the video stream in omnidirectional detection scenes. The authors of [8] developed a fully connected convolutional network for predicting spatial significance based on motion information from video frames. Obtaining long-term information based on the Spatio-temporal coherence of the main foreground plans and the previous objectivity is solved in [9]. The use of a correlation filter in conjunction with a multiparticle correlation tracker with triangular structure constraints is shown in [10]. The problem of object detection is solved in many works, but despite this, a number of promising questions remain from the point of view of further research.

In [1], the authors use a neural network approach as the main method for object detection. It is proposed to use a new end-to-end convolutional neural network for detecting multimodal objects, which solves the problem of detecting pixel values by combining the neural network functions. In [2], a context-sensitive detection module in convolutional neural network architecture is presented, which detects observable objects while simultaneously creating connections between each pixel of the image and its local and global contextual pixels. However, this method is used only on those classes of objects that are included in training samples. The combined method proposed in [13] assumes insignificant projective distortions of the object. At the same time, the search parameters are adapted to the corresponding projective changes in the properties of the object's image; therefore, this method is inapplicable for performing the search operation. The deterministic method from [3] has shown resistance to projective distortions, but this method has high computational complexity.

From this point of view, probabilistic methods are the most suitable methods for performing a detection operation in a video stream [12]. The probabilistic methods are characterized by initial information about the object's initial position in the first frame. In addition, the use of such filters allows maintaining constant computational complexity for any number of measurements.

This paper proposes applying a probabilistic method for detecting a moving object using the Kalman filter as an example.

## 2   Materials and Methods

### 2.1   *Theoretical Analysis of Kalman Filter*

The Kalman filter is a recursive filter that estimates the state vector of a dynamical system based on a series of incomplete and noisy measurements. The algorithm for implementing the Kalman filter performs repeated actions of prediction and adjustment of the system's state under consideration.

The mathematical filtering model is set as follows: at the input of the system, there are initial values $\hat{x}_{k-1}$ and $P_k^-$ – prediction of the state of the system at the previous moment in time and prediction of the error, respectively. Then the first phase of the prediction is to calculate:

1.   System state predictions

$$\hat{x}_k^- = F\hat{x}_{k-1} + Bu_{k-1}, \tag{1}$$

where $\hat{x}_k^-$ - is the prediction of the state of the system at the current moment in time, $F$ is the transition matrix between states (dynamic model of the system), $\hat{x}_{k-1}$ is the state of the system at the last moment in time, $B$ is the matrix for applying the control action, $u_{k-1}$ - control action at the last moment in time.

2.   Predictions of error covariance

$$P_k^- = FP_{k-1}F^T + Q, \tag{2}$$

Where $P_k^-$ is the prediction of the error, $P_{k-1}$ is the error at the last moment in time, $Q$ is the covariance of the process noise.

The second phase of adjustment requires the definition of the following indicators:

1.   Kalman gain

$$K_k = P_k^- H^T \left(HP_k^- H^T + R\right)^{-1}, \tag{3}$$

Where $K_k$ is the Kalman gain, $H$ is the measurement matrix representing the measurement-state ratio, $R$ is the measurement noise covariance;

2.   Updating the estimate based on the $z_k$ dimension

$$\hat{x}_k = \hat{x}_k^- + K_k\left(z_k - H\hat{x}_k^-\right), \tag{4}$$

where $z_k$ is the measurement at the current time.

3.   Update error covariance

$$P_k = (I - K_kH)P_k^-, \tag{5}$$

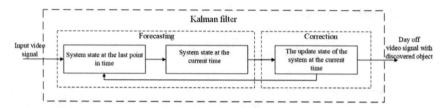

**Fig. 1** Generalized diagram of the Kalman filtering process

Where $I$ is the identity matrix.

The described filtering process can be generalized by scheme 1. The recurrent principle of the filter operation, shown in Fig. 1, allows for constant computational complexity for any number of observations. In this case, the search for a moving object is performed according to the scheme shown in Fig. 2. The principle of operation of the block for predicting new locations of existing tracks is described by formulas (1)–(2). Formulas (3)–(5) determine the operation of the track update blocks and the update of unassigned tracks. A track is understood as determining the location of a moving object (several objects) in time using a camera.

Consider the numerical implementation of the above algorithm. To do this, we will solve the problem of probabilistic assessment of the state of the object's position and update this assessment in real time using two prediction steps and correction of the obtained estimate.

The position of the object will be evaluated by two input arguments: the initial position $x_0$ and the speed of movement $v$ as

$$x_{k-1} = \begin{bmatrix} x_0 \\ \frac{dx}{dt} = \hat{x} \end{bmatrix}. \tag{6}$$

The movement is specified through acceleration:

$$u = a = \frac{d^2 x}{dt^2}. \tag{7}$$

The motion model and the observation model for the problem under consideration are as follows:

$$x_k = \begin{bmatrix} 1 & \Delta t \\ 0 & 1 \end{bmatrix} x_{k-1} + \begin{bmatrix} 0 \\ \Delta t \end{bmatrix} u_{k-1} + w_{k-1} \tag{8}$$

$$y_k = \begin{bmatrix} 1 & 0 \end{bmatrix} x_k + v_k \tag{9}$$

Consider predicting the state of an object at the next moment in time. Since the prediction is given by (1), then, in general, in matrix form, we can write:

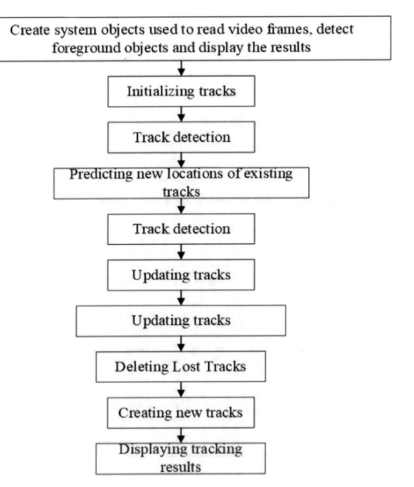

**Fig. 2** Scheme of detecting a moving object in a video stream based on the Kalman filter

$$\begin{bmatrix} \check{x}_{ks} \\ \check{x}_{kv} \end{bmatrix} = \begin{bmatrix} F_{11} & F_{12} \\ F_{21} & F_{22} \end{bmatrix} \cdot \begin{bmatrix} x_{k-1_{11}} \\ x_{k-1_{21}} \end{bmatrix} + \begin{bmatrix} b_1 \\ b_2 \end{bmatrix} \cdot u_{k-1} \tag{10}$$

We write the prediction of covariance (2) in matrix form as:

$$\begin{bmatrix} \check{P}_{11} & \check{P}_{12} \\ \check{P}_{21} & \check{P}_{22} \end{bmatrix} = \begin{bmatrix} F_{11} & F_{12} \\ F_{21} & F_{22} \end{bmatrix} \cdot \begin{bmatrix} P_{k-1_{11}} & P_{k-1_{12}} \\ P_{k-1_{21}} & P_{k-1_{22}} \end{bmatrix} \cdot \begin{bmatrix} F_{11} & F_{12} \\ F_{21} & F_{22} \end{bmatrix}^T + \begin{bmatrix} Q_{11} & Q_{12} \\ Q_{21} & Q_{22} \end{bmatrix} \tag{11}$$

From (3) follows a matrix notation for calculating the Kalman gain value:

$$\begin{bmatrix} K_{11} \\ K_{21} \end{bmatrix} = \begin{bmatrix} \check{P}_{11} & \check{P}_{12} \\ \check{P}_{21} & \check{P}_{22} \end{bmatrix} \cdot \begin{bmatrix} h_{11} \\ h_{12} \end{bmatrix}^T \cdot \left( [h_{11} \quad h_{21}] \cdot \begin{bmatrix} \check{P}_{11} & \check{P}_{12} \\ \check{P}_{21} & \check{P}_{22} \end{bmatrix} \begin{bmatrix} h_{11} \\ h_{12} \end{bmatrix}^T + R \right)^{-1} \tag{12}$$

Let be

$$\mathcal{T} = [h_{11} \quad h_{21}] \cdot \begin{bmatrix} \check{P}_{11} & \check{P}_{12} \\ \check{P}_{21} & \check{P}_{22} \end{bmatrix} \begin{bmatrix} h_{11} \\ h_{21} \end{bmatrix}^T + R \tag{13}$$

Then $S = \mathcal{T}^{-1}$.

Since the matrix $H$ takes the values $H = [1 \; 0]$, then (13) can be written in the form:

$$\mathcal{T} = h_{11} \cdot \check{P}_{11} \cdot h_{11} + R \tag{14}$$

Taking into account (13), we rewrite the Kalman gain:

$$\begin{bmatrix} K_{11} \\ K_{21} \end{bmatrix} = \begin{bmatrix} \check{P}_{11} & \check{P}_{12} \\ \check{P}_{21} & \check{P}_{22} \end{bmatrix} \cdot \begin{bmatrix} h_{11} \\ h_{12} \end{bmatrix}^T \cdot S, \tag{15}$$

For position correction according to (4), we have a matrix notation of the form:

$$\begin{bmatrix} \hat{x}_{ks} \\ \hat{x}_{kv} \end{bmatrix} = \begin{bmatrix} \check{x}_{ks} \\ \check{x}_{kv} \end{bmatrix} + \begin{bmatrix} K_{11} \\ K_{21} \end{bmatrix} \cdot \left( y_k - [h_{11} \quad h_{12}] \cdot \begin{bmatrix} \check{x}_{ks} \\ \check{x}_{kv} \end{bmatrix} \right) \tag{16}$$

Correction of covariance according to (5) is specified in the vector form of representation:

$$\begin{bmatrix} \hat{P}_{11} & \hat{P}_{12} \\ \hat{P}_{21} & \hat{P}_{22} \end{bmatrix} = \left( 1 - \begin{bmatrix} K_{11} \\ K_{21} \end{bmatrix} \cdot [h_{11} \quad h_{21}] \right) \cdot \begin{bmatrix} \check{P}_{11} & \check{P}_{12} \\ \check{P}_{21} & \check{P}_{22} \end{bmatrix} \tag{17}$$

Consider the hardware implementation of the described algorithm for calculators to predict the system's state and covariance.

## 3   Results and Discussion

In this work, we have built a Kalman filter architecture using high-speed CSA and KSA adders. The complete architecture of the Kalman filter is shown in Fig. 3, but in this paper we consider 2 blocks: calculator $\hat{x}_k^-$ and calculator $P_k^-$ are characterizing the parameters of predicting the state of the system and covariance errors.

Three terms are fed to the input of the CSA adder, one of which is fed to the bit carry. The output of this adder outputs two values: sum (S) and bit carry (C). The advantage of this adder is the high computation speed due to parallel computations for each bit. The Kogge-Stone prefix adder allows you to add two numbers. The advantage of this adder is the fast summation speed. However, the use of such an adder entails an increase in the number of hardware costs.

In the proposed architecture, the input will receive data with a width of 8 bits. After performing the necessary calculations, the variables $\hat{x}_k^-$ and $P_k^-$ will be 18 and

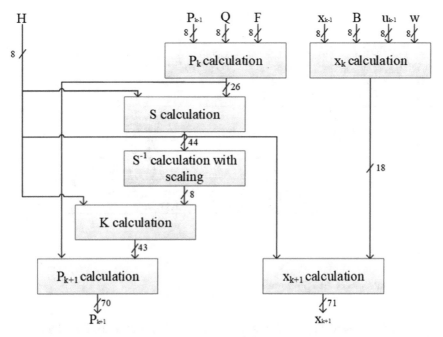

**Fig. 3** Kalman filter architecture

26 bits, respectively. The architectures of these calculators are presented in Fig. 4 and Fig. 5.

Let us illustrate the operation of the described algorithm using software and hardware modeling using the example of video from archived data.

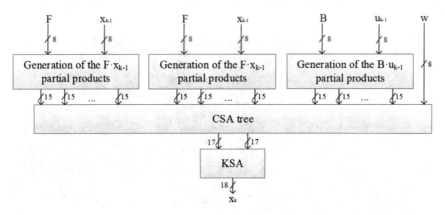

**Fig. 4** Architecture of the system state calculator $\hat{x}_k^-$

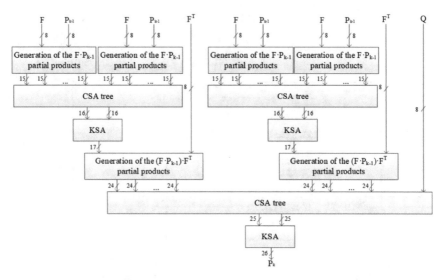

**Fig. 5** Architecture of the $P_k^-$ - covariance error calculator

For software modeling, used archived data in the form of a video file with a frame size of $720 \times 1280$ and an input rate of up to 30 frames per second. The simulation was carried out using the Matlab R2021a software package.

The result of the operation of the described algorithm is 2 new video files with selected moving objects from the running video file. Figure 6 shows an example of the program's operation for detecting a moving object from a video taken from the data archive [14].

As a result of the program execution, sections of the video data stream frames with moving objects were detected. Other objects do not overlap moving objects during

**Fig. 6** Detection of a moving object in a video stream

**Table 1** Results of hardware simulation of the designed device

| Approach | Delay, ns | LUT Slices, units | Power, W |
|----------|-----------|-------------------|----------|
| Original | 28.429    | 2355              | 190.158  |
| Proposed | 26.979    | 3742              | 449.503  |

the entire video sequence, and projective transformations associated with camera tilt do not occur. However, there are affine transformations related to the scalability and rotation of moving objects.

The hardware modeling of the Kalman filter architecture was performed in the VHDL using the Xilinx Vivado 2019.2 environment. The device is implemented on a Defense-Grade Virtex7 motherboard with xq7vx980t-2Lrf1930 parameters. The results of the hardware simulation are presented in Table 1. Original means the use of built-in operations of addition and multiplication in VHDL. The proposed architecture uses the CSA and KSA adders.

The table shows that the operations for predicting the state of the system and covariance errors in the proposed architecture are performed 5.4% faster than in the original one. However, they require more equipment and capacity costs by 37% and 58%, respectively.

Thus, the proposed architecture of calculators for predicting the state of the system and covariance errors can reduce the time spent.

# 4 Conclusion

The article substantiates the use of a probable model on the example of the Kalman filter in solving the problem of detecting a moving object in a video data stream. The hardware architecture of the filter for the prediction computers of the state and covariance of the system has been developed. The hardware simulation results showed that the developed architecture could be applied to computer vision systems, where the time delay for computations is a critical indicator. The hardware implementation of the algorithm has shown its applicability for detecting moving objects subject to affine transformations.

Further research will develop an architectural implementation of calculators for correcting the system's state and hardware comparison of the work of the set architectures not only in the positional number system but also in the Residue Number System (RNS).

**Acknowledgements** Research in section 2 was supported by North-Caucasus Center for Mathematical Research under agreement №. 075-02-2021-1749 with the Ministry of Science and Higher Education of the Russian Federation. Research in section 3 was supported by Russian Foundation for Basic Research (project №. 19-07-00130 A).

# References

1. Zhang Q, Huang N, Yao L, Zhang D, Shan C, Han J (2020) RGB-T salient object detection via fusing multi-level CNN features. IEEE Trans Image Process 29:3321–3335
2. Ren Q, Lu Sh, Zhang J, Hu R (2021) Salient object detection by fusing local and global contexts. IEEE Trans Multimedia 23:1442–1453
3. Metzler ChA, Lindell DB, Wetzstein G (2021) Keyhole imaging: non-line-of-sight imaging and tracking of moving objects along a single optical path. IEEE Trans Comput Imaging 7:1–12
4. Tu Zh, Li Zh, Li Ch, Lang Y, Tang J (2021) Multi-interactive dual-decoder for RGB-thermal salient object detection. IEEE Trans Image Process 30:5678–5691
5. Li J, Su J, Xia Ch, Tian Y (2020) Distortion-adaptive salient object detection in 360° omnidirectional images. IEEE J Sel Top Signal Process 14(1):38–48
6. Yu H, Zheng K, Fang J, Guo H, Wang S (2020) A new method and benchmark for detecting co-saliency within a single image. IEEE Trans Multimedia 22(12):3051–3063
7. Liu J, Zhoub Sh, Wu Y, Chen K, Ouyang W, Xu D (2021) Block proposal neural architecture search. IEEE Trans Image Process 30:15–25
8. Fang Y, Zhang Ch, Huang H, Lei J (2019) Visual attention prediction for stereoscopic video by multi-module fully convolutional network. IEEE Trans Image Process 28(11):5253–5265
9. Li Y, Chen Ch, Hao A, Qin H (2020) Accurate and robust video saliency detection via self-paced diffusion. IEEE Trans Multimedia 22(5):1153–1167
10. Ruan W, Chen J, Yi Wu, Wang J, Liang Ch, Hu R, Jiang J (2019) Multi-correlation filter with triangle-structure constraints for object tracking. IEEE Trans Multimedia 21(5):1122–1134
11. He X, Johansson KH, Fang H (2021) Distributed design of robust Kalman filters over corrupted channels. IEEE Trans Signal Process 69:2422–2434
12. Lim J, Kim H-S, Park H-M (2020) Interactive-multiple-model algorithm based on minimax particle filtering. IEEE Signal Process Lett 27, 36–40
13. Hua J, Li Ch (2019) Distributed robust Bayesian filtering for state estimation. IEEE Trans Signal Inf Process Over Netw 5(3):428–441
14. https://disk.yandex.ru/i/iGCuf0kn0PtzNQ

# Solving the Euler Problem for a Flexible Support Rod Base on the Finite Difference Method

Khusen Kulterbaev⬤, Madina Lafisheva⬤, and Lyalusya Baragunova⬤

**Abstract** The article focuses on the non-classical problem of the stability loss in a rectilinear rod with a flexible support. The mathematical model employed to study bifurcation consists of a basic differential equation of rod bending enhanced with boundary conditions. Through the finite difference method, they are reduced to a system of algebraic equations with a square matrix. There is a view offered at rods with constant and variable cross sections. Critical forces taken as unknown values are contained in the characteristic equation of the matrix, of which they are extracted numerically and graphically with the Matlab computing system. The identified critical forces were verified with tests on the well-known Euler problem as well as by comparing the results of two examples. There are conclusions offered, whichi are of practical value.

**Keywords** Critical force · Differential equations of longitudinal bending · Boundary conditions · Algebraic equations system

## 1 Introduction

The stability loss of the rectilinear equilibrium shape of a centrally compressed straight rod (Fig. 1) is called longitudinal bending, which is the simplest and yet one of the most important engineering issues related to construction and machine-building structures [1].

The main problem of identifying the critical force of a compressed rod was solved by Euler in the 18th Century while he was in close cooperation with the St. Petersburg Academy of Sciences. However, that time the issue appeared irrelevant due to compressed rods being mostly of stone, of short length, and had large cross sections. After about a century, though, circumstances changed. The industrial Revolution in

K. Kulterbaev (✉) · M. Lafisheva
NCFU, North-Caucasus Center for Mathematical Research, Stavropol, Russia
e-mail: kulthp@mail.ru

L. Baragunova
Kabardino-Balkaria State University H.M. Berbekov, Nalchik, Russia

© The Author(s), under exclusive license to Springer Nature Switzerland AG 2022   143
A. Tchernykh et al. (eds.), *Mathematics and its Applications in New Computer Systems*,
Lecture Notes in Networks and Systems 424,
https://doi.org/10.1007/978-3-030-97020-8_13

Europe, America and Russia resulted in a rapid introduction of metal into construction—first cast iron, and later steel. Such long metal pillars could carry significant
compressive loads with relatively small transverse dimensions. In addition, they are
easy to be manufactured in large numbers and do not require much to be erected.
Euler's problem gradually became classic and got in demand. Technological needs
have resulted in a large number of various calculation schemes [2, 3].

While in the classical scheme a compressed rod had articulated supports only,
nowadays the compressed rod design schemes feature a wide variety of supports,
cross-sectional shapes and loading methods.

To identify the critical forces, analytical methods were used at the initial stage.
The complexity of calculation schemes resulted in a need to perform large transformations, solving transcendental equations and their systems. Apart from the
mentioned shortcomings, there were some others detected. Analytical methods are
not adjusted to identifying critical forces for rods with variable cross-sections along
the axis, stepped, in cases of simultaneous application of several concentrated forces,
distributed loads, supports not only at the ends, yet at other points as well.

These circumstances have lately stimulated the use of numerical methods and
computing complexes relying on computer technologies [4–6]. In Russia, the numerical method of finite differences was developed and implemented in the works by
A.A. Samarsky and his followers. At the early stages and further on, the development
of numerical methods was due to research and academic publications [7–12], which
focus on general and special issues of employing numerical methods in calculations
and design of construction and machine-building units.

In this paper, identifying the critical forces is done based on the Matlab calculation
system as well as on the finite difference method for two cases of rods: with constant
and variable cross sections along the axes.

## 2 Mathematical Model

Next, we will take a look at a straight rod with a pivotally supported left end and a
right end resting on an elastic spring, viewing it as a basic task for working through
the algorithm for applying the finite difference method (Fig. 1).

During that, the rod is loaded at the ends by centrally applied compressive F
forces. The curved axis of the rod in the supercritical state is known to be described
by a fourth-order differential equation

$$[b(x)v''(x)]'' + Fv''(x) = 0, \quad b(x) = EJ(x), \; x \in (0, l). \tag{1}$$

Here we see generally accepted designations used.

For a rod of constant cross-section along the axis, it would be a good idea to
simplify Eq. (1) through dividing it by $b = $ const, which will finally produce

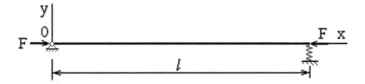

**Fig. 1** Calculation scheme

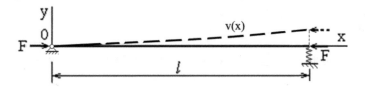

**Fig. 2** Rod in deformed condition

$$v''''(x) + kv''(x) = 0, \quad k = F/b, \ b = EJ, \ x \in (0, l). \tag{2}$$

Equations (1), (2) need boundary conditions to be attached to them. At the left end, the articulated support matches a deflection and bending moment, which equal zero

$$v(0) = 0, \ M(0) = 0, \ M(0) = bv''(0) = 0 \ \Rightarrow \ v''(0) = 0. \tag{3}$$

Identifying the boundary conditions at the right end will take using a diagram of a rod in a supercritical deformed state (Fig. 2) rather than Fig. 1.
Here follow boundary conditions

$$Q(l) = bv'''(l), \mathrm{M}(l) = 0. \tag{4}$$

Analyzing these, it is to be kept in mind that the section rotation angle here is very small. This means that the effect of the F force on the transverse Q force can be neglected. The transverse force here is due to a stretched spring with the C stiffness coefficient. Given that, the first equation in (4) is to appear as

$$bv'''(l) - cv(l) = 0. \tag{5}$$

Equations (1)–(5) make up mathematical models of problems for two rod variants.

## 3 Constant Cross-Section Rod

Further on, it will take translating the analytical mathematical model described above into a numerical one by employing the finite difference method. For this, let us replace the area of the argument x continuous change in (2) with the area of discrete change (Fig. 3)

$$L_h = \{x_i = (i-1)\,h, \quad i = 1,\, 2,\, \ldots,\, n\}.$$

with a step of $h = l/(n-1)$. Rather than a $v(x)$ continuous function, we will be considering a $v_i \approx v(x_i)$ grid function.

The derivatives entering the mathematical model are replaced by finite-difference ones with the $O(h^2)$ accuracy. The basic Eq. (2) for a rod with a constant cross-section will assume the following appearance

$$v_{i-2} + \alpha v_{i-1} + \beta v_i + \alpha v_{i+1} + v_{i+2} = 0, \quad i = 3, 4, \ldots n - 2. \tag{6}$$

$$\alpha = -4 + kh^2, \ \beta = 6 - 2kh^2.$$

Here it adds on the boundary conditions corresponding to the articulated support of the left end looking as

$$v_1 = 0, \quad 2v_1 - 5v_2 + 4v_3 - v_4 = 0. \tag{7}$$

At the right end (4), (5) produce

$$-v_{n-3} + 4v_{n-2} - 5v_{n-1} + 2v_n = 0, \, 3v_{n-4} - 14v_{n-3} + 24v_{n-2}$$
$$-18v_{n-1} + \gamma v_n = 0, \gamma = 5 - 2\frac{h^3c}{b}. \tag{8}$$

As we rewrite (6)–(8) in a matrix-vector form, it will give us

$$Bv = 0, \tag{9}$$

where B is a square matrix of n order, $v$ is a column-vector with $v_i$, $i = 1, 2, \cdots, n$ components.

In the expanded form, the equation appears as

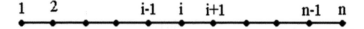

Fig. 3  Grid area

$$
\begin{pmatrix}
1 & & & & & & \\
2 & -5 & 4 & -1 & & & \\
1 & \alpha & \beta & \alpha & 1 & & \\
\cdots & \cdots & \cdots & \cdots & \cdots & \cdots & \cdots \\
& & 1 & \alpha & \beta & \alpha & 1 \\
& & 3 & -14 & 24 & -18 & \gamma \\
& & & -1 & 4 & -5 & 2
\end{pmatrix}
\cdot
\begin{pmatrix}
v_1 \\ v_2 \\ v_3 \\ \cdots \\ v_{n-2} \\ v_{n-1} \\ v_n
\end{pmatrix}
=
\begin{pmatrix}
0 \\ 0 \\ 0 \\ \cdots \\ 0 \\ 0 \\ 0
\end{pmatrix}.
\tag{10}
$$

The zero elements of the B matrix are not included.
Equation (10) has an obvious solution

$$
v = (0, 0, \ldots, 0)^{\mathrm{T}},
$$

which is of no interest. A non-zero solution is possible in case of a zero determinant of the B matrix.

$$
det\,(B) = 0.
\tag{11}
$$

The F values in case of which Eq. (11) holds, are, actually, the critical forces of the compressed rod. Identifying these values for large n values in simple ways is not possible. A reasonable solution in this case is using a numerical-graphical method for identifying critical forces. Its point lies in the fact that the finite difference method is employed in combination with computer software and calculation systems, and then the $F - det\,(B)$ graph is constructed [12]–[14].

The elements of the B matrix contain the longitudinal F force. The roots of this algebraic equation are identified visually by using graphs displayed on the monitor.

The $F_\kappa$, $\kappa = 1, 2, 3$ ... values, which correspond to the intersection points of the abscissa axis and the graph, are, actually, the critical forces. They are easily and accurately read from the monitor screen when examining the graphs. It is to be noted that the high accuracy of identifying the coordinates of the points on the F axis is due to the "magnifying glass" in the calculation systems, which allows multiplying the fragments of drawings.

*Example 1.* To test the proposed algorithm, we shall take a rod with articulated supports at both ends and with conditional initial data

$$
\pi^2 E J / l^2 = 1.
$$

The exact values of the critical forces spectrum elements in this case are the elements of the following set

$$
F_\kappa = k^2, \ k = 1, 4, 9, 16 \ldots.
$$

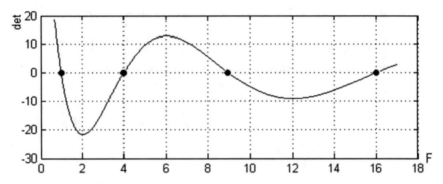

**Fig. 4** Values of critical forces

The program designed in the Matlab system offered a result appearing as it is in Fig. 4.

It is obvious that the critical forces read from the figure exactly coincide with those calculated by Euler's classical formula.

This example of testing the algorithm and the corresponding computer program relying on the well-known problem related to the critical forces of a constant cross-section rod allows stating that the finite difference method can be used to solve complex problems with no need to involve complex analytical methods.

*Example 2.* Now we shall consider a real steel rod with a design scheme based on Fig. 2. Here below come the physical and mechanical parameters:

$$E = 2{,}1\,10^8 \text{ kN/m}, l = 4\text{m}, J = 5 \cdot 10^{-8}\text{m}^4, c = 2 \cdot 10^{-4}\text{kN/m}.$$

There was a program developed within the Matlab system with a graph obtained (see Fig. 5). The first three critical force values were read from it.

$$F_K = \{6{,}477; 25.908; 58.292\} \text{ kN}.$$

Euler's analytical method allows finding the critical force, yet only in rods featuring constant cross-section. In case of rods with variable cross-section, it is a more complicated issue, since there are systems of transcendental equations that have to be solved. This example shows that this problem can be easily solved employing the numerical-graphical method combined with the Matlab system.

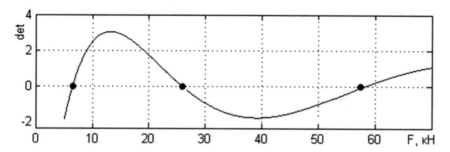

**Fig. 5** Critical force values for a rod of constant cross-section

## 4    Variable Cross-Section Rod

In variable cross-section rods, the axial moment of inertia will be represented in the main Eq. (1) by a longitudinal coordinate function. The finite-difference representation of this equation will be significantly different from Eq. (6). For additional boundary conditions, the finite-difference Eqs. (7), (8) will reveal no change.

Applying the same procedures for replacing derivatives with their finite difference analogues will produce a finite difference equation similar to (6)

$$
\begin{aligned}
&b_{i-1}v_{i-2} + p_i v_{i-1} + q_i v_i + r_i v_{i+1} + b_{i+1}v_{i+2} = 0, \quad i = 3, 4, \ldots, n-2. \\
&p_i = \alpha_i + Fh^2, \quad q_i = \beta_i - 2Fh^2, \quad r_i = \gamma_i + Fh^2, \\
&\alpha_i = -2b_{i-1} - 2b_i, \quad \beta_i = b_{i-1} + 4b_i + b_{i+1}, \quad \gamma_i = -2b_1 - 2b_{i+1}.
\end{aligned} \tag{12}
$$

(7), (8), (13) represent a homogeneous system of linear algebraic equations.

$$
Bv = 0, \tag{13}
$$

which we will put in a matrix-vector form

$$
\begin{pmatrix}
1 & & & & & \vdots & & & & & & \\
2 & -5 & 4 & -1 & & \vdots & & & & & & \\
b_2 & p_3 & q_3 & r_3 & b_4 & \vdots & & & & & & \\
& b_3 & p_4 & q_4 & r_4 & b_5 & \vdots & & & & & \\
\cdots & \cdots & \cdots & \cdots & \cdots & \cdots & \cdots & \cdots & \cdots & \cdots & \cdots & \cdots \\
& & & & & \vdots & b_{n-3} & p_{n-2} & q_{n-2} & r_{n-2} & b_{n-1} & \\
& & & & & \vdots & 3 & -14 & 24 & -18 & \gamma & \\
& & & & & \vdots & & -1 & 4 & -5 & 2 &
\end{pmatrix}
\cdot
\begin{pmatrix}
v_1 \\ v_2 \\ \vdots \\ \vdots \\ \vdots \\ \vdots \\ v_{n-1} \\ v_n
\end{pmatrix}
=
\begin{pmatrix}
0 \\ 0 \\ \vdots \\ \vdots \\ \vdots \\ \vdots \\ 0 \\ 0
\end{pmatrix}
$$

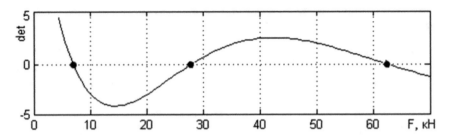

**Fig. 6** Critical force values for a rod of variable cross-section

The critical forces in this case, too, will be identified based on Eq. (11), its roots to be found in the same graphical way on the computer.

*Example 3.* A steel rod of circular cross-section features the following physical and mechanical parameters:

$$E = 2{,}110^8 \text{ kN/m}^2, \ l = 4\text{m}, \text{c} = 2 \cdot 10^{-4} \text{ kN/m}, \ J = 5 + 0,5 \sin \pi / l \text{m}$$

The program developed in Matlab produced the result to be seen in Fig. 6. This makes it easy to read the values of the critical forces

$$F_{\kappa} = \{\ 7{,}052;\ 27{,}792;\ 62{,}498\}\ \kappa H.$$

It is notable that the outcomes of the two solved examples feature little difference. The explanation for it is simple: the initial data of the examples, apart from the axial moment of the cross section, are similar. The axial moment in the third example is set as a variable value, which depends on the spatial coordinate x, yet slightly different from the constant cross-section in the second example. This choice was made for a reason, in order to have a check of the calculation correctness obtained via different algorithms and computer programs. Indeed, the critical forces described above differ slightly, which allows claiming the obtained results as reliable.

## 5  Conclusions

1.  The numerical-graphical method employed to identify the critical forces of compressed rods is a simple and universal means of identifying the critical forces of compressed rods.
2.  The numerical-graphical method has the property of universality and does not entail the need to use transcendental and Bessel's functions.
3.  The examples reveal that the algorithms of the finite difference method are easily implemented within the Matlab calculation system when dealing with stability problems.

# References

1. VI Feodosiev (2003) Strength of materials. MSTU im. N.E.Bauman, Moscow, Russia
2. Maslennikov AM (2016) Dynamics and stability of structures. Textbook and workshop for universities. Yurayt, Moscow, Russia
3. Gorshkov AG, Troshin VN, Shalashilin VI (2002) Strength of materials. Fizmatlit, Moscow, Russia
4. Makarov EG (2009) Resistance of materials using computing systems: In: 2 books the main course. Higher School, Moscow, Russia
5. Korobko VI, Korobko AV (2008) Dynamics and stability of rod systems. Textbook. ASV Publishing House, Moscow, Russia
6. Gorodetsky AS, Barabash MS, Sidorov VN (2016) Computer modeling in problems of structural mechanics. ASV Publishing House, Moscow, Russia
7. Verzhbitsky VM (2002) Fundamentals of numerical methods. Higher School, Moscow, Russia
8. Formalev VF, Reviznikov DL (2006) Numerical methods, 2nd edn. Fizmatlit, Moscow, Russia
9. Ilyin VP, Karpov VV, Maslennikov AM (2005) Numerical methods for solving problems in structural mechanics. ASV Publishing House, Moscow, Russia
10. Zolotov AB, Akimov, PA, Sidorov, VN (2008) Mathematical methods in structural mechanics (with the foundations of the theory of generalized functions). ASV Publishing House, Moscow, Russia
11. Zolotov AB, Akimov PA, Sidorov VN, Mozgaleva ML (2009) Numerical and analytical methods for calculating building structures. ASV Publishing House, Moscow, Russia
12. Kulterbaev KhP, Baragunova LA, Shogenova MM, Senov KhM (2018) About a high-precision graphoanalytical method of determination of critical forces of an oblate rod. In: Proceedings 2018 IEEE international conference "quality management, transport and information security, information technologies" (IT & QM & IS), 24–28 September, pp 794–796
13. Kulterbaev KhP, Baragunova LA (2012) Implementation of the problem of eigenvalues of a compressed-stretched bar on a computer. In: Computer technologies: Materials of the All-Russian scientific and technical conference, pp 90–94
14. Kultebaev KhP (2021) The nonclassical problem of stability of the compressed bar. In: E3S web of conferences 281. 01031 CATPID-2021, Part 110. 10.1051/ e3sconf/202128101031

# Multiscale Model Reduction for the Poroelasticity Problems Using Embedded Fracture Model

Aleksei Tyrylgin⬤, Maria Vasilyeva⬤, and Anatoly Alikhanov⬤

**Abstract** In this paper, we present the application of an generalized multiscale finite element method in numerical simulation of poroelasticity problems in fractured media. Mathematical model contains a coupled system of equations for displacement and pressure, where for fractures we use an embedded fracture model. The most important feature of mathematical models of poroelasticity is that the equations of the system are coupled. Fine grid approximation is constructed based on the finite element method for the displacements and finite volume approximation for the pressure in fractured media. To construct structured coarse grid approximation a generalized multiscale finite element method is used, where we solve local spectral problem for construction of the multiscale basis functions for displacement and pressures. Numerical results are presented for the two and three-dimensional model problem with different number of the multiscale basis functions. We compute relative $L_2$ error between the multiscale solution with the fine-scale solutions by choosing different numbers of multiscale basis functions.

**Keywords** Poroelasticity · Generalized multiscale finite element method · Fractured media · Numerical simulation · Embedded fracture model · Finite volume approximation

A. Tyrylgin (✉)
Multiscale Model Reduction Laboratory, North-Eastern Federal University, Yakutsk, Republic of Sakha (Yakutia), Russia
e-mail: tyrylgin.aa@s-vfu.ru

M. Vasilyeva
Department of Mathematics and Statistics, Texas A&M University, Corpus Christi, TX, USA
e-mail: vasilyeva@tamucc.edu

A. Alikhanov
North-Caucasus Center for Mathematical Research, North-Caucasus Federal University, Stavropol, Russia

© The Author(s), under exclusive license to Springer Nature Switzerland AG 2022
A. Tchernykh et al. (eds.), *Mathematics and its Applications in New Computer Systems*,
Lecture Notes in Networks and Systems 424,
https://doi.org/10.1007/978-3-030-97020-8_14

153

# 1 Introduction

The coupling of flow and deformation has attracted many researchers across various disciplines due to its profound implications in science and engineering. Poroelastic effects in fractured media is necessary for many real world applications, for example, reservoir geomechanics, which have crucial roles in reservoir compaction, ground subsidence, gas reservoirs, and underground waste disposal [1, 3, 5]. The basic mathematical model of poroelasticity is construct by the coupling of equations of pressure and deformation via Biot model [4]. Fine grid approximation depends on the mesh construction, where the discrete fracture model is used for conforming fracture and porous matrix grid [2, 10, 12] and in the case of separate independent construction of the fracture and porous matrix grids, an embedded fracture model is used [11, 13].

Embedded fracture model is used to process a complex mesh of fractures on a fine grid. EFM can use meshes for the porous matrix that are independent of the fracture meshes. For accurate numerical solution of the flow and deformation using embedded fracture model, we should use a fine grid which leads to the large discrete system of equations. To reduce the size of the discrete system, a multiscale methods are used [6–9]. In this work, we develop a Generalized Multiscale Finite Element method for solution of the poroelasticity problems in fractured media with embedded fracture model. The construction of the offline basis functions is given separately for pressure and displacement fields and involves the solution of the local spectral problems. The offline basis functions are selected using eigenvectors that correspond to dominant eigenvalues. We numerically investigate presented method for two and three—dimensional model problems in fractured and heterogeneous poroelastic media.

The work is organized as follows. In Sect. 2, we present the mathematical model and a fine grid approximation of the poroelasticity problem with embedded fracture model. In Sect. 3, a coarse grid approximation is constructed using Generalized Multiscale Finite Element Method. Numerical results for two—dimensional model poroelasticity problems are presented in Sect. 4. Finally, we describe conclusions.

# 2 Mathematical Model and Fine Grid Approximation

Let $\Omega$ is computation domain for the porous matrix and $\gamma$ is the lower dimensional fracture domain. We consider a mathematical model of flow and geo-mechanics in fractured poroelastic medium that described by a following system of equations for displacements, pressure in porous matrix and fractures

$$
\begin{aligned}
-\mathrm{div}\boldsymbol{\sigma}\left(u\right) + \alpha \, \mathrm{grad} \, p_m &= 0, x \in \Omega, \\
c_m \frac{\partial p_m}{\partial t} - \mathrm{div}\left(\mathrm{k_m} \, \mathrm{grad} \, p_m\right) + r_{mf}\left(p_m - p_f\right) &= f_m, x \in \Omega, \\
c_f \frac{\partial p_f}{\partial t} - \mathrm{div}\left(\mathrm{k_f} \, \mathrm{grad} \, p_f\right) + r_{fm}\left(p_m - p_f\right) &= f_f, x \in \gamma,
\end{aligned}
\tag{1}
$$

with a linear relation between stress $\sigma$ and strain $\varepsilon$ tensors

$$\sigma(u) = \lambda \varepsilon_v I + 2\mu\varepsilon(u), \quad \varepsilon(u) = 0.5\big(\nabla u + (\nabla u)^T\big),$$

where $\lambda, \mu$ are the Lames parameters, $u$ is the displacements, $p$ is the pressure, $\alpha$ is the Biot coefficient, $f$ is the source term, $k_\alpha = \kappa_\alpha = \nu$, $\nu$ is the viscosity, $\kappa_\alpha$ is the permeability for $\alpha = m, f$, $r_{\alpha\beta} = \eta_{\alpha\beta} r$, $r$ is the transfer coefficient, $\eta_{\alpha\beta}$ is the geometric factors, $c_\alpha$ is the compressibility for $\alpha = m, f$. Note that, we neglect the gravitational forces, suppose that effect of the mechanics to the flow is relatively small and consider effect of the porous matrix pressure to the displacements. In this work, we use the two-dimensional problem for illustration of the presented method and consider an implicit scheme for approximation of time with given time step $\tau$. Let $\mathcal{T}_h = \cup_i \varsigma_i$ be a fine scale finite element partition of the domain $\Omega$ and $\mathcal{E}_\gamma = \cup_l u_l$ is the fracture mesh. $N_f^m$ is the number of cells in $\mathcal{T}_h$, $N_f^f$ is the number of cell for fracture mesh $\mathcal{E}_\gamma$.

For approximation of the flow problem, we use a finite volume approximation on the structured fine grids and obtain the following discrete system

$$c_m \frac{p_{m,i} - \breve{p}_{m,i}}{\tau} |\varsigma_i| + \sum_j T_{ij}\big(p_{m,i} - p_{m,j}\big) + \sum_l q_{il}\big(p_{m,i} - p_{f,l}\big) = f_m|\varsigma_i|, \forall i = 1, N_f^m$$

$$c_f \frac{p_{f,l} - \breve{p}_{f,l}}{\tau} |u_l| + \sum_n W_{ln}\big(p_{f,l} - p_{f,n}\big) + \sum_i q_{il}\big(p_{m,i} - p_{f,l}\big) = f_f|u_l|, \forall l = 1, N_f^f$$

$$(2)$$

where $T_{ij} = \frac{k_m |E_{ij}|}{\Delta_{ij}}$ ($|E_{ij}|$ is the length of interface between cells $\varsigma_i$ and $\varsigma_j$, $\Delta_{ij}$ is the distance between mid point of cells $\varsigma_i$ and $\varsigma_j$), $W_{ln} = k_f/\Delta_{ln}$ ($\Delta_{ln}$ is the distance between point and $|u_l|$ is the volume of the cells and $\varsigma_i$ and $u_l$, $q_{il} = r$ if $\varsigma_i \cap u_l \neq 0$ and equals zero otherwise. Here $(\breve{p}_m, \breve{p}_f)$ are solutions from the previous times step.

For displacement, we use Galerkin method with linear basis functions.

$$\int_\Omega \sigma(u) : \varepsilon(v)dx - \int_\Omega \alpha p_m \mathcal{I} \cdot \varepsilon(v)dx = 0, \tag{3}$$

Therefore, we have a following computational algorithm for solution on the fine grid in the matrix form.

Solve pressure system for $p = \big(p_m, p_f\big)^T$ :

$$\left(\frac{1}{\tau}M + A\right)p = F, \tag{4}$$

where

$$M = \begin{pmatrix} M_m & \\ & M_f \end{pmatrix}, F = \begin{pmatrix} F_m + \dfrac{1}{\tau} M_m \breve{p}_m \\ F_f + \dfrac{1}{\tau} M_f \breve{p}_f \end{pmatrix} \quad A = \begin{pmatrix} A_m + Q & -Q \\ -Q & A_f + Q \end{pmatrix},$$

$$M_m = \{m_{ij}^m\}, \quad m_{ij}^m = \begin{cases} c_m |\varsigma_i| & i = j \\ 0 & i \neq j \end{cases}, \quad M_f = \{m_{ln}^f\}, m_{ln}^f = \begin{cases} c_f |\iota_l| & l = n, \\ 0 & l \neq n \end{cases}$$

and $A_m$, $A_f$ are the transmissibility matrices, $Q$ is the transfer term matrix between porous matrix and fractures, $F_m = \{f_i^m\}$, $f_i^m = f_m |\varsigma_i|$, $F_f = \{f_i^f\}$, $f_i^m = f_f |\iota_l|$.

Solve is displacements system $u$:

$$Du = B, \tag{5}$$

where $D = [d_{ij}] = \int_\Omega \sigma(\psi_i) : \varepsilon(\psi_j) dx$ is the elasticity stiffness matrix, $B = [b_j] = \int_\Omega \alpha p_m \cdot \varepsilon(\psi_j) dx$.

## 3   Coarse Grid Approximation Using GMsFEM

For coarse grid approximation of the poroelasticity problems in fractured porous media, we use the Generalized Multiscale Finite Element Method (GMsFEM). GMsFEM contains following steps: (1) coarse grid and local domains construction; (2) generation of the projection matrix using local multiscale basis functions; (3) construction of the coarse grid system using projection matrix; (4) solution of the coarse scale problem and reconstruction of the fine grid solution.

For construction of the multiscale basis functions for pressures and displacements, we solve a spectral problem in the local domain $\omega_i$

- pressures, $\phi_j^{\omega_i} = (\phi_{m,j}^{\omega_i}, \phi_{f,j}^{\omega_i})$

$$A^{\omega_i} \phi_j^{\omega_i} = \lambda_{p,j} S^{\omega_i} \phi_j^{\omega_i},$$

- displacements, $\Phi_j^{\omega_i} = (\Phi_{x,j}^{\omega_i}, \Phi_{y,x}^{\omega_i})$

$$D^{\omega_i} \Phi_j^{\omega_i} = \lambda_{u,j} G^{\omega_i} \Phi_j^{\omega_i},$$

where $D^{\omega_i}$ and $A^{\omega_i}$ are the restrictions of the global matrices $D$ and $A$ to the local domain $\omega_i$. Here

$$S^{\omega_i} = \begin{pmatrix} S_m^{\omega_i} & 0 \\ 0 & S_f^{\omega_i} \end{pmatrix},$$

$$S_m^{\omega_i} = \{s_{ij}^{\omega_i,m}\}, \quad s_{ij}^{\omega_i,m} = \begin{cases} k_m |\varsigma_i| & i = j \\ 0 & i \neq j \end{cases}, \quad S_f^{\omega_i} = \{s_{ln}^{\omega_i,f}\}, \quad s_{ln}^{\omega_i,f} = \begin{cases} k_f |\iota_l| & l = n \\ 0 & l \neq n \end{cases},$$

and $G^{\omega_i} = [g_{ij}] = \int_{\omega_i} (\lambda + 2\mu) \psi_j dx$.

We form the multiscale spaces for pressures and displacements using eigenvectors $\{\phi_1^{\omega_i}, \phi_2^{\omega_i}, \ldots, \phi_{L_p}^{\omega_i}\}$ and $\{\Phi_1^{\omega_i}, \Phi_2^{\omega_i}, \ldots, \Phi_{L_u}^{\omega_i}\}$ corresponding to the first smallest $L_p$ and $L_u$ eigenvalues, where $\lambda_{p,1} \leq \lambda_{p,2} \leq \cdots \leq \lambda_{p,L_p}$ and $\lambda_{u,1} \leq \lambda_{u,2} \leq \cdots \leq \lambda_{u,L_u}$. Furthermore, for obtaining conforming basis functions we use linear partition of unity functions $x^{\omega_i}$. We construct transition matrices $R_u$ and $R_p$ from a fine grid to a coarse grid and use it for reducing the dimension of the problem

$$R_u = \left\{ \chi^{\omega_1} \Phi_1^{\omega_1}, \chi^{\omega_1} \Phi_2^{\omega_1}, \ldots, \chi^{\omega_1} \Phi_{L_u}^{\omega_1}, \ldots, \chi^{\omega_{N_c}} \Phi_1^{\omega_{N_c}}, \chi^{\omega_{N_c}} \Phi_2^{\omega_{N_c}}, \ldots, \chi^{\omega_{N_c}} \Phi_{L_u}^{\omega_{N_c}} \right\},$$

$$R_u = \left\{ \chi^{\omega_1} \phi_1^{\omega_1}, \chi^{\omega_1} \phi_2^{\omega_1}, \ldots, \chi^{\omega_1} \phi_{L_p}^{\omega_1}, \ldots, \chi^{\omega_{N_c}} \phi_1^{\omega_{N_c}}, \chi^{\omega_{N_c}} \phi_2^{\omega_{N_c}}, \ldots, \chi^{\omega_{N_c}} \phi_{L_p}^{\omega_{N_c}} \right\},$$

where $\chi^{\omega_i}$ is linear partition of unity functions, $L_p$ and $L_u$ are the number of basis functions for pressure and displacements, $N_c$ is the number of vertices of a coarse grid.

Then the system of equations can be translated into a coarse grid, and we have following computational algorithm in the matrix form

Solve pressure system for $p_c = \left( p_{c,m}, p_{c,f} \right)^T$ :

$$\left( \frac{1}{\tau} M_c + A_c \right) p_c = F_c, \tag{6}$$

where $M_c = R_p M R_p^T$, $A_c = R_p A R_p^T$ and $F_c = R_p F$.

Solve displacements system for $u$ :

$$D_c u_c = B_c, \tag{7}$$

where $D_c = R_u D R_u^T$ and $B_c = R_u B$.

After obtaining of a coarse-scale solution, we can reconstruct fine-scale solution $u_{ms} = R_u^T u_c$ and $p_{ms} = R_p^T p_c$. Next, we present numerical results for a test problems and compare multiscale solutions for pressures and displacements ($p_{ms}$ and $u_{ms}$) with the reference fine grid solutions ($p$ and $u$).

## 4   Numerical Results

– In this section, we consider poroelasticity problem in fractured porous media. We consider two and three-dimensional problems in $\Omega = [0, 1] \times [0, 1]$ and $\Omega = [0, 10] \times [0, 10] \times [0, 5]$ :

– *Two—dimensional model problem.* Fine grid contains 40,401 vertices and 40,000 cells. Fracture grid contains 686 vertices and 684 cells. For study of the presented multiscale method, we consider two coarse grids $10 \times 10$ and $20 \times 20$;

– *Three—dimensional model problem.* Fine grid contains 35,301 vertices and 32,000 cells. Fracture grid contains 6204 vertices and 11,140 cells. Coarse grid—10 × 10 × 5.

## 4.1   Two-Dimensional Problems

We simulate a two-dimensional problem in fractured porous media for two test problems. We set parameters of model problem $E = 10$, $v = 0.3$, $k_m = 10^{-5}$, $k_f = 1$ for *Test* 1. For *Test* 2, we use a heterogeneous coefficient, where elasticity modulus and heterogeneous permeability are presented in Fig. 1. We set $\alpha = 1$, $\sigma = k_m$, $c_m = 0.1$ and $c_f = 0.01$. The calculation is performed by $T_{max} = 100$ with time step $\tau = 10$. Computational grids and fracture distribution are presented in Fig. 1.

In Figs. 2 and 3, distribution of pressure and displacement along $X$ and $Y$ directions at final time for 8 multiscale basis functions in homogeneous and heterogeneous media are presented. In Table 1, we present the relative $L_2$ errors for different number of the multiscale basis functions on the coarse grid 10 × 10 and 20 × 20 between multiscale and fine grid solutions. In the Table 1 for *Test* 1, the results show that

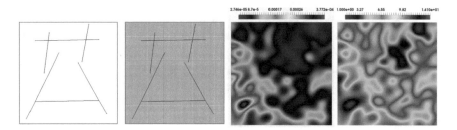

**Fig. 1** Fracture distribution, fine grid, heterogeneous permeabilities $K$ and elasticity parameter $E$ (from left to right) for two-dimensional problem

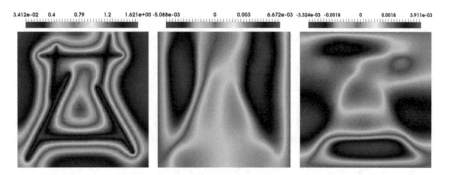

**Fig. 2** Distribution of pressure, displacement along X and Y directions at the last moment of time for 8 multiscale basis functions for two-dimensional problem (Test 1)

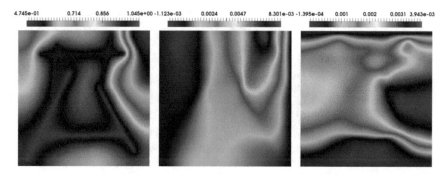

4.745e-01    0.714    0.856    1.045e+00 -1.123e-03    0.0024    0.0047    8.301e-03 -1.395e-04    0.001    0.002    0.0031 3.943e-03

**Fig. 3** Distribution of pressure, displacement along X and Y directions at the last moment of time for 8 multiscale basis functions for two-dimensional problem (Test 2)

**Table 1** Relative errors in % for displacement and pressure with different numbers of multiscale basis functions for two-dimensional problems (*Test 1*(left) and *Test* 2(right)). $L_p = L_u = L$.

| Test 1 L | 10 × 10 | | 20 × 20 | | Test 2 L | 10 × 10 | | 20 × 20 | |
|---|---|---|---|---|---|---|---|---|---|
| | $e^u$ | $e^p$ | $e^u$ | $e^p$ | | $e^u$ | $e^p$ | $e^u$ | $e^p$ |
| 1 | 4.850 | 1.840 | 1.794 | 0.896 | 1 | 13.461 | 10.781 | 3.195 | 4.065 |
| 2 | 1.933 | 1.332 | 0.535 | 0.550 | 2 | 4.285 | 7.350 | 1.037 | 2.477 |
| 4 | 1.120 | 0.648 | 0.372 | 0.292 | 4 | 3.850 | 4.208 | 1.082 | 1.278 |
| 8 | 0.395 | 0.488 | 0.136 | 0.195 | 8 | 1.103 | 3.502 | 0.274 | 0.992 |
| 12 | 0.289 | 0.330 | 0.077 | 0.144 | 12 | 1.009 | 1.628 | 0.126 | 0.552 |

4 multiscale basis functions are enough to achieve good results with 1.120% of relative $L_2$ error for displacement and 0.372% of relative $L_2$ pressure and in the case of *Test* 2 4 multiscale basis functions are sufficient with 1.103 % relative $L_2$ error for displacement and 0.274% relative $L_2$ error for pressure.

## 4.2   Three-Dimensional Problems

We simulate a three-dimensional model problem in fractured porous media for two test problems. We set parameters of model problem $E = 10, v = 0.3, k_m = 10^{-3}, k_f = 1$ for Test 1. For Test 2, we use a heterogeneous coefficient for elasticity modulus and heterogeneous permeability are presented in Fig. 4. We set $\alpha = 1, \sigma = 0.001, c_m = 0.02$ and $c_f = 0{:}002$. The calculation is performed by $T_{max} = 50$ with time step $\tau = 5$. Computational grids and fracture distribution are presented in Fig. 4.

In Figs. 5 and 6, distribution of pressure and displacement along $X, Y$ and $Z$ directions at final time for 12 multiscale basis functions in homogeneous and heterogeneous media are presented. In Table 2, we present the relative $L_2$ errors for different

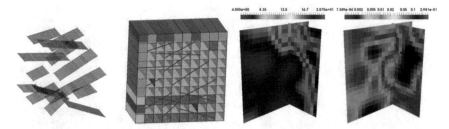

**Fig. 4** Fracture distribution, fine grid, heterogeneous permeabilities $K$ and elasticity parameter $E$(from left to right) for three-dimensional problem

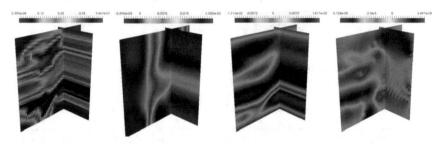

**Fig. 5** Distribution of pressure, displacement along $X, Y$ and $Z$ directions at the last moment of time for 12 multiscale basis functions for three-dimensional problem (Test 1)

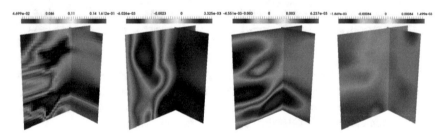

**Fig. 6** Distribution of pressure, displacement along $X, Y$ and $Z$ directions at the last moment of time for 12 multiscale basis functions for three-dimensional problem (Test 2)

**Table 2** Relative errors in % for displacement and pressure with different numbers of multiscale basis functions for three-dimensional problems (*Test 1*(left) and *Test* 2(right)). $L_p = L_u = L$.

| Test 1 L | $10 \times 10 \times 5$ | | Test 2 L | $10 \times 10 \times 5$ | |
|---|---|---|---|---|---|
| | $e^u$ | $e^p$ | | $e^u$ | $e^p$ |
| 1 | 48.820 | 2.781 | 1 | 69.051 | 6.320 |
| 2 | 7.513 | 2.289 | 2 | 24.456 | 5.005 |
| 4 | 6.710 | 1.947 | 4 | 8.624 | 2.725 |
| 8 | 2.793 | 1.859 | 8 | 3.862 | 2.16 |
| 12 | 0.856 | 1.453 | 12 | 1.039 | 1.783 |

number of the multiscale basis functions on the coarse grid $10 \times 10 \times 5$ between multi-scale and fine grid solutions. For the three-dimensional problem, we also observe good convergences for the poroelasticity problem in fractured media. Presented results of the poroelasticity problem show that proposed multiscale method can provide good accuracy for a small number of multiscale basis functions for embedded fracture model.

# 5   Conclusion

In this paper, we presented the poroelasticity problems in fractured porous media. The fine grid approximation is based on the finite volume approximation for pressure in a fractured medium and the finite element method for displacement. The coarse grid approximation is developed using a structured coarse grid and is based on a generalized multiscale finite element method. The efficiency of the method has been tested on a two and three-dimensional model problem. We compared a relative error between multi- scale and fine-scale solutions for different number of multiscale basis functions. The proposed multiscale method provides a good accuracy for two and three-dimensional problems for poroelasticity problems by using embedded fracture model.

**Acknowledgements** TA's work is supported by the mega-grant of the Russian Federation Government №. 14.Y26.31.0013 and by North-Caucasus Center for Mathematical Research under agreement №. 075-02-2021-1749 with the Ministry of Science and Higher Education of the Russian Federation. MV's work is supported by the mega-grant of the Russian Federation Government №. 14.Y26.31.0013. AA's work is supported by North-Caucasus Center for Mathematical Research under agreement №. 075-02-2021-1749 with the Ministry of Science and Higher Education of the Russian Federation.

# References

1. Akkutlu IY, Efendiev Y, Vasilyeva M, Wang Y (2017) Multiscale model reduction for shale gas transport in a coupled discrete fracture and dual-continuum porous media. J Nat Gas Sci Eng 48:65–76
2. Akkutlu IY, Efendiev Y, Vasilyeva M, Wang Y (2018) Multiscale model reduction for shale gas transport in poroelastic fractured media. J Comput Phys 353:356–376
3. Akkutlu IY, Fathi E (2012) Multiscale gas transport in shales with local kerogen heterogeneities. SPE J 17(04):1002–1011
4. Biot MA (1941) General theory of three-dimensional consolidation. J Appl Phys 12(2):155–164
5. Dean RH, Gai X, Stone CM, Minkoff SE (2006) A comparison of techniques for coupling porous flow and geomechanics. SPE J 11(01):132–140
6. Efendiev Y, Hou TY (2009) Multiscale finite element methods: theory and applications, vol 4. Springer, Heidelberg
7. Efendiev Y, Galvis J, Hou TY (2013) Generalized multiscale finite element methods (GMsFEM). J Comput Phys 251:116–135

8. Hajibeygi H, Bonfigli G, Hesse MA, Jenny P (2008) Iterative multiscale finite-volume method. J Comput Phys 227(19):8604–8621
9. Jenny P, Lee SH, Tchelepi HA (2005) Adaptive multiscale finite-volume method for multiphase flow and transport in porous media. Multiscale Model Simul 3(1):50–64
10. Karimi-Fard M, Durlofsky LJ, Aziz K (2004) An efficient discrete-fracture model applicable for general-purpose reservoir simulators. SPE J 9(02):227–236
11. Lee SH, Lough MA, Jensen CL (2001) Hierarchical modeling of flow in naturally fractured formations with multiple length scales. Water Resour Res 37(3):443–455
12. Li L, Lee SH (2008) Efficient field-scale simulation of black oil in a naturally fractured reservoir through discrete fracture networks and homogenized media. SPE Reserv Eval Eng 11(04):750–758
13. Tene M, Al Kobaisi MS, Hajibeygi H (February 2015) Algebraic multiscale solver for flow in heterogeneous fractured porous media. In: SPE reservoir simulation symposium. OnePetro

# Error Correction Method in Modular Redundant Codes

Viktor Berezhnoy ⓘ

**Abstract** Correction codes, which can detect and correct errors that occur during data storage and processing, are necessarily used in various computing devices to ensure noise immunity. The article discusses modular redundant codes using a non-positional number system in residue classes. They show good results when being used in digital signal processing, cloud storage, data storage systems. The corrective abilities of codes in the residual class system appear when additional control bases are introduced into their composition. As a result, the overall range of data presentation is increased, which makes it possible to detect information distortion. Error numbers move from the working range of data to the residual one. Moreover, different types of errors lead to the movement of numbers in different areas of the excess range, known as error intervals. The article proposes to use the results on the distribution of error intervals to develop a new method for detecting and correcting errors. The method makes it possible to simultaneously translate data from modular representation into positional representation and correct distorted information, which significantly reduces the time spent on detecting and correcting errors. The proposed method is fast and simple to implement. Comparison of the temporal characteristics of the well-known projection method with the developed method shows the significant advantage of the latter.

**Keywords** Modular arithmetic · Correction codes · Residue number system · Detecting errors · Correcting errors

## 1  Introduction

The correction codes to detect and correct errors can be used almost in all digital storage and information processing devices. They are mostly used in various information and computer networks when transmitting packets [1], as well as in information storage devices when reading data from them [2].

V. Berezhnoy (✉)
North Caucasus Center for Mathematical Research, North-Caucasus Federal University, 355017 Stavropol, Russia
e-mail: vvberezhnoi@ncfu.ru

© The Author(s), under exclusive license to Springer Nature Switzerland AG 2022  163
A. Tchernykh et al. (eds.), *Mathematics and its Applications in New Computer Systems*,
Lecture Notes in Networks and Systems 424,
https://doi.org/10.1007/978-3-030-97020-8_15

Arithmetic in residual classes (modular arithmetic) [3] has one of the approaches to creating correcting codes, which, using redundancy in the data representation, makes it possible to create codes for detecting and correcting errors (modular redundant codes).

Modular redundant codes are considered to ensure the reliability of information transfer in systems based on DS-CDMA technology [4], to improve the reliability and security of cloud storage [5], when designing reliable hybrid memory systems [6].

From a mathematical point of view, a modular redundant code is formed from $n$ residuals when dividing the number $A$ into coprime modules, which are the bases $p_i, i = [\overline{1, n}]$, of the residual class system (RNS).

Any code that is required to have the ability to detect and correct an error is characterized by the presence of two groups of numbers, information and control. The information group includes numbers that form the numerical value of the encoded value, and the control group includes numbers that are additionally entered to detect and correct possible distortions during transmission. These additional numbers are redundant in terms of a numerical value, and they increase the overall length of the code, which, of course, ultimately somewhat reduces the channel capacity in serial transmission or increases the number of channels in parallel transmission. However, this redundancy provides the code with the ability to detect and correct errors.

It is necessary to add redundant (control) modules to the original (working) set of bases to give correcting properties to the system code of residual classes. To do this, in a system with bases $p_1, p_2, \ldots, p_n$ and the operating range $R = p_1 p_2 \ldots p_n$, at least one control base $p_{n+1}$ is introduced which is mutually simple with any of the accepted bases and data (numbers) are represented in a system of $n + 1$ bases. The range $P = R p_{n+1}$ is called the full range of the system with one reference base. All data values with which the computing device operates must lie in the range $[0, R)$. An error is considered to be such a distortion of the data, which translates their values into the range $[R, P)$.

Correcting abilities of redundant RNS codes increase both with an increase in the value of the control base $p_{n+1}$, and with an increase in the number of check bases $p_{n+1}, p_{n+2}, \ldots, p_{n+k}$ [3].

The properties of modular redundant codes are well analyzed in the literature [7–9]. The basis of most of the approaches that allow correcting errors in all RNS modules is the projection method [8]. The essence of which is the sequential exclusion of each base from the RNS and the identification of the one for which the error occurred. This approach increases the time spent on error detection and correction by $n + k$ times, which significantly reduces its efficiency.

One of the more effective methods for detecting and correcting errors is the syndromic decoding method [10]. It is based on the method of extending the RNS bases using the generalized positional number system (GPRS) and calculating the error syndromes $\delta i$, which determine the error base $\delta_i$ and the magnitude of the error $\Delta$. This method is $n$ times more time efficient than the projection method. However, it also requires additional costs for base expansion and translation into positional number system.

This article proposes a new method for error detection and correction that allows to detect and correct an error in the process of converting modular redundancy code to positional representation, which insignificantly increases the time spent on error correction as opposed to making the transmission error-free.

## 2 Materials and Methods

### 2.1 Modular Redundancy Codes

Data packets transmitted over computer networks, as well as information stored and processed in computer systems, are usually represented by combinations of binary code that can be interpreted as numbers. Using this fact, any number can be represented in modular redundant code, representing a number in a non-positional number system RNS.

Let a set of integer positive bases $p_1, p_2, \ldots, p_n$ be given. Then any number A can be represented by non-negative remainders after dividing it into bases $A = (a_1, a_2, \ldots, a_n), a_i = A \bmod p_i = |A|_{p_i}$.

Modular arithmetic is based on the Chinese remainder theorem (CRT) [3], which states the uniqueness of the representation of a number in the RNS in the interval $[0, R)$, where $R$ is the working range of representing numbers in the RNS, while the modules $p_i$ are pairwise mutually prime. The value of the residues $a_i$ modulo $p_i$ is an integer in the interval $[0, p_i - 1]$. One of the important consequences of CTO is the key method for translating numbers from RNS into positional representation:

$$A = \sum_{i=1}^{n} a_i B_i (mod\ R), \tag{1}$$

where $B_i = \frac{R}{p_i} \cdot \left| \left( \frac{R}{P_i} \right)^{-1} \right|_{p_i}$ is orthogonal bases of RNS. This approach served as the main set of methods that implement both the operation of transferring a number from RNS to positional representation, and other important operations in RNS.

### 2.2 Corrective Abilities of RNS

The corrective properties of the RNS can be seen when the control (redundant) modules are introduced into it [11]. The numbers $p_j, j = n + 1, n + 2, \ldots, n + r$, which satisfy the conditions of the CTO, are taken as redundant modules, i.e. mutually prime among themselves and throughout the entire set of original non-redundant (working) modules.

Error correction methods in RNS are based on the basic principle: "corrupted" bases change the final value of the number so that it ceases to belong to the working range $[0, R\text{-}1]$. Tracking the value of the number presented in the RNS, the presence of an error can be detected.

*Example 1.* Let us choose the base system $p_1 = 2$, $p_2 = 3$, $p_3 = 5$, $p_4 = 7$, for which the working range is $R = 2 \cdot 3 \cdot 5 \cdot 7 = 210$ and the control base $p_5 = 11$. Then the full range is defined as: $P = 210 \cdot 11 = 2310$. Let us calculate the orthogonal bases of the system:

$$B_1 = (1, 0, 0, 0, 0) = 1155, \ B_2 = (0, 1, 0, 0, 0) = 1540,$$

$$B_3 = (0, 0, 1, 0, 0) = 1386, \ B_4 = (0, 0, 0, 1, 0) = 330,$$

$$B_5 = (0, 0, 0, 0, 1) = 210.$$

The number transmitted was $A = 19 = (1, 1, 4, 5, 8)$, instead of which the number received turned out to be $\tilde{A} = (1, 1, 4, 6, 8)$. To detect an error, we calculate the value $\tilde{A}$:

$$\tilde{A} = 1 \cdot 1155 + 1 \cdot 1540 + 4 \cdot 1386 + 6 \cdot 330 + 8 \cdot 210 (mod\ 2310)$$

$$= 11899 (mod\ 2310) = 349 > 210.$$

Since the number received $\tilde{A} > R$, then it is incorrect, and an error occurred during transmission in one or more residuals. This detects the presence of an error in the transmission of the number. Moreover, in this case, we can confidently assert that any error made can be detected, since no working base exceeds the control base of the system in its size [3].

To localize distorted modules and correct errors, it is necessary to determine the module for which the error $p_i$ occurred and the magnitude of the error itself (error depth) $\Delta a_i$.

In work [12], a method was proposed for investigating the distribution of errors in the full range of RNS, which makes it possible to determine the limits of error intervals in which numbers containing errors get into when they are converted from a modular code to a positional one. Using this information, we will develop a method for detecting and correcting errors in RNS numbers.

## 2.3 Development of a Method for Error Correction in Modular Redundancy Code

To detect errors in the RNS, it is necessary to ensure the required redundancy of the modular code, the value of the control base $p_{n+1}$ must exceed the value of any of the working modules $p_i$ [11]:

$$p_{n+1} > p_i \left(i = \overline{1, n}\right). \tag{2}$$

Let the bases $p_1, p_2, \ldots, p_n, p_{n+1}$ and the number $A = (\alpha_1, \alpha_2, \ldots, \alpha_n, \alpha_{n+1})$ in the RNS be given. Then the translation of the number into the positional number system can be carried out through orthogonal bases according to expression (1). The appearance of an error number $\tilde{a}_i$ for any reason brings the number $\tilde{A}$ out of the working range $[0, \overline{R - 1}]$ into the excess range $[R, \overline{P - 1}]$. Mathematically, the effect of an error of depth $\Delta a_i$ modulo $p_i$ can be described by the expression:

$$\tilde{A} = [A + \Delta a_i B_i](mod\ P). \tag{3}$$

To detect an error, it is enough to determine how the number $\tilde{A}$ exceeds the lower limit $R$ of the excess range $[R, \overline{P - 1}]$.

However, to localize the error base and the depth of the error, it is necessary to determine in what error interval $K_i^j = [0, \overline{R - 1}]$, where $i \left(i = \overline{1, n}\right)$ is a number of the base and $j = \Delta a_i \left(j = \overline{1, p_i - 1}\right)$ is the depth of the error, the number $\tilde{A}$ falls [12]. In this case, the condition of non-intersection of error intervals must be met. Then, from the value of the distorted number $\tilde{A}$, one can determine not only the distorted remainder $\tilde{a}_i$, but also the depth of the error $\Delta a_i$ i and correct the error by performing the operation

$$A = [\tilde{A} - \Delta a_i B_i](mod\ P). \tag{4}$$

Let us determine the requirements for the value of the control bases for the localization of single errors.

## 2.4 Error Correction with One Control Base

We will consider the case with one control base. In [13], the correction capabilities of the code of the residual class system with one control base were estimated and the lower limit of redundancy was determined for guaranteed localization and correction of single errors. It has been proven that the value of the control base should be selected from the condition:

$$p_{n+1} > p_n p_{n-1}, \tag{5}$$

where $p_n$ and $p_{n-1}$ are the largest operating modules of the RNS in absolute terms.

In this case, the numbers $\tilde{A}$ distorted by errors will be in error intervals $K_i^j$, which will not intersect with each other.

The analysis of the distribution of errors over intervals of the full range of RNS with one redundant basis, carried out in [12], allows us to make the following statements:

1.  the distribution of errors within the excess range of the ordered system of residual classes has an uneven character, symmetric with respect to the middle of the excess range, and is a function of the RNS bases;
2.  for the implementation of the possibility of correcting any single error on a working basis, a necessary and sufficient condition is not to intersect error intervals at working bases;
3.  with an increase in the value of the control base, the "last" intersection of error intervals is observed with errors on the largest bases $p_{n-1}$ and $p_n$.

Thus, using the error intervals $K_i^j$ and providing condition (5) for one control base, it is possible to detect, localize and correct the error for any of the RNS work modules.

To localize the error, it is necessary and sufficient to determine the error interval $K_i^j$, in which the value of the distorted number $\tilde{A}$ falls. The error interval $K_i^j$ will indicate not only the base $p_i$, along which the error occurred, but also unambiguously determine the depth of the error $\Delta a_i$. To correct this error in the number presented in the remainders $(a_1, a_2, \ldots, \tilde{a}_i, \ldots, a_n, a_{n+1})$ it is necessary to subtract the error depth $\Delta a_i$ from the error number $\tilde{a}_i$, and for a number in the positional number system (PNS) it is necessary to use the expression (4).

*Example 2.* Let the system of bases of the RNS be given $p_1 = 2, p_2 = 3, p_3 = 5, p_4 = 17$, and $p_1, p_2, p_3$ are the information bases of the system, and $p_4$ is the control one. Then the working range R $= 2 \cdot 3 \cdot 5 = 30$ and the full range P $= 2 \cdot 3 \cdot 5 \cdot 17 = 510$, and the orthogonal bases of the system:

$$B_1 = 255, B_2 = 340, B_3 = 306, B_4 = 120.$$

We will find the error intervals by the formula (3):

$$K_1^1 = [255 \bmod 510, 255 + 29 \bmod 510] = [255, 284];$$
$$K_2^1 = [340 \bmod 510, 340 + 29 \bmod 510] = [340, 369];$$
$$K_2^2 = [680 \bmod 510, 680 + 29 \bmod 510] = [170, 199];$$
$$K_3^1 = [306 \bmod 510, 306 + 29 \bmod 510] = [306, 335];$$
$$K_3^2 = [612 \bmod 510, 612 + 29 \bmod 510] = [102, 131];$$
$$K_3^3 = [918 \bmod 510, 918 + 29 \bmod 510] = [408, 437];$$
$$K_3^4 = [1224 \bmod 510, 1224 + 29 \bmod 510] = [224, 253].$$

As can be seen, the error intervals do not intersect, due to the fact that $p_4 = 17 > 3 \cdot 5$.

Let the initial number be $A = 23 = (1, 2, 3, 6)$. We suppose that an error has occurred with the base $p_3 = 5$, and the error number $\tilde{A}$ has the form $\tilde{A} = (1, 2, 4, 6)$. Let us convert this number in PNS:

$$\tilde{A} = (1 \cdot 255 + 2 \cdot 340 + 4 \cdot 306 + 6 \cdot 120) \, mod \, 510$$
$$= 2879 \, mod \, 510 = 329.$$

Obviously, $\tilde{A} = 329 \in K_3^1 = [306, 335]$. Therefore, the error occurred at the third base and the error depth $\Delta a_3 = 1$. To correct the error, subtract the error depth from the error remainder $\tilde{a}_3$:

$$\tilde{a}_3 - \Delta a_3 = 4 - 1 = 3 = \alpha_3.$$

We get the correct number $A = (1, 2, 3, 6)$. We can also correct the number already converted to the positional number system using the expression (4).

$$A = (329 - 1 \cdot 306) \, mod \, 510 = 23 \, mod \, 510 = 23.$$

However, the use of only one control base allows you to detect and correct errors only in the working modules of the RNS, since errors in the control module cover the entire excess range $\left[ R, \overline{P - 1} \right]$ and, accordingly, the error intervals in the control module are superimposed at intervals in working modules.

## 2.5   Correction of Errors with Two Control Bases

To localize and correct the error $\Delta a_i$ for any module, including the control one, it is necessary to use two control bases $p_{n+1}, p_{n+2}$ in absolute terms exceeding any working base $p_i$. In this case, the error intervals for all modules, both working and control, will not intersect with each other, which will ensure the localization of the error at one of the RNS bases.

*Example 3.* We will consider the base system $p_1 = 2, p_2 = 3, p_3 = 5, p_4 = 7, p_5 = 11$, where modules $p_1, p_2, p_3$ are working ones, $p_4, p_5$ are control ones. Let us determine the error intervals for this set by expression (3).

Orthogonal bases for a given base system

$$B_1 = 1155, B_2 = 1540, B_3 = 1386, B_4 = 330, B_5 = 210.$$

The results are presented in the Table 1.

**Table 1** Limits of error intervals for a set of working modules {2, 3, 5} with control bases {7, 11}

| Modules with a error | $a_i$ | Limits $K_i^s$ | | $K_i^s$ |
|---|---|---|---|---|
| | | 0* | $(M-1)^*$ | |
| $p_1 = 2$ | 1 | 1155 | 1184 | $K_1^1$ |
| $p_2 = 3$ | 1 | 1540 | 1569 | $K_2^1$ |
| | 2 | 770 | 799 | $K_2^2$ |
| $p_3 = 5$ | 1 | 1386 | 1415 | $K_3^1$ |
| | 2 | 462 | 491 | $K_3^2$ |
| | 3 | 1848 | 1877 | $K_3^3$ |
| | 4 | 924 | 953 | $K_3^4$ |
| $p_4 = 7$ | 1 | 330 | 359 | $K_4^1$ |
| | 2 | 660 | 689 | $K_4^2$ |
| | 3 | 990 | 1019 | $K_4^3$ |
| | 4 | 1320 | 1349 | $K_4^4$ |
| | 5 | 1650 | 1679 | $K_4^5$ |
| | 6 | 1980 | 2009 | $K_4^6$ |
| $p_5 = 11$ | 1 | 210 | 239 | $K_5^1$ |
| | 2 | 420 | 439 | $K_5^2$ |
| | 3 | 630 | 659 | $K_5^3$ |
| | 4 | 840 | 869 | $K_5^4$ |
| | 5 | 1050 | 1079 | $K_5^5$ |
| | 6 | 1260 | 1289 | $K_5^6$ |
| | 7 | 1470 | 1499 | $K_5^7$ |
| | 8 | 1680 | 1709 | $K_5^8$ |
| | 9 | 1890 | 2009 | $K_5^9$ |
| | 10 | 2100 | 2129 | $K_5^{10}$ |

Analyzing the Table 1, we can say that there are no intersections of error intervals and, therefore, it is possible to detect, localize and correct any single error in any of the modules, including the control ones. Let us consider an example.

*Example 4.* We represent the number $A = 23$ in the RNS $A = 23 = (1, 2, 3, 2, 1)$. If an error has occurred, the number has taken the value $A* = (1, 2, 3, 4, 1)$. Let us convert the number to PNS.

$$A* = (1 \cdot 1155 + 2 \cdot 1540 + 3 \cdot 1336 + 4 \cdot 330 + 1 \cdot 210) \, mod \, 2310 = 683.$$

As $683 > P = 30$, that number is wrong. According to the table, we determine to which error interval the number 683 belongs. This interval is $K_4^2$. Therefore, the error occurred at the base $p_4 = 7$, the error depth is $\Delta a_4 = 2$.

Having these values known, it is easy to carry out the correction according to the expression (4).

$$A = (683 - 2 \cdot 330) mod \, 2310 = 23 \, mod \, 2310 = 23.$$

Since the error intervals $K_i^j$ do not depend on the number, but only on the RNS bases, they can be calculated in advance, stored in memory and used to localize errors.

## 3 Results

Using the considered method of localization and error correction, it is possible to propose a method for error correction in a redundant modular code.

1. For a given base system of the RNS, the error intervals $K_i^j = [0, \overline{R-1}]$ are calculated using expression (3). The limits of the error intervals are stored in memory.
2. The number A in the RNS is converted to the PNS by any of the known methods [3, 10], for example, by orthogonal bases (1) and is compared with the error intervals $K_i^j$. If the number A is correct, then it will be in the working range and will not fall into any error interval. If the number A contains an error, then it will fall into some error interval $K_i^j$ and it is necessary to localize and correct the error.
3. The error intervals determine the error base $p_i$ and the depth of the error $\Delta a_i$.
4. Conduct the correction using expression (4).

To implement this method, the following functional diagram can be proposed (Fig. 1).

The input of block 1, which performs the functions of the RNS-PNS converter, receives a number in the modular code $(\alpha_1, \alpha_2, \ldots, \alpha_n, \alpha_{n+1}, \alpha_{n+2})$. Block 1 translates the number from the modular code to the positional one. As a result, at the output of block 1, a binary number A is formed, which moves to block 2. Block 2 is a memory for storing pre-calculated constants $\Delta \alpha_i B_i$. The value of the number A in block 2 determines the memory cell in which the corresponding constant $\Delta \alpha_i B_i$ is stored, for the case if the number A contains an error or the value 0 is returned if the number A is correct. Constants $\Delta \alpha_i B_i$ or 0 go to the next block 3. Block 3 represents a modulo P subtractor that corrects the error according to expression (4). If the number A is correct, then a zero value is subtracted from it, which allows the number A to pass to the output of the device without making any changes to it.

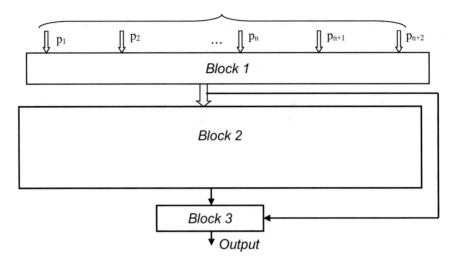

**Fig. 1** Functional diagram of the error correction device

## 4 Discussion

This method and its hardware implementation are significantly more efficient than the projection method [8]. Let us carry out a comparative analysis of the time spent by these methods.

As a characteristic for estimating time costs, we will use a dimensionless time step that is necessary to perform one operation of bitwise summation of binary numbers modulo 2. The conversions arising during the usual summation add clock signals according to the number of bits of the summed number. This means that it takes 8 clock cycles to add an 8-bit number normally.

As RNS modules, we will use a set of optimal modules for the 4-byte model {37, 41, 43, 47, 53, 59}, with two control bases {61, 63}.

*Projection Method.* To calculate the projections for each base, it is necessary to convert the number from RNS to PNS for 7 bases out of 8. Let us calculate the number of clock cycles for such calculations. To perform multiplication $a_i B_i$, which multiplies the $n = log_2 p_i$ for $a_i$ and $N$ for $B_i$. This operation requires $n - 1$ additions of N-bit numbers. For the chosen base system $n = 6$ and the number of modules in projection 7, $N = 42$. When multiplying $a_i B_i$, 5 additions of 42-bit numbers $5 * 42 = 210$ are performed. The resulting products are summed modulo the full range $P$. To speed up this operation, we will simultaneously add 2 44-bit numbers $(a_i B_i + a_{i+1} B_{i+1}) mod P$. Since there are 7 modules in this RNS, then at the first stage 6 numbers are summed up in pairs, in the second 4 numbers in pairs and in the third stage 2 numbers. Then the number of additions is $3 * (42 + 42) = 252$. In total, $210 + 252 = 462$ clock cycles.

For 8 bases of RNS, calculating all projections will require $8 * 462 = 3696$ clock cycles. In addition, initially it is also required to translate the number into the PNS in order to find out if there is an error in the number A, and then conduct one more translation after correcting the error. In this case for $n = 6$, $N = 48$, $528 * 2 = 1056$ number of clock cycles will be required. In total, in the worst case (when all 8 projections are found), it is necessary to spend $3696 + 1056 = 4752$ clock cycles.

*The Proposed Method.* Let us determine similar time characteristics for the method considered in the article and the proposed error correction device shown in Fig. 1.

It also requires translation from RNS to PNS numbers. It is performed similarly to the translation in the projection method and will require, as already known from previous calculations, 528 clock cycles. The selection of the constants $\Delta a_i B_i$ s carried out in one cycle of memory access. To correct the error, subtraction $\tilde{A} - \Delta a_i B_i (mod\ R)$, is performed, which takes $2N = 48 * 2 = 96$ clock cycles.

Then the total number of clock signals required to correct errors will be $528 + 1 + 96 = 625$.

Thus, the proposed method allows to reduce the time spent on error correction by 7.6 times in comparison with the projection method for a 4-byte data representation model.

# 5   Conclusion

The error correction method proposed in the article uses the distribution of the limits of error intervals [12] to determine the distorted RNS base $p_i$ and the error depth $\Delta a_i$. It significantly reduces the time spent on determining the error due to the fact that for a given base system of the RNS, the limits of the error intervals $K_i^j$ and, accordingly, the values $\Delta a_i$ are calculated in advance and stored in memory. The error correction is carried out already in the positional representation of the number A, which also makes it possible to exclude the additional costly operation of the RNS-PNS transformation.

Thus, the proposed method is less time consuming and can be proposed for error correction in a redundant modular code when transmitting messages in information computing networks, protecting data when reading from storage devices, and processing information in parallel modular computing devices.

Further research will be aimed at accelerating the conversion operation from modular code to positional code and determining the hardware costs for the implementation of the proposed error correction method.

# References

1. Zarei B, Muthukkumarasay V, Wu X (2013) A residual error control scheme in single-hop wireless sensor networks. In: 2013 IEEE 27th international conference on advanced information networking and applications (AINA), Barcelona, Spain Spain, pp 197–204. https://doi.org/10. 1109/AINA.2013.101
2. Yaakobi E, Grupp L, Siegel PH, Swanson S, Wolf JK (2012) Characterization and error-correcting codes for TLC flash memories. In: 2012 international conference on computing, networking and communications (ICNC), pp 486–491. https://doi.org/10.1109/ICCNC.2012. 6167470
3. Szabo NS, Tanaka RI (1967) Residue arithmetic and its applications to computer technology. McGraw-Hill
4. Hanzo L, Yang LL, Kuan EL, Yen K (2003) Single-and multi-carrier DS-CDMA: multi-user detection, space-time spreading, synchronisation, standards and networking. Wiley, Hoboken
5. Tchernykh A, Schwiegelsohn U, Talbi EG, Babenko M (2016) Towards understanding uncertainty in cloud computing with risks of confidentiality, integrity, and availability. J Comput Sci
6. Haron NZ, Hamdioui S (2011) Redundant residue number system code for fault-tolerant hybrid memories. ACM J Emerg Technol Comput Syst 7(1)
7. Tay TF, Chang CH (2017) Fault-tolerant computing in redundant residue number system. In: Embedded systems design with special arithmetic and number systems. Springer, Cham, pp 65–88
8. Goh VT, Siddiqi MU (2008) Multiple error detection and correction based on redundant residue number systems. IEEE Trans Commun 56(3):325–330
9. Ding C, Pei D, Saiomaa A (1996) Chinese remainder theorem: applications in computing, coding, cryptography. World Scientific, Singapore
10. Chervyakov NI, Lyakhov PA, Babenko MG, Garyanina AI, Lavrinenko IN, Lavrinenko AV, Deryabin MA (2016) An efficient method of error correction in fault-tolerant modular neuro-computers. Neurocomputing 205:32–44. ISSN 0925-2312. https://doi.org/10.1016/j.neucom. 2016.03.041
11. Yang LL, Hanzo L (2001) Redundant residue number system based error correction codes. In: Proceedings of 54th vehicular technology conference, pp 1472–1476
12. Berezhnoy V, Kuchukova E (2020) Method for analyzing the corrective ability of residue number system. In: International conference engineering and telecommunication (En&T), pp 1–5. https://doi.org/10.1109/EnT50437.2020.9431282
13. Chervyakov NI, Berezhnoi VV, Olenev AA, Kalmykov IA (1994) Minimizing code redundancy for a system of residue classes with a single check base. Eng Simul Link Disabl 12(1):95–102

# Comparative Analysis of the Mathematical Literacy Level in First-Year Students of the North-Caucasus Federal University, Based on the International PISA Studies

Olga Rozhenko⬥, Anna Darzhania⬥, Viktoria Bondar⬥, Marine Mirzoyan⬥, and Olga Skvortsova⬥

**Abstract** The article examines the results of testing first year students of the North-Caucasus Federal University under the international program for assessing the quality of education PISA (Program for International Student Assessment) using the methods of mathematical statistics. The aim of the research is to carry out a comparative analysis of the level of mathematical literacy of university students in the light of modern educational standards in the field of mathematical education.

**Keywords** PISA · Mathematical education · International research · Competency · Mathematical literacy

## 1 Introduction

The leading international organizations operating in the field of education assessment have been doing global comparative studies focusing on the quality of education for more than twenty years now [1, 2]. These studies are aimed at carrying out comparative analyses of the current status and transformations occurring in the education systems of various countries, as well as at assessing the efficiency of strategic decisions taken in education [3]. It is to be noted that such comparison is not made in an abstract way, based on various sources of literature; nor are they based on the analysis of the outcomes obtained from well-known mathematics contests where selected students only are involved. The value of these studies resides in the comparative analysis being carried out relying on the results of works that are representative samples of schoolers from different countries [4, 5]. This present study employs a uniform

O. Rozhenko (✉)
North Caucasus Center for Mathematical Research, North-Caucasus Federal University, 355017 Stavropol, Russia
e-mail: r.o.d@mail.ru

A. Darzhania · V. Bondar · M. Mirzoyan · O. Skvortsova
North-Caucasus Federal University, Stavropol, Russian Federation

© The Author(s), under exclusive license to Springer Nature Switzerland AG 2022　　　175
A. Tchernykh et al. (eds.), *Mathematics and its Applications in New Computer Systems*,
Lecture Notes in Networks and Systems 424,
https://doi.org/10.1007/978-3-030-97020-8_16

toolkit, developed in view of international priorities in the field of education. This paper offers a view at mathematical statistics methods used to study the results of testing first-year students of the North-Caucasus Federal University within the PISA international program (Program for International Student Assessment). The purpose of this study is to conduct a comparative analysis of the mathematical literacy level among university students in the light of modern academic standards in the field of mathematical education [6, 7].

The system of education in our country is a source of concern nowadays since there are a number of relevant problems related to the integration of the Russian education into the international education environment [3, 7]. The issue of objective assessment regarding the quality of education has been growing for a long time, whereas an effective solution to it can be found only taking into account the data from international research focusing on the quality of education [8].

From the year 2000 on, our country has been actively involved in various international programs for education quality research. Russia is currently a member of international studies like TIMSS (Trends in Mathematics and Science Study); PIRLS (Progress in Inter-national Reading Literature Study); TALIS (Teaching and Learning International Survey); PIAAC (The Program for the International Assessment of Adult Competencies). The studies mentioned above are aimed at obtaining efficiency assessment for education systems in various countries all over the world, as well as at updating education itself based on the results obtained [9, 10].

In particular, from 2000 on, Russia has been involved in PISA—the most significant international study on assessing the quality of education, where fifteen-year-old schoolers are tested in a number of areas. One of the most important areas of PISA research is identifying the level of mathematical literacy in students.

Mathematical literacy is to be the focus of special attention in the next cycle of the PISA study for the year 2022. In the same year, this study will focus for the first time on schoolers' creative thinking. Within the study, mathematical literacy is defined as "an individual's ability to formulate, apply and interpret mathematics in a variety of contexts". Besides, the concept of mathematical literacy incorporates mathematical reasoning, ability to use mathematical concepts, facts, procedures and tools to describe, explain and predict various phenomena.

Mathematical literacy is known to be one of the most important factors behind academic studies in most areas in a university. Students majoring in economic, engineering, natural science, and computer areas should have a high level of mathematical literacy and mathematical culture, which is important to master the chosen professions and shape appropriate general professional competencies that would allow solving fundamental and applied problems within the respective areas of human activity.

The age of students joining the first year of university training usually lies within the range of 16–18. Given that, there was definite interest taken in carrying out an experiment aimed at testing mathematical literacy in freshmen based on international PISA tests, in order to check skills like critical thinking, creativity, perseverance, information management, systemic thinking, the ability to design mathematical reasoning.

## 2   Materials and Methods

As mentioned earlier, Russia is to host the next round of the international PISA study in 2022. In April 2021, the preparatory measures were already taken for the PISA-2022 study, where around three thousand schoolchildren representing five areas of the country coming from 58 academic institutions took part.

As an experiment, in order to test mathematical literacy, over a hundred first-year students of the North Caucasus Federal University (NCFU) joined the testing employing the *PISA for Schools* toolkit developed by the Organization for Economic Cooperation and Development (OECD).

The purpose of this work is to study the results of the PISA tests completed by the NCFU students, as well as to see how much these results are consistent with global monitoring indicators in mathematical education, based on mathematical statistics and correlation analysis methods.

The general test that the students took, included 41 assignments. Statistical processing embraced two stages. The initial stage focused on testing the hypothesis concerning the normal distribution of test results. The major stage implied a comparison of the results subject to several criteria relevant within the PISA study, in particular, a comparative analysis was carried out following the territorial principle and the students' respective places of residence. The study also included analysis of the differences in terms of the test performance based on a comparative analysis of the average results demonstrated by schoolchildren from OECD countries, NCFU students and Russian schoolchildren.

## 3   Results

Within the experiment, 126 first-year students of non-mathematical areas studying in the NCFU were tested using demonstration tests in the field of the PISA mathematical literacy, as presented on the website of the Center for Education Quality Assessment. The results of PISA international studies, as well as the data obtained through the test, allowed carrying out a study on the level of mathematical literacy among the university's first-year students.

During the experiment, a hypothesis was proposed at the initial stage, stating that the results of the tests completed by first-year students follow the normal distribution law. To test this hypothesis, Pearson's criterion was employed. Experimental data was used to calculate $n^T$ theoretical frequencies, as well as the observed attribute value was found based on the following formula:

$$\chi^2_{obs} = \sum\nolimits_{i=1}^{7} \frac{\left(n_i^T - n_i\right)^2}{n_i^T} = 3.36, \tag{1}$$

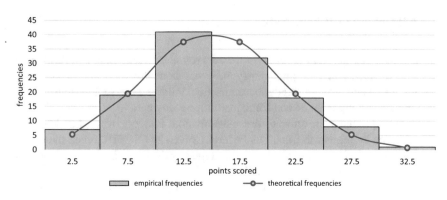

**Fig. 1** Empirical frequencies curve and theoretical frequencies diagram

where $\chi^2_{obs}$ is the observed value of the attribute; $n_i^T$ are theoretical frequencies; $n_i$ are empirical frequencies. Given the set significance value of $\alpha = 0.05$ within the critical chi-square chart, the attribute critical value was found:

$$\chi^2_{cr}(m - r - 1; \alpha) = \chi^2_{cr}(4; 0.05) = 9.5, \tag{2}$$

where k = m-r-1 is the number of degrees of freedom;

m is the number of intervals;

r is the number of the evaluated parameters within the distribution law in question. For a normal distribution law, r = 2.

Table and Fig. 1 contain the results of testing the hypothesis of the normal distribution of the results obtained through the PISA test carried out among the NCFU students.

The obtained data suggest that the experimental data and the hypothesis of the test results distribution according to the normal law do not contradict each other.

This means that given the significance level set at $\alpha = 0.05$, there is no reason to reject the null hypothesis (Table 1).

The next stage of the experiment implied a comparison of the test completion by the NCFU students and the average results demonstrated while completing similar tests by schoolchildren from OECD countries and from Russia. The results of this comparative analysis can be seen in Fig. 2.

The experimental data processing produced correlation coefficients for each test task between each group—first-year NCFU students, schoolchildren of OECD countries and schoolchildren of Russia. The correlation coefficient between the results of Russian schoolchildren and university students was found to be 0.747, whereas the correlation coefficient between the results demonstrated by the OECD schoolchildren and university students was 0.919.

The obtained data allow arriving at the conclusion revealing that there is a closer statistical relationship detected between the results of the tests completed by 1st year university students and schoolers of the OECD countries (the correlation being appr.

**Table 1** Testing the hypothesis of normal distribution of the PISA test results among the NCFU students

| Average sample value, | Dispersion, $D_x$ | Intervals | Empirical frequencies, $n_i$ | Theoretical frequencies, $n_i^T$ | $\frac{(n_i^T - n_i)^2}{n_i^T}$ |
|---|---|---|---|---|---|
| | | 01–05 | 7 | 5.28 | 0.59 |
| | | 05–10 | 18 | 19.45 | 0.01 |
| | | 10–15 | 42 | 37.48 | 0.32 |
| | | 15–20 | 33 | 37.46 | 0.8 |
| | | 20–25 | 17 | 19.45 | 0.12 |
| | | 25–30 | 8 | 5.26 | 1.42 |
| | | 30–35 | 1 | 0.74 | 0.09 |
| $\overline{x_s} = 15$ | 38,1804 | Σ | **126** | **126** | **3.36** |

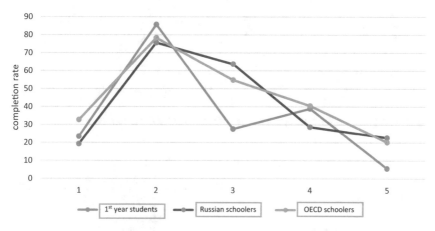

**Fig. 2** Comparison of the test results from NCFU 1$^{st}$ year students with those of Russian schoolers and schoolers from the OECD countries

0.93) than between the results of the NCFU students and those of Russian schoolers (correlation—0.75).

In order to verify the PISA study data based on the territorial feature, several segments were created to group together the test results. The first segment was included residents of the Stavropol Region; the second segment included students coming from the Krasnodar Region; the third segment was made up by students from the Karachai-Circassian Republic, while residents of a number of other North-Caucasus areas (Dagestan, Ingushetia, Kabardino-Balkarian and Chechen Republics) entered the fourth segment. The fourth segment being made up through integrating residents of several areas was due to the small number of people coming from these parts of the Russian Federation. Figure 3 shows the results of the test completed by representatives of each of the above segments.

Sample averages and linear correlation coefficients between different groups were identified and calculated for each of the groups through the stage of statistical analysis, with the results to be seen in Table 2 and Table 3.

The experiment revealed that the highest results were achieved in the first two territorial segments, namely school graduates in the Stavropol and Krasnodar Regions. Residents of the North-Caucasus Republics featured slightly lower results: Ingushetia, Dagestan, the Chechen and Kabardino-Balkarian Republics. The lowest

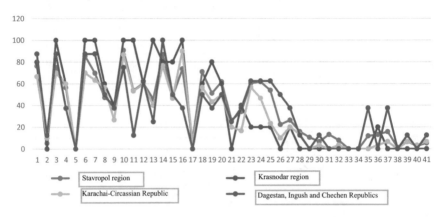

**Fig. 3** Completion of tests by representatives of various territorial groups (segments)

**Table 2** General distribution of test results through the territorial segments

| Area | Sample average | Dispersion | Number of students |
|---|---|---|---|
| Ingushetia, Dagestan, Kabardino-Balkarian and Chechen Republics | 35.37 | 913.30 | 8 |
| Karachai-Circassian Republic | 31.06 | 828.58 | 30 |
| Krasnodar Region | 36.59 | 1598.05 | 5 |
| Stavropol Region | 36.71 | 852.04 | 76 |

**Table 3** Table of correlation between the results of the tests and the residence area

| Territorial location | Territorial location | | |
|---|---|---|---|
| | Stavropol Region | Krasnodar Region | Karachai-Circassian Republic |
| Dagestan, Ingushetia, Kabardino-Balkarian and Chechen Republics | 0.89 | 0.69 | 0.81 |
| Karachai-Circassian Republic | 0.97 | 0.86 | |
| Krasnodar Region | 0.86 | | |

results were observed in the third territorial segment, namely, school graduates of the Karachai-Circassian Republic.

It should be noted that the samples on a territorial basis were also checked for the homogeneity degree, where the general population dispersions were compared for all four segments. As for the null hypothesis, we accepted it as the hypothesis of statistical homogeneity of samples. The proposed competing hypothesis was the following: $H_1 : D(X) \neq D(Y)$. The observed value of the criterion was found by the formula as follows:

$$F_{obs} = \frac{s_1^2}{s_2^2},\tag{3}$$

где $F_{obs}$ is the observed value of the criterion; $s_1^2$ and $s_2^2$ are the higher and the lower dispersion values, respectively.

Using the tables, critical values of the criterion were identified. The research showed that experimental data suggest that statistically significant differences are observed only between the results of the test completion in the first (Stavropol Region) and the fourth segments (students from Dagestan, Ingushetia, Kabardino-Balkarian and Chechen republics).

The international PISA study revealed that the mathematical literacy of Russian teenagers depends directly on the population size of their area of residence. Residents of larger places always showed higher results through the study. In particular, the results demonstrated by schoolers from Moscow are much above those obtained by schoolchildren from provincial areas. Schoolchildren from Moscow have results that are part of the top ten globally in many competencies. This explains the interest in analyzing the relationship between the place of permanent residence and the success in completing the tests.

All the tested students were divided into two groups depending on their place of residence: the first group included residents of urban areas, whereas the second group were rural residents (in villages, auls and farms). The results of the respective studies can be seen in Fig. 4 and Table 4.

The obtained data allowed concluding that school graduates' level of mathematical literacy depends directly on the population size of the respective residential area, i.e. achievements in completing tests are definitely higher among graduates coming from urban areas. This outcome is fully consistent with both Russian and global trends.

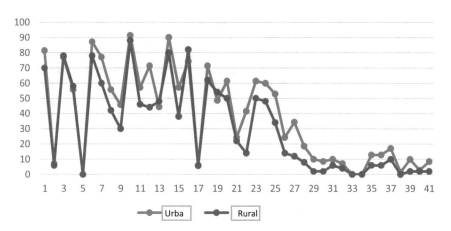

**Fig. 4** Test completion depending on students' residence—urban VS. rural

**Table 4** Test completion success depending on residence—urban VS. rural

| Area type | Average rate of entire test completion | Share in entire number | Number of students | Correlation coefficient |
|---|---|---|---|---|
| City | 38.7 | 58 | 72 | 0,96 |
| Villages, auls and farms | 30,71 | 42 | 54 | |

## 4  Discussion and Conclusion

The test aimed to identify the level of mathematical literacy among 1$^{st}$ year students majoring in non-mathematical areas of NCFU, which was carried out subject to the PISA international tests, allows the following conclusions:

1.  The experimental data offers no reason to reject the normal distribution hypothesis regarding the test outcomes demonstrated by first-year students of the NCFU.
2.  The closest correlation was observed between the outcomes shown by 1$^{st}$ year NCFU students and students of OECD schools (correlation coefficient—0.918), unlike that between Russian schoolchildren and the NCFU students (correlation coefficient—0.746).
3.  Graduates coming from schools of the Stavropol and Krasnodar Regions demonstrated the highest indicators in respective mathematical competencies if compared to students from other territorial segments who were part of the experiment.
4.  The experiment offered proof to a relationship between the population size and the successful completion of the PISA test.

The obtained results of the comparative analysis involving the level of mathematical literacy in the NCFU first-year students are fully consistent with the results of research carried out with the PISA tools and focusing on education quality, both in Russia and around the world.

The PISA testing approach, like most international studies, cannot but feature a number of downsides and specific issues, just like it does not allow a completely reliable objective assessment of the education level in a particular country. However, these studies cannot be ignored, since they are mainly aimed at identifying weak spots in the education system of the country as a whole. The results of such studies are viewed as a reflection of the available competencies that are of global fundamental importance in the modern world, and which are trends in the international system of education.

**Acknowledgements** The work is supported by North-Caucasus Center for Mathematical Research under agreement №. 075-02-2021-1749 with the Ministry of Science and Higher Education of the Russian Federation.

# References

1. Orechova M (2020) Influence of international large-scale assessments on national education policies. Vilnius Univ Open Ser 3:66–76. https://doi.org/10.15388/SRE.2020.6
2. Teltemann J, Jude N (2019) Assessments and accountability in secondary education: international trends. Res Comp Int Educ 14:249–271. https://doi.org/10.1177/1745499919846174
3. Breakspear S (2012) The policy impact of PISA: an exploration of the normative effects of international benchmarking in school system performance. https://doi.org/10.1787/5k9fdfqffr28-en
4. Hanushek E, Wössmann L (2006) Does educational tracking affect performance and inequality? Differences-in-differences evidence across countries. Econ J 116:63–63. https://doi.org/10.1111/j.1468-0297.2006.01076.x
5. Michel A (2017) The contribution of PISA to the convergence of education policies in Europe: MICHEL. Eur J Educ 52. https://doi.org/10.1111/ejed.12218
6. Kubova-Semaka J (2020) An integral approach to the meaning of competence. Vilnius Univ Open Ser 3:120–135. https://doi.org/10.15388/SRE.2020.11
7. Glaesser J (2018) Competence in educational theory and practice: a critical discussion. Oxf Rev Educ 45:1–16. https://doi.org/10.1080/03054985.2018.1493987
8. Yusupova N, Skudareva, G (2020) Quality accordingly PISA: from math teachers' continuing education to students' mathematical literacy, pp 2817–2831. https://doi.org/10.3897/ap.2.e2817
9. Kelly P, Kotthoff H-G (2017) PISA, national and regional education policies and their effect on mathematics teaching in England and Germany. Res Comp Int Educ 12:174549991772428. https://doi.org/10.1177/1745499917724286
10. Sellar S, Lingard B (2014) The OECD and the expansion of PISA: new global modes of governance in education. Br Educ Res J. https://doi.org/10.1002/BERJ.3120

# Implementation of the Learning Procedure with Equalized Errors on the Receiving Side of the Information Transmission System

**Alexander Malofey**◉**, Oleg Malofey**◉**, Alexander Zhuk**◉**, Alexander Troshkov**◉**, and Victoria Bondar**◉

**Abstract** There is an effect of combining an indirect method for assessing the quality of received symbols with code methods for detecting errors. This combination is applicable at the level of the first and second decision circuits in the information transmission system, where it is possible to implement algorithms for improving the probabilistic characteristics of majority methods of protecting information from errors by combining them with indirect methods of increasing fidelity. A method for assessing the quality of received symbols was chosen based on the use of a quality detector, as it significantly increases the reliability of reception with sufficient ease of implementation. On the basis of the calculations, the advantage of combining code and indirect methods is proved and an algorithm for majority decoding based on this statement is proposed. An algorithm for improving the quality of symbol-by-symbol reception by introducing the dependence of the level of the erasing channel threshold on the value of the interference acting in the communication channel and achieving its optimality has been developed. Variants of the technical implementation of these algorithms are proposed.

**Keywords** Indirect methods · Uncertainty zone · Decision scheme · Training procedure · Function minimization · Error correction · Decoding algorithm

A. Malofey (✉)
Stavropol Branch of the Krasnodar University of the Ministry of Internal Affairs of Russia, Stavropol, Russia
e-mail: skandin@mail.ru

O. Malofey · A. Zhuk
North Caucasus Federal University, Stavropol, Russia

A. Troshkov
Stavropol State Agrarian University, Stavropol, Russia

V. Bondar
North Caucasus Center for Mathematical Research, North-Caucasus Federal University, 355017 Stavropol, Russia

© The Author(s), under exclusive license to Springer Nature Switzerland AG 2022
A. Tchernykh et al. (eds.), *Mathematics and its Applications in New Computer Systems*,
Lecture Notes in Networks and Systems 424,
https://doi.org/10.1007/978-3-030-97020-8_17

# 1 Introduction

It is known that the size of the zone of uncertainty, the decision-making of the first decision circuit (DC), affects the noise immunity of reception [1]. This is due to the fact that the level of the erasure channel threshold and the size of the erasure zone are unambiguously related to the probabilities of transformation and false erasure, and if these parameters are incorrectly selected, the effect of using the method for assessing the quality of received symbols may turn out to be negative.

# 2 Materials and Methods

One of the methods for increasing the noise immunity of reception in conditions of uncertainty, proposed by Chase [1–3], consists in ordering the received symbols according to their reliability and choosing from among them 5–10 the least reliable ones for special processing, which consists in enumerating all their possible values with simultaneous decoding of the received combination after each performed substitution. For a binary input alpha-whit and the number of selected characters 5–10, the number of options that the decoder should consider is $2^5$–$2^{10}$. This procedure is called "chasing" and is mainly used to decode Bose-Chowdhury-Hawkingham codes (BCH codes) in a communication channel (CC) with a fairly low noise level. Chasing is a general technique by which the performance of almost any block code decoding algorithm can be improved at the cost of additional computation time. At the same time, an increase in the number of chased symbols by 2–3 makes it necessary to speed up the decoder by 4–8 times so that the final information delivery time remains unchanged [4].

However, this method of making non-rigid decisions, in essence, does not solve the problem of choosing and regulating the threshold level in the process of receiving information in real CC of automated control systems (ACS) exposed to powerful weakly correlated interference, due to the likelihood of an excessively large number of erased characters, or replacing prohibited ones. combinations allowed during chasing. In this case, errors in the final result of decoding are due to the non-optimal choice of the threshold in the first DC.

Some methods for choosing the threshold level of the CC binary erasure channel (BEC) are considered in [5] and are reduced to solving the problem of system optimization. Among them, as the most applicable at the present time, include methods for optimizing the threshold level according to the Neumann-Pearson criteria, as well as the minimum and maximum [6, 7]. However, the practical application of these methods in real CC does not allow achieving the theoretically calculated positive effect, and the impact of non-stationary interference in the CC makes it difficult to regulate the threshold level during the reception process.

Thus, the methods of increasing the noise immunity of reception, based on the use of the erasure signal, require solving the problem of optimizing the threshold level of the first DC, that is, minimizing the probabilities of transformation and false erasure for any changes in the characteristics of the CC [8]. The mathematical formulation of the problem is reduced to the solution of the automatic classification problem, which consists in assigning the vector $x$ to one of the s classes of the set $\{w_i\}$, and the components of the vector $x$ are "information" about the observed object.

$$f(x) = w^T x + \text{const},\qquad(1)$$

## 3   Results

The simplest non-trivial mathematical formulation of this problem arises in the case when $s = 2$ (due to the discreteness of the signal taking the values "O" or "1") and the linear decision function assigns $x$ to the class $w_1$ for $f(x) > 0$ and to the class $w_2$ for $f(x) < 0$ with an admittedly small number of incorrect classifications [9], while the procedure for calculating linear decision functions should be adaptive (that is, respond to any changes in the noise environment in the CC), quickly converge to the local minimum errors and be performed without any a priori assumption regarding the form of the statistical distribution of the vectors $x$ in each class $w_i$ (that is, the parameters of the demodulated signal unknown in advance) [10, 21]. Let us call it a "learning procedure with equalized errors" or "VO-procedure", and the word "learning" indicates its adaptive nature.

Let $x$ be a $d$-dimensional feature vector $x = (x_1,..., x_d)$, where $-\infty < x_i < \infty$, and let $y$ be an additional feature vector $y = (x_0, x_1, \ldots, x_d)^T$. Let $X(n)$ and $Y(n)$ be vector random variables that are functions of the step number $n$ and generated by the training sequence in the corresponding spaces $x$ and $y$. In particular, $Y(n)$ is an additional feature vector at step $n$. Let $w(n)$ be the class to which $Y(n)$ belongs, and let $\{w_i\}$ be the alphabet of classes. Due to the discreteness (two-valuedness) of the demodulated signal, this alphabet has a length of two: $\{w\} = \{w_1, w_2\}$. Let $w$ be the weight vector $(w_1, \ldots, w_d)^T$. Let $V(n)$ be a vector random variable generated by the learning process in the space $v$ and which is also a function of the step number.

The additional vector in the proportional learning procedure is defined recursively as follows:

$$V(n + 1) = V(n) + r_n Q(n),\qquad(2)$$

where $\rho_n$ is the value of the $n$-th step and

$$Q(n) = \begin{cases} Y(n), & \text{if } V(n)^T Y(n) < 0 \text{ and } w(n) = w_2; \\ -Y(n), & \text{if } V(n)^T Y(n) > 0 \text{ and } w(n) = w_1; \\ 0, & \text{in other cases.} \end{cases}\qquad(3)$$

In principle, (2) is a kind of gradient descent [13, 14], that is, a method in which the minimum of the loss function $J(v)$ is sought. The usual procedure for finding such a minimum is reduced to determining the recursive function $V(n)$ converging (stochastically) to zero of the gradient of the loss function $J(v)$. Such a recursive function can be obtained by stochastic approximation methods [16]:

$$J(v) = E(-v^T Q | V = v), \qquad (4)$$

We now obtain another expression for (4) that leads directly to the VO-learning procedure: let $f_i(x) = P(w_i) p(x|w_i)$, where $P(w_i)$—is the prior probability that $w = w_i$ and $p(x|w_i)$ is the conditional distribution density $X = x$ at $w = w_i$. We call $p(x|w_i)$ the density conditional in the class, and $f_i(x)$—the component density. Let $\vartheta_i$ be the solution domain corresponding to the class $w_i$.

Using this notation, (4) can be written as:

$$J(v) = |w|[M_1(v) + M_2(v)], \qquad (5)$$

where $|w|$—vector length $w$,

$$M_1(v) = P(w = w_1, y \in \Re_2) E\left(\frac{v^T y}{|w|} | V = v, w = w_1, y \in \Re_2\right); \qquad (6)$$

$$M_2(v) = P(w = w_2, y \in \Re_1) E\left(\frac{-v^T y}{|w|} | V = v, w = w_2, y \in \Re_1\right). \qquad (7)$$

It is easy to show what $\frac{v^T y}{|w|}$ is the distance between $x$ and the boundary of the region $\vartheta_2$, which is a hyperplane. This distance is positive for $x \in \vartheta_2$. Similarly, $\frac{-v^T y}{|w|}$ it represents the distance between $x$ and the boundary of the region $\vartheta_1$, which is a hyperplane. This distance is positive for $x \in \vartheta_1$. For the case of two classes, both boundaries coincide. Thus, $M_i(v)$ is the first moment of the tail (error) of the function $f_i(x)$.

Although this learning procedure is asymptotically accurate for linearly divisible class-conditional densities $\{p(x|w_i)\}$ [9], its asymptotics can differ significantly from the minimum error probability in the case of overlapping class-conditional densities. On the other hand, the minimum of $M_1(v) + M_2(v)$ occurs for some value of $v$, denoted by $v_e$, which is often very close to the minimum error probability value of $v$.

Let us denote by $v_p$ the value of $v$ that provides the minimum probability of error. Experience [15] shows that $v_e$ and $v_p$ are often very close. Indeed, when $f_1(x)$ and $f_2(x)$ are symmetric with respect to each other and, therefore,

$$f_2(x) = f_1(b - x), \qquad (8)$$

where $b$—is the centroid of $f_1(x) + f_2(x)$, then $v_e$ and $v_p$ coincide.

Introduction $|w|$ into the expression for $J(v)$ as a factor (5) makes it possible to substantially separate the points of minima of the function $J(v)$ and $v(p)$, even when $v_e$ and $v_p$ coincide. The use of (5) also leads to a shift of $W(n)$ towards small values of the vector $w$ at $n \to \infty$, since $J(v) = 0$ at $v = 0$. As a result, the direction of the vector $W(n)$ at $n \to \infty$ often becomes insufficiently defined. Thus, the asymptotic behavior of the proportional-increment learning procedure can be unsatisfactory in cases where the class-conditional densities overlap.

To overcome this drawback of the training procedure with proportional increment, the loss function is used in the form

$$J(v) = M_1(v) + M_2(v), \tag{9}$$

which leads to the VO-training procedure described in the article.

We obtain a VO-learning procedure from the loss function (9) using the methods of gradient descent and stochastic approximation [16–18]. Suppose that a continuous differentiable function $J(v)$ has a unique minimum at $v^*$ and it has no local minima. The basic gradient descent procedure for such a function $J(v)$ can be represented as a recursive equation.

$$v_{n+1} = v_n - \rho_n \nabla J(v_n), \tag{10}$$

where

$$\nabla J(v) = gradient \; from J(v) = \begin{bmatrix} \vdots \\ \frac{dJ}{dv_i} \\ \vdots \end{bmatrix} \tag{11}$$

Then, for any sufficiently small $\rho_n$, the sequence $\{J(v^*)\}$ will be monotonically decreasing and converging as $n \to \infty$ to $J(v^*)$. The gradient descent method was extended [15] to the case of "noisy" functions, that is, functions $J(v)$ depending on one or more random variables. This extension leads to a stochastic approximation.

Stochastic approximation is associated with finding the extrema and zeros of the regressive function, that is, the average value of a random variable as a function that depends on a parameter. Suppose that $J(v)$ is a regression function and that, in addition, one can find a random variable $Z$ depending on $v$ such that

$$E(Z|v) = -\nabla J(v). \tag{12}$$

Also assume that $J(v)$ has a unique minimum at $v^*$. If $E(Z|v_n)$ were known, then $v^*$ could be found using gradient descent (10):

$$v_{n+1} = v_n + \rho_n E(Z|v_n). \tag{13}$$

Usually for small $n$ $E(Z|v_n)$ is unknown. In this case, with the method of stochastic approximation in expression (13), the replacement of $E(Z|v_n)$ by the value $Z$ obtained at step $n$, that is, $Z(n)$, is used. Under this condition, $v_n$ becomes a random variable, which we denote by $V(n)$. Thus, (13) is transformed to the form

$$V(n + 1) = V(n) + \rho_n Z(n). \tag{14}$$

It follows from the theory of stochastic approximation that the stochastic convergence of $\{V(n)\}$ to $v^*$ depends on the choice of $\rho_n$, $Z(n)$ and registration functions [16]. For example, if $\rho_n$ decreases too fast as $n$ increases, then $V(n)$ does not converge to $v^*$. It follows from the theory of stochastic approximation that the stochastic convergence of $\{V(n)\}$ to $v^*$ depends on the choice of $\rho_n$, $Z(n)$ and registration functions [16]. For example, if $\rho_n$ decreases too quickly as $n$ increases, then $V(n)$ does not converge to $v^*$.

Bearing in mind that the training procedure works in an additional weight space, we will consider $V(n)$ in (14) as an additional weight vector at step $n$ and $Z(n)$ as an appropriately chosen function depending on $Y(n)$, $V(n)$ and $w(n)$. For the appropriate choice of $Z(n)$ in the case of the usual stochastic approximation, one would have to look for a random variable $Z$ satisfying (12). However, the choice of $Z$ requires special care [16] in the case when $J(v)$ is a loss function for the VO-learning procedure (9), because in this case $J(v)$ does not depend on |v|. This independence leads to ambiguity in the position of the minimum of the function $J(v)$. Indeed, if $J(v)$ is a minimum at $v$, then this is also a minimum at $v^*$ for all real values of $a$. In this case, $Z$ should be chosen such that the convergence properties of $V(n)$ do not depend on $|V(n)|$, that is, depend only on $\frac{V(n)}{|V(n)|}$.

In [15], the results of modeling the VO-learning procedure by the Monte Carlo method are presented. Below is one of the examples of modeling. The feature space in the example is two-dimensional, respectively, the additional feature weight vector $V(n)$ is three-dimensional. After calculating $V(n)$, it was normalized by dividing by $|W(n)|$. Result vector

$$\widetilde{V}(n) = \frac{V(n)}{|W(n)|} \tag{15}$$

was built in space $\widetilde{V}_0 \widetilde{V}_1$. The training was 1000 steps long.

Example: $f_1(x)$ and $f_2(x)$ are Gaussian distributions with mean values

$$\mu_1 = \begin{pmatrix} -\dfrac{1}{2\sqrt{2}} \\ -\dfrac{1}{2\sqrt{2}} \end{pmatrix}, \quad \mu_2 = \begin{pmatrix} \dfrac{3}{2\sqrt{2}} \\ \dfrac{3}{2\sqrt{2}} \end{pmatrix}.$$

The covariance matrices for these densities are the same:

**Fig. 1** Densities and
hyperplanes as an example

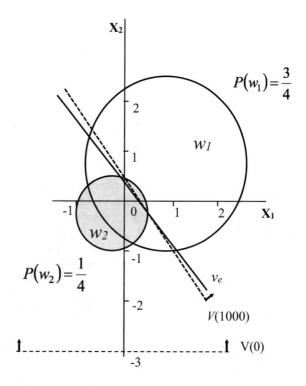

$$\underline{\sum}_1 = \underline{\sum}_2 = \begin{pmatrix} 1 & 0 \\ 0 & 1 \end{pmatrix}.$$

The prior probabilities are $P(w_1) = \dfrac{1}{4}$ и $P(w_2) = \dfrac{3}{4}$ respectively.

These densities are shown in Fig. 1. The circles correspond to the values of x satisfying.

$$f_i(x) = e^{-1} \max_x = [f_1(x) + f_2(x)], \ i = 1, 2.$$

The starting vector is $V(0) = \begin{pmatrix} -3 \\ 0 \\ 1 \end{pmatrix}$. The sequence $\rho_n$ is

$$\rho_n = 0, 1\sqrt{\dfrac{25}{25 + s(n)}}, \tag{16}$$

where $s(n)$ is defined as follows. Let be

$$e(n) = \begin{cases} 1 \text{ at } Z(n) \neq 0 \\ 0 \text{ at } Z(n) = 0 \end{cases} \tag{17}$$

then

$$s(n) = \begin{cases} 0, & \text{for } \sum_{k=0}^{n} e(k), \\ \left( \sum_{k=0}^{n} e(k) \right) - 1 \text{ in other cases.} \end{cases} \tag{18}$$

Figure 1 shows the positions of the hyperplane $V(n)$ for $n = 0$ and $n = 1000$. To illustrate the discrepancy between $V(1000)$ and $v_e$, the figure shows a hyperplane corresponding to the VO-weight vector $v_e$. The trajectory $\widetilde{V}(n)$ in space $\widetilde{V}_0\widetilde{V}_1$ is shown in Fig. 2 for $n = 0, 25, 50, ..., 1000$. Some points are marked with corresponding n values. The constructed trajectory makes it possible to judge the speed of convergence of the training procedure under consideration.

The asymptotic error probability for the component densities in the considered example, attained by the VO-weight vector $v_e$, is 0,131; the minimum error probability achieved by the hyperplane is 0,127; the asymptotic error probability achieved

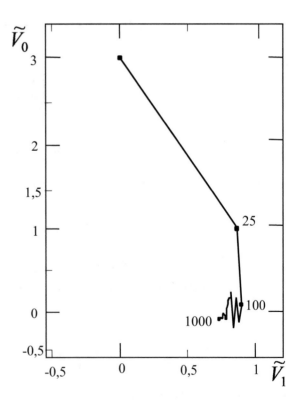

**Fig. 2** The trajectory of convergence in the example

using the training set is 0,137, which indicates a good asymptotic behavior of the training VO-procedure. The probability of error in the example was 0,064 and 0,073, respectively, for classes $w_1$ and $w_2$.

# 4  Conclusion

The above-described VO-procedure was applied for communication systems, in particular, with the aim of optimizing the threshold level of the first DC BEC by generating, on the basis of the VO-procedure, a control action proportional to the change in the input parameters of the demodulated signal [19]. At the same time, the probabilities of transformation and false erasure of the symbol are minimized when the interference environment in the CC changes, which increases the noise immunity of the reception as a whole.

An example of a technical implementation is a device for adaptive majority decoding of duplicated telemechanical signals [20]. The block diagram of a device that technically implements the VO-learning procedure, given by Eqs. (2) and (3), is shown in Fig. 3.

Here the mathematical expression (3) looks like this:

$$Q(n) = \begin{cases} Y(n), & V(n)^T Y_n < \Pi \\ -Y(n), & V(n)^T Y_n > \Pi \end{cases} \tag{19}$$

where

$V(n)$—gain of amplifier 1 with adjustable gain (URKU);

$\rho_n$—the step with which this procedure converges to a minimum of the probability of error and is implemented by the gain of amplifier 4 with a constant gain (UPKU), set manually or automatically, depending on the nature of the non-stationary interference in the communication channel;

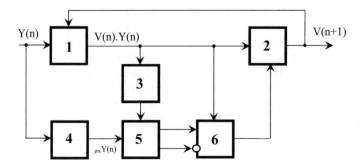

**Fig. 3** Block diagram of a device that implements the VO-learning procedure

$Y(n)$—samples of channel signals distorted by interference and entering the input of this device (interference analyzer);

П—the value of the threshold of block 3 and the convergence of which to the minimum of the error probability is proved.

The device works as follows.

The input signal $Y(n)$ is fed to the inputs of the URKU (block 1) and the UPKU (block 4). The gain of the URKU has the values $V(n)$, $V(n + 1)$ at different times, while the URKU implements the action $V(n) \cdot Y(n)$. The UPKU has a gain $\rho_n$, which determines the step of convergence of the procedure, and implements the action $\rho_n \cdot Y(n)$. The signal from the output of the URKU goes to the input of the comparator (block 3), which analyzes the fulfillment of the conditions $V(n)Y(n) < 0$, or $V(n)Y(n) > 0$. When the first inequality is fulfilled, the switch (block 5), controlled by the signal from the output of the comparator, passes the amplified signal from the output of the UPPC actually, or inversely when the second inequality is fulfilled. Thus, the condition $\pm \rho_n Y(n)$ is realized. For the general case, the comparator compares not with the logical zero level, but with the reference voltage initially set when a signal is applied to the input of the device in the absence of interference in the CS. The adder (block 6) performs the action $V(n) \pm \rho_n Y(n)$, after which the obtained value $V(n + 1)$ is compared in the comparison unit (subtractor) (block 2) with the value $V(n)$. The mismatch signal, proportional to the difference of the compared values, is fed to the input of the URKU through the figurative connection of the feedback and regulates the coefficient of its amplification, which realizes the dependence $V(n + 1) = f[V(n)]$. In addition, the resulting control action is the sought-for VO-learning procedure, which has all the previously described advantages.

According to the specified algorithm, depending on the level of interference in the communication channel, a signal is generated at the output of the device, the level of which is proportional to the change in the interference acting at a given time in the communication channel. This signal adjusts the "erase" bandwidth of the quality detector.

A further increase in noise immunity in accordance with the proposed method can be achieved if, according to the law of VO-learning, one regulates not only the threshold level, but also the volume of the uncertainty zone. With the deterioration of the noise environment in the spacecraft, that is, the deviation of the decisive hyperplane in space, its volume must undoubtedly be increased (Fig. 4). In addition, the introduction into the device instead of the UPKU of the second URKU, with the coefficient $\rho$, regulated by the VO-procedure, will make it possible to implement the "ravine" method [16] of gradient descent. In this case, the step of convergence of the procedure will be variable and the greater, the greater the deviation of the hyperplane $V(n)$ from the asymptotic value. The convergence of the procedure to a minimum of errors will accelerate.

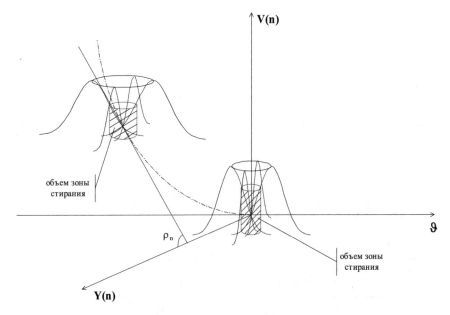

**Fig. 4** Change in the position of the erasure zone in space, caused by the VO-training procedure

To optimize the threshold level of the first decision circuits, a mathematical gradient descent method should be applied, which leads to a learning procedure with equalized errors. The simplicity of the technical implementation of this procedure allows it to be applied in the first decisive schemes of the information transmission system without significantly complicating them and increasing the decision-making time.

Thus, the VO-learning procedure makes it possible to achieve a minimum of the loss function of various processes. Its implementation makes it possible to obtain a control action that changes the parameters of the system and reduces them to the asymptotic value of the input action (Figs. 1 and 2).

The area of possible application of the considered VO-procedure is wide and includes many areas of activity in which the need for automatic classification arises: navigation; medical diagnostics; aerial reconnaissance; space photography; psychological learning theory; automatic control systems; economy [15]. In particular [15], the VO-training procedure was used to analyze medical X-ray images in order to detect tumors on the X-ray images. Moreover, the convergence of the procedure even for the case of seven-dimensional feature vectors turned out to be small [15], which once again proves its advantages.

**Acknowledgements** The work is supported by North-Caucasus Center for Mathematical Research under agreement №. 075-02-2021-1749 with the Ministry of Science and Higher Education of the Russian Federation.

# References

1. Berlekamp ER (1980) Error-correcting coding technique. TIIER 68(5):24–58
2. Chase D (1973) A combined and modulation approach for communication over dispersive channels. IEEE Trans Commun 3:159–174
3. Chase D (1972) A class of algorithms for decoding block codes with channel measurement information. IEEE Trans Inform Theory IT-18(1):170–182
4. Malofey OP, Malofey AO, Kuchukov VA, Malofey MS, Kharechkina YuO, Kharechkin AN (08 June 2019) A modified device for correcting errors, taking into account the erasure signal. Patent for invention RU 2711035 C1, 14.01.2020. Application № 2019114114
5. Tolstyakov VS (1972) Detection and correction of errors in discrete devices. Sov. Radio., Moscow
6. Shuvalov VP (1972) Indirect error detection methods. Communication, Moscow
7. Levin BR (1975) Fundamentals of statistical radio engineering. Book 2. Sov. Radio., Moscow
8. Malofey OP, Malofey AO, Shangina AE (2017) Analysis of algorithms for noise-resistant coding and increasing their efficiency in complex systems. Sci Innov Technol (2):43–52
9. Nilsson N (1967) Learning machines. Mir, Moscow
10. Sklansky J (1965) Threshold training of two-mode signal detection. IEEE Trans Inf Theory IT-11(3):353–362
11. Rosenblatt F (1958) The perceptron: a probabilistic model for information storage and organization in the brain. Psychol Rev 65(6):386–408
12. Motzkin TS, Shoenberd IJ (1954) The relaxation method for linear inequalities. Con J Math 6:393–404
13. Widrow B, Hoff ME (1960) Adaptive switching circuits. In: IRE WESCON convention record, Part 4, no. 8, pp 96–104
14. Ho YC, Kashyap RL (1965) An algorithm for linear inequalities and its applications. IEEE Trans Electron Comput EC-14(10):683–688
15. Wassel GN, Sklansky J (1976) Adaptive nonparametric classifier. TIIER (8):52–62
16. Vazan M (1973) Stochastic approximation. Mir, Moscow
17. Duda R, Hart P (1976) Pattern recognition and scene analysis. Mir, Moscow
18. Vasiliev FP (1988) Numerical methods for solving extreme problems. Science, Moscow
19. Malofey AO (02 September 2016) Increasing the noise immunity of reception using a modified majority decoding algorithm, which implements correction taking into account the erasure signal. In: The collection: actual problems of the formation of professional competence among cadets and students of universities of the Ministry of Internal Affairs of Russia. Electronic collection of materials of the interuniversity round table, Stavropol, RU, pp 205–209
20. Malofey AO, Malofey OP et al (2007) Pat. 2309553 Russian Federation, MPK[7] N 03 M 13/00. Error correction device with an extended set of decision rules and taking into account the adaptive erasure signal. No 2005126769, Appl. 24.08.2005, publ. 27.10.2007, Bul. No 10, 24 p ill
21. Malofey OP, Malofey AO, Shangina AE (2018) On the issue of optimization of hardware costs in infotelecommunication systems. In: Proceedings of the 2018 IEEE conference of Russian young researchers in electrical and electronic engineering, ElConRus pp 342–346

# A Feature Fusion Based Deep Learning Model for Deepfake Video Detection

Sk Mohiuddin⬚, Shreyan Ganguly⬚, Samir Malakar⬚, Dmitrii Kaplun⬚, and Ram Sarkar⬚

**Abstract** In the recent past, several tools developed for facial image manipulation utilizing techniques like FaceSwap and Deepfake with implausible success. Such tools enable to edit faces present in a video with a minimum effort and the outcomes are incredibly close to real faces. Despite their handful number of uses, they possess many unpleasant impacts like creating fake news and defaming celebrities on social media. Hence, pinpointing whether a video contains manipulated faces in it becomes a task of utmost importance. To this end, we have designed a face manipulation detection technique that can figure out whether a target video contains manipulated faces in it or not. To do this, we first crop out the target face from video frames. Next, we extract deep features using MesoInception from RGB as well as their YCbCr and HSV versions, which are then concatenated to form a single feature vector. On the top of this new feature vector, we add the classification layer to classify the images into fake and real. To evaluate our model performance, we use two publicly available video datasets, namely Celeb-DF (V2) and Faceforensis++. The experimental results lay out the effectiveness of the proposed model over some recently published models.

**Keywords** Deepfake · Face manipulation · Deep learning · Celeb-DF (V2) · Faceforensics++

S. Mohiuddin (✉) · S. Malakar
Department of Computer Science, Asutosh College, Kolkata, India
e-mail: myselfmohiuddin@gmail.com

S. Ganguly
Department of Construction Engineering, Jadavpur University, Kolkata, India

D. Kaplun
Department of Automation and Control Processes, Saint Petersburg Electrotechnical University "LETI", Saint Petersburg, Russia
e-mail: dikaplun@etu.ru

R. Sarkar
Department of Computer Science and Engineering, Jadavpur University, Kolkata, India

© The Author(s), under exclusive license to Springer Nature Switzerland AG 2022     197
A. Tchernykh et al. (eds.), *Mathematics and its Applications in New Computer Systems*,
Lecture Notes in Networks and Systems 424,
https://doi.org/10.1007/978-3-030-97020-8_18

# 1 Introduction

Detection of deepfake videos/images is considered as a very crucial task because they possess widespread reprobate sides on many social media. The paramount concerns raised by deep-fake videos is in case of face identity theft and the defamation of celebrity, where their face are found to be swapped with another individual's face or they are found to mimic some fake expressions. Many organizations of social media (for example, Google and Facebook) are facing criticism due to the increase of sharing such fake images and videos. Hence, they are trying to find some suitable solutions to fight against this. They have made several open challenges for the researchers by providing sample free video datasets aiming at overcoming such obstinate circumstances.

Unsurprisingly, computer graphics-based face manipulation is visually detectable by human bare eyes while zooming the images. Also such manipulations leave some signs in case of videos. However, modern artificial intelligence (AI) based synthesis media rarely leave any perceptible artifacts. Figure 1 depicts the differences among manipulated face images generated by computer graphics based, early and modern generation AI aided face manipulation methods. Generative adversarial networks (GANs) are the largely used backbone for creating such AI based face manipulations. GAN based applications, like FaceApp,[1] FaceSwap,[2] etc., are freely available on the internet for the entertainment purpose. However, some users are applying its evil side.

Researchers around the world have tried to detect above mentioned face manipulations by either extracting handcrafted features from perceivable visual artifacts [6, 7, 10] or by employing deep learning aided models where almost non-perceptible artifacts are found [1, 14, 17]. Durall et al. [6] have used Discrete Fourier Transform (DFT) to investigate abnormal behavior in the synthesis images which are then fed

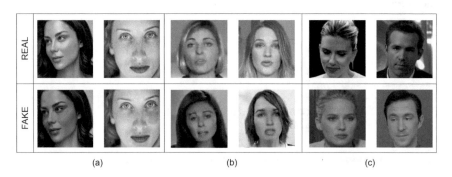

**Fig. 1** Samples of manipulated faces with their corresponding real faces: It shows manipulated faces generated by the **a** computer graphics based method, **b** early generation AI-based method, and **c** modern AI-based method

---

[1] https://apps.apple.com/gb/app/faceapp-ai-face-editor/id1180884341.

[2] https://github.com/MarekKowalski/FaceSwap.

into the classical machine learning method for the classification purpose. To extract pixel correlation information, the Expectation Maximization (EM) algorithm have been applied on each channel of the input images by Guarnera et al. [7]. Extracted features are passed through a naive classifier to distinguish forged frames from real ones. Unlike the preceding methods, Koopman et al. [10] have tried to detect manipulation patterns from camera captured noises where photo response non-uniformity (PRNU) patterns are applied on a group of equal size cropped faces extracted from videos. Average correlation and variations in correlation scores among the groups are then passed through Welch's t-test to identify manipulated faces. These methods are fast in terms of computation time, however fail if no dominant visual artifacts present in the manipulated faces.

As a result, researchers have shifted their approach from traditional handcrafted feature engineering to the deep learning based approaches. Such approaches become predominant over handcrafted based methods as they can learn better features from the training samples that boost the end results. Hence, several Convolutional Neural Network (CNN) based models on AI based face manipulation detection could be found in literature. For example, Shang et al. [14] have built a CNN architecture that utilizes two modules stacked in a steady progression. One module captures the relation of a pixel with its closest ones and the other module searches for the relation between the original and the manipulated faces. They utilize a high-resolution network (HRNet-w30) [17] in both the modules with different setups. To focus on mesoscopic properties of an image, Afchar et al. [1] have proposed a model made out of two diverse CNN structures: i) MesoNet, a CNN model comprising 4 convolutional layers followed by a fully connected layer, and ii) MesoInception-4 where first two layers of MesoNet are replaced by Inception module [18]. The concept of an unnatural motion of facial components is brought into notice by Amerini et al. [2] where PWC-Net [16], a CNN model, is used to extract optical flow matrices from the face images first and then feed these into VGG16 [15] based semi-trainable CNN model to filter out authentic videos from forged ones. To learn adaptively editing features from images Bayar and Stamm [3] have designed their own CNN model. The method is designed to detect universal image manipulations rather than detecting a specific type of manipulation. In another work [13], Rossler et al. have conducted a comprehensive study with existing deepfake detection methods and some state-of-the-art CNN models, and concluded that XceptionNet [4] performs the best as a backbone of the CNN network on the dataset prepared by them. Recently, Nirkin et al. [12] have proposed a method where they have looked for face recognition similarity between a tightly segmented face and the face context (excluding the tightly segmented face) and the artifacts left by the deep faking methods. Thus, they use three deep learning based models, and the confidence scores from classifiers are stacked and passed through deep learner to distinguish real faces from fake ones.

The performance of the existing CNN based methods is satisfactory when trained and tested on videos from the same dataset where face manipulation is made by some specific method. Intuitively, if the characteristic of artifacts left by the GAN based face faking method is known to the detector then it performs well. However, the task becomes critical if some unknown artifacts appear in the test images and most of the existing models fail in such scenarios [11, 12]. Also, He et al. [8] have shown that concatenating deep features extracted from different color space images helps in generalizing a deepfake image classification method. Motivated by these facts, we propose a feature fusion based deep learning model that performs better then many state-of-the-art face manipulation models for both intra-dataset and inter-dataset model training and testing setups. Our method uses MesoInception-4 proposed by Afchar et al. [1] as a feature extractor. The CNN model is employed to generate three different feature vectors from three color spaces of an image. Next, we fuse these features and pass through a classification model to distinguish a real video from a fake video.

## 2 Proposed Model

In this work, we propose a novel deep learning based deepfake detection model which takes a target video as input and classifies it either as fake or real. The proposed model primarily consists of two major modules - (i) Cropping of faces from video frames and (ii) feature fusion aided deep learning based classification module. The architecture of the proposed model is shown in Fig. 2. In this model, information from different color spaces of the input images is considered an important factor for the classification task. We explain the modules of the proposed model in the following subsections.

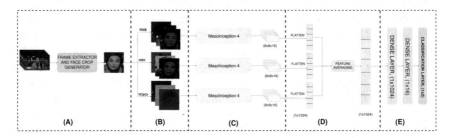

**Fig. 2** The architecture of the proposed model. The initial module (see block **A**) generates the cropped face images from the frames of an input video. This module considers three image formats of the face crops (see block **B**). For feature extraction (see block **C**), MesoInception-4 takes three different color space variants of the face crops. As a deep feature fusion operation, we take average of the three obtained features (see block **D**). Lastly, the model includes multi-layer perceptron neural network and a classification layer (see block **E**)

## 2.1   Generation of Cropped Faces from Video Frames

This is the initial module of the proposed model, which takes a video as the input, and returns the cropped face of the subject in the video. To accomplish this task, selection of frames from the video is the primary task. We choose to extract P-type frames in this work as they are compressed moderately in the videos. The selected P-type frame is passed through a multi-mask cascading neural network (MTCNN) [9] based face detector algorithm to detect subjects' faces in the frame. We use this network because of its lightweight architecture which uses a multi-task learning mechanism, hence works well with faces of all scales. The used method also performs better in different lighting conditions. After detecting the bounding box around the face region using MTCNN, we increase the size of the detected bounding box by 10% to include some background information around the face region. Cropped faces are resized to (256 × 256) to satisfy the dimensional requirement of the deep learning model described in the later section.

## 2.2   Deep Feature Fusion Based Classification

The cropped face obtained using the previous module is in RGB image format. This is then converted to two different color spaces - HSV, and YCbCr. In the deepfake videos, a considerable inconsistency is visible in facial color as a left-over artifact of manipulation techniques. We try to exploit this fact to classify deepfake videos by representing an input image with different color spaces in this work. Also, the studies [5, 8] show that complementary and useful features can be obtained from a particular image by using its differently processed forms and feeding such information to a CNN aided feature extractor model. Inspired by these facts, we use three different forms of face crops as the input to the CNN model, and finally fuse such differently extracted features to obtain the final features. In this work, we choose to use MesoInception-4 [1] as the CNN model to generate feature maps, and averaging operation as a feature fusion technique.

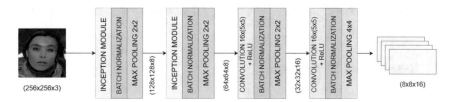

**Fig. 3** Architecture of MesoInception-4 blocks used to generate feature maps in the proposed model

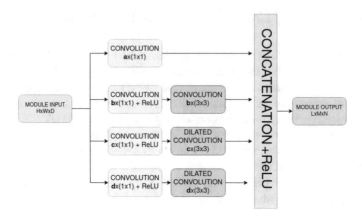

**Fig. 4** Architecture of inception modules present in MesoInception-4 blocks, which are useful in generating feature maps of multi-scale target objects

MesoInception-4 is an improved version of MesoNet [1] designed for deepfake video/image detection. MesoInception-4 architecture is similar to MesoNet architecture which comprises four convolution layers followed by batch normalization and pooling layers. However, in the case of MesoInception-4, the first two convolution layers of MesoNet are replaced with inception blocks as shown in Fig. 3. The inception blocks are useful in generating important feature maps of multi-scale target objects. This is because it tries to generate an aggregated feature map from kernels of different sizes as shown in Fig. 4.

MesoInception-4 generates feature maps of dimension $(8 \times 8 \times 16)$. We further flatten the obtained feature maps, to form feature vectors each of dimension $(1 \times 1024)$. Getting three feature vectors from three variants of each input image, we average the vectors to form a single feature vector of dimension $(1 \times 1024)$. Averaging the feature vectors helps in reducing the feature dimension by 3 times keeping the information of each feature vector almost intact, hence reducing the parameters needed to train the model, thereby making the model light.

Finally, we attach a couple of dense layers after that with a dropout of 0.5. We end up the model with a softmax classification layer of dimension $(2 \times 1)$ which implies the confidence score of the model determining if a face crop image is real or fake.

## 3 Results and Discussion

It has already been mentioned that in this work, we develop a feature fusion aided deep learning model to examine whether a questioned video contains face manipulated contents made by some face faking methods or not. The model first extracts frames from the questioned video and then crops out the face region from each frame. Next, the cropped face images are converted to YCbCr and HSV image formats which along with the RGB version are passed through MesoInception-4 to extract the feature maps. These three types of feature vectors are fused to generate a single feature vector that has been used in the classification model. In the following subsections, we first describe the data preparation method and then discuss the experimental protocols used here and the obtained results by this model.

## 3.1 Dataset Preparation

For evaluating the performance of the proposed model, we classify a questioned video into either real or fake. The videos where faces are manipulated using the faceswap via deep learning based method i.e., we only consider videos belonging to the deepfake category. We collect deepfake video samples from two popularly used public datasets, namely Celeb-DF (V2) [11] and FaceForensics++ [13]. The Celeb-DF (V2) dataset contains 5639 synthesis videos of different celebrities that were generated using 590 real videos collected from YouTube with varying genders, ages and backgrounds. Thus this dataset contains in total 6,229 videos (5639 deepfake videos and 590 real videos) out of which 518 videos were present as test samples (178 real and 340 fake). Here, it is worth in mentioning that all the fake videos of this dataset fall under the deepfake category.

The FaceForensics++ contains 1000 real videos generated from a collection of 977 YouTube videos, selected in such a way that all videos mostly contain frontal face without occlusions which in turn help the researchers to generate almost realistic forged videos. All the 1000 real videos were manipulated using four face manipulation methods namely, DeepFake, Face2Face, FaceSwap and NeuralTexture. So, the dataset contains 5000 videos in total. The facial manipulation methods used in this dataset are of two categories, namely facial expression manipulation (Face2Face and NeuralTexture) and facial identity manipulation videos (DeepFake and FaceSwap) [13]. We consider fake videos generated by deepfaking as fake videos while Pristine category videos (i.e., the original videos) are considered as real videos since our method targets detection of identity manipulation via a deep learning method. The distribution of training and test sets of videos is shown in Table 1 where we also mention the number frames per class used in our experiments. During model training we use 10% of the training samples as validation samples.

**Table 1** Distribution of training and test videos and frames for each dataset

| Dataset | #Real videos in | | #Fake videos in | | #Real frames in | | #Fake frames in | |
|---|---|---|---|---|---|---|---|---|
| | Training | Test | Training | Test | Training | Test | Training | Test |
| Celeb-DF (V2) | 412 | 178 | 5299 | 340 | 4000 | 178 | 4000 | 340 |
| FaceForensics++ | 800 | 200 | 800 | 200 | 8000 | 2000 | 8000 | 2000 |

## 3.2 Experimental Setups

The proposed model is developed in Python using Tensorflow framework. It is trained and tested on the Google Colab platform which provides Nvidia Tesla K80 GPU consisting of 4992 CUDA cores. We train our model for 150 epochs (step size/epoch = 40) with batch size of 32, learning rate of 0.001 and Adam optimizer. The training is performed from scratch.

Face tampering using faceswap via deep learning model sometimes leaves behind imperceptible artifacts that might help identifying the fake faces. However, use of different face manipulation methods generates different such artifacts. Hence, it is not guaranteed that a deepfake detection method would perform closely irrespective of deepfaking techniques employed for face manipulation. Inspired by this fact, we follow two different experimental setups. In the first case, we use samples from the same dataset for model training and performance evaluation. We call this evaluation strategy as an intra-dataset evaluation. In this setup, the model knows or can learn the artifacts that may present in test samples. In the second case, the model is trained and evaluated on samples from different datasets. In this scenario, the model is not aware of artifacts present in the test samples and this evaluation strategy is termed as the inter-dataset evaluation.

## 3.3 Experimental Results

The proposed model is evaluated on the said datasets following both the experimental setups mentioned above. The reported results are obtained at frame level. Area Under the Curve (AUC) score is used as an evaluation parameter following the state-of-the-art methods. The results obtained by the proposed method are shown in Table 2. Our method provides 85.25% and 81.81% AUC scores on Celeb-DF (V2) and FaceForensics++ datasets respectively following the intra-dataset experimental protocol as shown in Table 1. When evaluated using the inter-dataset protocol, the proposed method provides 67.44% and 64.64% AUC scores on Celeb-DF (V2) and FaceForensics datasets respectively.

## 3.4 Comparison with Past Methods

To have a fair comparison on the current datasets' splits, we estimate the performances of some recent methods [1, 13] at our end using the codes made available by the researchers. The comparative results are recorded in Table 2. The results in this table indicate that the proposed method outperforms the methods that used MesoNet-4 and MesoInceptionNet-4 as their backbone architecture. It can be observed that the Xception based model outperforms proposed model while considering intra-dataset

**Table 2** Performance comparison of proposed method with some past methods. Here D1 is the Celeb-DF (V2) dataset and D2 is the FaceForensics++ dataset

| Work Ref. | Method | Intra-dataset | | Inter-dataset | |
|---|---|---|---|---|---|
| | | D1 | D2 | D1 | D2 |
| Afchar et al. [1] | Meso-4 | 69.75 | 80.20 | 57.29 | 62.17 |
| Afchar et al. [1] | MesoInception-4 | 81.79 | 77.76 | 63.78 | 61.04 |
| Rossler et al. [13] | XceptionNet | 84.13 | **82.31** | 60.87 | 50.45 |
| **Proposed** | Stacking based | **85.25** | 81.81 | **67.44** | **64.64** |

evaluation strategy on FaceForensics++ dataset, however, for other cases our method outperforms the rest of the methods considered here for comparison.

## 4 Conclusion

Detection of face manipulated images/videos consisting of face manipulated frames is the utmost prerequisite to cut off the spreading of image/video based fake news or to reduce the defaming of famous people among others. Here we propose a feature fusion aided deep learning model for detecting videos consisting of manipulated faces. We use multiple color space variants of an input image to generate different feature maps using a CNN model, which results in mostly complementary features. We fuse the generated features by average operation and then the obtained feature vector is passed through the classification layer. The proposed method outperforms many recently published methods, which is confirmed after evaluating the model performance on two benchmark face manipulation datasets - FaceForensics++ and Celeb-DF. In this work, MesoInception-4 has been used to generate features, which proves to be quite cost effective considering its light architecture. However, other CNN models can be used to generate features, which is in the scope of further research. Also, other challenging video datasets can be considered in future to ensure the robustness of the proposed model.

## References

1. Afchar D, Nozick V, Yamagishi J, Echizen I (2018) Mesonet: a compact facial video forgery detection network. In: 2018 IEEE international workshop on information forensics and security (WIFS). IEEE, pp 1–7
2. Amerini I, Galteri L, Caldelli R, Del Bimbo A (2019) Deepfake video detection through optical flow based CNN. In: Proceedings of the IEEE/CVF international conference on computer vision workshops
3. Bayar B, Stamm MC (2016) A deep learning approach to universal image manipulation detection using a new convolutional layer. In: Proceedings of the 4th ACM workshop on information hiding and multimedia security, pp 5–10

4. Chollet F (2017) Xception: deep learning with depthwise separable convolutions. In: Proceedings of the IEEE conference on computer vision and pattern recognition, pp 1251–1258

5. Dey S, Roychoudhury R, Malakar S, Sarkar R (2021) Screening of breast cancer from thermogram images by edge detection aided deep transfer learning model. Multimedia Tools Appl 1–19

6. Durall R, Keuper M, Pfreundt F-J, Keuper J (2019) Unmasking deepfakes with simple features. arXiv preprint arXiv:1911.00686

7. Guarnera L, Giudice O, Battiato S (2020) Deepfake detection by analyzing convolutional traces. In: Proceedings of the IEEE/CVF conference on computer vision and pattern recognition workshops, pp 666–667

8. He P, Li H, Wang H (2019) Detection of fake images via the ensemble of deep representations from multi color spaces. In: 2019 IEEE international conference on image processing (ICIP), pp 2299–2303

9. Jiang B, Ren Q, Dai F, Xiong J, Yang J, Gui G (2018) Multi-task cascaded convolutional neural networks for real-time dynamic face recognition method. In: International Conference in Communications, Signal Processing, and Systems. Springer, Cham, pp 59–66

10. Koopman M, Rodriguez AM, Geradts Z (2018) Detection of deepfake video manipulation. In: The 20th Irish machine vision and image processing conference (IMVIP), pp 133–136

11. Li Y, Yang X, Sun P, Qi H, Lyu S (2020) Celeb-DF: a large-scale challenging dataset for deepfake forensics. In: Proceedings of the IEEE/CVF conference on computer vision and pattern recognition, pp 3207–3216

12. Nirkin Y, Wolf L, Keller Y, Hassner T (2021) Deepfake detection based on discrepancies between faces and their context. IEEE Trans Pattern Anal Mach Intell

13. Rossler A, Cozzolino D, Verdoliva L, Riess C, Thies J, Nießner M (2019) Faceforensics++: learning to detect manipulated facial images. In: Proceedings of the IEEE/CVF international conference on computer vision, pp 1–11

14. Shang Z, Xie H, Zha Z, Yu L, Li Y, Zhang Y (2021) PRRNet: pixel-region relation network for face forgery detection. Pattern Recogn 116:107950

15. Simonyan K, Zisserman A (2014) Very deep convolutional networks for large-scale image recognition. arXiv preprint arXiv:1409.1556

16. Sun D, Yang X, Liu M-Y, Kautz J (2018) PWC-net: CNNs for optical flow using pyramid, warping, and cost volume. In: Proceedings of the IEEE conference on computer vision and pattern recognition, pp 8934–8943

17. Sun K, Zhao Y, Jiang B, Cheng T, Xiao B, Liu D, Mu Y, Wang X, Liu W, Wang J (2019) High-resolution representations for labeling pixels and regions. arXiv preprint arXiv:1904.04514

18. Szegedy C, Liu W, Jia Y, Sermanet P, Reed S, Anguelov D, Erhan D, Vanhoucke V, Rabinovich A (2015) Going deeper with convolutions. In: Proceedings of the IEEE conference on computer vision and pattern recognition, pp 1–9

# Search for an Object with a Recognition Error by a Desktop Grid

**Evgeny Ivashko and Ilya Chernov**

**Abstract** Scientific computing problems include a "needle in a haystack" problem. One searches for a unique target independently examining objects from a finite set. If there is a possible error in an object examination, one should have a strategy for re-examination the objects. High-throughput computing systems such as Desktop Grids commonly use so-called replication to increase reliability and throughput of a computing process. In this paper we present and analytically prove the optimal strategy of re-examination and derive an explicit formula for the expected number of examinations needed to find the target. We also show connection between replication in Desktop Grids and re-examination strategy in "needle in a haystack" problems.

**Keywords** Desktop grid · Volunteer computing · Unreliable search · Urn problem · Needle in a haystack

## 1 Introduction

In this paper, we consider the following search problem, which is called "needle in a haystack" problem. The problem is as follows. There is a target object among many useless objects. One draws blindly and examines objects sequentially one by one, trying to find the target as soon as possible. The target object may be mistakenly recognised as useless with some positive probability. Useless objects are always recognised correctly. Examined objects are available for further examination and the

I. Chernov—These authors contributed equally to this work.

E. Ivashko (✉) · I. Chernov
Institute of Applied Mathematical Research, Karelian Research Centre of RAS, Petrozavodsk, Russia
e-mail: ivashko@krc.karelia.ru

I. Chernov
e-mail: chernov@krc.karelia.ru

Petrozavodsk State University, Petrozavodsk, Russia

© The Author(s), under exclusive license to Springer Nature Switzerland AG 2022
A. Tchernykh et al. (eds.), *Mathematics and its Applications in New Computer Systems*,
Lecture Notes in Networks and Systems 424,
https://doi.org/10.1007/978-3-030-97020-8_19

number of examinations of each object is known. The goal is to minimise the average number of examinations before the target is found.

Problems of this type are common in scientific computing: finding a password having its hash, finding integer roots of equations, searching for counterexamples of conjectures, just to name a few.

Since the objects are examined independently, high-throughput computing systems such as Desktop Grid are commonly used to solve such kind of problems in case of a huge number of such objects and/or computationally intensive examination of the objects.

Desktop Grids use idle time of non-dedicated geographically distributed computing nodes connected by a relatively low-speed (Internet or LAN) network. In general, Desktop Grids have the server-client architecture and focuses on the efficient execution of a large number of loosely-coupled tasks [1].

Desktop Grids have huge computing potential at low cost, and great scalability. However, Desktop Grids have some peculiarities compared to other computational systems (such as computing clusters and classical GRID systems): low reliability of computing nodes, huge hardware and software heterogeneity, and lack of trust to computing nodes.

Desktop Grid project use so-called replication mechanism that orders to solve each task two or more times (the factor of replication) in order to reduce the risk of wrong answer, but at the cost of efficiency loss. An important question is what factor of replication to use. Higher factor reduces the possibility of wrong answer acceptance but also significantly reduces effective computing power because more computing nodes do the same work (see for example [2]).

In "needle-in-a-haystack" problems, the search is over as soon as the target is found. Error risk can be ignored because it is quite easy to check the answer using trusted computing resources or redundant computing. Besides, missing the target due to the wrong answer is possible and this risk cannot be ignored, so further checks are necessary. This risk can be made arbitrarily low by high replication factor, but this is quite costly.

For "needle-in-a-haystack" problems with the single target, we provide an answer what the optimal replication factor is, and give a formula for the expected number of checks. This formula could be used to develop a software estimator that would give researchers an idea of time needed to finish an experiment.

This paper presents an extended version of work [3] enriched with proves of mathematical statements. The mathematical model of this problem is described in Sect. 2, the optimal strategy including an explicit formula for the expected number of checks needed to find out the target, and numerical experiments are given in Sects. 3 and 4, respectively. Finally, in Sect. 5 we discuss several other related problems based on the initial one and make the concluding remarks.

## 2   The Mathematical Model

We consider the following urn system. There is the countable number of urns numbered by $i = 0, 1, \ldots$, the urn $0$ is called the first urn. An urn $i$ contains $k_i$ marbles, the total amount of marbles is $n$. Initially, all marbles are in the first urn. All marbles are white except one which is black and called the *target*. One *move* is choosing an urn $i$, drawing a marble randomly, and determining its colour. White marbles are identified correctly, while the black one can be missed with probability $q$. The examined marble is put into the next urn $i + 1$.

In this statement the problem belongs to the class of urn problems [4]. In different statements the search problem is considered by Craswell [5], Pelc [6], Marks and Zaman [7], and other researchers. Note, that in [8] we considered the more general problem statement with the cost of examination $C_{m,s}$ of a ball which depends on its number $m$ and the box number $s$.

Let us denote the sequence (distribution) of marbles by $\{k\} = \{k\}_{i=0}^{\infty} = k_0, k_1, \ldots$. Drawing a marble (without finding the target) from an urn $s$ (a *move*) changes this distribution to $k_0, \ldots, k_{s-1}, k_s - 1, k_{s+1} + 1, k_{s+2}, \ldots$ denoted by $\{k\}^s$. Note that moves can be commuted provided that both are possible: $\{k\}^{s,t} = \{k\}^{t,s}$ if $k_s > 0$ and $k_t > 0$.

Define the probability of correct identification of the target be $p = 1 - q$. Note that the probability that the target reaches without being found the urn $j$ is $q^j$. The probability that a specific urn contains the target is defined by the following Theorem 1.

**Theorem 1.** The probability that the target is in the urn $j$, provided that the distribution of marbles is $\{k\}$, is equal to

$$P_j(\{k\}) = \frac{k_j q^j}{\sum\limits_{i=0}^{\infty} k_i q^i}.$$

*Proof.* The probability that the target is in a chosen urn is the conditional probability: it should have been drawn the given number of times and all examinations should have failed. Denote by $A_j$ the event that the target has reached the urn $j$, and by $\bar{B}$ the event that the target is still not found. Then the probability that the target is in the urn $j$ is

$$(A_j \mid \bar{B}) = \frac{P(A_j \cap \bar{B})}{P(\bar{B})}. \tag{1}$$

Here the probability that the target is in the urn $j$ under the condition that the target is still not found is

$$P(A_j \cap \bar{B}) = \frac{k_j}{n} q^j. \tag{2}$$

The probability $P(\bar{B})$ that the target is still not found is defined as the sum of all the probabilities that the target is in one of the urns:

$$P(\bar{B}) = \frac{k_0}{n} + \frac{k_1}{n}q + \frac{k_2}{n}q^2 + \cdots = \sum_{i=0}^{\infty} \frac{k_i}{n}q^i = \frac{1}{n}\sum_{i=0}^{\infty} k_i q^i. \qquad (3)$$

Substituting formulae (2) and (3) into (1), we have

$$P_j(\{k\}) = P(A_j \mid \bar{B}) = \frac{k_j q^j}{\sum_{i=0}^{\infty} k_i q^i}. \qquad (4)$$

This completes the proof.                                                                    □

For early target finding one should determine a strategy as a sequence of moves until the target is found. Formal definition: having the distribution $\{k\}$ of the marbles in the urns a strategy is a rule of choosing the urn $j$. Having in mind non-zero probability of failing in identification of the target the strategy has to generate an infinite sequence of moves which breaks off when and only when the target is finally found.

Obviously each move changes the distribution of marbles in the urns $\{k\}$ and the probabilities distribution $P_j(\{k\})$. Note that the distribution $\{k\}$ of the marbles in the urns specifies all properties of the considered system at any time. Therefore the system has the Markov property, i.e. the state of the system does not depend on the previous sequence of moves and depends only on the current distribution of the marbles in the urns.

We assume that initially all the marbles are in the first urn, i.e. initial distribution is $k_0 = n$ and $k_i = 0$ for any $i > 0$.

The next section is devoted to the optimal strategy derivation and analytical prove. We also derive an explicit formula determining the expected number of checks needed to find out the target. But first let us prove that the expected number of moves is finite for at least one strategy (to be proved optimal later).

**Lemma 1.** Assume that $p > 0$. There are strategies that provide the finite expected number of moves needed to find the target.

*Proof.* Consider a strategy that checks each non-empty urn at least once in $T$ moves. Here $T$ is the same for all urns. Then the target is examined at least once in every $Tn$ moves. It can be found after each $i$-th examination with probability $pq^{i-1}$, so the number of moves is at most $Tni$. This distribution is the geometric distribution with the expectation $Tn/p$, which is finite provided that $p > 0$. This value is the estimation of the expected number of moves of the chosen strategy, which is, therefore, finite.

□

In particular, the optimal strategy provides the finite number of moves on the average. Note that there exist strategies that provide a finite expectation, though are not subject to the lemma. Also, there are strategies with infinite expectation, revealed by the next lemma.

**Lemma 2.** Assume that all marbles are initially in the first urn. Any strategy with a finite expected number of moves needed to find the target examines each urn exactly $n$ times.

*Proof.* Assume that the strategy examines an urn $i$ at most $m < n$ times. There are chances to miss the target with a positive probability: if the target is in the urn already examined $m$ times. Then the target would never be found, so the number of the moves becomes infinite with positive probability.

On the other hand, an urn can not remain non-empty after $n$ examinations because all marbles would have been already in the next urn.                                    □

For an arbitrary initial distribution $k_i$ of marbles among the urns $i = 0, 1, \ldots$, the lemma reads as follows: all urns are examined exactly $n$ times except for the first one, which is examined exactly $k_0$ times.

The strategy we are going to prove optimal examines the first non-empty urn, so that urns are emptied and excluded one by one in the ascending order.

## 3  Solution of the Optimisation Problem

First, let us derive the probability to find the target checking the urn $j$.

**Lemma 3.** The probability to find the target checking the urn $j$ is equal to

$$\hat{P}_j(\{k\}) = \frac{pq^j}{\sum_{i=0}^{\infty} k_i q^i}.$$

*Proof.* As the marbles are drawn randomly with the uniform distribution of probability, we need three independent events with known probabilities to occur simultaneously: the target must be in the urn $j$ ($P_j(\{k\})$), it must be selected from $k_j$ possibilities ($1/k_j$), and, selected, it must be recognised as the target ($p$):

$$\hat{P}_j(\{k\}) = \frac{k_j q^j}{\sum_{i=0}^{\infty} k_i q^i} \cdot \frac{1}{k_j} \cdot p = \frac{pq^j}{\sum_{i=0}^{\infty} k_i q^i}.$$

Let us consider changing the probability $\hat{P}_j(\{k\})$ after a move. Moving a marble from an urn $j$ to the next urn $j + 1$ (i.e. drawing a marble without finding the target), one remains the same numerator while the denominator is added the quantity $q^{j+1} - q^j = -pq^j$ and, therefore, strictly decreases.                                    □

The following lemma is used to derive the optimal strategy.

**Lemma 4.** The optimal strategy checks all non-empty urns: if $k_i > 0$ for some $i$, then the optimal sequence of moves for this distribution contains $i$ at least once (actually, not less than $k_i$ times).

*Proof.* Assume the contrary: $k_i > 0$ for some $i \geq 0$, and the optimal sequence of moves for this distribution does not contain the urn $i$. Then the probability $P_i(\{k\})$ that the target is in this urn is strictly positive; in this case, the target will never be found. This means that the average number of moves cannot be finite, and this completes the proof.                                                                        $\square$

Lemma 4 implies, in particular, that the sequence of moves of the optimal strategy contains exactly $k_0$ zeros, i.e., all marbles of the urn 0 must be checked, sooner or later.

Let us denote as $E(\{k\})$ the expected number of moves needed to find a target, having initial marbles distribution $\{k\}$ and using the optimal strategy. Now let us prove the following lemma.

**Lemma 5.** Leading zeros in the distribution of marbles do not change the expected number of moves needed to find the target, i.e.,

$$E(0, k_1, \dots) = E(k_1, k_2, \dots).$$

*Proof.* Note that the only available information to choose a strategy is the distribution of probabilities that the target is in an urn. Let the distribution of marbles be $k_0 = 0, k_1, k_2, \dots$. Then, for any $j > 0$:

$$P_j(0, k_1, k_2, \dots) = \frac{k_j q^j}{\sum\limits_{i=0}^{\infty} k_i q^i} = \frac{k_j q^j}{\sum\limits_{i=1}^{\infty} k_i q^i} = \frac{k_{j-1} q^{j-1}}{\sum\limits_{i=0}^{\infty} k_i q^i} = P_{j-1}(k_1, k_2, \dots).$$

This implies $E(0, k_1, \dots) = E(k_1, k_2, \dots)$.                                                $\square$

For the sake of brevity, denote

$$\Sigma = \sum_{i=0}^{\infty} k_i q^i.$$

Also, denote the expected number of moves needed to find the target, having initial marbles distribution $\{k\}$, making the first move $s$, and using after that the optimal strategy as $E^s(\{k\})$. Let us consider relations between $E(\{k\})$ and $E^s(\{k\})$.

**Lemma 6.** The quantity $E^s(\{k\})$ is expressed as follows:

$$E^s(\{k\}) = 1 + E(\{k\}^s)\frac{\Sigma - pq^s}{\Sigma}.$$

*Proof.* Checking the urn $s$ takes one move; the game is over if the target is found $(\hat{P}_s(\{k\}))$ or continues with the new distribution of marbles $\{k\}^s$ if:

- either the target was in the urn but another marble was drawn: the probability is $P_s(\{k\})\frac{k_s-1}{k_s}$;
- or the target was drawn and not recognised $(P_s(\{k\})\frac{1}{k_s}q)$;
- or the urn did not contain the target at all $(1 - P_s(\{k\}))$.

The expected number of moves is subject to the recurrent equation

$$E^s(\{k\}) = 1 + 0 \cdot \hat{P}_s(\{k\})$$

$$+ E(\{k\}^s)\left(P_s(\{k\})\frac{1}{k_s}q + P_s(\{k\})\frac{k_s-1}{k_s} + 1 - P_s(\{k\})\right)$$

$$= 1 + E(\{k\}^s)\frac{q^s(1-p) + q^s(k_s-1) + \Sigma - k_s q^s}{\Sigma}$$

$$= 1 + E(\{k\}^s)\frac{\Sigma - pq^s}{\Sigma}.$$

This completes the proof. □

Now we are ready to present the optimal strategy which is quite simple: for any distribution of marbles, check the first non-empty urn.

**Theorem 2.** The optimal strategy is $S(\{k\}) = z, k_z > 0, k_i = 0$ if $i < z$.

*Proof.* Having Lemma 5, it is sufficient to prove the statement for the case $k_0 \geq 1$.

Next, let us assume that the optimal strategy is different, so it generates the sequence of moves in which some urn $s > 0$ is checked before the first urn (the urn 0). Consider the alternative strategy, which generates sequence with these moves swapped: first we check the first urn, then the urn $s$. We are going to show that this alternative strategy is better in the sense of fewer expected moves up to finding the target, and so that the optimal strategy is to exhaust the first urn first.

Let the distribution of marbles before these two moves be $\{k\}$. Consider the expected numbers of moves for the strategies starting with moves $(0, s)$ and $(s, 0)$ (denoted by $E^{0,s}(\{k\})$ and $E^{s,0}(\{k\})$, respectively). We make these two moves using Lemma 6:

$$E^{0,s}(\{k\}) = 1 + \frac{\Sigma - p}{\Sigma}E(\{k\}^0) = 1 + \frac{\Sigma - p}{\Sigma}\left[1 + \frac{\Sigma - p - pq^s}{\Sigma - p}E(\{k\}^{0,s})\right]$$

$$= 1 + \frac{\Sigma - p}{\Sigma} + \frac{\Sigma - p - pq^s}{\Sigma}E(\{k\}^{0,s}),$$

$$E^{s,0}(\{k\}) = 1 + \frac{\Sigma - pq^s}{\Sigma}E(\{k\}^s) = 1 + \frac{\Sigma - pq^s}{\Sigma}\left[1 + \frac{\Sigma - pq^s - p}{\Sigma - pq^s}E(\{k\}^{s,0})\right]$$

$$= 1 + \frac{\Sigma - pq^s}{\Sigma} + \frac{\Sigma - pq^s - p}{\Sigma}E(\{k\}^{s,0}).$$

Comparing $E^{0,s}$ and $E^{s,0}$, we have:

$$E^{0,s} - E^{s,0} = \frac{\Sigma - p}{\Sigma} - \frac{\Sigma - pq^s}{\Sigma} < 0.$$

Therefore, we have shown that the alternative strategy is better than the optimal one, which contradicts optimality. So, the optimal strategy can never check an urn $s > 0$ before the first urn (the urn 0).

Note that we used commuting of moves and, therefore, assumed that $k_s > 0$, $k_0 > 0$. However, $k_0 > 0$ as was noted at the beginning of the proof, and $k_s > 0$ because the optimal strategy selects this urn for drawing a marble. So, changing the order of urn checks is possible.                                                                 □

**Note.** *The strategy proved optimal is optimal for any initial distribution $\{k\}$ of marbles over urns and for any error probability $q$. This is by no means intuitive, because even for poor recogniser and/or most marbles already checked, it is still optimal to proceed with the unchecked ones.*

Now let us evaluate the expected number of moves $E(\{k\})$ for two distributions of marbles: $k_0 > 0$, $k_i = 0$ for $i > 0$ and $k_0 > 0$, $k_1 > 0$, $k_i = 0$ for $i > 1$ (i.e., only the first urn is non-empty and only two first urns are non-empty). There is a reason for that: according to Theorem 2, the first non-empty urn should be checked until it becomes empty, leading empty urns are discarded. So, only these two distributions can possibly appear from the initial state $k_0 = n$, $k_i = 0$ for $i > 0$.

**Theorem 3.** The expected number of moves

$$E(n, 0, 0, \dots) = \frac{n+1}{2} + \frac{q}{p}n$$

provided that the optimal strategy is used.

*Proof.* There are two cases: either the target is obtained in the first urn, or it is missed, the first urn becomes empty, so the target is obtained during the same game, recurrently. Probability of the first case is the probability $p$ of the correct identification of the target, probability of the other case is $q = 1 - p$. The expected number of moves provided that the target is identified correctly is $(n+1)/2$. In the second case, it equals $n + E(n, 0, \dots)$: all marbles from the first urn have been tested and then the new game is started. Therefore, the optimal number of moves is subject to the equation

$$E(n, 0, 0, \dots) = p\frac{n+1}{2} + q\left(n + E(n, 0, 0, \dots)\right).$$

This can be reduced to

$$E(n, 0, 0, \dots) = \frac{n+1}{2} + \frac{q}{p}n.$$

This completes the proof.                                                     □

**Theorem 4.** The expected number of moves

$$E(k, m, 0, \dots) = \frac{kp}{k+mq} \cdot \frac{k+1}{2} + \frac{nq}{k+mq}\left(k + \frac{n+1}{2} + \frac{q}{p}n\right)$$

provided that the optimal strategy is used (here $n = k + m$, $k > 0$, $m \geq 0$).

*Proof.* Employing the same idea as in Theorem 3, we have:

$$
\begin{aligned}
E(k, m, 0, \dots) &= \frac{k}{k+mq}\left(p\frac{k+1}{2} + q\left(k + E(n, 0, 0, \dots)\right)\right) \\
&\quad + \frac{mq}{k+mq}\left(k + E(n, 0, 0, \dots)\right) \\
&= \frac{kp}{k+mq} \cdot \frac{k+1}{2} + \frac{(k+m)q}{k+mq}\left(k + E(n, 0, 0, \dots)\right) \\
&= \frac{kp}{k+mq} \cdot \frac{k+1}{2} + \frac{nq}{k+mq}\left(k + \frac{n+1}{2} + \frac{q}{p}n\right).
\end{aligned}
$$

This completes the proof.                                                     □

In the next section, we present results of numerical experiments, comparing the optimal strategy to two other strategies.

## 4 Numerical Experiments

Table 1 provides the results of numerical experiments to compare the optimal strategy with two heuristics: the "proportional" strategy and the "best probability" strategy.

The "proportional" strategy is choosing an urn randomly with the probability proportional to the number of marbles in the urn, i.e.,

$$P(\text{choose urn } j) = \frac{k_j}{\sum\limits_{i=0}^{\infty} k_i}.$$

The "best probability" strategy is choosing the urn with the highest probability of containing the target, i.e.,

$$j = \arg\max_i P_i(\{k\}).$$

To model the expected number of moves according to the strategies, we developed a simulator. The modelling results are averaged on 1 million iterations using the standard C++ pseudo-random generator `rand()`. Table 1 contains also the theoretical value of $E$ to estimate the computation error.

**Table 1** Numerical experiments. Average number of moves before finding the target, 100 marbles, $10^6$ iterations

| Strategy/p | 0.98 | 0.95 | 0.9 | 0.8 | 0.55 |
|---|---|---|---|---|---|
| Optimal theoretical | 52.54 | 55.76 | 61.61 | 75.5 | 132.32 |
| Optimal | 52.539 | 55.75993 | 61.6158 | 75.5151 | 132.344 |
| Largest probability | 53.0234 | 57.6061 | 65.3259 | 82.9889 | 150.362 |
| Proportional | 102.041 | 105.24 | 111.06 | 124.933 | 181.825 |

Note also, that all three strategies are linear with respect to the number of marbles $n$, i.e., increasing the number of marbles ten times increases the expected number of moves also ten times, for all strategies.

## 5  Conclusion

In this paper, we presented the original mathematical model of a "needle in a haystack" problem, i.e. search for an object with a recognition error probability. The problem is directly connected to high-throughput computing using Desktop Grids (volunteer computing, in particular).

The paper presents the optimal strategy of searching for the target. The strategy has the trivial form (which is to check the first non-empty urn), but it is not intuitively clear. In the case of large number of marbles one should continue to search for the target in the urn even if the probability that this urn contains the target is vanishingly low. In this paper, we analytically prove the optimality of this strategy and give explicit formulae for the expected number of moves needed to find the target.

As for Desktop Grids, the important result is that no replication mechanism (a conventional mechanism of Desktop Grids) should be used while solving "needle in a haystack" problem because this contradicts the optimal strategy.

## References

1. Ivashko E, Chernov IA, Nikitina N (2018) A survey of desktop grid scheduling. IEEE Trans Parallel Distrib Syst 1 (2018). https://doi.org/10.1109/TPDS.2018.2850004
2. Rumyantsev A, Chakravarthy S, Morozov E, Remnev S (2018) Cost and effect of replication and quorum in desktop grid computing. In: Dudin A, Nazarov A, Moiseev A (eds) Information technologies and mathematical modelling. Queueing theory and applications. Springer, Cham, pp 143–156
3. Ivashko E, Chernov I (2021) Search for an object with a recognition error probability in a desktop grid environment. In: 2021 Ivannikov ISP RAS open conference (ISPRAS)

4. Johnson NL, Kotz S (1977) Urn models and their application: an approach to modern discrete probability theory. Wiley series in probability and mathematical statistics. Wiley, New-York
5. Craswell KJ (1973) How to find a needle in a haystack. Two-Year College Math J 4(3):18–22
6. Pelc A (1989) Searching with known error probability. Theoret Comput Sci 63:185–202
7. Marks C, Zaman T (2016) A multi-urn model for network search. arxiv e-prints
8. Chernov I, Ivashko E (2020) Risky search with increasing complexity by a desktop grid. In: Voevodin V, Sobolev S (eds) Supercomputing. Springer, Cham, pp 622–633

# Numerical Methods for Solving the Robin Boundary Value Problem for a Generalized Diffusion Equation with a Non-smooth Solution

Nikki Kedia, Anatoly A. Alikhanov, and Vineet Kumar Singh

**Abstract** Solutions of Robin boundary value problem for a generalized diffusion equation with a non-smooth solution are studied. The Caputo derivative in the generalized sense has been discretized by using a difference scheme of order $(2 - \alpha)$ on a non-uniform mesh with $0 < \alpha < 1$ in the temporal direction. Test example shows how the grading of the mesh is essential for non-smooth solution and using such kind of mesh generate stronger results.

**Keywords** Generalized L1 scheme · Generalized Fractional derivative ·
Non-uniform mesh

## 1 Introduction

Fractional Calculus has gained considerable attention of researchers lately. The reason being so many effective applications of fractional calculation in various fields of science and engineering [1–4]. The time-fractional diffusion equation is a kind of linear integro-differential equation. Since, it is not always possible to find the analytical solution of such equations, hence the application of numerical methods come into the picture. In this manuscript a generalized time-fractional diffusion equation is considered which takes into account the memory of the procedure which was first described by Boltzmann in 1874 and 1876 [5, 6]. The meaning of memory is that the outcome of any process will depend not only on the present time but also on the previous time layers. These are described by memory functions. In fractional calculus, these functions represent the kernel of an integro-differential operator known as power-law memory.

N. Kedia (✉) · V. K. Singh
Department of Mathematical Sciences, Indian Institute of Technology (Banaras Hindu University), Varanasi, India
e-mail: nikki.kedia.rs.mat18@itbhu.ac.in

A. A. Alikhanov
North-Caucasus Center for Mathematical Research, North-Caucasus Federal University, Stavropol, Russia

© The Author(s), under exclusive license to Springer Nature Switzerland AG 2022
A. Tchernykh et al. (eds.), *Mathematics and its Applications in New Computer Systems*,
Lecture Notes in Networks and Systems 424,
https://doi.org/10.1007/978-3-030-97020-8_20

In the rectangle $Q = \{(x,t) : 0 \le x \le 1, 0 \le t \le T\}$ consider the following Robin boundary value problem:

$$\partial_{0,t}^{\alpha,\psi(t)} u = \frac{\partial}{\partial x}\left(k(x,t)\frac{\partial u}{\partial x}\right) - q(x,t)u + f(x,t), \quad x \in (0,1), \quad t \in (0,T]. \quad (1)$$

$$\begin{aligned}
k(0,t)u_x(0,t) &= \beta_1(t)u(0,t) - \mu_1(t), \\
-k(1,t)u_x(1,t) &= \beta_2(t)u(1,t) - \mu_2(t),
\end{aligned} \quad (2)$$

$$u(x,0) = u_0(x). \quad (3)$$

where,

$$\partial_{0,t}^{\alpha,\psi(t)} u(x,t) = \frac{1}{\Gamma(1-\alpha)} \int_0^t \frac{\psi(t-\xi)}{(t-\xi)^\alpha} \frac{\partial u}{\partial \xi}(x,\xi)d\xi, \quad (4)$$

is the generalized Caputo fractional derivative of order $\alpha$, $0 < \alpha < 1$ where $\psi(t)$ is the weight function with $\psi(t) \in C^2[0,T]$, $\psi(t) > 0$ and $\psi'(t) \le 0$ for all $t \in [0,T]$, $0 < c_1 \le k(x,t) \le c_2$ and $q(x,t) \ge 0$ for all $(x,t) \in Q$. One of the most widely used difference approximation of the Caputo fractional derivative (4) is the $L1$ method [1, 7]. For Reimann Liouville fractional derivative the L1 method was developed by Langlands and Henry [8] and for the Caputo fractional derivative the $L1$ method was devised by Lin and Xu [9]. There has been a huge amount of research done into the L1 approach for solving the subdiffusion problem. However, with the exception of a few, the solution was presumed to be sufficiently smooth in the majority of those works. The occurrence of singularity in the derivatives of the solution $u(x,t)$ at $t = 0$ as indicated in [10, 11] might cause the solution to be non-smooth in a closed interval even for a homogeneous problem with smooth initial data. Stynes et al. established in [12] that the L1 scheme's rate of convergence on a uniform mesh was $\mathcal{O}(N^{-\alpha})$ and on a non-uniform mesh was $\mathcal{O}(N^{\alpha-2})$. Huang and Stynes have published a number of recent papers based on this technique [13–15]. Alikhanov invented the L1 technique for generalized Caputo fractional derivative on the uniform grid for smooth and non-smooth solutions in [16, 17], by separating the problem into two parts: one smooth but unknown, and the other known but non-smooth. When utilising this method for non-smooth solutions, a convergence order of $2 - \alpha$ is achieved at the final time, and one in the domain. It achieved a convergence order of $2 - \alpha$ for smooth solutions. In [18] some fast and parallel numerical methods were considered for accelerating the implementation of numerical schemes on Caputo fractional derivative in the generalized sense. Also, in [19] the L1 scheme was developed for multiterm generalized Caputo fractional derivative on uniform mesh for smooth solutions.

The manuscript is arranged as: Sect. 2 presents the approximation of the generalized Caputo derivative by L1 formula and also a difference scheme is developed. Stability and convergence of the scheme for non-smooth solution has been established in Sect. 3. Numerical experiments for the test example are performed in Sect. 4 which validates the theoretical results. Some final observations are made in Sect. 5.

## 2 Approximation of the Generalized Caputo Derivative by L1 Formula and a Difference Scheme

### 2.1 Derivation of Formula

Let $u(x, t)$ be the exact solution of the problem (1)–(3). The interval $[0, T]$ is divided into sub-intervals with $0 = t_0 < t_1 \ldots < t_N = T$ and $\tau_n = t_n - t_{n-1}$, $1 \leq n \leq N$ where $\tau_n$ is the time step size. The formula for the approximation of the Caputo derivative $\partial_{0,t_n}^{\alpha, \psi(t)} v(t)$ with $(0 < \alpha < 1,\ \psi(t) > 0,\ \psi'(t) \leq 0,\ \psi(t) \in C^2[0, T])$ is given as:

$$\partial_{0,t_n}^{\alpha, \psi(t)} v(t) \approx \Delta_{0,t_n}^{\alpha, \psi(t)} v = \sum_{s=0}^{n-1} c_s^n (v(t_{s+1}) - v(t_s)), \tag{5}$$

where,

$$c_s^n = \frac{1}{\Gamma(2 - \alpha)} \left[ \psi(t_n - t_s - \frac{\tau_{s+1}}{2}) a_s^n + (\psi(t_n - t_{s+1}) - \psi(t_n - t_s)) b_s^n \right],$$

with

$$a_s^n = \frac{(t_n - t_s)^{1-\alpha} - (t_n - t_{s+1})^{1-\alpha}}{\tau_{s+1}},$$

$$b_s^n = \frac{1}{\tau_{s+1}^2} \left( \frac{1}{2 - \alpha} [(t_n - t_s)^{2-\alpha} - (t_n - t_{s+1})^{2-\alpha}] - \frac{\tau_{s+1}}{2} [(t_n - t_s)^{1-\alpha} + (t_n - t_{s+1})^{1-\alpha}] \right).$$

The details of the derivation can be found in the paper [20].

### 2.2 Difference Scheme

In the rectangle $Q$ consider the mesh as $v_{h\tau} = v_h \times v_\tau$, with

$$v_h = \{x_i = ih : i = 0, 1, \ldots, M, hM = l\} \text{ and } v_\tau = \{t_n : 0 = t_0 < t_1 < t_2 < \ldots < t_{N-1} < t_N = T\}.$$

the assigned difference scheme to the model (1)–(3) is:

$$\Delta_{0,t_n}^{\alpha, \psi(t)} \varsigma_i = \wedge \varsigma_i^n + \phi_i^n, \quad n = 1, \ldots, N, \ i = 0, 1, 2, \ldots, M, \tag{6}$$

$$\varsigma(x, 0) = u_0(x) \quad i = 0, 1, 2, \ldots, M, \tag{7}$$

where

$$\wedge \varsigma = \begin{cases} \frac{2}{h}(a_1 \varsigma_x - \kappa_1 \varsigma), & i = 0, \\ (a \varsigma_{\bar{x}})_x - d\varsigma, & 1 \le i \le M - 1, \\ -\frac{2}{h}(a_M \varsigma_{\bar{x}} + \kappa_2 \varsigma), & i = M, \end{cases}$$

and

$$\phi = \begin{cases} \frac{2}{h}\bar{\mu}_1, & i = 0, \\ f, & 1 \le i \le M - 1, \\ \frac{2}{h}\bar{\mu}_2, & i = M. \end{cases}$$

with $\quad ((a\varsigma_{\bar{x}})_x - d\varsigma)_i = \dfrac{a_{i+1}\varsigma_{i+1} - (a_{i+1} + a_i)\varsigma_i + a_i\varsigma_{i-1}}{h^2} - d_i\varsigma_i, \quad i = 1, \ldots,$ $M - 1,$

$$\varsigma_{\bar{x},i} = \frac{\varsigma_i - \varsigma_{i-1}}{h}, \quad \varsigma_{x,i} = \frac{\varsigma_{i+1} - \varsigma_i}{h}, \quad a_i^n = k(x_{i-1/2}, t_n), \quad d_i^n = q(x_i, t_n),$$

$$\kappa_1 = \beta_1 + 0.5hd_0, \quad \kappa_2 = \beta_2 + 0.5hd_M, \quad \bar{\mu}_1 = \mu_1 + 0.5hf_0, \quad \bar{\mu}_2 = \mu_2 + 0.5hf_M.$$

The non-uniform mesh used here for the case of non-smooth solution is defined by $t_n = T(n/N)^r$ for $n = 0, 1, \ldots, N$, where the constant mesh grading is being chosen by the user. When $r = 1$, the mesh is uniform. Also, $\tau_n = t_n - t_{n-1}$ for $n = 1, 2, \ldots, N$.

## 3 Stability and Convergence

In this section we prove that the difference scheme has the order of approximation as $\mathcal{O}(N^{\alpha-2} + h^2)$.

**Lemma 3.1.** [20–22] For $n \in \mathbb{N}$, $\{a_s^n | 0 \le s \le n - 1\}$ and $\alpha \in (0, 1)$, the following result holds:

$$a_{n-1}^n > a_{n-2}^n > \ldots > a_s^n > a_{s-1}^n > \ldots a_0^n > \frac{1 - \alpha}{(t_n - t_0)^\alpha}$$

and

$$b_{n-1}^n > b_{n-2}^n > \ldots > b_s^n > b_{s-1}^n > \ldots > b_0^n > 0.$$

**Corollary 3.1.** [20] For $n \in \mathbb{N}$, $\{c_s^n | 0 \le s \le n - 1\}$, $\alpha \in (0, 1)$ and $\psi(t) \in C^2[0, T]$, where $\psi(t) > 0$, $\psi'(t) \le 0 \; \forall t \in [0, T]$, the following results hold:

$$c^n_{n-1} > c^n_{n-2} > \ldots > c^n_s > c^n_{s-1} > \ldots c^n_0 > \frac{\psi(t_n - t_0 - \frac{\tau_1}{2})}{\Gamma(1-\alpha)(t_n - t_0)^\alpha} > \frac{\psi(T)}{\Gamma(1-\alpha)T^\alpha}.$$

**Proof.** For the proof of this Theorem please refer to the paper [20].

**Theorem 1.** The difference scheme (6)–(7) is unconditionally stable and its solution satisfies the following inequality:

$$||\varsigma^n||_0^2 \leq ||\varsigma^0||_0^2 + \frac{\Gamma(1-\alpha)T^\alpha}{\psi(T)\epsilon_0} \max_{0 \leq n \leq N-1} \left( ||f^{n+1}||_0^2 + \bar{\mu}_1^2(t_{n+1}) + \bar{\mu}_2^2(t_{n+1}) \right). \quad (8)$$

where $[\varsigma, v] = \sum\limits_{i=1}^{M-1} \varsigma_i v_i h + 0.5\varsigma_0 v_0 h + 0.5\varsigma_M v_M h$, $||\varsigma||_0^2 = [\varsigma, \varsigma]$ and $\epsilon_0 = \dfrac{min(c_1, \kappa_1, \kappa_2)}{2}$.

**Proof.** Taking the inner product of (6) with $\varsigma^n$, we get

$$[\varsigma^n, \Delta^{\alpha, \psi(t)}_{0, t_n} \varsigma] - [\wedge \varsigma^n, \varsigma^n] = [\varsigma^n, \phi], \quad (9)$$

Since,

$$[\varsigma^n, \Delta^{\alpha, \psi(t)}_{0, t_n} \varsigma] \geq \frac{1}{2} \Delta^{\alpha, \psi(t)}_{0, t_n} ||\varsigma||_0^2,$$

$$[\varsigma^n, \wedge \varsigma^n] = 0.5h\varsigma_0(\wedge \varsigma)_{i=0} + \sum_{i=1}^{M-1}(\varsigma_i(\wedge \varsigma)_i)h + 0.5h\varsigma_M(\wedge \varsigma)_{i=M}$$

Putting the value of $\wedge \varsigma$ for $i = 0$ and $M$, we get

$$[\varsigma^n, \wedge \varsigma^n] = \varsigma_0 a_1 \varsigma_{x,0} - \kappa_1(\varsigma_0)^2 + \sum_{i=1}^{M-1}(\varsigma_i(\wedge \varsigma)_i)h - \varsigma_M a_M \varsigma_{\bar{x},M} - \kappa_2 \varsigma_M^2$$

$$\sum_{i=1}^{M-1}(\varsigma_i(\wedge \varsigma)_i)h = \sum_{i=1}^{M-1} \varsigma_i(a\varsigma_{\bar{x}})_{x,i}h - \sum_{i=1}^{M-1} d_i \varsigma_i^2 h$$

$$\sum_{i=1}^{M-1} \varsigma_i (a\varsigma_{\bar{x}})_{x,i} h = \sum_{i=1}^{M-1} \varsigma_i a_{i+1} \varsigma_{\bar{x},i+1} - \sum_{i=1}^{M-1} \varsigma_i a_i \varsigma_{\bar{x},i}$$

$$= \sum_{i=0}^{M-1} \varsigma_i a_{i+1} \varsigma_{\bar{x},i+1} - \varsigma_0 a_1 \varsigma_{\bar{x},1} - \sum_{i=1}^{M} \varsigma_i a_i \varsigma_{\bar{x},i} + \varsigma_M a_M \varsigma_{\bar{x},M}$$

$$= \sum_{i=1}^{M} \varsigma_{i-1} a_i \varsigma_{\bar{x},i} - \varsigma_0 a_1 \varsigma_{\bar{x},i} - \sum_{i=1}^{M} \varsigma_i a_i \varsigma_{\bar{x},i} + \varsigma_M a_M \varsigma_{\bar{x},M}$$

$$= -\sum_{i=1}^{M} a_i (\varsigma_{\bar{x},i})^2 h - \varsigma_0 a_1 \varsigma_{\bar{x},i} + \varsigma_M a_M \varsigma_{\bar{x},M}$$

Putting the value of $\sum_{i=1}^{M-1} (\varsigma_i (\wedge\varsigma)_i) h$ in $[\varsigma^n, \wedge\varsigma^n]$ one arrives at

$$-[\varsigma^n, \wedge\varsigma^n] \geq c_1 \|\varsigma_{\bar{x}}\|_1^2 + \kappa_1 (\varsigma_0^n)^2 + \kappa_2 (\varsigma_M^n)^2$$

where $\|\varsigma_{\bar{x}}\|_1^2 = \sum_{i=1}^{M} (\varsigma_{\bar{x},i})^2 h$. Again,

$$[\varsigma^n, \phi] = 0.5\varsigma_0 \phi_0 h + \sum_{i=1}^{M-1} \varsigma_i \phi_i h + 0.5\varsigma_M \phi_M h = \varsigma_0 \bar{\mu}_1 + (\varsigma, f) + \varsigma_M \bar{\mu}_2$$

$$\leq \epsilon (\varsigma_0^n)^2 + \frac{1}{4\epsilon} \bar{\mu}_1^2 + \epsilon \|\varsigma\|_0^2 + \frac{1}{4\epsilon} \|f\|_0^2 + \epsilon (\varsigma_M^n)^2 + \frac{1}{4\epsilon} \bar{\mu}_2^2$$

Now, using the values of $[\varsigma^n, \wedge\varsigma^n]$ and $[\varsigma^n, \phi]$ in (9) we get

$$\frac{1}{2} \Delta_{0,t_n}^{\alpha,\psi(t)} \|\varsigma\|_0^2 + c_1 \|\varsigma_{\bar{x}}\|_1^2 + \kappa_1 (\varsigma_0^n)^2 + \kappa_2 (\varsigma_M^n)^2 \leq \epsilon \|\varsigma\|_0^2 + \epsilon (\varsigma_0^n)^2 + \epsilon (\varsigma_M^n)^2 + \frac{1}{4\epsilon} \left( \|f\|_0^2 + \bar{\mu}_1^2 + \bar{\mu}_2^2 \right),$$

Also, using the inequality

$$\|\varsigma\|_0^2 \leq \|\varsigma_{\bar{x}}\|_1^2 + \varsigma_0^2 + \varsigma_M^2$$

and taking $2\epsilon_0 = min(c_1, \kappa_1, \kappa_2)$ we get

$$\Delta_{0,t_n}^{\alpha,\psi(t)} \|\varsigma\|_0^2 \leq \frac{1}{\epsilon_0} \left( \|f\|_0^2 + \bar{\mu}_1^2 + \bar{\mu}_2^2 \right). \tag{10}$$

The inequality (10) can be written in the form

$$c_{n-1}^n \|\varsigma^n\|_0^2 \leq \sum_{s=1}^{n-1} (c_s^n - c_{s-1}^n) \|\varsigma^s\|_0^2 + c_0^n \|\varsigma^0\|_0^2 + \frac{1}{\epsilon_0} \left( \|f\|_0^2 + \bar{\mu}_1^2 + \bar{\mu}_2^2 \right),$$

From Corollary 3.1 we have $c_0^n \geq \dfrac{\psi(T)}{\Gamma(1-\alpha)T^\alpha} = m_2$ (let), we get

$$c_{n-1}^n \|\varsigma^n\|_0^2 \leq \sum_{s=1}^{n-1}(c_s^n - c_{s-1}^n)\|\varsigma^s\|_0^2 + c_0^n \left( \|\varsigma^0\|_0^2 + \frac{1}{m_2\epsilon_0}\left( \|f\|_0^2 + \bar{\mu}_1^2 + \bar{\mu}_2^2 \right) \right),$$

$$(11)$$

Denote $E = \|\varsigma^0\|_0^2 + \dfrac{1}{m_2\epsilon_0} \max_{0\leq n\leq N-1} \left( \|f\|_0^2 + \bar{\mu}_1^2 + \bar{\mu}_2^2 \right)$ The inequality (11) is reduced to

$$c_{n-1}^n \|\varsigma^n\|_0^2 \leq \sum_{s=1}^{n-1}(c_s^n - c_{s-1}^n)\|\varsigma^s\|_0^2 + c_0^n E \tag{12}$$

It is obvious that at $n = 1$ we have

$$\|\varsigma^n\|_0^2 \leq \|\varsigma^0\|_0^2 + \frac{1}{m_2\epsilon_0} \max_{0\leq n\leq N-1} \left( \|f^{n+1}\|_0^2 + \bar{\mu}_1^2(t_{n+1}) + \bar{\mu}_2^2(t_{n+1}) \right), \tag{13}$$

Let us prove (13) for $n = 2, 3, \ldots$ by using mathematical induction method. So, let us assume (13) is true for all $n = 1, 2, 3, \ldots, k$;

$$\|\varsigma^n\|_0^2 \leq E, \quad n = 1, 2, \ldots, k.$$

From (12) at $n = k + 1$ one has

$$c_k^{k+1}\|\varsigma^n\|_0^2 \leq \sum_{s=1}^{k}(c_s^{k+1} - c_{s-1}^{k+1})\|\varsigma^s\|_0^2 + c_0^{k+1} E$$

$$\leq \sum_{s=1}^{k}(c_s^{k+1} - c_{s-1}^{k+1})E + c_0^{k+1} E = c_k^{k+1} E$$

Hence, we prove (13) for all n.

As a result, using the value of $m_2$ one arrives at Eq. (8). Hence, our theorem is proved.

From the a priori estimate (8) it follows that the solution of the difference scheme (6)–(7) converges to the solution of the differential problem (1)–(3) with the rate equal to the approximation error order $\mathcal{O}(N^{\alpha-2} + h^2)$.

## 4 Numerical Results

In this section, the maximum error and Convergence Order (CO) for the domain have been calculated using the following norm $||.||_{Q(v_{h\tau})}$, where $||w||_{Q(v_{h\tau})} = \max\limits_{(x_i,t_j)\in v_{h\tau}} |w|$ and $w = \varsigma - u$ is the error, where $\varsigma$ is the approximate solution. In time, CO$=$ $\log_{\frac{N1}{N2}} \frac{||w_1||}{||w_2||}$ and in space, CO$= \log_{\frac{h1}{h2}} \frac{||w_1||}{||w_2||}$. $|| \cdot ||_0$ is the definition of the $L_2$-norm.

**Example 4.1.** We examine a test problem with non-smooth solution. Let $u(x, t) = t^\beta \sin(\pi x)e^{-bt}$ be the exact solution of (1)–(3) with $\psi(t) = e^{-5t}$, $0 < \beta < 1$, the coefficients $q(x, t) = 1 - sin(xt)$, $k(x, t) = 2 - \cos(xt)$, $T = 1$, $\mu_1(t) = -\pi t^\beta e^{-bt}$, $\mu_2(t) = \pi t^\beta e^{-bt}(\cos(t) - 2)$, $\beta_1(t) = 2 - sin(t)$ and $\beta_2(t) = 2 - cos(3t)$.

The numerical results of this example are shown and illustrated below.

- Due to the presence of a singularity in the first derivative of the solution at $t = 0$, the exact solution is non-smooth. As can be seen in Table 1, a uniform mesh yields poor outcome in this scenario.
- For different values of $\alpha$ and $\beta$ with fixed $h = 1/1000$ in the temporal direction, maximum norm error and CO estimated in the domain are provided in Table 1.
- The results for uniform and non-uniform meshes are compared, revealing that the non-uniform grid produces CO about $\mathcal{O}(N^{\alpha-2})$, confirming the theoretical findings.

**Table 1** For $h = 1/1000$, CO and maximum norm error calculated in the domain for Example 4.1

| $\alpha$ | $\beta$ | N | Uniform | | Non-uniform | |
|---|---|---|---|---|---|---|
| | | | $||w||_{Q(v_{h\tau})}$ | CO | $||w||_{Q(v_{h\tau})}$ | CO |
| 0.3 | 0.3 | 40 | 2.23e-02 | – | 7.74e-04 | – |
| | | 80 | 1.93e-02 | 0.2133 | 2.76e-04 | 1.4891 |
| | | 160 | 1.69e-02 | 0.1885 | 9.39e-05 | 1.5547 |
| | | 320 | 1.49e-02 | 0.1779 | 3.12e-05 | 1.5885 |
| 0.5 | 0.5 | 40 | 1.85e-02 | – | 1.30e-03 | – |
| | | 80 | 1.43e-02 | 0.3787 | 5.42e-04 | 1.2866 |
| | | 160 | 1.10e-02 | 0.3761 | 2.12e-04 | 1.3524 |
| | | 320 | 8.40e-03 | 0.3818 | 8.02e-05 | 1.4047 |
| 0.7 | 0.7 | 40 | 1.10e-02 | – | 2.10e-03 | – |
| | | 80 | 7.00e-03 | 0.6500 | 9.61e-04 | 1.1500 |
| | | 160 | 4.50e-03 | 0.6932 | 4.35e-04 | 1.1442 |
| | | 320 | 2.90e-03 | 0.6390 | 1.93e-04 | 1.1719 |
| 0.9 | 0.9 | 40 | 6.00e-03 | – | 4.00e-03 | – |
| | | 80 | 3.10e-03 | 0.9324 | 2.00e-03 | 1.0045 |
| | | 160 | 1.60e-03 | 0.9597 | 9.71e-04 | 1.0315 |
| | | 320 | 8.21e-04 | 0.9664 | 4.70e-04 | 1.0492 |

**Table 2**  For $N = 5000$, $L_2$-norm, maximum norm error and CO in space calculated in the domain.

| $\alpha$ | $\beta$ | h | $\max\limits_{0 \leq n \leq N} \|w^n\|_0$ | CO in $\|\cdot\|_0$ | $\|w\|_{Q(v_{h\tau})}$ | CO in $\|\cdot\|_{Q(v_{h\tau})}$ |
|---|---|---|---|---|---|---|
| 0.3 | 0.3 | 1/10 | 1.50e-03 | – | 2.10e-03 | – |
|  |  | 1/20 | 3.74e-04 | 2.0030 | 5.28e-04 | 2.0030 |
|  |  | 1/40 | 9.35e-05 | 1.9977 | 1.32e-04 | 1.9977 |
| 0.5 | 0.5 | 1/10 | 9.00e-04 | – | 1.30e-03 | – |
|  |  | 1/20 | 2.24e-04 | 2.0049 | 3.17e-04 | 2.0048 |
|  |  | 1/40 | 5.59e-05 | 2.0052 | 7.90e-05 | 2.0050 |
| 0.7 | 0.7 | 1/10 | 5.89e-04 | – | 8.33e-04 | – |
|  |  | 1/20 | 1.46e-04 | 2.0129 | 2.06e-04 | 2.0121 |
|  |  | 1/40 | 3.57e-05 | 2.0324 | 5.06e-05 | 2.0309 |
| 0.9 | 0.9 | 1/10 | 4.01e-04 | – | 5.70e-04 | – |
|  |  | 1/20 | 9.51e-05 | 2.0777 | 1.35e-04 | 2.0723 |
|  |  | 1/40 | 2.00e-05 | 2.2446 | 2.88e-05 | 2.2301 |

- Thus, from Table 1 it is clear that uniform mesh does not work for non-smooth solutions.
- Table 2 shows the maximum norm error and CO in space for Example 4.1 when $N = 5000$ is fixed and $h = 1/M$ is varied.

## 5  Conclusion

A scheme for a generalised diffusion problem with a non-smooth solution and Robin boundary conditions is presented in this article. The major goal of this study is to demonstrate that a non-uniform mesh is needed to handle the singularity in the derivative of a solution at $t = 0$. It is clear that a non-uniform grid achieves significantly better outcome than a uniform mesh. For the non-smooth solution with respect to the $L_2$-norm, the stability and CO of $\mathcal{O}(N^{\alpha-2})$ has been devised. A second order of convergence has been developed in the spatial direction. Theoretical results are validated by numerical test example.

**Acknowledgements**  The first author sincerely thanks the financial support from the Ministry of Education, Govt. of India, under Junior Research Fellow (JRF) scheme. The study of the second author was supported by the Russian Science Foundation grant No. 22-21-00363, https://rscf.ru/project/22-21-00363/.

# References

1. Oldham K, Spanier J (1974) The fractional calculus theory and applications of differentiation and integration to arbitrary order. Elsevier
2. Podlubny I, Chechkin A, Skovranek T, Chen Y, Jara BMV (2009) Matrix approach to discrete fractional calculus II: partial fractional differential equations. J Comput Phys 228(8):3137–3153
3. Hilfer R (2000) Applications of fractional calculus in physics. World Scientific
4. Kilbas A (2006) Theory and applications of fractional differential equations
5. Boltzmann L (1878) Zur theorie der elastischen nachwirkung. Ann Phys 241(11):430–432
6. Boltzmann L (2012) Theory of elastic aftereffect [zur theorie der elastischen nachwirkung], vol 1. Cambridge University Press, Cambridge, pp 616–644
7. Zhang Y-N, Sun Z-Z, Liao H-L (2014) Finite difference methods for the time fractional diffusion equation on non-uniform meshes. J Comput Phys 265:195–210
8. Langlands T, Henry BI (2005) The accuracy and stability of an implicit solution method for the fractional diffusion equation. J Comput Phys 205(2):719–736
9. Lin Y, Xu C (2007) Finite difference/spectral approximations for the time-fractional diffusion equation. J Comput Phys 225(2):1533–1552
10. Luchko Y (2012) Initial-boundary-value problems for the one-dimensional time-fractional diffusion equation. Fract Calc Appl Anal 15(1):141–160
11. Sakamoto K, Yamamoto M (2011) Initial value/boundary value problems for fractional diffusion-wave equations and applications to some inverse problems. J Math Anal Appl 382(1):426–447
12. Stynes M, O'Riordan E, Gracia JL (2017) Error analysis of a finite difference method on graded meshes for a time-fractional diffusion equation. SIAM J Numer Anal 55(2):1057–1079
13. Huang C, Stynes M, Chen H (2021) An $\alpha$-robust finite element method for a multi-term time-fractional diffusion problem. J Comput Appl Math 389:113334
14. Huang C, Stynes M (2020) $\alpha$-robust error analysis of a mixed finite element method for a time-fractional biharmonic equation. Numer Algorithms 87:1–18
15. Huang C, Stynes M (2020) Superconvergence of a finite element method for the multi-term time-fractional diffusion problem. J Sci Comput 82(1):1–17
16. Alikhanov AA (2017) A time-fractional diffusion equation with generalized memory kernel in differential and difference settings with smooth solutions. Comput Methods Appl Math 17(4):647–660
17. Alikhanov AA (2017) A difference method for solving the Steklov nonlocal boundary value problem of second kind for the time-fractional diffusion equation. Comput Methods Appl Math 17(1):1–16
18. Gu X-M, Huang T-Z, Zhao Y-L, Lyu P, Carpentieri B (2021) A fast implicit difference scheme for solving the generalized time-space fractional diffusion equations with variable coefficients. Numer Methods Partial Differ Equ 37(2):1136–1162
19. Khibiev AK (2019) Stability and convergence of difference schemes for the multi-term time-fractional diffusion equation with generalized memory kernels. J Samara State Tech Univ Ser Phys Math Sci 23(3):582–597
20. Kedia N, Alikhanov AA, Singh VK (2021) Stable numerical schemes for time-fractional diffusion equation with generalized memory kernel. Appl Numer Math 172:546–565
21. Soori Z, Aminataei A (2019) A new approximation to Caputo-type fractional diffusion and advection equations on non-uniform meshes. Appl Numer Math 144:21–41
22. Gao G-H, Sun Z-Z, Zhang H-W (2014) A new fractional numerical differentiation formula to approximate the Caputo fractional derivative and its applications. J Comput Phys 259:33–50

# Asymptotic of Calculation of Hydro-Mechanical Fields of a Marine Mobile Object by the Finite Volume Method

**Svetlana Osmukha**⬤ **and Gennady Zelenkov**⬤

**Abstract** The article describes the study of methods for describing hydrodynamic coefficients using finite volumes, as well as numerical methods based on the study of data from the aerotube experiment. The methods of converting the data of air tube experiments when working out the elements of the fixed assets of the fleet in the bases of orthogonal functions (bursts) investigated and described. Based on the obtained mathematical models of processes using the methods of multi-turn engineering design, methods for constructing computer models in a structure that provides for an open process of developing and approving project elements proposed. An urgent task is to develop methods for reducing costs associated with the design and creation of marine infrastructure facilities and fixed working capital. The relevance of the tasks follows from the expensive cycles of project development, the labor costs of highly qualified specialists.

**Keywords** Bernstein's polynomial · Coordinate transformation · Numerical simulation

## 1 Introduction

Liquid friction related to the parameters of the geometry of marine moving objects. Depending on the indicators of the purpose of the profiles streamlined by the flow, they can have a variety of configurations, in some cases; on the contrary, poorly streamlined forms performed, in order to obtain static stability of marine mobile objects. If we consider a finite volume—a conditional parallelepiped, which we replace with a computational model, we also assume that this volume occupies the same volume of liquid.

The aim of the research is to develop methodological foundations of digital data transformation platforms in the tasks of structural-parametric synthesis of multi-turn engineering design in the field of complex theory of water transport [7].

S. Osmukha (✉) · G. Zelenkov
Admiral F. F. Ushakov Maritime State University, Novorossiysk, Russia
e-mail: rusalsvetik@mail.ru

© The Author(s), under exclusive license to Springer Nature Switzerland AG 2022    229
A. Tchernykh et al. (eds.), *Mathematics and its Applications in New Computer Systems*,
Lecture Notes in Networks and Systems 424,
https://doi.org/10.1007/978-3-030-97020-8_21

The solution of the scientific and technical problem of digital transformation of the production system, reconstruction and modernization of the fixed assets of the fleet is based on theoretical research, including the stages of: aerotube experiments; descriptions by methods of structural-parametric synthesis of a mathematical model; creation of a basic digital model in a mathematical orthogonal basis; study of the properties of models; stability and transformation of the model by means of compression mapping technology in the space of bursts.

Aerotube experiment—aims to obtain sets of hydrodynamic coefficients in the conditions of structural selection and model refinement through multi-turn engineering design for the tasks of production, operation and modernization potential of a marine mobile object [10].

## 2 Materials and Methods

### 2.1 A Subsection Sample

For any set of finite volumes of a conditional parallelepiped, the frontal part can distinguished; more often, it takes a spherical shape. Models of finite volume components of a complex parallelepiped shown in Fig. 1.

The calculation of coordinates performed in the global coordinate system and each of the associated with each finite volume—local, or connected.

Global system—$O(x, y, z)$, any related $O'_n(x_n, y_n, z_n)$.

The basis system: $O = \begin{vmatrix} i & j & k \\ x_m^n & y_m^n & z_m^n \\ x_{m+1}^{n+1} & y_{m+1}^{n+1} & z_{m+1}^{n+1} \end{vmatrix}$.

The main vector of the set;

$$R = a_0 + a_1 x + a_2 y + a_3 z; \tag{1}$$

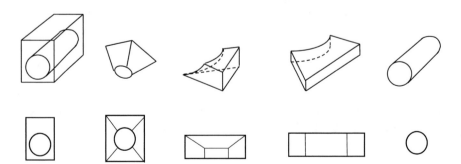

**Fig. 1** Representations of finite volumes and their interpretation

recalculation of coordinates to another system:

$$R' = a_1 x_1' + a_2 y_1' + a_3 z_1' - b;$$ (2)

where is every linear combination:

$$x_i^j, y_i^j, z_i^j, \text{индексы } i = 1 \cdots 3 \text{ и} = 1 \cdots 3, \forall\{i, j\}, i \neq j;$$

$$b = a_0(a_1, a_2, a_3).$$ (3)

Formulas for converting coordinates to curved ones known [1]:
–   the equation of the sphere:

$$\xi = \sqrt{x^2 + y^2 + z^2}; \eta = -\frac{y}{x}; \varsigma = -\frac{z}{x};$$ (4)

where $-1 \leq \eta \leq 1; -1 \leq \varsigma \leq 1;$
–   the prism equation;

$$\xi = (x^2 + y^2 + z^2)/\sqrt{x^2 + z^2}; \eta = 2y/(x^2 + y^2 + z^2); \varsigma = -\frac{z}{x};$$ (5)

where $0 \leq \eta \leq \frac{1}{r}; -1 \leq \varsigma \leq 1; r -$ radiusofthesphere;
–   the equation of the circle:

$$\xi = \sqrt{x^2 + z^2}; \eta = y; \varsigma = -\frac{z}{x};$$ (6)

where $-1 \leq \varsigma \leq 1;$
–   cylinder equation:

$$\xi = x; \eta = \sqrt{y^2 + z^2}; \varsigma = arctg(z/y) + k\pi;$$ (7)

where $-\pi \leq \varsigma \leq \pi.$

Approximation of hydrodynamic coefficients on grid differences allows scaling, i.e. compression and stretching to any given surface of the sphere.

Linear combinations can used to calculate forces and moments on finite volumes [2]:

$$\frac{1}{2}(\rho div(V) + \begin{vmatrix} u \\ v \\ w \end{vmatrix} \cdot \Delta\rho = 0;$$

For a particular set of coordinates:

$$
\begin{aligned}
&\frac{\partial u}{\partial t} + \frac{1}{2}\left(u\frac{\partial u}{\partial x} + v\frac{\partial u}{\partial y} + w\frac{\partial u}{\partial z} + \frac{\partial u^2}{\partial x} + \frac{\partial uv}{\partial y} + \frac{\partial uw}{\partial z}\right) - \frac{1}{\rho Re}\left(\frac{\partial^2 u}{\partial x^2} + \frac{\partial^2 u}{\partial y^2} + \frac{\partial^2 u}{\partial z^2}\right) \\
&\quad = -\frac{1}{\rho k M^2}\frac{\partial p}{\partial x} + \frac{1}{2}u div(V) + \frac{1}{3\rho Re}\frac{\partial}{\partial x}div(V); \\
&\frac{\partial v}{\partial t} + \frac{1}{2}\left(u\frac{\partial v}{\partial x} + v\frac{\partial v}{\partial y} + w\frac{\partial v}{\partial z} + \frac{\partial uv}{\partial x} + \frac{\partial v^2}{\partial y} + \frac{\partial vw}{\partial z}\right) - \frac{1}{\rho Re}\left(\frac{\partial^2 v}{\partial x^2} + \frac{\partial^2 v}{\partial y^2} + \frac{\partial^2 v}{\partial z^2}\right) \\
&\quad = -\frac{1}{\rho k M^2}\frac{\partial p}{\partial y} + \frac{1}{2}v div(V) + \frac{1}{3\rho Re}\frac{\partial}{\partial y}div(V); \\
&\frac{\partial w}{\partial t} + \frac{1}{2}\left(u\frac{\partial w}{\partial x} + v\frac{\partial w}{\partial y} + w\frac{\partial w}{\partial z} + \frac{\partial uw}{\partial x} + \frac{\partial vw}{\partial y} + \frac{\partial w^2}{\partial z}\right) - \frac{1}{\rho Re}\left(\frac{\partial^2 w}{\partial x^2} + \frac{\partial^2 w}{\partial y^2} + \frac{\partial^2 w}{\partial z^2}\right) \\
&\quad = -\frac{1}{\rho k M^2}\frac{\partial p}{\partial z} + \frac{1}{2}w div(V) + \frac{1}{3\rho Re}\frac{\partial}{\partial z}div(V); \\
&\frac{\partial T}{\partial t} + \frac{1}{2}\left(u\frac{\partial T}{\partial x} + v\frac{\partial T}{\partial y} + w\frac{\partial T}{\partial z} + \frac{\partial uT}{\partial x} + \frac{\partial vT}{\partial y} + \frac{\partial wT}{\partial z}\right) - \frac{k}{\rho Re}\left(\frac{\partial^2 T}{\partial x^2} + \frac{\partial^2 T}{\partial y^2} + \frac{\partial^2 T}{\partial z^2}\right) \\
&\quad = -\frac{k-1}{\rho}p div(V) + \frac{1}{2}T div(V) \\
&\qquad + \frac{k(k-1)M^2}{\rho Re}\left[2\left(\left(\frac{\partial u}{\partial x}\right)^2 + \left(\frac{\partial v}{\partial y}\right)^2 + \left(\frac{\partial w}{\partial z}\right)^2\right) + \left(\frac{\partial u}{\partial y} + \frac{\partial v}{\partial x}\right)^2 \right. \\
&\qquad \left. + \left(\frac{\partial u}{\partial z} + \frac{\partial w}{\partial x}\right)^2 + \left(\frac{\partial v}{\partial z} + \frac{\partial w}{\partial x}\right)^2 - \frac{2}{3}(div(V))^2\right];
\end{aligned}
\tag{8}
$$

in the model (8): $V = (u, v, w)$—velocity vector; $p, \rho, T$— pressure, density and temperature functions.

The general solution of Eq. (8) is:

$$
p = \rho T.
\tag{9}
$$

The transition to the local coordinate system performed through the replacement of variables [3]:

$$
\begin{aligned}
\frac{\partial w}{\partial x} &= \frac{\partial w}{\partial \xi}\xi_x + \frac{\partial w}{\partial \eta}\eta_x + \frac{\partial w}{\partial \zeta}\zeta_x; \\
\frac{\partial w}{\partial y} &= \frac{\partial w}{\partial \xi}\xi_y + \frac{\partial w}{\partial \eta}\eta_y + \frac{\partial w}{\partial \zeta}\zeta_y; \\
\frac{\partial w}{\partial z} &= \frac{\partial w}{\partial \xi}\xi_z + \frac{\partial w}{\partial \eta}\eta_z + \frac{\partial w}{\partial \zeta}\zeta_z.
\end{aligned}
\tag{10}
$$

Higher derivatives, when considering a finite segment of the Fourier series:

$$
\begin{aligned}
\frac{\partial^2 w}{\partial x^2} + \frac{\partial^2 w}{\partial y^2} + \frac{\partial^2 w}{\partial z^2} &= \frac{\partial w}{\partial x} = \frac{\partial^2 w}{\partial \xi^2}\nabla\xi\nabla\xi + \frac{\partial^2 w}{\partial \eta^2}\nabla\eta\nabla\eta + \frac{\partial^2 w}{\partial \zeta^2}\nabla\zeta\nabla\zeta \\
&\quad + 2\frac{\partial^2 w}{\partial \xi\partial \eta}\nabla\xi\nabla\eta + 2\frac{\partial^2 w}{\partial \xi\partial \zeta}\nabla\xi\nabla\zeta + 2\frac{\partial^2 w}{\partial \zeta\partial \eta}\nabla\zeta\nabla\eta + \frac{\partial w}{\partial \xi}(\xi_{xx} + \xi_{yy} + \xi_{zz}) \\
&\quad + \frac{\partial w}{\partial \eta}(\eta_{xx} + \eta_{yy} + \eta_{zz}) + \frac{\partial w}{\partial \zeta}(\zeta_{xx} + \zeta_{yy} + \zeta_{zz});
\end{aligned}
\tag{11}
$$

where: $\nabla \xi = (\xi_x, \xi_y, \xi_z); \nabla \eta = (\eta_x, \eta_y, \eta_z); \nabla \zeta = (\zeta_x, \zeta_y, \zeta_z);$ a $\nabla \xi \nabla \eta, \nabla \eta \nabla \zeta, \nabla \xi \nabla \zeta$—scalar products of vectors.

In numerical modeling, the values of the faces are set based on the stress-strain state of the elements of the equipment complex, and the values of the flow velocities $u = v = w = 0$, values $\rho = T = \text{const}, p = p(0)$.

## 2.2 Calculations on Bernstein Polygonal Grids

The shape of the elements of surfaces, in the process of flow, considered through the algebra of Bernstein polynomials [4].

Tensor product:

$$Q(u, w) = \sum \sum B_{i,j} J_{n,i}(u) K_{m,j}(w) \tag{12}$$

where: $J_{n,i}(u)$ и $K_{m,j}(w)$ basic (grid) functions in parametric coordinates.

Functions:

$$J_{n,i}(u) = \binom{n}{i} u^i (1 - u)^{n-i},$$

$$K_{m,j}(w) = \binom{m}{j} w^i (1 - w)^{m-j},$$

$$\binom{n}{i} = \frac{n!}{i!(n-i)!},$$

$$\binom{m}{j} = \frac{m!}{j!(m-j)!}, \tag{13}$$

for the coordinates $u$ and $w$, a uniform grid is given in equal increments, in matrix form:

$$Q(u, w) = [U][N][B][M]^T[W]. \tag{14}$$

In (14):

$$[U] = [u^n, u^{(n-1)} \dots 1];$$
$$[W] = [w^n, w^{(n-1)} \dots 1];$$

$$[N] = \begin{bmatrix} \binom{n}{0}\binom{n}{i}(-1)^i & \cdots & \binom{n}{i}\binom{n-n}{i-n}(-1)^0 \\ \vdots & \ddots & \vdots \\ \binom{n}{0}\binom{n-i}{i-1}(-1)^0 \cdots & & 0 \end{bmatrix}; \qquad (15)$$

$$[B] = \begin{bmatrix} B_{0,0} & \cdots & B_{0,m} \\ \vdots & \ddots & \vdots \\ B_{n,0} & \cdots & B_{n,m} \end{bmatrix}; \qquad (16)$$

similarly:

$$[M] = \begin{bmatrix} \binom{m}{0}\binom{m}{j}(-1)^j & \cdots & \binom{m}{j}\binom{m-j}{j-m}(-1)^0 \\ \vdots & \ddots & \vdots \\ \binom{m}{0}\binom{m-j}{j-1}(-1)^0 \cdots & & 0 \end{bmatrix} \qquad (17)$$

The Bezier surface constructed by (14) shown in the Fig. 2.

The uniform grid of the characteristic polyhedron shown in Fig. 3 allows the approximation of the Bezier surface approximation function along the vertices.

The approximation is of interest:

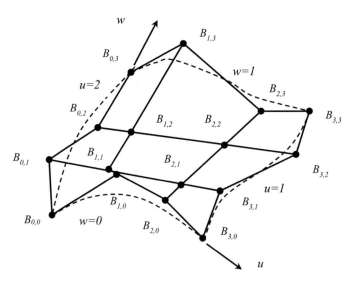

**Fig. 2** Bezvier surface with vertices constructed from a characteristic polyhedron

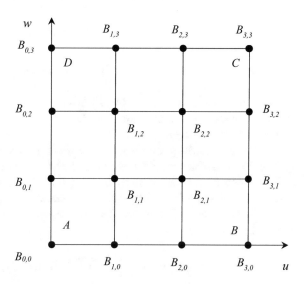

**Fig. 3** A regular mesh of a Bezier surface with vertices of a characteristic polyhedron

- with a step from vertex to vertex with a given parameter of the approximation step;
- step change depending on the detail of the approximation, especially in places where the smoothness of the polynomial is critical in terms of the approximation of the surface to the hydromechanical variables.

Family of derivatives (14):

$$Q_u(u, w) = [U][N][P][N]^T[W], \tag{18}$$

$$Q_w(u, w) = [U][N][P][N]^T[W'], \tag{19}$$

$$Q_{u,w}(u, w) = [U'][N][P][N]^T[W'], \tag{20}$$

$$Q_{u,u}(u, w) = [U''][N][P][N]^T[W'], \tag{21}$$

$$Q_{w,w}(u, w) = [U][N][P][N]^T[W''], \tag{22}$$

with partial derivatives:

$$[U'] = [3u^2\ 2u\ 1\ 0], \tag{23}$$

$$[W']^T = [3w^2\ 2w\ 1\ 0], \tag{24}$$

$$\left[U''\right] = [6u\,2\,0\,0],\tag{25}$$

$$\left[W''\right]^{T} = [6w\,2\,0\,0].\tag{26}$$

The construction of surfaces by (18–26) makes it possible to obtain a parametric approximation by parametric curves to the surface under study, by methods of cubic (bicubic) Bezier splines.

However, a fairly good result of constructing a surface by methods of parametric Bezier splines has computational difficulties associated with the "stitching" of various surfaces constructed by Bezier splines. *The transition from one surface to another provides an intractable effect of thickening computational grids.*

The solution is seen in the application of the decomposition of Bezier surfaces in parametric directions:

$$Q(u,\,w) = \sum\nolimits_{i=1}^{n+1} \sum\nolimits_{j=1}^{m+1} B_{i,j} N_{i,k}(u) M_{j,l}(w),\tag{27}$$

где $N_{i,k}(u)$, $M_{j,l}(w)$—orthogonal basis functions.
Indeed, let:

$$N_{i,1}(u) = \begin{cases} 1,\, x_i \le u < x_{i+1}, \\ 0 \end{cases}\tag{28}$$

$$N_{i,k}(u) = \frac{(u - x_i)N_{i,k-1}(u)}{x_{i+k-1} - x_i} + \frac{(x_{i+k} - u)N_{i+1,k-1}(u)}{x_{i+k} - x_{i+1}},\tag{29}$$

$$M_{j,l}(w) = \begin{cases} 1,\, y_j \le w < y_{j+1}, \\ 0 \end{cases}\tag{30}$$

$$M_{j,l}(w) = \frac{(w - y_j)M_{j,l-1}(w)}{y_{j+l-1} - y_j} + \frac{(y_{j+l} - w)M_{j+1,l-1}(w)}{y_{j+l} - y_{j+1}}\tag{31}$$

where $x_i$, $y_j$—coordinates of the nodal vector.

Let's say that $B_{i,j}$—the vertices of the computational grid of the surface, which can have different intervals, are ideally strictly rectangular.

The polygonal surface of a B-spline is the defining polyhedron of a surface defined by sets of polynomials of a certain smoothness.

**Theorem 2.1** *If the surface of the defining B spline is invariant with respect to the affine transformation, then each of these invariants is generated by the transformation of the defining grid.*

Let a uniform grid given:

$$P(x,\,y,\,z) = \{x_i + \Delta x,\, y_i + \Delta y,\, z_i + \Delta z\}\forall\{x = y = z\},$$

The coordinates define the cylindrical surface:

$$\xi = x; \eta = \sqrt{y^2 + z^2}; \varsigma = arctg(z/y) + k\pi;$$

where—$\pi \le \varsigma \le \pi$.

Then, there is a transformation of the scale of the cylinder of the form:

$$P^*(\xi^*, \eta^*, \varsigma^*) = P^*\big(P(x_i, y_i, z_i)\big) \blacksquare \tag{32}$$

As follows from Theorem 2.1, the step of setting the grid can be parametric, and then the description of the surface made by means of parametric derivatives:

$$Q_u(u, w) = \sum_{i=1}^{n} \sum_{j=1}^{m} B_{i,j} N'_{i,k}(u) M_{j,l}(w), \tag{33}$$

$$Q_w(u, w) = \sum_{i=1}^{n} \sum_{j=1}^{m} B_{i,j} N_{i,k}(u) M'_{j,l}(w), \tag{34}$$

$$Q_{uw}(u, w) = \sum_{i=1}^{n} \sum_{j=1}^{m} B_{i,j} N'_{i,k}(u) M'_{j,l}(w), \tag{35}$$

$$Q_{uu}(u, w) = \sum_{i=1}^{n} \sum_{j=1}^{m} B_{i,j} N''_{i,k}(u) M_{j,l}(w), \tag{36}$$

$$Q_{ww}(u, w) = \sum_{i=1}^{n} \sum_{j=1}^{m} B_{i,j} N_{i,k}(u) M''_{j,l}(w). \tag{37}$$

The parametrized Bezier surface is well describe by the basic functions related to the Rises basis. Description of Bezier surfaces, in this case it is convenient to produce in systems of bursts (wavelets) [5].

## 3 Results

The study of methods for describing hydrodynamic coefficients using finite volumes gives good results in solving systems of differential equations for given surfaces that make up the profile of a marine mobile object. The problem seems to be a set of different angular moments that are difficult to describe by a set of surfaces streamlined by a liquid.

Numerical methods based on the study of data from the aerotube experiment have a significant computational difficulty, consisting in choosing a scaling grid. A regular grid can considered as a large-scale one if an approximation of the flow of "small" structural elements is required, but then an estimate of the spectrum of the dynamics of the vortex flow is required.

The solution to this problem seen by using sets of decomposition bases or a bank of decompositions in an orthogonal basis obtained by compression and transformation of the basis.

Strictly speaking, use the basis of bursts, it is possible to decompose the spectrum of vortices separated by the difference in the flow velocity of the profiles of the surfaces of a marine mobile object.

## 4   Discussion and Conclusion

The scientific novelty of the work provided by the theoretical result obtained for the first time—asymptotic by Bernstein polynomials in the problems of approximation of hydrodynamic coefficients [6] in the cycle of multi-turn engineering design of ships, marine mobile objects and special outboard equipment when studying in the basis of Morley surges [9].

A new, for the first time proposed practical result—computational methods of multi-turn engineering design when working out the tasks of designing marine mobile objects, special marine equipment and outboard equipment related to the field of fixed assets of the fleet [7].

Numerical methods based on the study of data from the aerotube experiment have a significant computational difficulty, consisting in choosing a scaling grid. A regular grid can considered as a large-scale one if an approximation of the flow of "small" structural elements is required, but then an estimate of the spectrum of the dynamics of the vortex flow is required [8]. The solution to this problem seen by using sets of decomposition bases or a bank of decompositions in an orthogonal basis obtained by compression and transformation of the basis. Strictly speaking, use the basis of burst; it is possible to decompose the spectrum of vortices separated by the difference in the flow velocity of the profiles of the surfaces of a marine mobile object.

## References

1. Heuberger P, Van Den Hof P, Wahlberg B (2005) Modelling and identification with rational orthogonal basis functions
2. Dantsevich IM, Lyutikova MN, Novikov AY, Osmukha SA (2020) Analysis of a nonlinear system dynamics in the Morlet wavelet basis. In: IOP Conference Series Materials Science and Engineering, vol 873, p 012035, 1–8 (2020). https://doi.org/10.1088/1757-899X/873/1/012035
3. Zaslonov VV, Golovina AA, Popov AN (2020) Creating a crewless ship in the framework of the technological paradigm of the Russian Federation. Lecture Notes in Networks and Systems, vol 115, pp 468–474. https://doi.org/10.1007/978-3-030-40749-0_56
4. Slavič J, Mihalec M, Javh J, Boltežar M (2017) Morlet-wave damping identification: Application to high-speed video. In: Conference Proceedings of the Society for Experimental Mechanics Series vol 10B. Springer, New York, pp 27–30

5. Andersen Bachelor K (2016) Development of a time-domain modeling platform for hybrid marine propulsion systems
6. Bertram V (2012) Practical ship hydrodynamics. Elsevier, Amsterdam
7. Lyutikova MN, Dantsevich IM, Pankina SI (2021) The intelligent underwater laboratory. In: IOP conference series: earth and environmental science this link is disabled, vol 872, no 1, p 012003
8. Yusupov OK, Ibadullaev KK, Yusupova AK (2016) One application way of programming methods in the physical problem solution. Sci World Int Sci J I(5(33)):43–46
9. John J. Benedetto Sampling Theory and Wavelets. http://www.prometheus-inc.com/asi/multimedia1998/papers/benedetto.pdf
10. Osmukha SA (2019) Simulation of the distribution of hydrodynamic coefficients in the fluid flow according to the aerotube experiment. Nat Tech Sci (2):39–41

# A Deep Feature Selection Method for Tumor Classification in Breast Ultrasound Images

Payel Pramanik⬡, Souradeep Mukhopadhyay⬡, Dmitrii Kaplun⬡, and Ram Sarkar⬡

**Abstract** Breast cancer is a serious concern worldwide amongst women. The malignant growth in breast is a life-threatening issue which must be taken care of and treated at an early stage of its formation. Identification of such tumors in the breast is very crucial and important. In the present work, we introduce a deep learning based method for breast tumor classification from breast ultrasound tumor images. We combine the concept of transfer learning (TL) along with a statistical method which involves a total of three phases. At first, we extract features from ultrasound tumor images with a fine-tuned TL model, next we rank these features based on the correlation coefficient values that we get after applying the Spearman Rank Correlation Coefficient method in the second phase. At the last phase, we select the top-ranked features which are most important and relevant, and classify the unseen breast ultrasound tumors using machine learning based classifiers. The proposed method achieves state-of-the-art classification accuracy of 98.72% on the test dataset with only 40% top-ranked extracted features by the TL model.

**Keywords** Breast cancer · Tumor classification · Deep learning · Feature selection · Ultrasound images

## 1 Introduction

Malignant growth in breast is the most typical sort of threat among females now-a-days. As indicated by the World Cancer Research Fund (WCRF), in the year 2018 itself, millions of new malignant growth cases accounted for (WCRF, 2020). Such quite forbidding statistics motivate the right utilization of the recent technological

P. Pramanik (✉) · S. Mukhopadhyay · R. Sarkar
Department of Computer Science and Engineering, Jadavpur University, Jadavpur, India
e-mail: ppramanik07@gmail.com

D. Kaplun
Department of Automation and Control Processes, Saint Petersberg Electrotechnical University "LETI", Saint Petersburg, Russia
e-mail: dikaplun@etu.ru

© The Author(s), under exclusive license to Springer Nature Switzerland AG 2022      241
A. Tchernykh et al. (eds.), *Mathematics and its Applications in New Computer Systems*,
Lecture Notes in Networks and Systems 424,
https://doi.org/10.1007/978-3-030-97020-8_22

advances, so that an economical early conclusion of the sickness can be made. Advancements in Artificial Intelligence (AI) and Machine Learning (ML) can also be seen in the field of medical image analysis. Aside from mammography imaging, ultrasound is one more generally utilized imaging methodologies for performing radiological treatment for this type of life threatening disease. In recent years, various methods based on ML and Deep Learning (DL) have been exploited to learn from previously collected ultrasound images and make more appropriate models for incoming new set of data.

In this work, we introduce a DL based method where we combine the concept Transfer Learning (TL) along with a statistical method to perform classification of breast tumors from a publicly accessible dataset, called breast ultrasound image database (BUSI) and it achieves state-of-the-art accuracy. In the last few decades, a lot of researchers have contributed to this field. We discuss some of the past method along with their shortcomings on ultrasound image based breast cancer data. Details are given in Table 1. The rest of this paper is structured as follows. In Sect. 2, we first state the prerequisites for this research work, and then describe our proposed method in detail. Next, in Sect. 3, we discuss the metrics that we use to evaluate the proposed model and provide the experimental results, and analyze and discuss the results in Sect. 4. Finally, we conclude our work and mention some future scopes of this work in Sect. 5.

## 2 Materials and Methods

In this section, we discuss about the dataset used, the prerequisites and the detailed methodology we propose in this work. First, we briefly explain the BUSI database, then the Spearman Rank Correlation Coefficient (SCC) method that we apply here. Next we elaborate the method for the breast tumor classification from the BUSI database.

### 2.1 Dataset Description

In this work, we use the publicly available BUSI[1] (Breast Ultrasound Images) database, collected in 2018, introduced by [2]. This database consists of a total of 780 (benign: 437, malignant: 210 and normal: 133) breast ultrasound images (500 × 500 pixels), collected from 600 female patients aged between 25 and 75 years old using two LOGIQ E9 ultrasound systems. This database also contains the hand drawn ground truth mask generated with MATLAB software for each of the corresponding images. Few sample images along with their masks from all the three categories of this database are shown in Fig. 1.

---

[1] https://scholar.cu.edu.eg/?q=afahmy/pages/dataset.

**Table 1** Description of some past methods experimented on ultrasound image based breast cancer data

| Authors | Method | Database | Shortcomings |
|---------|--------|----------|--------------|
| Shi et al. [14] | Histogram, texture and fractal features from the area of interest are exploited and Support Vector Machine (SVM) classifier is used | BUSI (36 malignant and 51 benign) | Ignores one of the most precious features like size of tumor |
| Yang et al. [16] | Texture featuresformulated on GLCM (extracted from multi resolution less sensitive ranklet transformation) and SVM classifier are used | Acuson Siemens, California, USA (78 benign and 38 Malignant) | The accuracy is low. Computational cost is high |
| Cai et al. [4] | Local texture features and phase congruency, along with SVM classifier are used | Department of Ultrasound, Huashan Hospital in Shanghai, China (69 benign, 69 malignant) | All area of interest based features are ignored |
| Lo et al. [10] | Tumor-texture and speckle-texture features along with logistic regression classifier are exploited | Breast US database (21 malignant, 48 benign) | Texture features are only considered as discriminating features |
| Xiao et al. [15] | 3 TL models based on InceptionV3, Xception and Resnet50, a CNN model with three convolutional layers (CNN3) are used. InceptionV3 outperforms other methods in terms of result | TirdAfliated Hospital of Sun Yatsen University (688 malignant, 1370 benign cases) | The accuracy is low (89.44%) |
| Moon et al. [12] | Shape, texture and ellipsoid fitting based 3D features are extracted from contour analysis and logistic regression classifier is used | SomoVuScanStation (U-systems, SanJose, CA, USA) (76 benign and 71 malignant) | Less sensitive model. Accuracy is low |
| Sadad et al. [13] | Texture and shape-based hybrid features are extracted and Decision tree, KNN are exploited as classifier | OASBUD (52 malignant and 48 benign), BUSI (BaheyaHospital, Egypt) (210 malignant and 437 benign) | Class imbalance problem is ignored |
| Mishra et al. [11] | Various texture features are used. Then recursive feature elimination and SMOTE oversampling steps are followed. Different ML classifiers are explored | BUSI (437 benign, 210 malignant and 133 normal) | Computation time is high |

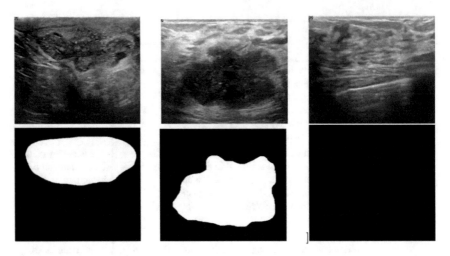

**Fig. 1** **i** Benign Ultrasound, **ii** Malignant Ultrasound **iii** Normal Ultrasound **iv** Benign Mask, **v** Malignant Mask **vi** Normal Mask

## 2.2 Spearman Rank Correlation Coefficient

SCC often known as the $\rho$ of Spearman, introduced by Charles Spearman in 1904 [5], is one of the oldest methods in statistics to measure correlation between two sets of data. Correlation determines if there exists a relation or dependencies between two random sample data. Let's say, $A_1, A_2, \ldots, A_n$ and $B_1, B_2, \ldots, B_n$ be two samples or variables (in this work features) each of size n. Then the $\rho$ of Spearman can be defined [5] as follows:

$$\rho = 1 - \frac{6 \sum diff_i^2}{n(n^2 - 1)} \tag{1}$$

where, $diff_i = Rank(A_i) - Rank(B_i)$. Rank $(A_i)$ defines the rank of $A_i$ compared to other values present in A for $i = 1, 2, \ldots, n$. Rank $(A_i) = 1$ if $A_i$ is the smallest value of A and Rank $(A_i) = n$ if $A_i$ is the largest value of A. Similarly, Rank $(B_i)$ denotes the rank of $B_i$. Repetition of values can be found in any sample, and in that case the average rank is assigned to such instances. For several such average rankings, Eq. 1 can be rewritten [5] as follows:

$$\rho = 1 - \frac{S_A + S_B - \sum diff_i^2}{2\sqrt{S_A S_B}} \tag{2}$$

where $S_A = \frac{n(n^2-1) - \sum_{i=1}^{i=g}(c_i^3 - c_i)}{12}$ g = total number of such groups that get the average rank and $c_i$ = the size of group i for the A variable. Similarly, $S_B = \frac{n(n^2-1) - \sum_{j=1}^{j=h}(c_j^3 - c_j)}{12}$ with h number of groups with average ranking and the size of $jth$ group is $c_j$ for sample B.

The value of $\rho$ lies between $-1$ and $+1$. A positive $\rho$ value signifies a positive correlation and a negative $\rho$ value means negative correlation between variables A and B. The value of $\rho = +1$ corresponds to a perfect positive correlation, whereas $\rho = -1$ corresponds to a perfect negative correlation between A and B. However, if A and B are not correlated at all then value of $\rho$ becomes 0.

## 2.3 Methodology

In this work, we propose a method for the identification of breast tumors from the ultrasound images which mainly involves three phases namely, (a) feature extraction: extraction of deep features from the input ultrasound images, (b) feature selection: finding statistically important and relevant features to obtain optimal feature subset, and (c) classification: training of an ML classifier to perform prediction on the test dataset.

**Deep Feature Extraction:** For feature extraction, we consider a TL model pre-trained on the ImageNet database as the base model, which we fine-tune using the dataset under consideration. We eliminate the last layer of the base model, and incorporate a Global Average Pooling (GAP) layer, followed by a dense layer of 1024 units with activation function as rectified linear activation unit (ReLU), and adjust the output layer as per the categories of the BUSI database with Softmax activation function. The structure of the fine-tuned TL model for deep feature extraction is shown in Fig. 2. We use the GAP layer as it is more native to the convolution structure because it enforces the correspondence between feature maps and categories, thereby showing the ability to capture the spatial information which is very important for image analysis. During training of the model, we train only the top layers (the incorporated layers) and freeze all the remaining layers of the base TL model.

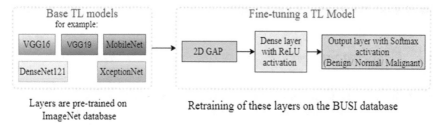

**Fig. 2** The structure of the fine-tuned TL model for deep feature extraction from the BUSI database. We consider five TL models namely, VGG16, VGG19, MobileNet, DenseNet121 and XceptionNet as individual base TL model for experimentation where we fine-tune each of them to perform feature extraction

**Selection of Statistically Important Features:** After extracting the deep features, we feed these features further into a statistic based method, called SCC, to obtain the most important features. We consider the correlation metric here, and based on correlation coefficient values we assign score to every feature and make the ranking of these extracted features accordingly. As we know redundant features get high correlation value due to high similarity so we reduce such redundant features by considering only the top-ranked features where ranking is done in increasing order of the assigned scores. After that we select top ranked features for the later phase of our method.

**ML Classifiers for Classification:** In this work, we use four different ML based classifiers namely SVM [7], Random Forest (RF) [8], K-Nearest Neighbors (KNN) [9], and Artificial Neural Network (ANN) [1] to perform prediction on the test dataset and to ensure the robustness of our method. We split the top-ranked feature set into 75% training, 10% validation and 15% test dataset. After that feed the training and validation sets as the input of these ML classifiers and performs prediction on the test dataset. The entire pipeline of the proposed method is shown in Fig. 3.

The proposed method achieves promising results for the breast tumor classification from the BUSI dataset. This is to be noted that this database has high class imbalance, however the success to deal with this problem lies in that fact that DL models can handle the class imbalance problems to some extent [3]. Further, the SCC method ranks the features as per their importance and discriminative capability, which intuitively eliminates the redundant features and provides an optimal feature set. This also intuitively reduces the computational time of the ML classifiers used for the task of classification.

## 3 Experimental Results

In this section, we mention the metrics for performance evaluation and then provide the figures and tables of the obtained results.

### 3.1 Performance Metrics

We evaluate our method using the following widely adopted metrics:

True Positive (TP): It refers to the positive class samples, correctly labeled by the classifier.

True Negative (TN): It refers to the negative class samples, correctly labeled by the classifier.

False Positive (FP): These are the negative class samples, incorrectly labeled as positive class samples.

False Negative (FN): These are the positive class samples, mislabeled as negative class samples.

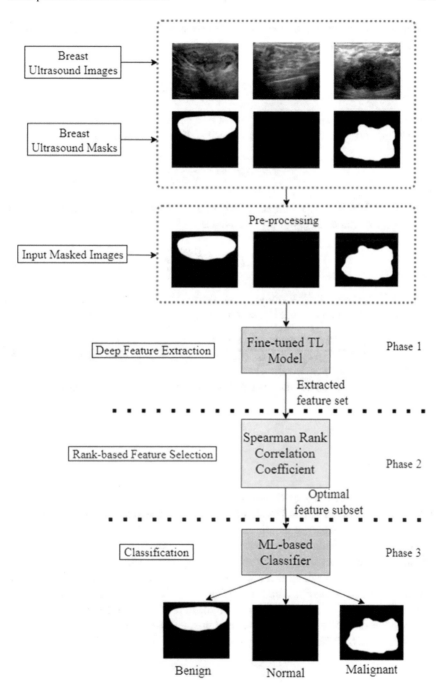

**Fig. 3** The pipeline of the proposed breast tumor classification model

Accuracy is the measure of ratio of correctly predicted labels to the total size of the dataset [9]. Accuracy is calculated as follows:

$$Accuracy = \frac{TP + TN}{TP + TN + FP + FN} \tag{3}$$

Precision or specificity is defined as the ratio of correctly predicted labels for a class to the total number of samples for the class [9]. It is calculated as follows:

$$Precision = \frac{TP}{TP + FP} \tag{4}$$

Recall or Sensitivity is defined as the ratio of true positive samples to the total positives from that class [9]. It is calculated as follows:

$$Recall = \frac{TP}{TP + FN} \tag{5}$$

F1-score is the weighted harmonic mean of precision and recall [9]. It gives equal weight to precision and recall. It is calculated as:

$$F1 = \frac{2 * Precision * Recall}{Precision + Recall} = \frac{2 * TP}{2 * TP + FP + FN} \tag{6}$$

Kappa score is a statistical measure to test inter-rater reliability [6] for categorical items.

## 3.2 Results

The experiments are conducted and evaluated on the said database with 75% training data, 10% validation data and 15% testing data. Initially, we experiment with different splits of training, testing and validation data during the training of the TL model. The experimental results are shown in Fig. 4. We find that the model achieves highest classification accuracy for the above mentioned splitting. We use the widely adopted Adam optimizer, and categorical cross-entropy as the loss function for the training purpose of the model. For smooth learning, we use step learning rate scheduler where the learning rate is reduced by a factor of 2 after the third epoch, where the total number of epochs is 30. We have implemented our method on a computer with NVIDIA Tesla K80 GPU.

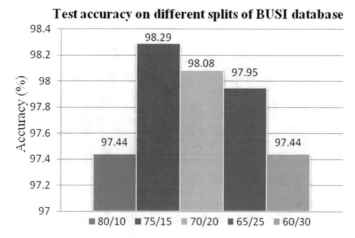

**Fig. 4** Test accuracy of the fine-tuned VGG19 model for different splits of training and test sets. For 75% training and 15% test data, the model achieves highest classification accuracy

**Table 2** Comparison of classification accuracies of different TL models on the BUSI database

| Fine-tuned CNN model | Training accuracy | Validation accuracy | Test accuracy |
|---|---|---|---|
| VGG19 | 0.9929 | 0.9683 | *0.9829* |
| VGG16 | 0.9875 | 0.9524 | 0.9808 |
| MobileNet | 0.9933 | 0.9104 | 0.9744 |
| DenseNet121 | 0.9883 | 0.9254 | 0.9744 |
| XceptionNet | 0.9883 | 0.9104 | 0.9573 |

**Table 3** Experimental results of different ML classifiers on top 40% ranked features extracted by the VGG19 model. All the results are evaluated on 15% test dataset

| ML classifier | Test accuracy | F1-score | Kappa score | Precision | Recall |
|---|---|---|---|---|---|
| SVM | 0.9872 | 0.9872 | 0.9781 | 0.9872 | 0.9872 |
| KNN | 0.9872 | 0.9872 | 0.9781 | 0.9872 | 0.9872 |
| ANN | 0.9829 | 0.9830 | 0.9709 | 0.9839 | 0.9829 |
| RF | 0.9829 | 0.9830 | 0.9709 | 0.9839 | 0.9829 |

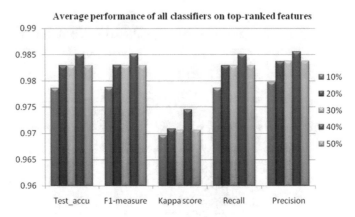

**Fig. 5** Comparative results of all four classifiers with different sets of selected features. Selection of features is done by SCC method where the results are obtained with different k values

**Table 4** Performance comparison of the proposed method with some past methods on the BUSI database

| Method | Accuracy | Precision | Recall | F1-score | Recall |
|---|---|---|---|---|---|
| Mishra et al. [11] | 0.974 | 0.958 | 0.96 | 0.959 | 0.9872 |
| Sadad et al. [13] | 0.966 | – | 0.943 | – | 0.9872 |
| Moon et al. [12] | 0.946 | 0.9 | 0.923 | 0.911 | 0.9829 |
| Proposed method | *0.987* | *0.987* | *0.987* | *0.987* | *0.9829* |

## 4   Result Analysis and Discussion

As mentioned earlier, in this work, we first extract deep features using TL models from the BUSI database. Initially, we experiment with five popular TL models and find that for the BUSI database VGG19 model yields comparatively higher classification accuracy with 98.29% on the test set. The comparative experimental results are shown in Table 2. We proceed further with this model for the second phase of our method and apply SCC method on the extracted feature set. We get a ranked feature set where ranking is done in increasing order based on the SCC score associated with each feature. Further, we select top $k$% features where $k \in \{10, 20, 30, 40, 50\}$ and feed it to various ML classifiers for the classification task as the last phase of this work. In doing this, we follow the same dataset split mentioned earlier and evaluate the performance for all the metrics with the reduced optimal feature set. We analyze these performances using Fig. 5. From this figure it is clear that with only top 40% features extracted by VGG19 model shows relatively more capability to classify the breast tumor images from the BUSI database. The outcomes of the experiment are recorded in Table 3.

From Table 3, it can be seen that the application of SCC method helps in increasing the classification accuracy along with all the other evaluation metrics on unseen test dataset. Not only that, we obtain highest classification accuracy with the optimal feature set, which is just 40% of the extracted features from TL model, thereby ensuring the presence of the most important and relevant features.

Finally, we compare the performance of our method with some recently published methods on the BUSI database and tabulate the same in Table 4. From Table 4, it is clear that the proposed model outperforms state-of-the-art models with a good margin. The gain in performance has to be attributed to use of a fine-tuned TL model and the elimination of the redundant features from the extracted deep feature set.

## 5  Conclusion

Breast cancer is one of the common and most threatening diseases amongst women worldwide. Proper identification of malignant growth in women's breast from ultrasound images can provide a timely care and treatment to a breast cancer patient. In this work, we introduce a DL based model where we combine the concept of TL along with a statistical method namely SCC for tumor classification from breast ultrasound images. We evaluate our method on the publicly accessible BUSI database with respect to various performance metrics. From the experimentation, it can be observed that the proposed method yields promising results on unseen test data and achieves a state-of-the-art result which suggests potential implementations of the proposed model in real diagnostics. However, this work has some limitations also as it focuses only on the region of interest (ROI) of the original ultrasound images i.e., the ground truth tumor masks. But the availability of such ROI masks is scarce in practical scenarios. Automatic segmentation from ultrasound images using some DL based approaches like UNet can be explored in the future.

## References

1. Theodoridis S, Koutroumbas K (eds) (2009) Pattern recognition, 4th edn. Academic Press, Boston, p iv
2. Al-Dhabyani W, Gomaa M, Khaled H, Fahmy A (2019) Dataset of breast ultrasound images. Data Brief 28:104863
3. Buda M, Maki A, Mazurowski M (2017) A systematic study of the class imbalance problem in convolutional neural networks. Neural Netw 106:10
4. Cai L, Wang X, Wang Y, Guo Y, Yu J, Wang Y (2015) Robust phase-based texture descriptor for classification of breast ultrasound images. Biomed Eng Online 14:12
5. Chambers LG (1989) 73.52 spearman's rank correlation coefficient. Math Gaz 73(466):331–332

6. Cohen J (1960) A coefficient of agreement for nominal scales. Educ Psychol Measur 20(1):37–46
7. Cristianini N, Ricci E (2008) Support vector machines. Springer, Boston, pp 928–932
8. Cutler A, Cutler D, Stevens J (2011) Random forests, vol 45, pp 157–176
9. Han J, Kamber M, Pei J (2012) Data mining concepts and techniques, 3rd edn
10. Lo C, Chang R, Huang C, Moon W (2015) Computer-aided diagnosis of breast tumors using textures from intensity transformed sonographic images. In: IFMBE proceedings, vol 47. Springer, Cham, pp 124–127. 1st Global Conference on Biomedical Engineering, GCBME 2014 and 9th Asian-Pacific Conference on Medical and Biological Engineering, APCMBE 2014; Conference date: 09-10-2014 Through 12-10-2014
11. Mishra AK, Roy P, Bandyopadhyay S, Das SK (2021) Breast ultrasound tumour classification: a machine learning-radiomics based approach. Expert Syst 38(7):e12713
12. Moon W, Shen Y, Huang C, Chiang L, Chang R (2011) Computer-aided diagnosis for the classification of breast masses in automated whole breast ultrasound images. Ultrasound Med Biol 37(4):539–548. Funding Information: This work was supported by a grant from the National Science Council of the Republic of China (NSC 99-2221-E-002-136-MY3) and was supported by the Converging Research Center Program through the Ministry of Education, Science and Technology, Republic of Korea (2010K001113)
13. Sadad T, Hussain A, Munir A, Habib M, Ali Khan S, Hussain S, Yang S, Alawairdhi M (2020) Identification of breast malignancy by marker-controlled watershed transformation and hybrid feature set for healthcare. Appl Sci 10(6):1900
14. Shi X, Cheng H, Hu L, Ju W, Tian J (2010) Detection and classification of masses in breast ultrasound images. Digit Signal Process 20:824–836
15. Xiao T, Liu L, Li K, Qin W, Yu S, Li Z (2018) Comparison of transferred deep neural networks in ultrasonic breast masses discrimination. BioMed Res Int 2018
16. Yang M-C, Moon WK, Wang Y-CF, Bae M, Huang CF, Chen J, Chang R-F (2013) Robust texture analysis using multi-resolution gray-scale invariant features for breast sonographic tumor diagnosis. IEEE Trans Med Imaging 32:08

# Reference Points Based RNS Reverse Conversion for General Moduli Sets

Alexander Stempkovsky⬤, Dmitry Telpukhov⬤, Ilya Mkrtchan⬤, and Alexey Zhigulin⬤

**Abstract** Residue Number System (RNS) is a non-positional number system which can lead to parallel arithmetic operations, thereby achieving high performance in hardware implementation. However, the complexity of the reverse conversion (from RNS to binary) significantly limits the efficiency of the RNS for most modern tasks. The article proposes a new approach to the implementation of one of the most difficult and important tasks of RNS – reverse conversion. A mixed approach based on both arithmetic calculations and a Look-up-Table (LUT) implementation is proposed. The advantage of the method lies in its flexibility, since it allows one to vary the proportion of computations performed in LUTs. This is achieved by recursively implementing LUTs using the proposed reference points method. The proposed approach results in ability to create the best solution for the given performance criteria and hardware constraints. Experimental results show that the proposed method significantly outperforms the method based on the Chinese Remainder Theorem in terms of speed at the expense of moderate hardware costs. At the same time, it is possible to create more compact solutions due to the loss in timing. The issues of optimization and design automation of reverse converters based on the reference points method require further more detailed study.

**Keywords** Residue number system · RNS · Reverse converter · Look-up-table · Chinese remainder theorem · Hardware implementation · Reference points

## 1 Introduction

The Residue Number System has been known and actively studied since the late 60s [1–3]. During this time it has found wide application in cryptography [4, 5], as well as in the hardware implementation of Digital Signal Processing (DSP) devices

A. Stempkovsky · D. Telpukhov (✉) · I. Mkrtchan · A. Zhigulin
Institute for Design Problems in Microelectronics, Moscow, Russia
e-mail: nofrost@inbox.ru

A. Stempkovsky
e-mail: stal09@ippm.ru

© The Author(s), under exclusive license to Springer Nature Switzerland AG 2022
A. Tchernykh et al. (eds.), *Mathematics and its Applications in New Computer Systems*,
Lecture Notes in Networks and Systems 424,
https://doi.org/10.1007/978-3-030-97020-8_23

[6, 7]. RNS introduces an additional level of parallelism - at the level of arithmetic operations. This allows long carry chains to be broken, thus speeding up computation and reducing power consumption.

However, there are current problems that significantly hinder the more active use of RNS. The main problem is the implementation of non-modular operations. These operations break parallelism by forcing access to all parallel channels at the same time. This includes operations such as comparison, rounding, sign determination, conversion to a positional system, and vice versa [8–10]. The standard approach is to reduce the number of non-modular operations in a device. However, it is often impossible to remove RNS-binary converter – the most difficult block to implement [11–13]. Thereby, each study that suggests a more efficient way to implement the reverse (RNS to binary) converter thus improves the perspective of the RNS and its possibilities for wider use.

Most of the methods for solving this problem are based on the use of special moduli sets [14]. Such moduli sets are close to the powers of two, opening up possibilities for using simpler and more efficient algorithms for the RNS-binary conversion problem. However, this approach severely limits the flexibility of RNS to achieve a high dynamic range when used in cryptography and digital signal processing. The article proposes a flexible method for the hardware implementation of RNS converters for general moduli sets.

## 2  Materials and Methods

### 2.1  Pure Look-up-Table Implementation

In this paper a new method of reverse conversion is proposed. It does not rely on a special set of moduli - any numbers can be used as long as they are coprime.

The most straightforward method of improving the timing on the critical path is to use a LUT (lookup table). For example, if a set of moduli is (3, 5, 7), then the dynamic range M would be 105 (from 0 to 104) and so would be the number of values needed to be stored. This approach would definitely improve the timing, but its main problem can be spotted easily - the bigger the M, the more values must be kept. In this case, value M correlates with the area of LUT.

To better understand the dependencies between contents of a table and its area and timing, an experiment was held: a set of 120 test LUTs was synthesized using Cadence tools. The results in the form of two 3d plots can be seen on Fig. 1. This experiment was made for a general LUT structure. Here, "data size in bits" can be interpreted as "output size in bits" and "LUT size in rows" can be interpreted as "number of values to store".

It can be seen that though the timing increase is close to logarithmic, its dimension is not sizable even considering large LUT. However, the area increase is something

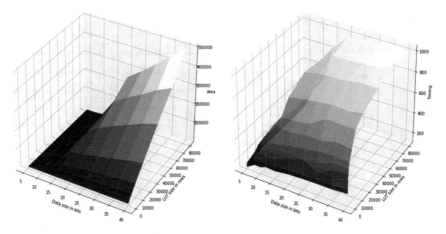

**Fig. 1** Area and timing dependences on LUT size and its data size

that should be addressed. The problem is - how to modify the design to lower the number of stored values without decreasing the dynamic range.

## 2.2 Reference Points Implementation

In order to overcome this issue, a method of reference points is introduced. Let a set of moduli be $(m_1, m_2, \ldots, m_n)$, where $m_n$, the largest module - is a reference-point module and $(x_1, x_2, \ldots, x_n)$, is a modular form of number $X$. The contents of a new LUT will consist of $\prod_{i=1}^{n-1} m_i$ values, called reference points, which correspond to $(x_1, x_2, \ldots, x_{n-1}, 0)$, and are represented by such $X$, so $x_n = X\%m_n = 0$. Operand $\%$ here denotes the operation of taking the remainder of a division. But the initial $(x_1, x_2, \ldots, x_n)$, however, can take values which are not stored in the table, so a certain correction should be made. In order to get the needed inputs for the LUT, modular subtraction $(x_i - x_n)\%m_i$ is performed, where $i \in [1, n-1]$. After gaining the value from the table, it should be added to $x_n$ to compensate for the initial correction and get the final $X$.

To illustrate this method, let's take $(2, 3, 5)$ as a set of moduli, 5 as a reference-point module and $X = (0, 2, 3)$ as a number which needs to be converted into a positional system. LUT for this moduli set will be as shown in Table 1. To find a reference point, let's perform modular subtraction and find needed values for LUT input: $in1 = (0-3)\%2 = 1, in2 = (2-3)\%3 = 2$. $in1 = 1$ and $in2 = 2$ correspond to the LUT output value of 5. Now, to get final result $X$, this output value must be added to $x_3 = 3$. The result is $X = 5 + 3 = 8$, which is the correct answer. This whole process is illustrated in more detail on Fig. 2 - here the values in the table are marked grey and values marked in red or white are not part of the table and are only shown for depictive purposes.

**Table 1** LUT for (2, 3, 5) moduli set with 5 as reference-point module

| Input | | Output |
|---|---|---|
| in1 | in2 | out |
| 0 | 0 | 0 |
| 1 | 2 | 5 |
| 0 | 1 | 10 |
| 1 | 0 | 15 |
| 0 | 2 | 20 |
| 1 | 1 | 25 |

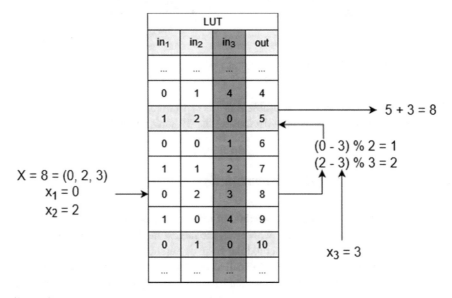

**Fig. 2** Illustrated example

Hardware implementation for the example above is shown on Fig. 3. By adding arithmetic modules, the LUT size was reduced by $m_n = 5$ times and it now contains 6 values instead of 30. It seems clear that this solution can be efficient for small amount of big modules, but yet is not good enough. For example, if the moduli set is (269, 271, 277) with 277 as reference-point module, the number of values to store would be 72,899, which is much better than 20,193,023, but still is a lot.

To reduce the amount of data even more, the gap between reference points should be higher. This goal can be achieved by taking not just one reference-point module, but the reference-point moduli subset. This leads us to the need for another LUT, the purpose of which is to compute the shift value, defined by the residues of reference-point moduli subset.

Let's use a (2, 3, 5, 7, 11) set of moduli to illustrate this approach. Example of such modification is shown on Fig. 4. LUT shift here is a reverse converter on its

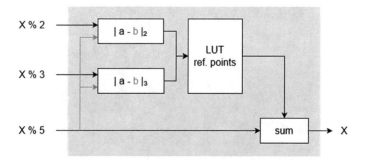

**Fig. 3** Hardware implementation of reverse converter in (2, 3, 5) basis

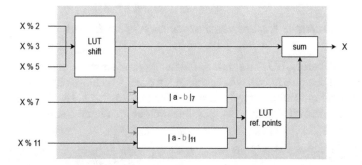

**Fig. 4** Hardware implementation of reverse converter with additional LUT

own, but for the small moduli set - (2, 3, 5). Such implementation reduces the number of data storage drastically - this is especially noticeable when using moduli sets with a large dynamic range.

Since LUT shift in this example can produce values up to 29, it is clear that number b will most certainly be bigger than a. For this reason, a modular subtractor would become more complex. There are two solutions for this. First one is to add a forward converter after the LUT. Expression (a - b)%m is the same as (a%m - b%m)%m, but in the second case it will be less costly in terms of hardware. The forward converter provides necessary value to be subtracted, and though this adds a new part to the design, it is more effective when considering large LUT shift. Hardware implementation of such design is illustrated on Fig. 5. Here, forward converter is operating on a (7, 11) modular basis.

Second option is aimed for the timing reduction and is pretty simple - the results of forward conversion for each value from LUT are precomputed and stored in this same LUT. In this case, the forward conversion module is not needed. It will increase the output dimension of the table, but if area reduction is not in the priority, then this option is acceptable.

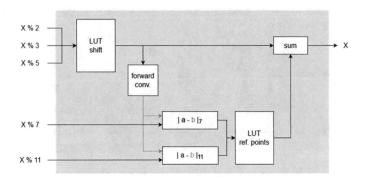

**Fig. 5** Solution with the use of forward converter

## 2.3 Hierarchical Reference Points Implementation

As mentioned before, LUT shift fulfills the functions of a reverse converter. It means, the described design can be used recursively in order to achieve either better area or better timing. For example, implementation for (2, 3, 5, 7, 11, 13, 17) set of moduli can be configured differently - two possible options are demonstrated on Fig. 6.

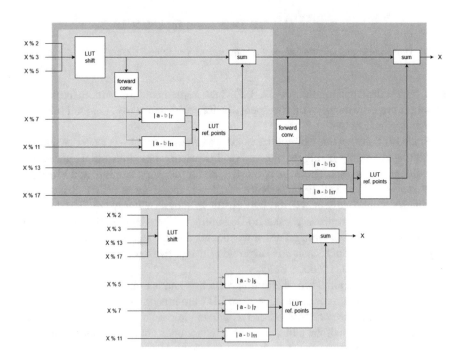

**Fig. 6** Reverse converter with different amount of layers

Here, the first implementation uses 3 layers - (2, 3, 5), (7, 11) and (13, 17), with LUT being the initial one. This configuration is aimed for area reduction, since the size of LUTs here is being cut by adding new layers. The second configuration is aimed for the timing reduction - it consists of a big LUT shift with pre recorded values for forward conversion. It also has a shorter critical path, in comparison to the first design.

## 3 Results

All the methods shown above can be used individually or in conjunction with one another to achieve the needed parameters of the circuit. Now, some of these possible configurations must be compared to other variants of reverse converters in order to confirm the efficiency of new methods. In this case, an implementation based on the CRT (Chinese remainder theorem) was used for comparison.

Searching for an optimal configuration is a complex task on its own and can be a subject of another research. Here, a set of 17 schemes was synthesized using Cadence tools with a Nangate standard cell library. A total of 3 sets of tests were conducted. Designs based on a Chinese remainder theorem and used for comparison will be called "CRT" for all tests.

In the first one, a (2, 3, 5, 7, 11, 13, 17) set of moduli was used. Designs marked 1 and 2 have 2 layers, configured as (2, 3, 13, 17) - first layer, (5, 7, 11) - second layer. Designs marked 3 and 4 have 3 layers, configured as (2, 3, 17) - first layer, (5, 13) - second layer, (7, 11) - third layer. Schemes 1 and 3 use a forward converter, schemes 2 and 4 use an extended LUT shift. Results are presented on Fig. 7.

In the second test, a (23, 29, 31, 37, 41) set of moduli was used. Designs marked 5 and 6 have 3 layers, configured as (23, 43) - first layer, (29, 41) - second layer and (31, 37) - third layer. Designs marked 7 and 8 have 2 layers, configured as (23, 29,

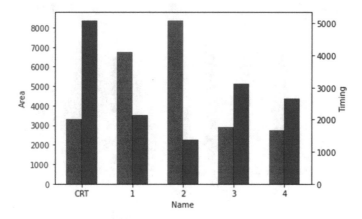

**Fig. 7** Results for (2, 3, 5, 7, 11, 13, 17) set of moduli

43) - first layer, (31, 37, 41) - second layer. Schemes 5 and 7 use a forward converter, schemes 6 and 8 use an extended LUT shift. Results are presented on Fig. 8.

The third test uses (2, 3, 5, 7, 11, 13, 17, 19, 23, 29) set of moduli. Designs marked 9 and 10 have 5 layers, configured as (2, 29) - first layer, (3, 23) - second layer, (5, 19) - third layer, (7, 17) - fourth layer, (11, 13) - fifth layer. Designs marked 11 and 12 have 4 layers, configured as (3, 5, 29) - first layer, (2, 7, 23) - second layer, (11, 19) - third layer, (13, 17) - fourth layer. At last, designs marked 13 and 14 have 3 layers, configured as (17, 5, 29) - first layer, (7, 23, 11) - second layer, (19, 13, 2, 3) - third layer. Schemes 9, 11, and 13 use a forward converter, schemes 10, 12, and 14 use an extended LUT shift. Results are presented on Fig. 9.

Results of both the first and the second set of tests show that new methods have up to 2 times better timing, but almost always lose in terms of area. The third set of tests shows that advantage in timing can be nullified when using a large amount of

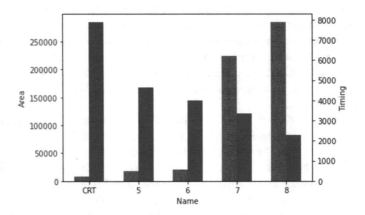

**Fig. 8**  Results for (23, 29, 31, 37, 41) set of moduli

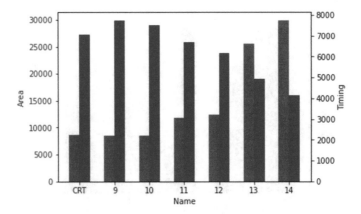

**Fig. 9**  Results for (2, 3, 5, 7, 11, 13, 17, 19, 23, 29) set of moduli

layers. It can also be seen that usage of extended LUT shift in the first layer instead of a forward converter can slightly improve timing in exchange for the small gain in area.

# 4 Conclusion

The article proposes a new LUT-based method for hardware implementation of RNS reverse converter for general moduli sets. The method of reference points is used to reduce the dimension of LUTs for the expense of additional arithmetic calculations. A hierarchical approach to the implementation of converters is proposed, which implies an iterative application of the reference points method to LUTs. This approach opens up a new optimization problem of choosing the best implementation in terms of depth of recursion and moduli set for each hierarchical level.

The research results have shown the prospects for flexible control of the area/delay due to the choice of the number of the hierarchical levels of the converter. Fewer levels are associated with the maximum area and best latency. Increasing the levels leads to a reduction in hardware costs at the expense of performance degradation.

Further work will focus on improving this approach. It is also planned to develop a computer-aided design system to automatically select the number and composition of hierarchical levels to achieve user-defined constraints.

**Acknowledgements** Research has been conducted with the financial support from the Russian Science Foundation (grant 17-19-01645).

# References

1. Szabo NS, Tanaka RI (1967) Residue arithmetic and its applications to computer technology, 1st edn. McGraw-Hill, New York
2. Akushskij IYA, YUdickij DI (1968) Mashinnaya arifmetika v ostatochnyh klassah. Sov. Radio, M., 440p
3. Garner H (1959) The residue number system. IRE Trans Electron Comput 8(6):140–147
4. Menezes AJ, Van Oorschot PC, Vanstone SA (2018) Handbook of applied cryptography. CRC Press, Boca Raton
5. Bajard JC, Imbert L (2004) A full RNS implementation of RSA. IEEE Trans Comput 53(6):769–774
6. Cardarilli GC, Nannarelli A, Re M (2007) Residue number system for low-power DSP applications. 2007 Conference record of the forty-first asilomar conference on signals, systems and computers. IEEE, pp 1412–1416
7. Amerbaev VM, Solovyev RA, Stempkovskiy AL, Telpukhov DV (2014) Efficient calculation of cyclic convolution by means of fast Fourier transform in a finite field. In: Proceedings of IEEE east-west design & test symposium (EWDTS 2014), pp 1–4. https://doi.org/10.1109/EWDTS.2014.7027043
8. Krasnobayev VA, Yanko AS, Koshman SA (2016) A Method for arithmetic comparison of data represented in a residue number system. Cybern Syst Anal 52(1):145–150

9. Telpukhov DV, Solovyev RA, Mkrtchan IA (2020) Hardware implementation of scaling in residue number system in application to convolutional neural networks. In: Information innovative technologies. Materials of the international scientific - practical conference, Prague, pp 165–169

10. Kong Y, Phillips B (2009) Fast scaling in the residue number system. IEEE Trans Very Large Scale Integrat Syst 17(3):443–447

11. Alia G, Martinelli E (1984) A VLSI algorithm for direct and reverse conversion from weighted binary number system to residue number system. IEEE Trans Circuits Syst 31(12):1033–1039

12. Amerbaev VM, Balaka ES, Solov'ev RA, Tel'puhov DV (2014) Postroenie obratnyh preobrazovatelej modulyarnoj arifmetiki s korrekciej oshibok na baze poliadicheskogo koda. Nejrokomp'jutery: Razrab Primen 9:30–36

13. Chervyakov NI et al (2017) Residue-to-binary conversion for general moduli sets based on approximate Chinese remainder theorem. Int J Comput Math 94(9):1833–1849

14. Cao B, Chang C, Srikanthan T (2007) A residue-to-binary converter for a new five-moduli set. IEEE Trans Circuits Syst I Regul Pap 54(5):1041–1049. https://doi.org/10.1109/TCSI.2007.890623

# Group Method of Data Separation in Wireless Sensor Networks

Yuriy Kocherov⬤, Mikhail Babenko⬤, and Dmitriy Samoylenko⬤

**Abstract** The paper proposes a reliable method of data transmission over wireless sensor networks. The proposed method is reliable because for data transmission over wireless sensor networks information is divided into parts using a group threshold data separation scheme. Data separation algorithms based on the residual class system are frequently used because they have high operation speed and low power consumption. The work shows that with the same number of parts, it is expedient to use a group data separation scheme. For the considered example, the increase in the average system uptime is 78%.

**Keywords** Wireless sensor networks · Threshold separation of data · Increased reliability · Residual class system · Chinese residual theorem

## 1 Introduction

Wireless sensor networks are a special class of networks that consist of a multitude of spatially distributed miniature devices (sensors) that are self-powered and united into a single network through a radio channel [1].

These miniature devices are used in various sectors of the national economy. There are three main types of sensors: sensors that measure physical, chemical, and biological processes.

Wireless sensor networks can include various types of sensors, such as: low sampling rate, seismic, magnetic, thermal, visual, infrared, radar, and acoustic sensors. Such sensors are used to monitor a wide range of environmental properties and situations.

Y. Kocherov (✉) · D. Samoylenko
North Caucasus Federal University, Stavropol, Russia
e-mail: kocherov_yra@mail.ru

M. Babenko
North Caucasus Center for Mathematical Research, North-Caucasus Federal University, 355017 Stavropol, Russia
e-mail: mgbabenko@ncfu.ru

Wireless sensor networks consist of an information collection network and a network for its distribution. The data collection network is used to obtain primary information about the state of an object or its properties and is transmitted to the storage using various methods of data distribution [2].

The following areas of application of wireless sensor networks are distinguished:

- military use,
- health monitoring,
- environmental applications (environmental monitoring),
- smart buildings (smart home technologies),
- commercial applications,
- security applications,
- detection of forest fires,
- detection of landslides,
- monitoring of water quality,
- industrial monitoring.

However, wireless sensor networks have following disadvantages:

- can be hacked by third parties
- since data is transmitted over radio channels, they are susceptible to interference,
- cannot be used for high-speed communication, as it is used for low-speed applications,
- the high cost of building these networks,
- when building these networks, it is necessary to take into account such problems as bandwidth, energy efficiency, cost of nodes, etc.

To solve problems such as increasing the reliability of wireless sensor networks and reducing the risk of hacking, it is proposed to use multichannel data transmission using algorithms for threshold data separation.

## 2   Materials and Methods

To efficiently use the bandwidth of the communication channels of the wireless sensor network and protect data, it is proposed to divide data into parts using wireless sensor networks and send them through different routes [3, 4].

The hreshold data separation scheme is a $(k, n)$ scheme where (is the total number of parts into which information is divided, is the minimum sufficient number of parts necessary to restore the original data). The advantage of this schemes is that the message is restored if there are $k$ or more number of pieces of data such that $k < n$ [5].

The concept of threshold data sharing was independently proposed in 1979 by Adde Shamir and George Blackley.

Shamir's threshold scheme is based on polynomial interpolation, which states that interpolation is impossible if there are fewer points known. Thus, through two points on the plane, an unlimited number of curves of the second degree can be constructed, and in order to construct the only true curve through them, a third one is needed.

George Blakely's scheme is based on the principle of vector data separation, according to which the intersection of linearly independent equations of the planes is a straight line and the intersection of linearly independent order is a point, the coordinate of which is shared data.

Unlike Shamir and Blakely's scheme, whose principle is based on calculating the equations of a polynomial or plane, there are other methods for dividing data, which are based on calculating the remainders of the integer division of information into a bases set.

The residual class system is a non-positional number system that is based on modular arithmetic. The numbers in the residual class system are represented as remainders after division by a series of coprime numbers. The mathematical substantiation of the functioning of the residual class system is based on the Chinese remainder theorem [6, 7].

The Mignott data scheme is based on a residual class system and allows a user, who has a certain permitted number of pieces of information, to restore it in a unique way.

The principle of operation of the scheme can be described by the following example. It is necessary to divide $S$ information among $n$ users in such a way that, having $k$ number of parts known, it would be possible to restore the original information, and having $k - 1$, it would not be possible.

This requires a sequence of natural numbers such that:

$$p_1 < p_2 < \ldots < p_n \text{ and } \coprod_{i=0}^{k-2} p_{n-i} < \coprod_{i=1}^{k} p_i.$$

Moreover, the following conditions must be met:

- any two numbers in the sequence must be mutually prime, i.e. $gnd\left(p_i, p_j\right) = (\forall : i, j = 0, 1, 2, \ldots, n; i \neq j)$;
- information must be in the range:

$$\alpha < S < \beta \text{ where } \alpha = \coprod_{i=1}^{k} p_i, {}_\text{a} \beta = \coprod_{i=1}^{k-2} p_{n-i},$$

$$\text{i.e.} p_1 \cdot p_2 \cdot \ldots \cdot p_k < S < p_{n-k+2} \cdot p_{n-k+3} \cdot \ldots \cdot p_n$$

Parts are calculated according $\alpha_i = S \mod p_i$ for all $i \in [1; n]$ and are distributed over the channels.

The Asmuth-Bloom scheme is a threshold secret sharing scheme built using a series of primes that allows data to be divided among the $k$ parties so that it can be restored by any participants.

To share the secret with the Asmuth-Bloom scheme, you need to choose a prime $q$ number larger than $S$.

The next step is to select $n$ mutually prime numbers $p_1 \cdot p_2 \cdot \ldots \cdot p_n$, satisfying the following conditions:

**Fig. 1** Structural diagram of dividing data into many subgroups

- $\forall i : q < p_i$;
- $\forall i : p_i < p_{i+1}$;
- $p_1 \cdot p_2 \cdot \ldots \cdot p_k < q \cdot p_{n-k+2} \cdot p_{n-k+3} \cdot \ldots p_n \, p_1 \cdot p_2 \cdot \ldots \cdot p_k < q \cdot p_{n-k+2} \cdot p_{n-k+3} \cdot \ldots \cdot p_n$
  [8].

In the next step, a random number $r$ should be chosen to and calculate $S' = S + q \cdot r$.

The parts of the secret are calculated using the formula $\alpha_i = S' \bmod p_i$. The following information. $\{q, p_i, \alpha_i\}$ is distributed among the participants.

From the review of data separation schemes, it can be seen that the use of data separation schemes based on the residual class system is preferable, since they are faster and less computationally complex [9].

To improve the reliability of wireless sensor networks, we propose to modify the data separation algorithm by dividing information into many subgroups [10] (Fig. 1).

The operation of the algorithm with division into many subgroups is presented below. It consists of two stages:

1) The information $S$ is divided into a set, consisting of $n$ parts of the "group leaders" $F_1, F_2, \ldots, F_n$.
2) Each "group leader" $F_1, F_2, \ldots, F_n$ is divided into its new set, consisting of $m$ parts $\left(F_{1_1}, F_{1_2}, \ldots, F_{1_m}\right)\left(F_{2_1}, F_{2_2}, \ldots, F_{2_m}\right) \ldots \left(F_{n_1}, F_{n_2}, \ldots, F_{n_m}\right)$.

The received pieces $n \times m$ of information $\left(F_{1_1}, F_{1_2}, \ldots, F_{1_m}\right)\left(F_{2_1}, F_{2_2}, \ldots, F_{2_m}\right)$ $\ldots \left(F_{n_1}, F_{n_2}, \ldots, F_{n_m}\right)$ are transmitted through wireless sensor networks.

## 3   Results

Any threshold data sharing scheme can be attributed to the majority system, as it can be regarded as a variant of a system with a parallel connection of a plurality of elements, the failure of which will occur if fewer than $k$ of $n$ parallel connected elements are operable.

In calculating the reliability of majority systems, the brute force method and combinatorial methods are used.

In the combinatorial method, according to formula (1), the probability of an event, when at least $k$ elements out of total $n$ number of elements are operable, is estimated:

$$P(t) = \sum_{i=k}^{n} C_i^n \cdot p(t)^i \cdot q(t)^{n-i} \tag{1}$$

where $p(t)$ is the probability of failure-free operation of one system element. $q(t) = 1 - p(t)$, $C_i^n = \frac{n!}{i!(n-i)!}$ is a binominal coefficient of $k$ from $n$.

Let us evaluate the performance of the system when $n = 5$ and $k = 3$ (Fig. 2). To estimate the distribution time of the failure-free operation of one element of the system, we take the exponential distribution law $p(t) = \lambda \cdot e^{-\lambda t}$ where $\lambda = 1$. Substituting all the value in the formula (1), we get:

$$P(t) = \sum_{i=3}^{5} \frac{5!}{i!(5-i)!} \cdot \left( \lambda \cdot e^{-\lambda t} \right)^i \cdot \left( 1 - \lambda \cdot e^{-\lambda t} \right)^{5-i}$$

We will also evaluate the performance of the system with proportionally increased $n = 25$ and $k = 15$ (Fig. 2). Substituting all the value in formula (1), we get:

$$P_x(t) = \sum_{i=15}^{25} \frac{25!}{i!(25-i)!} \cdot \left( \lambda \cdot e^{-\lambda t} \right)^i \cdot \left( 1 - \lambda \cdot e^{-\lambda t} \right)^{25-i}$$

From the graphs presented in Fig. 2 it can be seen that with a proportional increase in the values $n$ and $k$ the uptime decreases. Integrating $P(x)$ and $P_x(x)$ in the range $[0; \infty)$ we get the following results: $\int_0^\infty P(t)dt = 0,783$; $\int_0^\infty P_x(t)dt = 0,564$. For the considered example, the reduction in average system uptime is 72%.

We will consider the proposed group threshold data sharing scheme. To do this, we will consider a scheme of five groups of 5 elements in a way that at least three

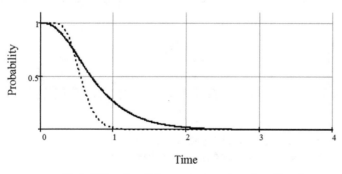

Fig. 2 System uptime graphs $P(x)$ and $P_x(x)$

elements of the total number of elements in the group remain operable, and the whole scheme works when there are less than three groups in working order.

Then the probability of uptime of any group can be estimated by the formula (2):

$$P_g(t) = \sum_{i=3}^{5} \frac{5!}{i!(5-i)!} \cdot \left(\lambda \cdot e^{-\lambda t}\right)^i \cdot \left(1 - \lambda \cdot e^{-\lambda t}\right)^{5-i} \tag{2}$$

Formula 3 can be used to estimate the probability of uptime of the scheme as a whole.

$$P_{ob}(t) = \sum_{i=3}^{5} \frac{5!}{i!(5-i)!} \cdot \left(P_g(t)\right)^i \cdot \left(1 - P_g(t)\right)^{5-i} \tag{3}$$

Figure 3 shows the graphs of the dependence of the time of failure-free operation of the group data separation scheme and the classical scheme for $n = 25$ and $k = 15$.

If $P_{ob}(x)$ and $P_x(x)$ are integrated in the range $[0;\infty)$ then we obtain the following results: $\int_0^\infty P_{ob}(t)dt = 0,719$; $\int_0^\infty P_x(t)dt = 0,564$.

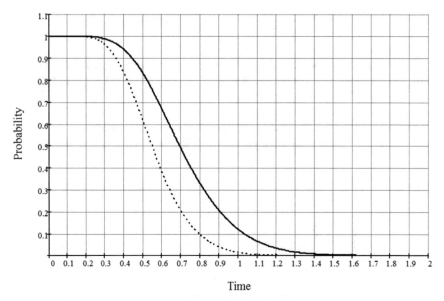

— Probability of no-failure operation of a group sheme
····· Probability of no-failure operation at n = 25 and k = 15

**Fig. 3** The graphs of the dependence of the time of failure-free operation of the Group data separation scheme $P_{ob}(t)$ and the classical scheme for $n = 5$ and $k = 3$

Thus, with the same number of data parts, it is advisable to use a group data separation scheme. For the considered example, the increase in average system uptime is 78%.

# 4   Conclusion

According to the results of the study, it can be said that in order to improve the reliability of wireless sensor networks, it is necessary to apply group schemes of threshold data separation.

This is ensured by the following factors:

– since the system of residual classes has corrective properties, the reliability of data transmission is increased. Moreover, the system of residual classes allows to detect errors in data transmission through wireless sensor networks. In addition, it allows you adaptively change the correcting properties of the code.
– the calculations carried out in the work shows that with the same number of data parts, it is advisable to use a group data separation scheme. For the example considered in the paper, the increase in the average system uptime is 78%.

# References

1. Sofi A, Jane Regita J, Rane B, Lau HH (2021) Structural health monitoring using wireless smart sensor network – an overview. Mech Syst Signal Process 163
2. Sidorenko AV (2015) The use of wireless sensor networks in the tasks of monitoring technogenic objects. Lect BSUIF 7(93):77–82
3. Jun S (2010) Method of encoding and transmission of multimedia data in wireless sensor networks. In: Measuring and computing equipment in technological processes, vol 2, pp 203–203
4. Anchitaalagammai JV, Muthumayil K, Subramaniam DK, Verma R, Muralikrishnan P, Visalaxi G (2021) An enhanced routing and lifetime performance for industrial wireless sensor networks. Intell Autom Soft Comput 31(3):1783–1792
5. Miranda-López V, Tchernykh A, Cortés-Mendoza JM, Babenko M, Radchenko G, Nesmachnow S, Du Z (2018) Experimental analysis of secret sharing schemes for cloud storage based on RNS. Commun Comput Inf Sci 796:370–383
6. Cherviakov NI, Babenko MG, Shabalina MN (2017). Development of a secure system for distributed data storage and processing in the clouds based on the concept of active security in RNS. In: Proceedings of 2017 20th IEEE international conference on soft computing and measurements, SCM (2017). IEEE, pp 558–560
7. Selianinau M (2021) Computationally efficient approach to implementation of the chinese remainder theorem algorithm in minimally redundant residue number system. Theory Comput Syst 65(7):1117–1140
8. Kocherov YN, Samoylenko DV, Tikhonov EE (2020) Safe storage of biometric data. In: 2020 international multi-conference on industrial engineering and modern technologies. FarEastCon, pp 1–5

9. Kocherov YN, Samoylenko DV (2021) Development of a reliable RTK communication method based on a group data separation method based on a residual class system. Izvestia SFU Tech Sci 1(218):218–235

10. Kocherov YN, Samoylenko DV, Koldaev AI (2018) Development of an antinoise method of data sharing based on the application of a two-step-up system of residual classes. In: 2018 international multi-conference on industrial engineering and modern technologies. FarEastCon, pp 1–5

# Data Processing in Problem-Solving of Energy System Vulnerability Based on In-memory Data Grid

**Sergey Gorsky⊙, Alexei Edelev⊙, and Alexander Feoktistov⊙**

**Abstract** Nowadays, data analysis is an integral part of large-scale scientific and applied experiments. Opportunities of modern computing environments allow us to move from operating with traditional storage systems within solving data-intensive problems to the in-memory data grid technologies. Such technologies improve the performance and scalability of data processing compared to external databases because of a faster random access memory and other hardware advancements. The considered data grid enables applications to cache data in the memory. Based on our practical experience, we discuss the advantages of applying the in-memory data grid technology to analyze the energy system vulnerability. The complexity of this problem is determined by considering possible combinations of simultaneous failures of energy system elements. We use open source-based Apache Ignite to support high-performance computing and data distribution. The study aims to evaluate the impact of the data grid scaling on the problem-solving quality criteria. We used the resources of the public access Irkutsk Supercomputer Center to carry out experiments.

**Keywords** HPC-cluster · Data processing · In-memory data grid · Experimental analysis

## 1 Introduction

Reducing overheads related to data analysis in solving scientific and applied problems is undoubted of great practical significance. In this regard, the research direction concerning the In-Memory Data Grid (IMDG) use brings special attention [1–3]. In

S. Gorsky (✉) · A. Feoktistov
Matrosov Institute for System Dynamics and Control Theory of SB RAS, Irkutsk, Russia
e-mail: gorsky@icc.ru

A. Feoktistov
e-mail: agf@icc.ru

A. Edelev
Melentiev Energy Systems Institute of SB RAS, Irkutsk, Russia
e-mail: flower@isem.irk.ru

A. Tchernykh et al. (eds.), *Mathematics and its Applications in New Computer Systems*,
Lecture Notes in Networks and Systems 424,
https://doi.org/10.1007/978-3-030-97020-8_25

particular, IMDG-based data processing is widely used in modeling various energy infrastructures [4–6].

IMDG is a data structure distributed among data grid nodes. As a rule, such data fully or partially reside in the Random Access Memory (RAM) to speed up computing. Computation speedup is achieved because of reducing the number of exchanges between memory and external storage [7].

The considered problem related to the vulnerability analysis focuses on studying the energy system response on possible failures of its elements. The vulnerability analysis undoubtedly has a combinatorial nature. In problem-solving, we form a set of combinations for potential simultaneous failures of system elements. Thus, high computing complexity and data-intensive processing distinguish carrying out such an analysis [8]. These distinctive features drive the use of High-Performance Computing (HPC).

Another feature of the problem under consideration is storing the data obtained during problem-solving in IMDG and providing access to this data grid for external applications for some time. Therefore, based on the problem dimension, it is required to perform the computing as quickly as possible and save their results in IMDG, minimizing the number of data grid nodes.

The rest of the paper is organized as follows: the next section addresses materials and methods used. In particular, we consider different data processing scenarios, an IMDG management system, and a workflow-based scientific application for problem-solving. Section 3 provides experimental analysis. Finally, Sect. 4 discusses the study results and concludes the paper.

## 2   Materials and Methods

### 2.1   Data Processing Scenarios

Within the considered problem-solving, we can use the following scenarios for data processing:

- External data storage use (Fig. 1a),
- Applying advanced read-through and write-through cache between an application and external data storage (Fig. 1b),
- Using advanced read and write cache only (Fig. 1c).

Fig. 1a demonstrates a traditional data processing scheme with synchronous read and write operations. This scheme ensures consistency and reliability in the data use. However, such data processing often becomes a bottleneck in the problem-solving process when a large number of transactions occur.

The data processing scheme with an intermediate layer based on IMDG is shown in Fig. 1b. In this case, IMDG uses a cache to exchange data frequently used by an application. It accesses an external database to retrieve required data missing in

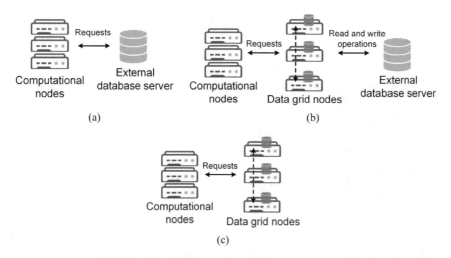

**Fig. 1** Traditional data processing scenario **a**, data processing scenario based on caching through IMDG **b**, scenario for full data processing based on IMDG **c**

the cache or store data from the cache unused for a long time. IMDG interacts with an external database asynchronously. However, in each specific case of using this scheme, ensuring effective data processing often requires additional research related to adapting the application's requests to the characteristics of data grid nodes and an external database [9].

When long-term data storage is not required, we exclude the use of an external database from the problem-solving process. Thus, we reduce the overhead of synchronizing data in the external database. Excluding an external database allows us to speed up data processing since reading and writing data is carried out directly into the cache. Figure 1c illustrates the data processing scheme for this case.

The latter scenario of data processing seems to be preferable. Preliminary comparative analysis confirms this conclusion [6]. At the same time, evaluating the computation scalability limits with a change in the number of computing nodes and nodes for IMDG requires an additional study.

## 2.2 Apache Ignite

In practice, a large spectrum the IMDG management systems are used. Among such systems are Hazelcast [10], Infinispan [11], Apache Ignite [12], and other tools [13]. Based on a comparative analysis of the functionality of the IMDG management systems [6], we have chosen Apache Ignite.

Open-source Java-based Apache Ignite also supports such programming languages as .Net, C++, and PHP. This IMDG management system provides the easy-to-use interface and specialized APIs that allow end-users to distribute computing and data processing among computing nodes and data grid nodes integrated within an Apache Ignite cluster. It also supports the standard ANSI SQL 1999. An Apache Ignite cluster scales horizontally by adding new nodes and vertically by increasing the RAM of nodes.

Thus, Apache Ignite allows us to implement the scenario of data processing selected in the previous section to the full.

An Apache Ignite cluster is a distributed data storage. Data processing is supposed to be performed on the same data grid node where data reside. Therefore, the critical issue is excluding or minimizing the data transfer across the cluster.

## 2.3 Workflow-Based Scientific Application for Solving the Problem of Energy System Vulnerability

To solve the problem of analyzing the energy system vulnerability, we have developed workflow-based scientific application using the Orlando Tools framework [14]. Figure 2 shows the problem-solving scheme (workflow) *s* formed in the application. Using this workflow, we find the energy system critical elements processing Monte Carlo simulation data in IMDG under the Apache Ignite control.

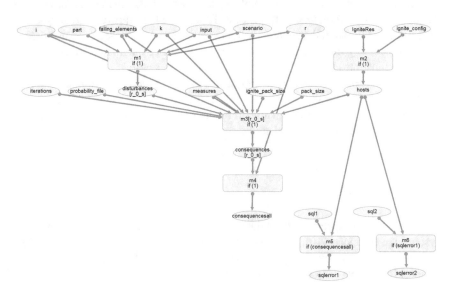

**Fig. 2** Workflow

The workflow includes six modules. Modules are launched in the asynchronous mode when data for their input parameters are ready. The following structure determines an additional condition for launching the module:

$$if(logic\_parameter),$$

where $logic\_parameter$ is a Boolean variable that takes values 0 or 1. The structure

$$if(1)$$

means the unconditional launch of a module when data for its input parameters are ready.

The module $m_1$ reads the initial information about the studied energy system from the database, generates a set of disturbance scenarios, and divides it into $r$ groups for further parallel processing. The structure of the energy system database consists of the following groups of tables:

- reference books of codes for system elements, energy technologies, and resources,
- initial data describing the spatially distributed infrastructure and production capacity of the energy system,
- rules for forming structural and functional sub-models of the energy system.

The module $m_2$ launches Apache Ignite on the dedicated node and then on data grid nodes through the common task queue of the HPC cluster. The module operation result is the parameter $hosts$ determining the addresses of data grid nodes.

The module $m_3$ simulates disturbances following the specified scenarios and forms variants of their consequences. At that, disturbance modeling is performed in parallel. Data related to disturbance scenarios and consequences resides on the Apache Ignite cluster. Each of the $r$ groups is processed by one instance of the $m_3$ module. Instances of the module $m_3$ generate reports on the simulated disturbances and their consequences.

The module $m_4$ collects and aggregates reports generated by the module $m_3$.

A preliminary analysis of the disturbance consequences is carried out by the modules $m_5$ and $m_6$. These modules implement SQL queries to IMDG. The parameters $consequencesall$ and $sqlerror1$ determine the conditions for their launching.

## 3 Experimental Analysis

This section presents the results of the IMDG use obtained from our experiments.

## 3.1   Computing Environment

We carry out experiments using the HPC-cluster resources of the public access Irkutsk Supercomputer Center [15]. The computing environment consists of the following components:

- 16 computing nodes with the following characteristics: 2 processors AMD Opteron 6276 (16 core, 2.3 GHz, 64 GB of RAM) for a node,
- 4 data grid nodes with the same characteristics,
- Dedicated node with the preinstalled Apache Ignite 2.9.1,
- Server of Orlando Tools with the preinstalled system software (Apache 2, MySQL, and PHP 5, as well as the Orlando Tools web interface, daemons, and databases) and various configuration files (application description, resource configuration, and Apache Ignite configuration, etc.).

The modules $m_1 - m_6$ are hosted on computing nodes. Apache Ignite is launched on data grid nodes.

## 3.2   Analysis of Computation Scaling

Within an analysis of computation scaling, we executed the workflow $s_1$ that includes the modules $m_1 - m_4$. Thus, $s_1$ is a part of $s$. The analysis of computation scaling determines the rational ratio of computing nodes and data grid nodes for the different observed parameters. We modeled only 3% of the total number of disturbance scenarios owing to the considered problem complexity.

In the experiments, the workflow makespan, average data recording time, average CPU utilization, and number of faults are selected as observed parameters.

Figure 3a shows the makespan of the workflow $s_1$. We can see a good computation speedup with an increase in computing nodes and data grid nodes. However, an exception is a case with 16 computing nodes and 1 data grid node. In this case, data write faults occurred, driving to the restart of modules (Fig. 3d). Therefore, module restarting led to an increase in the workflow makespan.

The change in the average recording time of data about disturbance scenarios during simulation modeling is shown in Fig. 3b. Figure 3c illustrates the average CPU utilization at various ratios of computing nodes and data grid nodes.

The observed parameters depend on the ratio of computing nodes and data grid nodes to a different extent. For example, the best makespan is achieved using 12 computing nodes and 1 data grid node (Fig. 3a). At the same time, the ratio of 6 computing nodes to 1 data grid node is rational while improving the average CPU utilization (Fig. 3c).

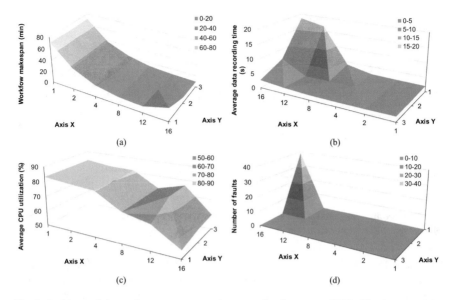

**Fig. 3** Problem-solving makespan **a**, average data recording **b**, average CPU utilization **c**, and the number of faults **d** vs. the number of computing nodes (axis X) and the number of data grid nodes (axis Y)

## 3.3 SQL Queries Processing

Next, we evaluate the Apache Ignite performance regarding the SQL queries processing. New tables are created and filled when SQL queries are executed. To speed up the SQL queries processing, the following settings concerning SQL tables are specified in the Apache Ignite configuration file (Fig. 4):

- Tables are configured to work in atomic (non-transactional) mode,
- Writing to tables is entirely asynchronous,
- Data of one table is split across nodes without creating backups,
- Logic structure of tables provides the location of data from different tables data concerning the same disturbance on the same cluster node through the declaration and use of affinity keys,
- Multithreaded execution of SQL queries is enabled in tables.

```
<bean class="org.apache.ignite.configuration.CacheConfiguration">
    <property name="name" value="Disturbances"/>
    <property name="backups" value="0"/>
    <property name="atomicityMode" value="ATOMIC"/>
    <!-- Set cache mode. -->
    <property name="cacheMode" value="PARTITIONED"/>
    <property name="writeSynchronizationMode" value="FULL_ASYNC"/>
    <!--The number of threads to process a SQL queries executed on the cache.-->
    <property name="queryParallelism" value="32"/>
```

**Fig. 4** Fragment of the Apache Ignite configuration file

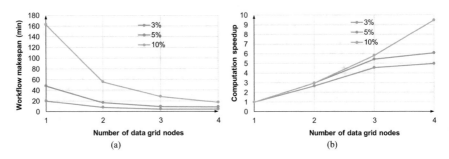

**Fig. 5** Average time of SQL queries processing vs. data set size

We executed the workflow $s_2$ that includes the modules $m_5$ and $m_6$. Thus, $s_2$ is also a part of $s$. The results of the computational experiments are shown in Fig. 5. Figure 5a demonstrates the makespan dependence of the modules $m_5$ and $m_6$ on the dimension of the solved problem (3%, 5%, and 10% of the total number of disturbance scenarios) and the number of nodes in the Apache Ignite cluster. The module makespan includes the overhead associated with launching modules from the common task queue of the HPC cluster. We can see that the makespan increase with growth in the dimension of the solved problem. Moreover, data processing time increases more rapidly in comparison with dimension growth.

At the same time, increasing the number of cluster nodes results in over-linear computation speedup (Fig. 5b). Such speedup can be assumed due to the separation of tables in IMDG. In this case, executing a SQL query requires processing a small amount of data. At the same time, when using only one data grid node, a much more significant amount of data is processed with each SQL query.

## 4  Discussion and Conclusions

The paper focused on the IMDG use in the problem-solving process to analyze the energy system vulnerability when processed data entirely resides in RAM. We analyzed the horizontal scaling of data processing by increasing the number of computing nodes and data grid nodes concerning the workflow makespan reduction. In addition, we also compared such observed parameters as the average data recording, the average CPU utilization, and the number of faults.

The experiment results have confirmed the possibility of improving the problem-solving quality criteria by scaling IMDG. At the same time, the intuitive reasons for limiting the scaling have become clear.

As future work, we assume to define formal evaluations of the possible rational scaling IMDG, considering the supposed number of the requests of applications and characteristics of computing nodes, data grid nodes, and interconnects. In addition, we plan to partially automize the data grid deployment based on the Ignite Apache in the cloud environment.

**Acknowledgements** The study was supported by the Ministry of Science and Higher Education of the Russian Federation (project no. FWEW-2021-0005 "Technologies for the development and analysis of subject-oriented intelligent group control systems in non-deterministic distributed environments"), as well as the Russian Foundation of Basic Research and Government of Irkutsk Region (project no. 20-47-380002).

# References

1. Arora I, Gupta A (2014) Improving performance of cloud based transactional applications using in-memory data grid. Int J Comput Appl 107(13):14–19
2. Siddiqa A, Karim A, Gani A (2017) Big data storage technologies: a survey. Front Inf Technol Electron. Eng 18(8):1040–1070
3. Salhi H, Odeh F, Nasser R, Taweel A (2017) Open source in-memory data grid systems: Benchmarking hazelcast and infinispan. In: Proceedings of the 8th ACM/SPEC on international conference on performance engineering, pp 163–164
4. Werth A, Andr'e A, Kawamoto D, Morita T, Tajima S, Tokoro M, Yanagidaira D, Tanaka K (2016) Peer-to-peer control system for DC microgrids. IEEE Trans Smart Grid 9(4):3667–3675
5. Zhou M, Feng D (2018) Application of in-memory computing to online power grid analysis. IFAC-PapersOnLine 51(28):132–137
6. Edelev A, Sidorov I, Gorsky S, Feoktistov AG (2020) Large-scale analysis of energy system vulnerability using in-memory data grid. In: ICCS-DE, pp 89–98
7. Zhu X, Qin X, Qiu M (2011) QoS-aware fault-tolerant scheduling for real-time tasks on heterogeneous clusters. IEEE Trans Comput 60(6):800–812
8. Gwalani H, Mikler AR, Ramisetty-Mikler S, O'Neill M (2017) Collection and integration of multi-spatial and multi-type data for vulnerability analysis in emergency response plans. In: Advances and new trends in environmental informatics. Springer, Cham, pp 89–101
9. Gai K, Qiu M, Liu M, Xiong Z (2018) In-memory big data analytics under space constraints using dynamic programming. Futur Gener Comput Syst 83:219–227
10. Johns M (2013) Getting Started with Hazelcast. Packt Publishing Ltd
11. Marchioni F, Surtani M (2012) Infinispan data grid platform. Packt Publishing Ltd
12. Bhuiyan S, Zheludkov M, Isachenko T (2017) High performance in-memory computing with apache ignite. Lulu.com
13. Top 15 In Memory Data Grid Platform. https://www.predictiveanalyticstoday.com/top-memory-data-grid-applications
14. Tchernykh A, Bychkov I, Feoktistov A, Gorsky S, Sidorov I, Kostromin R, Edelev A, Zorkalzev V, Avetisyan A (2020) Mitigating uncertainty in developing and applying scientific applications in an integrated computing environment. Program Comput Softw 46(8):483–502
15. Irkutsk Supercomputer Center of the SB RAS. http://hpc.icc.ru

# Wind Speed Recovering from Lidar Sensing Data by Solving of Inverse Problem

Nikolay Baranov[ORCID]

**Abstract** The wind characteristics measurements by use Doppler wind lidars is actively developing. Unlike measuring with anemometers, remote sensing allows you to get only an average value of the wind speed. In the case of essential spatial variability, the average value can differ significantly from the anemometer measurements. We consider the problem of improving the accuracy of determining wind characteristics from remote lidar sensing data in the case of a non-uniform wind field. Determination of the wind speed is considered as the inverse problem of recovering a function from the values of its integrals. The unknown function of the wind speed is represented in the form of its spline approximation. Various conditions of conjugate splines are considered. The analysis of the influence of the parameters of the problem on the quality of restoration is carried out. As a result, rational values of the parameters of the problem are obtained, which give a stable solution.

**Keywords** Wind speed reconstruction · Pulsed Doppler wind lidar · Radial speed · Spline approximation · Inverse problem

## 1 Introduction

The Doppler lidar remote sensing is the effective tool for the studying of the surface layer wind field. It is used in particular in the aviation and the wind energy [1–3]. A feature of lidar sensing is that we get the average value of the velocity in a certain area along the measurement direction. The size of this area depends on the pulse duration. In the case of a homogeneous wind field, such an estimate is accurate [4–7]. However, if the wind speed gradient changes in space, then the estimate will worse. Here we do not mean turbulent fluctuations, but we are talking specifically about the spatial variability of the wind field. In fact, in the case of a strongly inhomogeneous wind field, we must solve the inverse problem of the wind speed recovering from its particular integrals [8].

N. Baranov (✉)
Dorodnicyn Computing Centre, FRC CSC RAS, Moscow, Russia
e-mail: baranov@ians.aero

© The Author(s), under exclusive license to Springer Nature Switzerland AG 2022    281
A. Tchernykh et al. (eds.), *Mathematics and its Applications in New Computer Systems*,
Lecture Notes in Networks and Systems 424,
https://doi.org/10.1007/978-3-030-97020-8_26

Improving the quality of lidar measurements is the subject of ongoing research [9–11]. However, the problem of increasing the accuracy by solving the inverse problem of recovering a function has not yet been considered. Here we consider an approach to the problem of reconstructing the velocity function in the form of a spline function.

## 2 Materials and Methods

The initial data are the lidar measurements data of the wind field. We use the measurements data obtained with a pulsed Doppler wind lidar. The space resolution of a pulsed Doppler wind lidar is proportional to the pulse duration $\tau$ and is equal to

$$l = \frac{c\tau}{2},$$

где $c$ is the light speed [12].

Let $W(x)$ is a unknown radial speed function, which is defined on the interval $\left[x_0 - \frac{l}{2}; x_k + \frac{l}{2}\right]$. The measurements of the wind speed is performed at the points $x_i, i = 0, \ldots, k$, along a something direction at that

$$x_i - x_{i-1} = \alpha l,$$

where $\alpha \in \{0.25; 0.5; 0.75\}$ [13]. At each point $x_i, i = 0, \ldots, k$, we measure the average value of wind speed

$$\hat{W}_i = \frac{1}{l} \int_{x_i - \frac{l}{2}}^{x_i + \frac{l}{2}} W(x)dx. \tag{1}$$

The problem is to reconstruct the function $W(x)$ from the values of the particular integrals at the intervals $\left[x_i - \frac{l}{2}; x_i + \frac{l}{2}\right], i = 0, \ldots, k$.

We will solve the problem of the wind speed reconstruction in the form a spline approximation [14]

$$\tilde{W}(x) \approx \sum_{j=1}^{m} I_j(x)S_j(x), \tag{2}$$

where $S_j(x)$ is the cubic spline of the form

$$S_j(x) = a_{0j} + a_{1j}\left(x - z_{j-1}\right) + a_{2j}\left(x - z_{j-1}\right)^2 + a_{3j}\left(x - z_{j-1}\right)^3, \tag{3}$$

which is defined at the interval $\left[z_{j-1}; z_j\right], a_{0j}, a_{1j}, a_{2j}, a_{3j}$ are the spline coefficients and

$$I_j(x) = \begin{cases} 1, x \in [z_{j-1}; z_j], \\ 0, otherwise. \end{cases}$$

The nodes $z_j$, $j = 0, \ldots, m$, satisfy the conditions

$$z_0 = x_0 - \frac{l}{2}, \; z_m = x_k + \frac{l}{2}.$$

The integral (1) of the spline (3) defined at the interval $[z_{j-1}; z_j]$ is equal to

$$J_{ij} = \frac{a_{0j}}{l}\left(q_{ij}^{(2)} - q_{ij}^{(1)}\right) + \frac{a_{1j}}{2l}\left(\left(q_{ij}^{(2)}\right)^2 - \left(q_{ij}^{(1)}\right)^2\right)$$
$$+ \frac{a_{2j}}{3l}\left(\left(q_{ij}^{(2)}\right)^3 - \left(q_{ij}^{(1)}\right)^3\right) + \frac{a_{3j}}{4l}\left(\left(q_{ij}^{(2)}\right)^4 - \left(q_{ij}^{(1)}\right)^4\right), \tag{4}$$

where

$$q_{ij}^{(1)} = max\left(x_i - \frac{l}{2}; z_{j-1}\right),$$

$$q_{ij}^{(2)} = min\left(z_j; x_i + \frac{l}{2}\right),$$

if

$$x_i - \frac{l}{2} \le z_{j-1} < x_i + \frac{l}{2}$$

or

$$x_i - \frac{l}{2} < z_j \le x_i + \frac{l}{2},$$

and equals zero otherwise.

Then the condition that the integral of the spline approximation (2) on the each interval $\left[x_i - \frac{l}{2}; x_i + \frac{l}{2}\right]$, $i = 0, \ldots, k$, is equal to the given values (1) has the form

$$\sum_{j=1}^{m} J_{ij} = \hat{W}_i, i = 0, \ldots, k. \tag{5}$$

The linear equations system (5) for the coefficients of the spline (3) should be supplemented the equality conditions of the spline values at the nodes $z_j$, $j = 1, \ldots, m - 1$:

$$a_{0j} + a_{1j}(z_j - z_{j-1}) + a_{2j}(z_j - z_{j-1})^2 + a_{3j}(z_j - z_{j-1})^3 = a_{0j+1}, \tag{6}$$

The number of unknown coefficients equals $4m$ and the dimension of equations system (5), (6) is equal to $k + m$.

Therefore,

$$3m = k. \tag{7}$$

If the condition (7) is not fulfilled, we can use the additional conditions:

– the equality of the first derivatives at the nodes $z_j$, $j = 1, \ldots, m-1$,

$$S'_j(z_j) = S'_{j+1}(z_j), \tag{8}$$

then

$$a_{1j} + 2a_{2j}(z_j - z_{j-1}) + 3a_{3j}(z_j - z_{j-1})^2 = a_{1j+1}; \tag{9}$$

– the equality of the second derivatives at the nodes $z_j$, $j = 1, \ldots, m-1$,

$$S''_j(z_j) = S''_{j+1}(z_j), \tag{10}$$

then

$$a_{2j} + 3a_{3j}(z_j - z_{j-1}) = a_{2j+1}; \tag{11}$$

– the equality of the second derivative at the node $z_0$ to zero

$$S''_1(z_0) = 0, \tag{12}$$

then

$$a_{21} = 0; \tag{13}$$

– the equality of the second derivative at the node $z_m$ to zero

$$S''_m(z_m) = 0, \tag{14}$$

then

$$a_{2m} + 3a_{3m}(z_m - z_{m-1}) = 0. \tag{15}$$

The number of splines depends on the additional conditions used to solve the problem. If we use the additional conditions (8) or (10), we obtain

$$2m = k - 1. \tag{16}$$

If we use the additional conditions (8) and (10), we obtain

$$m = k - 2.$$

Thus, with the simultaneous fulfillment the conditions (8) and (10) at the conjugation points, it is always necessary to use additional (12) and (14) at the end nodes $z_0$ and $z_m$.

Solving the system of the linear Eqs. (5), (6) together with one or more systems (9), (11), (13), (15) we get the coefficients of splines $S_j(x)$.

## 3 Results

We carried out numerical simulation by varying the conditions for determining the spline coefficients and the number of measurement points. Table 1 shows which conjugation conditions are used depending on the number of measurement points.

Figures 1, 2 and 3 show some results of solving the problem for different values of the parameters. In the figures, the reference function $W(x)$ is shown in red. Blue color shows the measuring values $\hat{W}_i$. Other designations in the figures: «values»—used conditions (6); «values & 1st der»—used conditions (6), (8); «values & 1st & 2nd der»—used conditions (6), (8), (10); «values & 2nd der»—used conditions (6), (10).

In Fig. 4 an example of solving the inverse problem of recovering a function from its partial integrals for another standard function of polynomial type is shown.

**Table 1** Types of the conjugation conditions

| Conditions | Number of points | | | | | | | |
|---|---|---|---|---|---|---|---|---|
| | 7 ($k = 6$) | | 8 ($k = 7$) | | 9 ($k = 8$) | | 10 ($k = 9$) | |
| | Number of splines | Add. conditions | Number of splines | Add. conditions | Number of splines | Add. conditions | Number of splines | Add. conditions |
| (6) | 2 | – | 3 | (12), (14) | 3 | (12) | 3 | – |
| (6), (8) | 3 | (12) | 3 | – | 4 | (12) | 4 | – |
| (6), (10) | 3 | (12) | 3 | – | 4 | (12) | 4 | – |
| (6), (8), (10) | 4 | (12), (14) | 5 | (12), (14) | 6 | (12), (14) | 7 | (12), (14) |

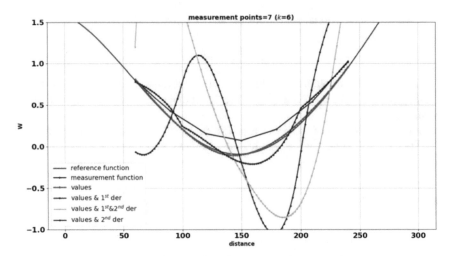

**Fig. 1** Reconstruction of the wind speed function for 7 measuring points

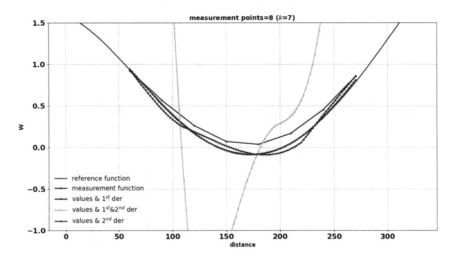

**Fig. 2** Reconstruction of the wind speed function for 8 measuring points

## 4 Discussion and Conclusion

In this paper, an approach to the problem of improving the lidar measurements quality of the wind field is considered. This approach consists in solving the inverse problem of recovering a function from the values of its partial integrals. The original function is approximated by a spline function. The presented results show that this approach allows one to improve the estimation of the function. To obtain a system of equations

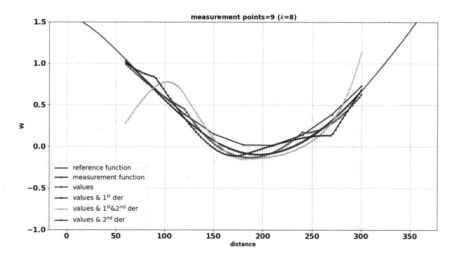

**Fig. 3** Reconstruction of the wind speed function for 9 measuring points

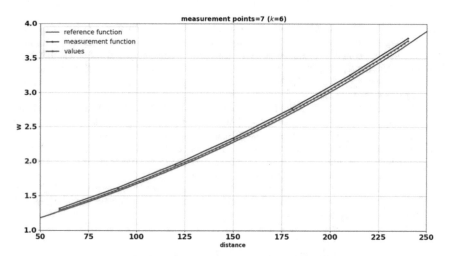

**Fig. 4** Reconstruction for the wind speed function of polynomial type

for determining the coefficients of the spline, additional conditions are used at the conjugation points of the splines.

Using only the conditions (6) (equality conditions of the spline values at the nodes) gives a stable solution if only $k = 3m$ (Figs. 1 and 4).

If it is required to use conditions (12) and (14) (the equality of the second derivative at the nodes $z_0$, $z_m$ to zero) to obtain a closed system of equations for determining the spline coefficients, then the solution is unstable.

If it is required to use only condition (12) to obtain a system of equations for determining the spline coefficients, this also degrades the quality of the solution (Fig. 3).

The conducted studies have shown that a good quality of the solution is achieved when using only conditions (6). For this case, the solution loses stability at a high dimension of the problem ($k > 20$).

If the number of points satisfies condition (16) and the conjugation condition for the first derivatives of the spline is used, a stable solution can be obtained for lower dimensions of the problem ($k < 10$).

Thus, the proposed approach can be used in remote wind sensing lidar systems to improve the quality of measurements.

# References

1. Menke R, Vasiljević N, Wagner J, Oncley SP, Mann J (2020) Multi-lidar wind resource mapping in complex terrain. Wind Energy Sci 5(3):1059–1073. https://doi.org/10.5194/wes-5-1059-2020
2. Kogaki T, Sakurai K, Shimada S, Kawabata H, Otake Y, Kondo K, Fujita E (2020) Field measurements of wind characteristics using LiDAR on a wind farm with downwind turbines installed in a complex terrain region. Energies 13:5135. https://doi.org/10.3390/en13195135
3. Liu Z, Barlow JF, Chan P-W, Fung JCH, Li Y, Ren C, Mak HWL, Ng E (2019) A review of progress and applications of pulsed doppler wind LiDARs. Remote Sens 11:2522. https://doi.org/10.3390/rs11212522
4. Mohandes M, Rehman S, Nuha H, Islam MS, Schulze FH (2021) Wind Speed predictability accuracy with height using LiDAR based measurements and artificial neural networks. Appl Artif Intell 35(8):605–622. https://doi.org/10.1080/08839514.2021.1922850
5. Oertel S, Eggert M, Gutsmuths C, Wilhelm P, Müller H, Többen H (2019) Validation of three-component wind lidar sensor for traceable highly resolved wind vector measurements. J Sens Sens Syst 8:9–17. https://doi.org/10.5194/jsss-8-9-2019
6. Zhou Z, Bu Z (2021) Wind measurement comparison of Doppler lidar with wind cup and L band sounding radar, Atmospheric Measurement Techniques, Discussion [preprint]. https://doi.org/10.5194/amt-2020-516
7. Liu H, Yuan L, Fan C, Liu F, Zhang X, Zhu X, Liu J, Zhu X, Chen W (2020) Performance validation on an all-fiber 1.54-μm pulsed coherent Doppler lidar for wind-profile measurement. Opt Eng 59(1):014109. https://doi.org/10.1117/1.OE.59.1.014109
8. Begmatov AH, Djaykov GM (2016) Numerical recovery of function in a strip from given integral data on linear manifolds. In: 2016 11th international forum on strategic technology (IFOST), pp 478–482. https://doi.org/10.1109/IFOST.2016.7884159
9. Klaas T, Emeis S (2021) The five main influencing factors on lidar errors in complex terrain. Wind Energy Science Discussion [preprint]. https://doi.org/10.5194/wes-2021-26
10. Kim H-G, Meissner C (2017) Correction of LiDAR measurement error in complex terrain by CFD: case study of the Yangyang pumped storage plant. Wind Eng 41(4):226–234. https://doi.org/10.1177/0309524X17709725
11. Risan A, Lund JA, Chang C-Y, Sætran L (2018) Wind in complex terrain—lidar measurements for evaluation of CFD simulations. Remote Sens 10(1):59. https://doi.org/10.3390/rs10010059
12. Simley E, Pao LY (2021) LIDAR wind speed measurements of evolving wind fields. University of Colorado Boulder, Colorado. Subcontract Report NREL/SR-5000-55516

13. Schroeder P, Brewer WA, Choukulkar A, Weickmann A, Zucker M, Holloway MW, Sandberg S (2020) A compact, flexible, and robust micropulsed doppler lidar. J Atmos Oceanic Technol 37(8):1387–1402. https://journals.ametsoc.org/view/journals/atot/37/8/jtechD190142.xml. Accessed 17 Nov 2021
14. Ezhov N, Neitzel F, Petrovic S (2018) Spline approximation, Part 1: basic methodology. J Appl Geodesy 12(2):139–155. https://doi.org/10.1515/jag-2017-0029

# A Simulation Device Model for Managing Access to Confidential Information Stored in a Cloud

**Alexander Troshkov**⊙**, Alexander Malofey**⊙**, Oleg Malofey**⊙**, and Victoria Bondar**⊙

**Abstract** The article offers a look at a model to be used for information resource protection in clouds, which allows managing authorized access to segments of the resources based on employees' biometric features. Ensuring a higher level of confidential data protection takes convenient access, which implies using biometric features. This process is impossible to be studied to the fullest with no model building. By using research methods like measurement and comparison subject to selected features, which characterize the management system operation, data storage and processing, ratios were obtained to be used for analyzing the outcomes of external effects caused on the recognition system, as well as for comparing incoming data. Such a study will result in developing a simulation device model, which is proposed to be employed for creating an access control system for data resources. The reliability of the model, as well as the calculation convergence between the impact and the system response were evaluated based on Fechner's mathematics. The integrated model can be incorporated into the data security employees' peripheral devices at automated workplaces, which will enhance significantly the durability of the protected data.

**Keywords** Simulation modeling · Biometric features · Algorithm · Personal identification and authentication · Information security · information resource

A. Troshkov
Stavropol State Agrarian University, Stavropol, Russia

A. Malofey (✉)
Stavropol branch of the Krasnodar University of the Ministry of Internal Affairs of Russia, Stavropol, Russia
e-mail: skandin@mail.ru

O. Malofey
North Caucasus Federal University, Stavropol, Russia

V. Bondar
North Caucasus Center for Mathematical Research, North-Caucasus Federal University, 355017 Stavropol, Russia

© The Author(s), under exclusive license to Springer Nature Switzerland AG 2022
A. Tchernykh et al. (eds.), *Mathematics and its Applications in New Computer Systems*,
Lecture Notes in Networks and Systems 424,
https://doi.org/10.1007/978-3-030-97020-8_27

# 1  Introduction

The currently available legislative and regulatory framework on personal biometric features to be used for data security technologies is not reliable enough and reveals systemic flaws. The major disadvantages include lack of consistency and methodology for building biometric identification/authentication to control access to protected data resources; lack of knowledge concerning all possible personal biometric features; imperfection in the biometric system model; a static approach to access control assessment; the planning of data security, at its best, indicates lists of employees and their access, whereas the access information resource is not fully coordinated, and there is no clear representation of the algorithm for working with sensitive papers [1]. The identified shortcomings allow claiming the need to develop a biometric identification/authentication system to restrict access to various types of data resources and exclude any unauthorized activities.

# 2  Materials and Methods

The purpose of this study is potential development of a simulation device model, which will serve to create an access control system for information resources (IR) containing confidential data. The empirical research method expands the potential of the model, includes various directions to solve technical problems employing a biometric system.

Research shows that a promising area for managing access to information resources lies in biometric systems [5, 7–14]. Employing an integrated biometric system to protect various information resources can rely on the designed model of an integrated biometric system [2, 3], which is to be seen in Fig. 1.

The model can be described as a specific closed technical space, which can be sued to solve the issue of access to information resources. The integrated biometric system is based on individual features of the human anthropometric and biometric system, which serve to identify signs, characteristics, and elements as described by vectors in the N-vector space. These vectors are chosen and determined by the selection and correction operations located in the operation-correction M area, while the same area connects the N-vector anthropo-biometric space with the reference vectors of the personal anthropo-biometric features (PABF) of space, S (see Fig. 2).

ABS- anthropo-biometric space (signs – $\mathbf{\Pi}$, feature – $\mathbf{F}$, elements – $\mathbf{E}$)

$$\underbrace{N\,(A,\,B)}_{}\quad \underbrace{MS\,(+,-)}_{} \tag{1}$$

Based on the cohesion model, an authentication algorithm was developed, which is to be seen in Fig. 3 [4].

The algorithm presented above will allow considering the sequence of actions at a certain set biometric feature. The of an authentication model based on biometric

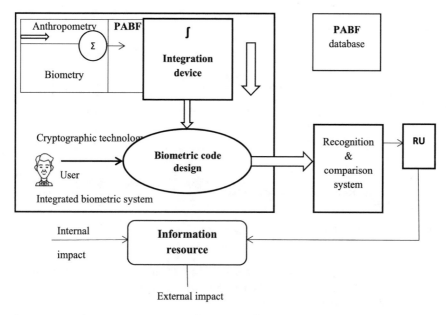

**Fig. 1** Integration of biometric system with peripheral devices

ABS- anthropo-biometric space (signs – **Π**, feature – **F**, elements – **E**).

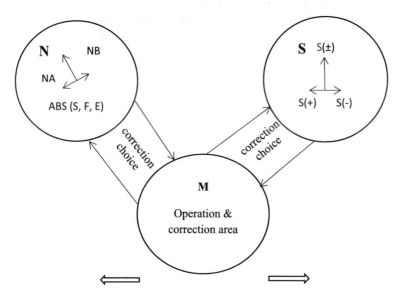

**Fig. 2** A Cohesion model of a vector anthropo-biometric space with PABF space vectors

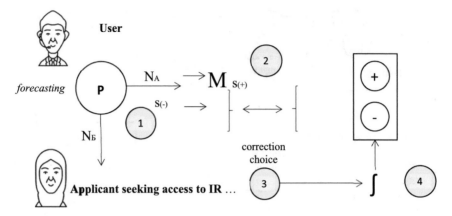

**Fig. 3** Working algorithm for an integrated biometric system

features allows designing models of threats to the data security and integrity posed by security violators (licensed by the Federal Service for Technical and Export Control (FSTEC) of the Russian Federation).

An analysis of scientific and technical sources [7, 8], guidance documents, enterprise information security plans, methods and techniques in the area of information protection allow identifying the criteria (indicators) of security violations or security violation systems, which are to be seen in Fig. 4.

A model like that is designed based on the guiding documents as envisaged in the Russian Government's Decree #1119 of 01/11/12 *On approving requirements for protection of personal data during their processing in personal data information systems.*

## 3  Results

Nowadays, as is obvious from various publications, information wars are underway on all continents, with the number of attacks on information resources having increased dozens of times. Figure 5 shows the main objects of attacks carried out by various kinds of security violators.

The criteria for security violations and the existing foci of attacks have pointed at the need to confirm the reliability of the studies carried out, and for this purpose, a solution to this problem has ben presented based on mathematical operations of the result convergence [5]. Figure 6 shows the result convergence diagrams by the criteria and by the indicators of violations. Convergence analysis of the availability probability of $P_{acc}$ of the IR, showed that the achievement control tends to $\rightarrow 1$ at the (.) access point. This suggests that the higher the indicator P, Z, D, T, C, the more likely is the probability of achieving the (.) access point tends to be $\leq 1$. The convergence of the results can be estimated by Fechner's mathematical method.

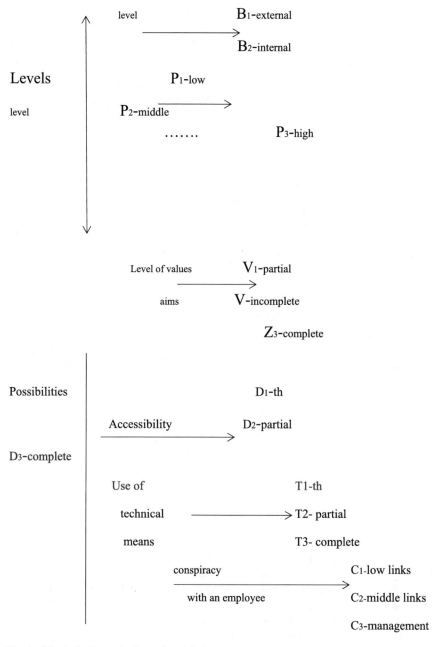

**Fig. 4** Criteria (indicators) of security violations

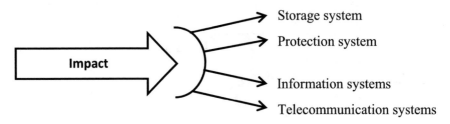

**Fig. 5** Focus of attack

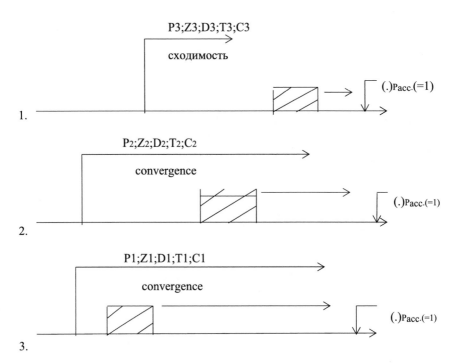

**Fig. 6** Convergence of P accessibility

There is no single approach to designing a security violation model (opinion expressed by research experts at the XIV conference on Information Security – 2018). In view of that, when designing such a model, there may be some improvement proposed regarding the quality of access control to segments of information resources. The design should be started with the life cycle and the value of information resources on the timeline to be seen in Fig. 7.

Figure 7 suggests that the violator will cause the greatest impact in zone A, i.e. where they strive to achieve the 1st probability of convergence (Fig. 6).

In other words, a specific attack on information resources in Zone A is to be qualified based on the following criteria:

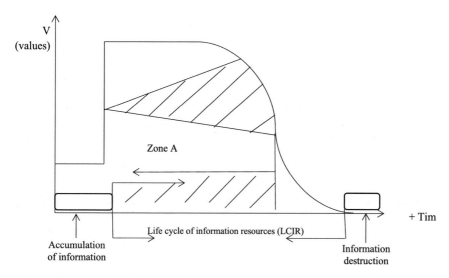

**Fig. 7** Life cycle and value of information resources

– Chasing cycles
– Resource knowledge level
– Use of technical means
– Direction of attacks
– Possible conspiracy
– IR life cycle

A visualization of M reiterating cycles (cycle motivation), the Y level of knowledge (extracted information), the T use of technical means (advance in research and technology), possible C collusion (the consent result), the life cycle (reference points in Fig. 7), can be seen in Fig. 8.

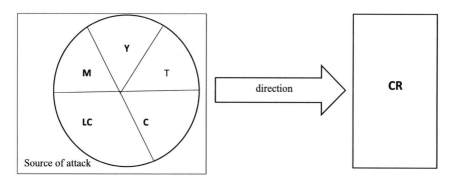

**Fig. 8** Reference points

The request motivation pursues the aims of M. The motivation determines, therefore, in what way the resources information will be obtained

$$M \rightarrow (t, f, p, i) \tag{2}$$

where t is priority, $f$ is dependence, p is the payability, $i$ is the ideologic content.

Further on, the motivation of the request for extracting the resource information will shape the level of the applied laws Y

$$Y \rightarrow (M, E, I, K) \tag{3}$$

where M is motivation, E is the level of education, I is the intelligence data, K is the knowledge of data protection.

In view of the motivation and the level of knowledge, a qualifying attribute is shaped regarding the use of technical means T

$$T \rightarrow (Res, \ Sk, \ Poss, \ Secr) \tag{4}$$

where Res is the use of advanced research & technology, Sk is skills in managing technology, Poss is the possibility and importance of delivering information, and Secr is secrecy of information delivery.

A special mention is to be made of possible conspiracy of employees Consp

$$Consp \rightarrow (E, Net, Psy) \tag{5}$$

where E is the level of education, Net is the personal connections network, and Psy is the psychological preparedness.

Here it would be useful to introduce a qualifying feature of the life cycle (LC IR) (.), which would describe the time and value-related indicator for fixing and marking the time scale.

$$LCIR \rightarrow (Value, \ T \ time) \tag{6}$$

Given the (2, 3, 4, 5, 6), Table 1 of the criteria-related features can be made up.

**Table 1** Criteria-related features

| Type | | | | | | | | | | | | | | | | |
|---|---|---|---|---|---|---|---|---|---|---|---|---|---|---|---|---|
| Criteria | t | $f$ | p | i | V | E | I | K | Re | Sk | Pos | Secr | E | Net | Psy | M |
| M | + | + | + | + | | | | | | | | | | | | |
| Y | | | | | + | + | + | + | | | | | | | | |
| T | | | | | | | + | | + | + | + | + | | | | |
| Consp | | | | | + | | | | | | | | + | + | + | |
| LCIR | + | | | | | | | | | | | | | | | |

Table 1 shows that if taken as a matrix, then there is only one intersection of values in the qualified attributes

$$M \leftrightarrow t \leftrightarrow LCIR \qquad (7)$$

As follows from the equations, the proposed model which incorporates the tools for analyzing biometric characteristics that are built into the peripheral devices of employees in charge of information security, it is hard to gain access to confidential IR. This entails enhanced IR durability and the protection of information circulating in the cloud in general.

## 4 Discussion

The presented results of a simulation device modeling feature a complete structure employing an evidence base relying on mathematical operations performed through modeling technical projects. The empirical research method allows judging the correctness of the chosen research streamline. The efficiency of the model for access and powers separation depends on how correctly was the choice made regarding a segment in the information space, on the choice of biometric characteristics in accordance with the classification assigned to the information, on the correctness of the presentation and comparison of the presented and declared biometric features. The described approach to evaluating the effectiveness of biometric access and information resource management has been implemented through a model that featuring an acceptable algorithm, which is good for the operation of offering users access to a given IR [6]. The correctness of mathematical operations has been proven by the convergence system based on Fechner's mathematics [2, 3]. The presented model can be further used to create an information platform and software for a decision support system by information security experts.

## 5 Conclusion

In view of all the above, the analysis, calculation and research allowed proposing a model of security threats, as well as they revealed that the employees' biometric characteristics are to be taken into account when shaping an algorithm for managing access to information cloud segments, which will protect information against unauthorized access.

**Acknowledgements** The team of the authors herewith would like to express their gratitude to research experts of the North-Caucasus Federal University as well as of the Stavropol State Agrarian University for valuable recommendations and advice offered through the activities on the design of the model.

# References

1. Krishtofik AM, Anischenko VV (2007) Metodologiia postroeniia sistem zaschity informatsii. Izvestiia IUFU. Tekhnicheskie nauki. Tematicheskii vypusk. «Informatsionnaia bezopasnost». Taganrog: Izdatelstvo TTI IUFU 2007 76:238s.
2. Troshkov AM, Troshkov MA (2012) Multi-mnogofaktornye biometricheskie kharakteristiki autentifikatsii lichnosti i sistema ikh zaschity. V- mezhdunarodnaia nauchno-tekhnicheskaia konferentsiia. g. Stavropol: Sev.Kav.GTU (Kislovodsk, 2–6 maia 2012g
3. Cherviakov NI, Golovko AN, Kondrashov AV (2010) Novyi metod i algoritm vypolneniia bazovykh operatsii v ellipticheskikh krivykh, ispolzuemykh v sistemakh kriptograficheskoi zaschity informatsii. / Infokommunikatsionnye tekhnologii 1:23–28
4. Troshkov AM, Troshkov MA (2012) Svidetelstvo o gosudarstvennoi registratsii programmy dlia EVM № 2012617031 «Informatsionnaia sistema autentifikatsii lichnosti po biometrich-eskim kharakteristikam». Zaiavka № 2012614575. Zaregistrirovano v Reestre programm dlia EVM 6
5. Briukhomitskii IUA (2008) Veroiatnostnyi metod klassifikatsii biometricheskikh parametrov lichnosti. Materialy KH Mezhdunarodnoi nauchno-prakticheskoi konferentsii «Informatsion-naia bezopasnost». Chast 1 – Taganrog. Izdatelstvo TTI IUFU 318s
6. Troshkov AM, Troshkov MA (2010) Zaschita kodirovannoi biometricheskoi informatsii na osnove svoistv struktury metall-okisel-poluprovodnik rotovoi polosti cheloveka. Sbornik nauchnykh trudov III NPK «Rossiiskaia tsivilizatsiia: proshloe, nastoiaschee i buduschee». «OOO Mir dannykh», Stavropol. 410s
7. Joshi JBD, Bertino E, Latif U, Ghafoor A (2005) A generalized temporal role-based access control model. IEEE Trans Knowl Data Eng 17(1):4–23. https://doi.org/10.1109/TKDE.2005.1
8. A wearable face recognition system on google glass for assisting social interactions. In: Computer vision - ACCV 2014 workshops Singapore, Singapore, November 1–2, 2014, Revised Selected Papers. Part III. P:419–433
9. Vasilyev VI, Lozhnikov PS, Sulavko AE, Fofanov GA, Samal S, Zhumazhanova S (2018) Flex-ible fast learning neural networks and their application for building highly reliable biometric cryptosystems based on dynamic features. IFAC-PapersOnLine 51(30):527–532. https://doi.org/10.1016/j.ifacol.2018.11.272
10. Sulavko AE, Volkov DA, Zhumazhanova SS, Borisov RV (2018) Subjects authentication based on secret biometric patterns using wavelet analysis and flexible neural networks. In: 2018 XIV international scientific-technical conference on actual problems of electronics instrument engineering (APEIE). - Novosibirsk, Russia. IEEE. - October 2, 2018. pp 218–227. https://doi.org/10.1109/APEIE.2018.8545676.
11. Defining and quantifying users' mental imagery-based BCI skills: a first step Fabien Lotte and Camille Jeunet Published 19 June 2018 © 2018 IOP Publishing Ltd Journal of Neural Engineering, 15(4) Citation Fabien Lotte and Camille Jeunet 2018 J. Neural Eng. 15 046030
12. Epifantsev BN (2017) Identifikatsionnyi potentsial polzovatelei kompiuternykh sistem v prot-sesse ikh professionalnoi deiatelnosti [Elektronnyi resurs] : monografiia/B.N. Epifantsev, A.E. Sulavko, A.S. Kovalchuk, N.N. Nigrei, S.S. ZHumazhanova, R.V. Borisov. – Omsk : SibADI, 2017. - 1 elektron. opt. disk (DVD-R). - Zagl. s etiketki diska
13. Osipov DL, Tebueva FB, Ryabtsev SS, Struchkov IV (2019) Development of a server software module for protected data sharing on the internet CEUR workshop proceedings. In: YSIP3 2019 - Proceedings of the young scientist's 3rd international workshop on trends in information processing
14. Gavrishev AA, Zhuk AP, Osipov DL (2016) An analysis of technologies to protect a radio channel of fire alarm systems against unauthorized access. In: SPIIRAS Proceedings 4 (47), pp 28–45

# Implementation of an Integro-Differential Model of a Singularly Loaded Rod by the Finite Element Method

Irina Oblasova⊙, Elena Timofeeva⊙, and Natalia Shiryaeva⊙

**Abstract** This article analyzes an integro-differential equation that models small deformations of an elastically supported console, as well as considers possible use of the finite element method for integro-differential models of a singularly loaded rod. For the proposed model, there is an algorithm for the approximate solution of the equation developed relying on a system of basic functions, the linear combination of which will be the desired approximate solution, while there are also test examples offered for exact solutions of some integro-differential equations and approximate solutions obtained through the finite element method. The examples serve proof to possible transfer of classical projection methods of approximate calculations to the model under study, while employing the developed algorithm of the finite element projection method and conducting a numerical experiment. The numerical experiment showed a reliable and fast convergence. The results obtained within this work allow separating approximate methods depending on the degree of their suitability and effectiveness in problems of unidimensional singular continuum elastic statics.

**Keywords** Mathematical model of a singularly fixed console · Integro-differential model · Integro-differential equation · Finite element method · FEM · Singularly loaded rod

## 1 Introduction

In monograph [10], we developed a mathematical model of a singularly fixed console, as well as studied its most significant properties, namely, the following integro-differential equation was constructed and analyzed.

I. Oblasova (✉) · E. Timofeeva
North-Caucasus Center for Mathematical Research, NCFU, Stavropol, Russia
e-mail: pravotor@list.ru

N. Shiryaeva
North-Caucasus Federal University, Stavropol, Russia

© The Author(s), under exclusive license to Springer Nature Switzerland AG 2022　　301
A. Tchernykh et al. (eds.), *Mathematics and its Applications in New Computer Systems*,
Lecture Notes in Networks and Systems 424,
https://doi.org/10.1007/978-3-030-97020-8_28

$$(pu'')'(x) + \int_0^x u(s)dQ(s) = F(x) + const,$$

assuming that $p$, $Q$ and $F$ are functions of limited variation, whereas the stroke means the usual derivative. The solutions were found within the class of continuously differentiable functions, whose second derivatives have limited by $[0, 1]$ variation, and the $(pu'')'$ derivatives have a finite by $[0, 1]$ change [3]. The genesis of such an equation is explained for a case where the equation is of a physical nature, arising from the problem of minimizing the energy functional for the console [5].

The results obtained in [6] allow substantiating numerical methods, as well as allow estimating the norms of integral operators that reverse the respective models, the convergence rate of iterative processes. To check possible transfer of classical projection methods of approximate calculations to the model investigated, let us construct an algorithm of the finite element projection method and carry out a numerical experiment.

Here we will consider an integro-differential equation

$$-(pu'')'(x) + \int_0^x u(s)dQ(s) = F(x) - F(0) - (pu'')'(0), \tag{1}$$

where $p$, $Q$ and $F$ are limited variation functions, while $\inf_{[0;1]} p > 0$ [9]. Of all the possible solutions for Eq. (1) we will be looking for one that satisfies the boundary conditions

$$u(0) = 0, \quad \frac{du}{dx}(0) = 0, \tag{2}$$

$$\frac{d^2u}{dx^2}(1) = 0, \quad \frac{d^3u}{dx^3}(1) = 0. \tag{3}$$

## 2 Materials and Methods

### 2.1 A Subsection Sample

For an approximate solution of Eq. (1), we will take a system of basic functions, whose linear combination will be the desired approximate solution. For this, we will break the interval $[0, 1]$ into equal parts by (nodal) points $\{x_k\}_{k=-3}^{k=n+3}$, taking $x_k = \frac{k}{h}$, where $h = \frac{1}{n}$. The points $x_{-3}, x_{-2}, x_{-1}, x_n, x_{n+1}, x_{n+2}, x_{n+3}$ have been introduced to ease the presentation. The basic functions shall be set by the formula (Fig. 1).

**Fig. 1** Graphs of $\phi_k(x)$
functions at $k = 0, 1, 2, n$

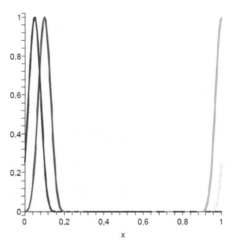

$$\phi_k(x) = \begin{cases} \dfrac{(x - x_k + 2h)^3}{4h^3}, & \text{for } x \in [x_k - 2h, x_k] \\[2mm] \dfrac{3x_k - 6x_k h + 4h^3}{4h^3} - \dfrac{3x_k(x_k - 4h)x}{4h^3} +, & \\[2mm] \dfrac{3(3c - 2h)x^2}{4h^3} - \dfrac{3x^3}{4h^3}, & \text{for } x \in [x_k - h, x_k] \\[2mm] -\dfrac{3x_k + 6x_k h - 4h^3}{4h^3} + \dfrac{3x_k(x_k + 4h)x}{4h^3} +, & \\[2mm] \dfrac{3(3c + 2h)x^2}{4h^3} + \dfrac{3x^3}{4h^3}, & \text{for } x \in [x_k, x_k + h] \\[2mm] -\dfrac{(x - x_k - 2h)^3}{4h^3}, & \text{for } x \in [x_k - 2h, x_k] \\[2mm] 0, & \text{for other } x. \end{cases} \qquad (4)$$

Rather than the desired $u(x)$ function, we will be looking for its values at the nodal points only, and, in view of that, in the equations we will be using not $u(x)$ function yet the function

$$v(x) = \sum_{i=-1}^{n+1} v_i \phi_i(x), \qquad (5)$$

where $v_i$ is its value at the $x_i$ nodal point. Subject to boundary conditions (2) we get

$$v_{-1} = v_0 = 0.$$

In this case, equality (5) appears as follows

$$v(x) = \sum_{i=1}^{n+1} v_i \phi_i(x), \tag{6}$$

We will integrate the integro-differential Eq. (1) along the [0, 1] segment by $\phi_k(x)$ function:

$$-\int_0^1 (pu'')'(x)d\phi_k(x) + \int_0^1 \left(\int_0^x u(s)dQ(s)\right)d\phi_k(x)$$

$$= \int_0^1 F(x)d\phi_k(x) - F(0) - (pu'')'(0)\int_0^1 d\phi_k(x) \tag{7}$$

The second integral in the left part, and the first integral in the right, we will integrate by parts:

$$\int_0^1 \left(\int_0^x u(s)dQ(s)\right)d\phi_k(x)$$

$$= \phi_k(x)\int_0^x u(s)dQ(s)|_{x=0}^{x=1} - \int_0^1 \phi_k(x)u(x)dQ(x) \tag{8}$$

$$= \phi_k(1)\int_0^1 u(s)dQ(s) - \int_0^1 \phi_k(s)u(s)dQ(s);$$

$$\int_0^1 F(x)d\phi_k(x) = F(x)\phi_k(x)|_{x=0}^{x=1} - \int_0^1 \phi_k(s)dF(s)$$

$$= F(1) \cdot \phi_k(1) - F(0) \cdot \phi_k(0) - \int_0^1 \phi_k(s)dF(s). \tag{9}$$

Here we will transform the first integral in the left part of equality (7)

$$\int_0^1 (pu'')'(x)d\phi_k(x) = \int_0^1 \phi_k'(x)d(pu'')(x)$$

$$= \phi_k'(x)(pu'')(x)|_{x=0}^{x=1} - \int_0^1 (pu'')(x)d\phi_k'(x) \tag{10}$$

This means that equality (7), in view of (8), (9) and (10), will appear as

$$-\int_0^1 (pu'')(x)d\phi_k'(x) + \phi_k'(1)(pu'')(1) - \phi_k'(0)(pu'')(0)$$

$$+\phi_k(1) \cdot \int_0^1 u(s)dQ(s) - \int_0^1 \phi_k(s)u(s)dQ(s) = F(1) \cdot \phi_k(1) - F(0) \cdot \phi_k(0)$$

$$-\int_0^1 \phi_k(s)dF(s) - \big(F(0) + (pu'')'(0)\big)(\phi_k(1) - \phi_k(0))$$

$$(11)$$

Then, after transformations

$$pu''(1)\phi_k'(1) - pu''(0)\phi_k'(0) - \int_0^1 pu''(x)d\phi_k'(x) - \int_0^1 u(s)\phi_k(s)dQ(s)$$

$$(12)$$

$$= -\int_0^1 \phi_k(s)dF(s) - \phi_k(1)(pu'')'(1) + \phi_k(0)(pu'')'(0)$$

or taking into account boundary conditions $u''(1) = u'''(1) = 0$,

$$\phi_k(0)(pu'')'(0) + pu''(0)\phi_k'(0) + \int_0^1 pu''(x)d\phi_k'(x) + \int_0^1 u(s)\phi_k(s)dQ(s)$$

$$(13)$$

$$= \int_0^1 \phi_k(s)dF(s)$$

Since derivative of $\phi_k''(x)$ is continuous, then

$$\int_0^1 pu''(s)d\phi_k'(s) = \int_0^1 pu''(s)\phi_k''(s)ds,$$

equality (13) is equivalent to

$$\phi_k(0)(pu'')'(0) + pu''(0)\phi_k'(0) + \int_0^1 pu''(x)\phi_k''(x)dx + \int_0^1 u(s)\phi_k(s)dQ(s)$$

$$= \int_0^1 \phi_k(s)dF(s) \tag{14}$$

We will put here, instead of $u(x)$, function (6). Since $\varphi_k(0) = \phi_k'(0) = 0$ for all $k \geq 1$, then for $k = 0, 1, 2, \ldots, n, n+1$ we get a system of $(n+1)$ equations in regard to $n+1$ unknowns $\{v_1, v_2, \ldots, v_{n+1}\}$.

$$\sum_{i=1}^{n+1} v_i \int_0^1 p\phi_i''(x)\phi_k''(x)dx + \sum_{i=1}^{n+1} v_i \int_0^1 \phi_i\phi_k dQ = \int_0^1 \phi_k dF \tag{15}$$

Here we will introduce notation

$$\langle \phi, \psi \rangle = \int_0^1 p\phi''\psi''dx + \int_0^1 \phi\psi dQ. \tag{16}$$

It is obvious that this is a bilinear symmetric functional in the space of continuous on $[0, 1]$ functions that have a second derivative summable with a square, and satisfying the conditions of $u(0) = u'(0) = 0$. Given the positivity of $p(x)$ function and non-decreasing $Q(x)$, it is also nondegenerate

$$\langle \phi, \psi \rangle = \int_0^1 p\phi''\phi''dx + \int_0^1 \phi\phi dQ > 0, \quad \langle \phi, \phi \rangle = 0 \Leftrightarrow \phi = 0. \tag{17}$$

and therefore, it can serve as a scalar product of functions. Then the coefficients of the Eq. (15)

$$A_{ij} = \langle \phi_i, \phi_j \rangle = A_{ji} \tag{18}$$

make up a Gram matrix of a system of linearly independent vectors $\phi_k$. The determinant of A matrix, therefore, is nonzero, so system (15) has a unique solution.

## 3   Results.

**Test Examples**
Here we will offer exact solutions for some integro-differential equations and approximate solutions obtained through the finite element method. To ensure simplicity of
calculations, we will assume $p(x) \equiv 1$.

**The External Force Consists of Concentrated Efforts**
Let $Q(x) \equiv const$ and $F(x)$ external force appears as

$$F(x) = \begin{cases} 0, & 0 \le x < 1/3; \\ -4, & 1/3 < x < 1/2; \\ -2, & 1/2 < x < 2/3; \\ 2, & 2/3 < x < 3/4; \\ 4, 2, & 3/4 < x < 4/5; \\ -0, 8, & 4/5 < x \le 1. \end{cases} \tag{19}$$

Direct calculation will show that influence function $K(x, s)$ shall appear as

$$K(x, s) = \begin{cases} \dfrac{3 \cdot x^2 \cdot s - x^3}{6}, & x < s; \\ \dfrac{3 \cdot x \cdot s^2 - s^3}{6}, & x \ge s. \end{cases}$$

Given that, the exact solution of the integro-differential equation

$$-(pu'')'(x) + \int_0^x u(s) dQ(s) = F(x) - F(0) - (pu'')'(0),$$

for $F(x)$, determined by equality (19) will appear as

$$u(x) = -4K(x, 1/3) + 2K(x, 1/2) + 4K(x, 2/3) + 2, 2K(x, 3/4) - 5K(x, 4/5),$$

with its graph to be seen in Fig. 2.

**Fig. 2** Exact solution

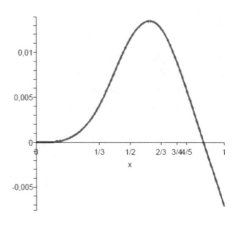

Let us break [0, 1] segment into ten equal parts and apply the above-described scheme of the finite element method. The obtained approximate solution will appear (precision up to thousandths)

$$
v(x) = \begin{cases}
0,102x^3, & 0 \le x \le 0,1; \\
+0,001x - 0,007x^2 + 0,124x^3, & x < 0,2; \\
-0,001 + 0,010x - 0,054x^2 + 0,202x^3, & x < 0,3 \\
0,014 - 0,134x + 0,427x^2 - 0,331x^3, & x < 0,4 \\
0,029 - 0,249x + 0,715x^2 - 0,571x^3, & x < 0,5 \\
-0,014 + 0,010x + 0,196x^2 - 0,225x^3, & x < 0,6 \\
-0,055 + 0,215x - 0,145x^2 - 0,036x^3, & x < 0,7 \\
-0,315 + 1,329x - 1,737x^2 + 0,722x^3, & x < 0,8 \\
0,045 - 0,023x - 0,047x^2 + 0,018x^3, & x < 0,9 \\
0,063 - 0,081x + 0,017x^2 - 0,006x^3, & x \le 1
\end{cases}
$$

Figure 3 shows graph $v(x)$. The exact and approximate solutions are to be seen in Fig. 4 (the thin line stands for the approximate solution). Figure 5 shows $(|u(x) - v(x)|)$ error rate.

Figure 6 depicts the exact and approximate solutions with segment [0, 1] broken in 100 equal parts (error rate in Fig. 7), whereas the time allocated to carry out the numerical experiment, which was done in the Maple 9.5 symbolic mathematics package, is equal to 3548 s $\approx$ 58 min [2].

**Fig. 3** Approximate
solution at N = 10

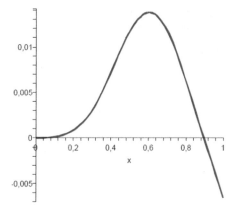

**Fig. 4** The exact and
approximate solutions with
the segment broken in 10
equal parts

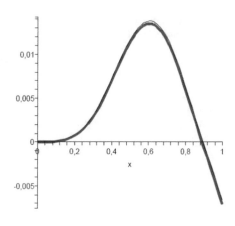

**Fig. 5** Error rate with the
segment broken in 10 equal
parts

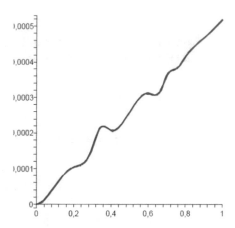

**Fig. 6** The exact and approximate solutions with the segment broken in 100 equal parts

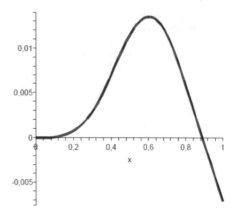

**Fig. 7** Error rate with the segment broken in 100 equal parts

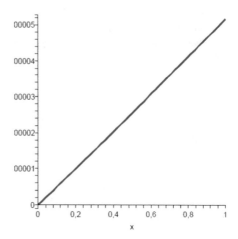

## The External Force has a Continuous Component

Now, let external force $F(x)$ appear as

$$F(x) = 1 - x + \begin{cases} 0, & 0 \leq x < 1/3; \\ -4, & 1/3 < x < 1/2; \\ -2, & 1/2 < x < 2/3; \\ 2, & 2/3 < x < 3/4; \\ 4, 2, & 3/4 < x < 4/5; \\ -0,8, & 4/5 < x \leq 1. \end{cases} \tag{20}$$

The exact solution $u(x)$ of integro-differential equation

$$-(pu'')'(x) + \int\limits_0^x u(s)dQ(s) = F(x) - F(0) - (pu'')'(0),$$

for $F(x)$, determined by equality (20), appears as

$$u(x)= \int\limits_0^1 K(x,s)(1-s)ds$$

$$-4K(x,1/3) + 2K(x,1/2) + 4K(x,2/3) + 2,2K(x,3/4) - 5K(x,4/5)$$

$$= \frac{1}{24}x^4 + \frac{1}{12}x^{24} - \frac{1}{120}x^5 - \frac{1}{12}x^3$$

$$-4K(x,1/3) + 2K(x,1/2) + 4K(x,2/3) + 2,2K(x,3/4) - 5K(x,4/5),$$

while its graph can be seen in Fig. 8.

By breaking the segment into 10 equal parts we will get the approximate solution $v(x)$, whose graph is to be seen in Fig. 9. The exact and approximate solutions are jointly presented in Fig. 10, while the error rate is in Fig. 11.

In case the segment is broken into 100 equal parts, we get the approximate solution $v(x)$, whose graph is to be seen in Fig. 13 (Fig. 12). The exact and approximate solutions can be both seen in Fig. 14 (the thin line shows the approximate solution), the error rate can be seen in Fig. 15.

**Fig. 8** Exact solution

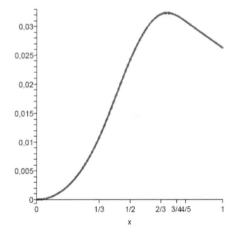

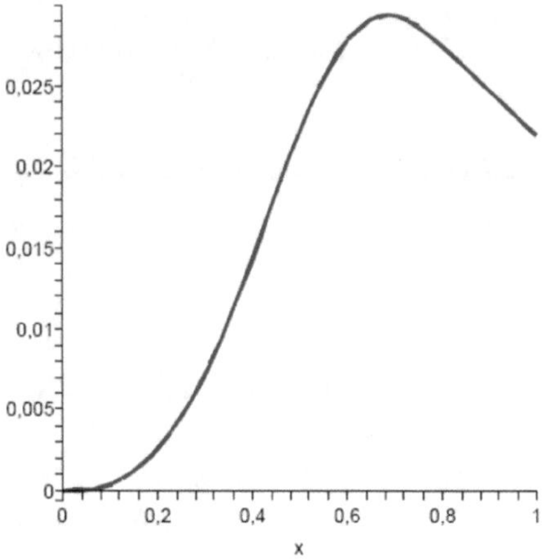

**Fig. 9** Approximate
solution at N = 10

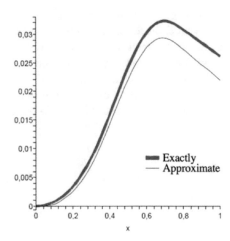

**Fig. 10** Exact and
approximate solutions at
breaking into 10 equal parts

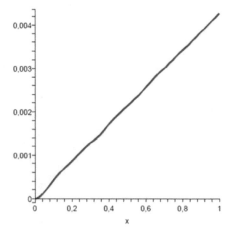

**Fig. 11** Error rate at
breaking into 10 equal parts

**Fig. 12** Exact solution

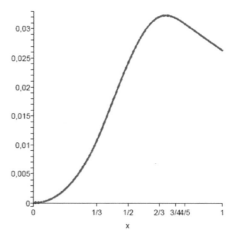

**Fig. 13** Approximate
solution at N = 100

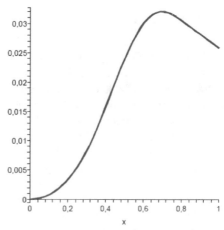

**Fig. 14** Exact and
approximate solutions at
breaking into 100 equal parts

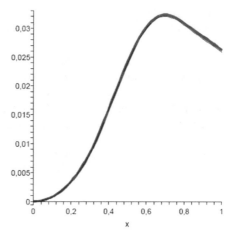

**Fig. 15** Error rate at
breaking into 100 equal parts

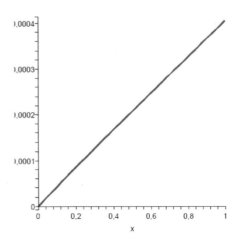

## 4  Conclusion

This article dwells on possible construction of an integro-differential equation of the considered type with strong features of approximate projection methods such as the finite element method. The approbation was done involving certain specific tasks.

From this stance, an algorithm was built in the beginning, which was done for the approximate solution of an integro-differential equation, which models deformations of a singularly loaded rod based on the finite element method. Standard functions for an ordinary equation with smooth coefficients were taken as the basic functions. There were two test examples offered for the built algorithm in case of $p(x) \equiv 1$ and $Q(x) \equiv 0$, which showed fast enough convergence. Since the influence function is known in these cases, then exact solutions could be obtained. The first case corresponds to the situation where the continuous component of the external force is zero, whereas the second one appears as $F_0(x) = 1 - x$. Cubic spline functions were taken as basic ones. Applying the algorithm produced approximate solutions for $n = 10$ and $n = 100$ (the graphs show exact $u(x)$ and approximate $v(x)$ solutions, as well as error rates $|u(x) - v(x)|$ at breaking the segment into 10 and 100 parts).

The numerical simulation completed for a number of model problems based on the projection (finite element method) method, whose results are in line with the real deformation process of a fixed elastic console with singularities, points at the reliability of the constructed model. The numerical experiment has proven a reliably high efficiency of the approach in question.

**Acknowledgements**  The work is supported by North-Caucasus Center for Mathematical Research under agreement №. 075-02-2021-1749 with the Ministry of Science and Higher Education of the Russian Federation.

# References

1. Belytschko T, Liu WK, Moran B (2001) Nonlinear finite elements for continua and structures, vol 16, p 650. Wiley, Chichester
2. Bhatti A (2005) Fundamental finite element analysis and applications: with mathematica and MATLAB computations, p 720. Wiley, New York
3. John F (1986) Partial differential equations, p 250. Springer, New York
4. Kudryavtsev OE (2010) Sovremennye chislennye metody resheniya integro-differentsialnykh uravnenii, voznikayuschikh v prilozheniyakh, p 141. Vuzovskaya kniga, Moskva
5. Lagnese JE, Leugering G, Schmidt EJPG (1994) Modelling analysis and control of dynamic elastic multi-link structures. Birkhauser, Boston
6. Oblasova IN (2013) Obosnovanie matematicheskoi modeli singulyarno zakreplennoi konsoli. Vestnik Severo-Kavkazskogo federalnogo universiteta. Nauchnyi zhurnal 2013, vol 1(34), pp 48–53. FGAOU VPO «Severo-Kavkazskii federalnyi universitet», Stavropol
7. Pokornyi YuV (1999) Integral Stiltesa i proizvodnye po mere v obyknovennykh differentsial-nykh uravneniyakh. Doklady RAN 364(2):167–169
8. Rektoris K (1985) Variatsionnye metody v matematicheskoi fizike i tekhnike, p 590. Mir, Moscow
9. Pazy A (1983) Semigroups of Linear Operators and Applications to Partial Differential Equations, p 280. Springer, Cham
10. Shiryaeva NV (2010) Matematicheskaya model singulyarno zakreplennoi konsoli v zadachakh teorii i praktiki. In: Kirikova VOI (ed) Obrazovatelno-innovatsionnye tekhnologii: teoriya i praktika: monografiya. Kniga 7, pp 106–121. VGPU, Voronezh

# Experience and Prospects of Implementing a Model of Mathematics Distance Learning in North-Caucasus Federal University Training e-Environment

**Elena Timofeeva**, **Nadezhda Timofeeva**, **Irina Oblasova**, **Natalia Shiryaeva**, **and Lusine Grigoryan**

**Abstract** The article offers a view at relevant issues of contemporary trends in employing remote learning technologies in academic environment, and transformation of higher education based on the example of the North-Caucasus Federal University, including through the given epidemic situation due to COVID-19 in the Russian Federation. The aim of the study is to analyze the implementation of the educational content developed in 2018 at the North-Caucasus Federal University in order to teach Mathematics via distance learning technologies (DLT), as well as to evaluate the experience of using the said remote model within the new academic paradigm, which is due to the coronavirus pandemic. There is a view at a modified distance learning model involving mixed learning, which has been introduced into the educational process and tested. Its implementation has relied on LMS Moodle (LMS—Learning Management System; Moodle—Modular Object-Oriented Distance Learning Environment). There is also a review of the virtual academic environment, the model of structure, its major principles and specific of its functioning and management, as well as a statistical analysis of the outcomes obtained through introducing the model in education. The article focuses on the main differences between the concepts of "distance learning" and "modified distance learning model", analyzes their strong and weak points, defines the major benefits of using the latter in the educational environment. The issue in question is view of from the stance of relevance and possible introduction of the said methodology for employing the potential contained in the distance learning technology. There is a special individual view offered at the issue mentioned above, as well as a possible way to resolve that.

E. Timofeeva (✉) · I. Oblasova
North-Caucasus Center for Mathematical Research NCFU, Stavropol, Russia
e-mail: teflena@mail.ru

I. Oblasova
e-mail: pravotor@list.ru

N. Timofeeva · N. Shiryaeva
North-Caucasus Federal University, Stavropol, Russia

L. Grigoryan
Stavropol State Pedagogical Institute, Stavropol, Russia

© The Author(s), under exclusive license to Springer Nature Switzerland AG 2022
A. Tchernykh et al. (eds.), *Mathematics and its Applications in New Computer Systems*,
Lecture Notes in Networks and Systems 424,
https://doi.org/10.1007/978-3-030-97020-8_29

**Keywords** Distance learning technologies · LMS Moodle · Modified distance learning model · Mixed learning

## 1 Introduction

Given the context of global informatization of higher education systems that the world is witnessing nowadays, there is a wide use of advanced information and communication technologies (ICT), which allow not to combine conventional learning tools only, yet also to significantly expand the list of those, while causing a significant impact on the information culture in the university educational environment [9, 10, 13, 16]. ICTs can change all internal elements of the education system—from the simplest content of education up to its administration and management. The introduction of distance learning into the university educational environment is a complex and multi-level process that takes theoretical understanding, the development of a system for psychological and didactic training of all the parties involved in education, as well as quality methodological support.

Apart from new soft- and hardware, there is also a need for newer teaching methods and principles of academic stuff presentation to achieve a proper fusion of technical capacity with methods for the presentation of the material itself. This serves proof to the urgency of the issue related to modernization of higher education involving distance technologies. The purpose of this work is to analyze the implementation of the developed educational content used to teach the discipline of *Mathematics* employing distance learning technologies at large and, namely, to evaluate the experience of employing the designed distance model within the new paradigm of education, which is due to the coronavirus pandemic.

The Decree by the Ministry of General and Professional Education of the Russian Federation (Decree #1050 of May 30, 1997) *On Launching an Experiment in the Area of Distance Learning* was the starting point for promoting distance learning in Russia [2, 3]. The outcomes produced by the experiment suggest that distance learning methods represent a promising streamline in terms of developing the national education system. As of today, from the legal stance, the distance learning methodology is subject to, above all, the Federal Law #273-FL of December 29, 2012 *On Education in the Russian Federation*, where Clause 13 envisages the implementation of degree programs involving various educational technologies, including distance ones [18].

The NCFU has not stayed away from the novelty, too. The initial sign of the preparation and transformation implying distance learning at the University was the Decree #816 of August 23, 2017 by the Ministry of Education and Science of the Russian Federation *On Approving the Procedure for Academic Institutions to Introduce Distance, Remote Learning Technologies when Running Degree Programs.*

## 2   Materials and Methods

### 2.1   A Review of the Development and Introduction Stages for the Modified Distance Learning Model

As representatives of the North-Caucasus Federal University attended the seminar on *Tools to Plan and Implement Mechanisms to Ensure Sustainable Economic Progress of the University*, this served the basis for the implementation of projects aimed at improving the economic efficiency of higher education institutions, which are accountable to the Ministry of Education and Science of Russia. This initiated the Project of *Introduction of Disciplines Delivered with Distance Learning Technologies to Students Majoring in Bachelor's Degree Programs*, which was successfully presented at the Plekhanov Russian University of Economics in October of 2017. In early November of 2017, the NCFU started the implementation of the roadmap within a pilot project on the introduction of disciplines delivered while using distance learning technologies. The educational environment of the designed modified distance learning model relies on three integrated components: technological, pedagogical, organizational and methodological.

The university's choice of the learning management system is the major factor determining effective arrangement of education while implementing a mixed learning model [19, 20]. As far as the NCFU is concerned, it is LMS Moodle that is used as such an information system. It ensures administrative, technical and methodological support for processes involving e-learning, and it is a software that is used to develop e-courses, place them, as well as to carry out e-learning based on distance learning technologies, to analyze students' activity, to grade their performance, etc. [7].

It is to be taken into account that when talking of e-learning, the methodology for creating and using interactive LMS elements in e-courses is of great importance. The main role here is to be played by the respectively selected learning model relying on DOT [1, 3, 5, 8]. The choice of the developed model for employing DOT in the NCFU was based on the analysis and review of the available approaches to classifying learning models involving DOT, as proposed by various experts [6]. Partial replacement of the conventional academic classroom load with electronic educational resources (EER) while running the mixed learning model can be carried out only provided there are MOOC-type courses (mass open online courses) involved in the respective discipline [1, 15, 17]. In this case, the course is represented by a course developed by a group of experts and hosted on the LMS Moodle, which is available to be used in the education delivered at the NCFU. The model presented in this work, if viewed from the pedagogical angle, is a copy of the pedagogical scenarios of the conventional full-time education with certain electronic forms incorporated into it.

The work [14] offers a detailed view at the major preparation staged and development levels for a distance course in Mathematics. Blocks 1, 2, 3, 4 of the algorithm to be seen in Fig. 1 reflect the content of the process and allow following and identifying the efficiency preparing the Mathematics e-course through various levels in view of the indicators that are already in effect in the said area.

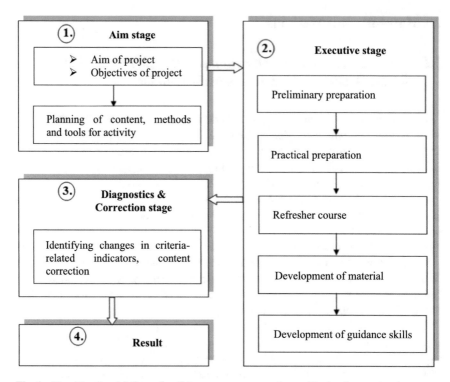

**Fig. 1**  Algorithm for a Mathematics distance course preparation and its development various stages

## 2.2  Analysis of the Modified Learning Model Implementation and the Course Content Specifics

As for the academic year of 2018–2019, subject to the roadmap embracing 10 training fields (350 students), the first stage of the project involving distance learning technologies in the Institute of Engineering (NCFU) [14] the workload in Mathematics to be delivered in the classrooms was cut by 50% (lectures). The developed model for distance teaching of Mathematics was developed and placed in the NCFU's LMS Moodle education management environment.

When analyzing the functional capacity of a mixed academic model involving distance learning technologies, the efficiency of the model appears an important aspect to be taken into consideration. One of the important criteria pointing at efficiency, which is to be mentioned here, is quality:

- the quality of educational services (educational and methodological material, the teacher's training, teaching technologies, educational process arrangement, evaluation material content, etc.);

- the quality of the product, which means the graduate's level of training (this is expressed in categories like knowledge, abilities, skills, and such indicators as competencies).

If we analyze the first of the selected indicators, then certain distinctive features of the developed e-resource are to be mentioned. The presentation of the theoretical material is arranged subject to a strict methodical sequence. Each lecture is well structured and divided into major issues. Video-lectures come along with slides executed with a font no less than 20, and include large-size and highlighted formulas, as well as image animation. Slides and presentations are to be available for individual viewing. The duration of each module is 12–15 min, whereas each module focuses on one issue only; the items from the respective lists are introduced gradually, one at a time. As far as tests are concerned, then both closed and open types of tests are used.

Based on pedagogical aspects, video lectures are recorded so as to special attention to the lecturer's speech culture: literacy in terms of phrase arrangements; easy-to-follow and clear presentation; expressiveness (i.e., intonation and tone, speech pace, pauses); correct use of special terminology with no excessive words. In case of distance learning, there are various ways to be used in terms of presenting educational material, arranging students' independent activities and knowledge monitoring. The teacher of Mathematics will use such means and types of activity that may promote students' acquisition of the required competencies to the fullest extent. Each issue highlighted within the lecture is to be presented with background information and a number of practical tasks, which is aimed at shaping a practical orientation within the studied issues. Practical tasks that are solved are aimed at mastering the types and methods of cognition that are employed in the respective field of knowledge and professional activity.

The conclusions presented following each lecture help students set the right emphasis within the covered material. The test tasks for independent completion, which follow each lecture, are to help verify the student's readiness to take up practical classes and undergo monitoring. The monitoring of the users' (students') activity relies on the frequency and the length of their work with the course and the respective modules.

In order to evaluate the effectiveness of the mixed learning model in education, there have been various types of reports from LMS Moodle diagnosed, as well as a statistical analysis carried out focusing on the effectiveness of research aimed at shaping skills and knowledge in Engineering students studying Mathematics [11]. Through the experiment, the learning of two groups—experimental and control, 50 persons each -was evaluated.

The results of the control as envisaged within the program, served the basis to compile tables that reflected the level of students' training [4, 11]. Table 1 shows the dynamics in the changes observed through the training of both the experimental and the control groups following the educational experiment. To ensure more visuality of the obtained results, there have been histograms designed (Fig. 2).

**Table 1** Comparative data on the level of training, experimental and control groups, prior to, and following the educational experiment (EE)

| Level | Control group respondents (%) | | Experimental group respondents (%) | |
|---|---|---|---|---|
| | Prior to EE | Following EE | Prior to EE | Following EE |
| High | 21.6 | 24.8 | 21.2 | 35.4 |
| Middle | 40.8 | 53.4 | 39.9 | 57 |
| Low | 37.6 | 21.7 | 38.9 | 7.6 |

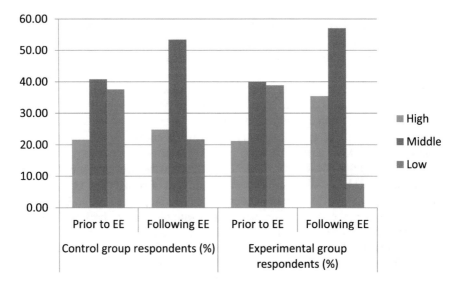

**Fig. 2** Histogram showing the training level dynamics, experimental and control groups

The uploaded charts and reports from LMS Moodle, while reflecting the average time spent in the system, as well as showing the average number of views and the days between visits, serve proof that students with low academic performance feature the worst indicators. A-students used the system more, even though their number of views was lower than that of C-students, who viewed the content more often, yet spent a shorter time on that. The analysis revealed a correlation between students' academic performance and their use of LMS Moodle.

The research allows concluding that active use of mixed learning technologies will not only boost learning effectiveness, yet is will also improve student academic performance. The active use of the educational environment, which is confirmed through the statistical data, also means that students acquired the required skills of independent distance learning, which will allow each of them follow the principle of life-long learning, which implies shaping and promoting everyone's personality through life. The significant positive dynamics observed in the experimental group due the implementation of the modified distance learning model serves a criterion to evaluate the efficiency and viability of experimental activities. This shows that a possible combination of conventional classes and distance learning involving the LMS Moodle system within the modified distance learning model will ensure its undeniable effect.

To study the impact that distance learning technologies cause on the quality of the basic degree program, the Department of Distance Technologies carried out a survey among first-year Engineering Institute students (January 2019), involving 70 students. The survey results revealed that the students described the impact of the distance learning technologies within the proposed disciplines as *positive* (15.94%) and *positive rather than negative* (40.58%).

## *2.3 Positive Aspects of Using the Available Content When Switching to Distance Learning*

The year 2020 brought whole world to serious challenge. The coronavirus pandemic made everyone change their typical approach to work, academic study, and relationships with colleagues, friends and closest ones. Given the changed conditions, from March on, universities began to implement a safer model of arranging training sessions following the imposed healthcare and epidemiological requirements, thus aiming to reduce the risks of the coronavirus spread, i.e., distance learning. The prevailing need in that case was the modernization of higher education institutions, the promotion of education oriented towards data and information in combination with flexibility and convenience through the process of knowledge acquisition.

The response from the NCFU academic system to the pandemic challenge turned out to be less painful due to the disciplines that were introduced in 2018, and delivered through distance learning technologies. As the disciplines in question involved electronic educational resources and digital platforms, this brought serious advantages. First, given a significant transformation of the education process, the LMS Moodle information system that was used in NCFU ensured its administrative, technical and methodological support to processes linked with e-learning involving distance learning technologies, analysis of students' activity, grading, etc. Second, the LMS functional tools (tests, tasks, forums, video-lectures), the use of which determines the level of the *learner—content, learner—teacher or learner—learner interaction*, whereas the resources ensured interactivity and communication through the entire process of education. Students' independent work involving using electronic content was arranged through Web-based learning support. Given the LMS, the feedback principle ensured customization of educational trajectories.

Nowadays, it is safe to say that in view of the pandemic, the developed content allowed the education system to maintain stability. This project proved a starting point for shaping a single data & educational environment for the university—the LMS Moodle system, which now serves a platform for placing the academic e-content of all the disciplines delivered today. This system allows the teacher to place academic e-content thus helping study the discipline in full, following the curriculum and the educational schedule, regardless of the mode through which the degree program is run (full-time, part-time, distance).

## 3 Results

The project presented here in this work, as implemented at the North-Caucasus Federal University, features certain specifics:

1. All courses are designed subject to the requirements of the federal state educational standards.
2. All courses meet the respective requirements for the learning outcomes of degree programs offered at the university.
3. Special attention shall be paid to the efficiency and quality of the content, as well as to procedures for evaluating learning outcomes.

Notable is that the course was developed based on unified topics for all areas studying Mathematics, and since September 2019 it has been implemented in all areas covered by the University's academic activities.

A significant proof to the success of the designed Mathematics e-resource is that it was awarded the top prize in the university contest of academic and methodological support for degree programs offered at the North-Caucasus Federal University.

## 4 Discussion

A comparison of distance learning and the modified distance learning model has revealed that, along with a large number of advantages, distance learning has certain disadvantages [12]. The developed modified distance learning model, however, which is based on the integration of electronic and classical modes compensates for these shortcomings. One of the disadvantages of distance learning, therefore, is the lack of direct academic emotional student-teacher contact, as well as a contact among the students themselves. The modified distance learning model, in turn, compensates for this via communication between students and the teacher in the process of discussing and studying the academic course topics in the classroom through lectures or practical skill-training workshops. The disadvantage of distance learning is a high level of self-organization, as well arranging the time to complete tasks and to master the content. In the modified distance learning model, the freedom in terms of choosing the time and the pace of studying is regulated by dates. In case of distance learning, there is a problem of student identification when checking knowledge, while in case the modified distance learning model is used, there are also classroom types of activities are employed along with tests carried out on the educational platforms to evaluate students' knowledge.

In view of the above, active introduction of distance learning into education should be based on an integrated approach to enhancing the quality of education through incorporating both electronic and classical types of training; on improving the psychological and pedagogical mechanisms to support students, as well as on the interaction between the university, the teaching staff and a team of IT infrastructure experts.

Paradoxical as it may seem, but under the social lockdown and self-isolation, which became the major global requirements for communication in 2020, we managed to facilitate significantly the transition to distance learning, promote interaction with students, enhance the efficiency of academic communication, attract students to discussing various discipline-related issues, build effective interaction with them—all this thanks to the distance learning course we developed.

## 5 Conclusion

The results of a comparative analysis of distance learning, and that of the proposed modified distance learning model, along with the results obtained through monitoring students' knowledge while testing the model, conclusion can be made suggesting that this method of making use of the potential contained in distance learning technology is promising on the whole. It allows solving one of the priority tasks faced by education—teaching students independent and meaningful use of the Internet technologies to design academic activities.

The model described herein will make an excellent addition to the full-time training mode, and can help significantly improve the quality and the efficiency of traditional education.

**Acknowledgements** The work is supported by North-Caucasus Center for Mathematical Research under agreement №. 075-02-2021-1749 with the Ministry of Science and Higher Education of the Russian Federation.

## References

1. Aleinikov BA, Aleinikov PB (2009) Struktura programmnogo obespecheniya, prednaznachennogo dlya provedeniya udalennykh zanyatii v rezhime realnogo vremeni. Vestnik Taganrogskogo gosudarstvennogo pedagogicheskogo instituta. Fiziko-matematicheskie i estestvennye nauki 1:85–88
2. Andreev AA (2014) MOOS v Rossii. Vysshee obrazovanie v Rossii, vol 6
3. Andreev AA, Soldatkin VI (1999) Distantsionnoe obuchenie: suschnost, tekhnologiya, organizatsiya, p 196. Izdatelstvo MESI, Moscow
4. Baraz V (2011) Metody statisticheskogo issledovaniya s primeneniem Excel: uch. Posobie, p 256. Izdatelstvo LAP Lambert Academic Publishing

5. Vaindorf-Sysoeva ME, Gryaznova TS, Shitova VA (2017) Metodika distantsionnogo obucheniya: ucheb. posobie dlya vuzov. M.E. Vaindorf-Sysoeva (red.) – M: Izdatelstvo IUrait, p 194. Obrazovatelnyi protsess, Seriya
6. Vaindorf-Sysoeva ME, Shitova VA (2013) O modelyakh primeneniya distantsionnykh obrazovatelnykh tekhnologii v sovremennom vuze. Vestnik Moskovskogo gosudarstvennogo gumanitarnogo universiteta im. M.A. SHolokhova. Pedagogika i psikhologiya 4:30–34
7. Gilmutdinov AKH, Ibragimov RA, Tsivilskii IV (2008) Elektronnoe obrazovanie na platforme MOODLE, p 169. KGU, Kazan
8. Gorbatiuk VF (2011) Modeli sistemy obucheniya v usloviyakh vnedreniya tekhnologii e-learning. Vestnik Taganrogskogo gosudarstvennogo pedagogicheskogo instituta. Fiziko-matematicheskie i estestvennye nauki 1:116–122
9. Kurbanov TK, Alieva UG, Abdulkhalimova MA, Surkhaev MA (2015) Model organizatsii informatsionnoi sistemy, resheniya zadach upravleniya v vuze. Nauka i Mir T 2 12(28): 92–94
10. Nikiforov OA, Glukhikh VR, Levkin GG (2015) Tendentsii primeneniya oblachnykh tekhnologii v obrazovatelnom protsesse. Innovatsionnaya ekonomika i obschestvo 1(7):80–86
11. Novikov DA (2004) Statisticheskie metody v pedagogicheskikh issledovaniyakh (tipovye sluchai), p 67. MZ-Press, Moscow
12. Polat ES (2004) Teoriya i praktika distantsionnogo obucheniya, p 416. Akademiya, Moscow
13. Surkhaev MA, Nimatulaev MM, Magomedov RM (2016) Modernizatsiya sistemy podgotovki buduschikh uchitelei v usloviyakh informatsionno-obrazovatelnoi sredy. Nauka i Mir T 3(2):96–97
14. Timofeeva EF, Grigoryan LA, Marchenko TV, Khalatyan KA (2019) A model of mathematics distance learning in university training e-environment. In: CEUR workshop proceedings: SLET 2019 - proceedings of the international scientific conference innovative approaches to the application of digital technologies in education and research, Stavropol-Dombay, 20–23 maya 2019 goda, – Stavropol-Dombay. CEUR-WS.org
15. Tretyakov VS, Larionova VA (2016) Otkrytye onlain kursy kak instrument modernizatsii obrazovatelnoi deyatelnosti v vuze. Vysshee obrazovanie v Rossii 7:55–67
16. Khutorskoi AV (2004) Distantsionnoe obuchenie i ego tekhnologii, p 416. Akademiya, Moscow
17. Uvarov AIU (2015) Zachem nam eti Muki. Informatizatsiya obrazovaniya 9:3–18
18. Rossiiskaya federatsiya (2018). Federalnyi zakon № 273-FZ ot 29.12.2012 «Ob obrazovanii v Rossiiskoi Federatsii» s izmeneniyami 2018 goda
19. Graf S, List B (2005) An evaluation of open source e-learning platforms stressing adaptation issues. In: Proceedings of the fifth IEEE international conference on advanced learning technologies, pp 163–165
20. Moore JL, Dickson-Deane C, Galyen K (2011) E-Learning, online learning, and distance learning environments: are they the same? Internet High Educ 14:129–135

# On Numerical Modeling of the Young's Experiment with Two Sources of Single-Photon Spherical Coordinate Wave Functions

**Alexandr Davydov** and **Tatiana Zlydneva**

**Abstract** Within the framework of the quantum mechanics of a photon, constructed by the authors in previous works, using the Maple environment, a numerical modeling was carried out of two-photon interference arising in the scheme of Young's experiment from two one-photon sources emitting simultaneously photons, the propagation of which is described by "spherical" diverging wave functions in the coordinate representation—the wave packets normalized to the total unit probability, which are a superpositions of six-component generalized eigenfunctions of the energy, momentum, and helicity operators, with a Gaussian isotropic distribution in photon momenta. In general, the curve graphically displaying the results obtained demonstrates a pronounced two-photon interference with characteristic maxima and minima and agrees well with an independently constructed curve modeled in a "quasi-classical" approach in terms of classical electrodynamics. It is concluded that the results obtained allow, in the future, using more powerful means of numerical analysis and calculations, to set the tasks of describing one- and two-photon phenomena observed in modern experiments, such as quantum cryptography and quantum computing, within the framework of the concept of photons as localized carriers of quantum information, using the wave function of a photon or a system of photons, including in an entangled state.

**Keywords** Two-photon interference · Wave function · Probability density · Wave packet · Quantum mechanics

## 1 Introduction

At present, areas related to single photons and their systems in an entangled state are rapidly developing, for example, quantum teleportation, quantum computers, and quantum cryptography. In the corresponding phenomena, the transfer of an individual photon is associated with the transfer of its certain localized state from one point

A. Davydov (✉) · T. Zlydneva
Nosov Magnitogorsk State Technical University, Magnitogorsk, Russia
e-mail: ap-dav@yandex.ru

© The Author(s), under exclusive license to Springer Nature Switzerland AG 2022        327
A. Tchernykh et al. (eds.), *Mathematics and its Applications in New Computer Systems*,
Lecture Notes in Networks and Systems 424,
https://doi.org/10.1007/978-3-030-97020-8_30

in space to another. Obviously, this localization should be described by the wave functions of a photon in a coordinate representation with all quantum–mechanical attributes, such as the probability density of detecting a photon, the equation (for it) of continuity, a quantum–mechanical equation that must be satisfied by the coordinate wave function of a photon.

However, despite many attempts to construct the photon wave function in the coordinate representation (see, for example, [2, 3, 6, 10, 13, 14, 18, 20]), there is an opinion [1, 4, 5, 12, 15, 17, 19], starting with [13], that it cannot be constructed, although in the momentum representation it is quite admissible and widely used, for example, in the substantiation of quantum electrodynamics. This situation, however, is somewhat "unusual", since for particles with mass the coordinate wave function can always be obtained from the wave function in the momentum representation by means of a Fourier transform.

In quantum electrodynamics, the transition probability amplitudes are used instead of wave functions, which are successfully applied for correctly description of the results of all experiments performed, such as the double-slit Young's experiment, and also equivalent to it, for example, with the Mach–Zehnder interferometer, both for single particles with a mass and for single photons [9, 11, 16].

In the works of the authors (see [7, 8] and references there), the wave function of a free photon, however, was constructed within the framework of quantum mechanics in the form of a six-component wave function, which is a superposition (integral, "wave packet") of generalized eigenfunctions of the operators of energy, momentum, and helicity in "bivector" representation. In the terminology of classical electrodynamics, these functions are "circularly polarized plane monochromatic waves" (see below formulas (2)–(3)). The six-component wave function of a photon in the momentum representation was also constructed (see formula (8)). The transition in this theory from the photon wave function in one representation to another will not be a problem: it is carried out by direct and inverse Fourier transform implemented by generalized eigenfunctions of the momentum operator $\hat{\mathbf{p}} = -i\hbar\nabla$. It is shown in theory (see [7, 8] and references there) that for a wave function normalized by one photon in the volume $V$ the relativistic invariance of the continuity equation is fulfilled.

In a number of papers (see [7, 8] and references there), single-photon interference was considered, demonstrated by modeling the photon wave function in the coordinate representation. Due to the difficulties encountered in the analytical calculation of the corresponding integrals, the final calculations were made approximately. However, for example, for the scheme of Young's experiment, formulas were obtained expressing interference similar to those that arise in classical electrodynamics. *In this article, we present the results of numerical modeling, without analytical approximations, which demonstrate explicit two-photon interference in a thought experiment from two single-photon sources emitting photons simultaneously from those points that correspond to the position of the slits in Young's experiment.*

## 2 Photon Wave Function in Coordinate Representation

In general, the photon wave function in the coordinate representation can be written [7, 8] as

$$\Psi(\mathbf{r}, t) = \int b(\mathbf{k}, +1)\, \Psi_{\mathbf{k},+1}(\mathbf{r}, t)\, d^3\mathbf{k} + \int b(\mathbf{k}, -1)\, \Psi_{\mathbf{k},-1}(\mathbf{r}, t)\, d^3\mathbf{k}, \quad (1)$$

where $\mathbf{k} = \mathbf{p}/\hbar$ is the "wave vector" of the photon; dimensionless "circularly polarized plane monochromatic waves"

$$\Psi_{\mathbf{k},+1}(\mathbf{r}, t) = (2\pi)^{-3/2} \mathbf{e}_{+1}(\mathbf{k})\, e^{i(\mathbf{kr} - kct)} \begin{pmatrix} 1 \\ 0 \end{pmatrix}, \quad (2)$$

$$\Psi_{\mathbf{k},-1}(\mathbf{r}, t) = (2\pi)^{-3/2} \mathbf{e}_{-1}(\mathbf{k})\, e^{i(\mathbf{kr} - kct)} \begin{pmatrix} 0 \\ 1 \end{pmatrix} \quad (3)$$

obey the orthonomization condition

$$\int d^3\mathbf{r}\, \Psi^{+}_{\mathbf{k}', \lambda'}(\mathbf{r}, t)\Psi_{\mathbf{k}, \lambda}(\mathbf{r}, t) = \delta_{\lambda\lambda'}\, \delta(\mathbf{k}' - \mathbf{k}), \quad (4)$$

forming a basis corresponding to formula (1), since they are generalized eigenfunctions of a complete set of mutually commuting operators of energy, momentum and helicity; the eigenvalues of the latter for a photon are known to be $\lambda = \pm 1$. In (2) and (3), the complex polarization vectors $\mathbf{e}_{+1}(\mathbf{k})$ satisfy a number of orthonormalization relations [14–20] and have the general form

$$\mathbf{e}_\lambda(\mathbf{k}) = [\mathbf{e}_I(\mathbf{k}) + i \lambda\, \mathbf{e}_{II}(\mathbf{k})]\big/ \sqrt{2}, \quad (5)$$

where are real unit mutually perpendicular vectors $\mathbf{e}_I(\mathbf{k})$ and $\mathbf{e}_{II}(\mathbf{k})$ form a right triplet of vectors with a vector $\mathbf{n} = \mathbf{k}/k$; to implement relation (4), the vector $\mathbf{e}_I(\mathbf{k})$ should not change its direction when the direction n changes to the opposite one.

The wave function (1) allows [7, 8] to calculate the probability density of photon detection normalized to unity

$$\rho_P(\mathbf{r}, t) = \Psi^{+}(\mathbf{r}, t)\Psi(\mathbf{r}, t). \quad (6)$$

In formula (1), the coefficients $b(\mathbf{k}, \lambda)$, generally speaking, can be specified arbitrarily (or calculated), but, in turn, must also satisfy the orthonormalization condition, which can be combined into a chain of general relations, taking into account (6)

$$\langle \Psi \mid \Psi \rangle \equiv \int d^3r \rho_P(\mathbf{r}, t) \equiv \int d^3k \, \rho_P(\mathbf{k}) = \int \sum_\lambda |b(\mathbf{k}, \lambda)|^2 d^3k = 1.$$

$$(7)$$

The photon wave function in momentum representation is [14–20] a six-component function

$$\psi^{(\pm)}(\mathbf{k}, t) = e^{\mp ikct} \left\{ b(\mathbf{k}, \pm 1) e_{\pm 1}(\mathbf{k}) \begin{pmatrix} 1 \\ 0 \end{pmatrix} + [b(-\mathbf{k}, \mp 1)]^* e_{\mp 1}(\mathbf{k}) \begin{pmatrix} 0 \\ 1 \end{pmatrix} \right\}. \quad (8)$$

As can be seen from (1) and (8), the wave functions of a photon in the coordinate and momentum representations are related to each other by the usual Fourier transform in quantum mechanics.

## 3  Two-Photon Interference of "Spherical Waves" from Two Point Sources in the Scheme of the Young Experiment

Let two point "isotropic" sources simultaneously emit single photons (each) in all directions with equal probability. Let us assume that the sources are located with respect to the screen fixing their hits in the same way as the "point" holes in Young's experiment. Let us choose the origin of coordinates in the middle of the distance $d$ between the sources, direct the $x$-axis from the "lower" source to the "upper" one, and the $z$-axis to the screen (Fig. 1).

A diverging spherical wave with a Gaussian momentum distribution is described by the following parametrization $b(\mathbf{k}, \lambda)$:

$$b(\mathbf{k}, +1) = b(\mathbf{k}, -1) = A \exp\left[ -\alpha^2(k - k_0)^2/2 - i\,\mathbf{kr}_0 \right], \quad (9)$$

where for the constant $A$ the normalization condition (7) gives the expression.

$$A = \frac{\alpha\sqrt{\alpha}}{\sqrt{2\pi\left[ \left(1 + 2\alpha^2 k_0^2\right)\left(1 + \mathrm{erf}(\alpha k_0)\right)\sqrt{\pi} + 2\alpha k_0 \exp\left(-\alpha^2 k_0^2\right) \right]}}. \quad (10)$$

Using the general formulas (see [7, 8] and references there) for the average values of physical quantities, the most important of them can be calculated using the impulse representation and formula (9). In particular, the average value of the momentum is obtained by exact zero, which corresponds to a spherical wave; the average value of the energy is $\overline{E} = \hbar k_0 c$, where $c$ is the speed of light in vacuum, and the average value of the square of the energy

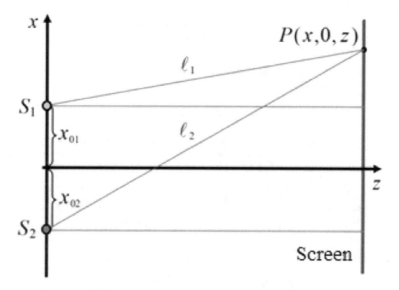

**Fig. 1** Scheme of a thought experiment of the type of Young's experiment with point single-photon sources

$$\overline{E^2} = 4\pi \, (A\hbar k_0 c)^2 \alpha^{-3} \left\{ \sqrt{\pi} \left( \alpha^2 k_0^2 + 3 + 3 \left( \alpha^2 k_0^2 \right)^{-1} \right) \left( 1 + \mathrm{erf}(\alpha \, k_0) \right) \right.$$
$$\left. + \exp \left( -\alpha^2 k_0^2 \right) \left( \alpha_0 + 5 \left( 2\alpha \, k_0 \right)^{-1} \right) \right\}.$$

The above formulas allow us to calculate the standard deviation of the photon energy in state (1), (9):

$$\Delta E = \sqrt{D_E} = \sqrt{\overline{E^2} - \left( \overline{E} \right)^2}. \tag{11}$$

This "theoretical" formula was used to find the value of the parameter $\alpha$, taking into account that the initial values are the average radiation wavelength $\lambda_0$ for which the parameter $k_0 = 2\pi/\lambda_0$, and the photon emission time $\tau$ associated with the "experimental" energy uncertainty value $\Delta E_{\mathrm{exp}}$ using the uncertainty relation:

$$\Delta E_{\mathrm{exp}} \tau = \hbar. \tag{12}$$

Substituting (11) into (12), we obtain an equation whose numerical solution gives the value $\alpha$. In this work, the parameter $\lambda_0$ was set from the microwave region available for the calculation (on an ordinary "home" computer), namely, $\lambda_0 = 1.5$ cm. The value of $\tau$ was chosen equal to $\tau = 0.200158$ ns. For these values, the parameter is $\alpha = 4.236338$ cm.

The average value of the helicity in the state (1), (9) is obviously zero, which also follows from the direct calculation; the average value of the radius vector of the photon detection point is equal to $\mathbf{r}_0$, appearing in (9). Therefore, when each source radiates separately from the other, for a photon from the "upper" source $\bar{r}_{01} = (x_{01}, 0, 0) = (d/2, 0, 0)$; from the "lower" source $\bar{r}_{02} = (x_{02}, 0, 0) = (-d/2, 0, 0)$.

In this work, we used the following vectors [7, 8]:

1)   at $0 \leq \theta \leq \pi/2$

$$e_I(\mathbf{k}) = \begin{pmatrix} e_{Ix} \\ e_{Iy} \\ e_{Iz} \end{pmatrix} = \begin{pmatrix} 1 - (1 - \cos\theta)\cos^2\varphi \\ -(1 - \cos\theta)\sin\varphi\cos\varphi \\ -\sin\theta\cos\varphi \end{pmatrix}. \qquad (13a)$$

$$e_{II}(\mathbf{k}) = \begin{pmatrix} e_{IIx} \\ e_{IIy} \\ e_{IIz} \end{pmatrix} = \begin{pmatrix} -(1 - \cos\theta)\sin\varphi\cos\varphi \\ \cos\theta + (1 - \cos\theta)\cos^2\varphi \\ -\sin\theta\sin\varphi \end{pmatrix}. \qquad (13b)$$

2)   at $\pi/2 \leq \theta \leq \pi$

$$e_I(\mathbf{k}) = \begin{pmatrix} e_{Ix} \\ e_{Iy} \\ e_{Iz} \end{pmatrix} = \begin{pmatrix} 1 - (1 + \cos\theta)\cos^2\varphi \\ -(1 + \cos\theta)\sin\varphi\cos\varphi \\ \sin\theta\cos\varphi \end{pmatrix}. \qquad (14a)$$

$$e_{II}(\mathbf{k}) = \begin{pmatrix} e_{IIx} \\ e_{IIy} \\ e_{IIz} \end{pmatrix} = \begin{pmatrix} (1 + \cos\theta)\sin\varphi\cos\varphi \\ \cos\theta - (1 + \cos\theta)\cos^2\varphi \\ -\sin\theta\sin\varphi \end{pmatrix}. \qquad (14b)$$

where $\theta$ and $\varphi$ define the vector $\mathbf{k}$ in the spherical coordinate system.

To describe the two-photon interference in the stated "formulation" of the thought experiment, we will assume, at the postulate level, that *the wave function of a system of simultaneously emitted photons is equal to the sum of wave functions of the form* (1) *of each photon:*

$$\Psi^{(\pm)}(\mathbf{r}, t) = \Psi_1^{(\pm)}(\mathbf{r}, t) + \Psi_2^{(\pm)}(\mathbf{r}, t). \qquad (15)$$

where the terms differ only by the vectors $\mathbf{r}_{01}$ and $\mathbf{r}_{02}$ of the positions of the sources of the positions of the sources that determine the difference of the coefficients $b(\mathbf{k}, \lambda)$ for them, according to (9). Therefore, the probability density of detecting any of these photons is determined by the formula (6), whereas if the source emits separately, the probability of a photon hitting the same point is determined by only one corresponding term in (15):

$$\rho_{P1}(\mathbf{r}, t) = \Psi_1^+(\mathbf{r}, t)\,\Psi_1(\mathbf{r}, t), \quad \rho_{P2}(\mathbf{r}, t) = \Psi_2^+(\mathbf{r}, t)\,\Psi_2(\mathbf{r}, t) \qquad (16)$$

As a result of integration over the azimuthal angle $\varphi$ of the vector $\mathbf{k}$, each term $\Psi_i(\mathbf{r}, t)$ in (15), where $i = 1, 2$, can be written as

$$\Psi_i(\mathbf{r},t) = \begin{pmatrix} u_i(\mathbf{r},t) \\ v_i(\mathbf{r},t) \end{pmatrix} = \begin{pmatrix} 1 \\ 0 \end{pmatrix} u_i(\mathbf{r},t) + \begin{pmatrix} 0 \\ 1 \end{pmatrix} v_i(\mathbf{r},t), \qquad (17)$$

where

$$u_i(\mathbf{r},t) = \begin{pmatrix} u_{ix}(\mathbf{r},t) \\ u_{iy}(\mathbf{r},t) \\ u_{iz}(\mathbf{r},t) \end{pmatrix} = \begin{pmatrix} v_{ix}(\mathbf{r},t) \\ -v_{iy}(\mathbf{r},t) \\ v_{iz}(\mathbf{r},t) \end{pmatrix}, \quad v_i(\mathbf{r},t) = \begin{pmatrix} v_{ix}(\mathbf{r},t) \\ v_{iy}(\mathbf{r},t) \\ v_{iz}(\mathbf{r},t) \end{pmatrix}, \qquad (18)$$

$$u_i(\mathbf{r},\ t) = A/(2\sqrt{\pi}) \int_0^\infty dk\, k^2 \exp\left[-\alpha^2(k - k_0)^2/2 - ikct\right] F_i(k), \qquad (19)$$

$$F_i(k) = \int_0^1 d\xi\, k^2 \begin{pmatrix} [J_0(\chi_i)(1+\xi) + J_2(\chi_i)(1-\xi)]\cos(kz\xi) \\ \mp[J_0(\chi_i)(1+\xi) - J_2(\chi_i)(1-\xi)]\sin(kz\xi) \\ 2J_1(\chi_i)\sqrt{1-\xi^2}\,\sin(kz\xi) \end{pmatrix}, \qquad (20)$$

где $J_0, J_1, J_2$—Bessel functions from argument $\chi_i = k(x - x_{0i})\sqrt{1 - \xi^2}$.

Figure 2 shows a graph of the probability density of detecting a photon (either of the two) with simultaneous radiation of both sources—in the form of the curve $\rho_P$ as a function of the $x$ coordinate of the observation point P on the screen (see Fig. 1) for radiation with the wavelength $\lambda_0 = 1.5$ cm and the duration $\tau = 0.200158$ ns at $x_{01} = -x_{02} = 1.5$ cm, $z = \ell = ct = 6.00$ cm, at time $t = 0.200$ ns.

Figure 2 shows that the $\rho_P$ curve demonstrates interference with three sharp peaks and two smoothed less intense outside peaks. In the center is the expected largest

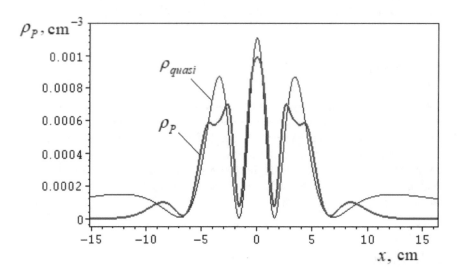

**Fig. 2** Two-photon interference from two isotropic radiation sources in the model (1), (9), (13), (14) in comparison with the quasi-classical description defined by formula (22)

interference peak. The shape of the middle peaks is slightly disturbed due to the specific parametrization (9) and the choice of the parameters $\lambda_0$ and $\tau$. However, the general trend in the behavior of the $\rho_P$ curve is in good agreement with the $\rho_{quasi}$ curve obtained using the "semiclassical" description of this thought experiment, under the assumption that each source emits electromagnetic radiation with an electric field strength given by the formula

$$\mathbf{E}_{quasi}(\mathbf{r}, t) = \mathbf{E}_0(\mathbf{r}/r) \, \exp\left[-(t - r/c)^2/\tau^2 - ik_0(ct - r)\right]/r, \qquad (21)$$

where the vector $\mathbf{E}_0 = \mathbf{E}_0(\mathbf{r}/r)$—depends only on the direction of the vector $\mathbf{r}$ and can be determined in absolute value from equating $\hbar k_0 c$ to all the electromagnetic energy from one act of source radiation passing through a sphere of radius $r$ for the time from $-\infty$ to $+\infty$. Thus, in absolute value it turns out to be equal to $E_0 = \sqrt{2\hbar k_0 \big/ \tau\sqrt{2\pi}}$. The "semiclassical" energy density can then be obtained as the density of electromagnetic energy divided by $\hbar k_0 c$, and the "probability amplitude" $A_{quasi}$—"by the taking the square root" of the probability density. Then for the upper and lower sources we get:

$$A_{i,\,quasi} = \exp\left[-(t - r_i/c)^2/\tau^2 - ik_0(ct - r_i)\right]\bigg/\left(r_i\sqrt{2\pi\, c\tau\sqrt{2\pi}}\right), \quad (22)$$

This expression plays the role of a one-component "semiclassical wave function" of a photon in the coordinate representation. Applying it to describe two-photon interference in the considered thought experiment, similarly to the above method for a six-component function, we obtain the $\rho_{quasi}$ curve shown in Fig. 2.

## 4 Conclusion

The numerical simulations carried out in this work generally demonstrate the possibility of applying the photon wave function in the coordinate representation to the description of interference phenomena in one-photon and two-photon experiments. The emerging nuances require clarification of the applied models and the use of computers with more RAM and higher performance. Taking this into account, the field of promising theoretical and applied research using the photon wave function seems to be very wide and varied.

## References

1. Berestetskii VB, Lifshitz EM, Pitaevskii LP (1982) Quantum electrodynamics, 2nd edn. Pergamon, NY

2. Bialynicki-Birula I (1996) The photon wave function. In: Eberly JH, Mandel L, Wolf E (eds) Coherence and quantum optics VII. Plenum Press, NY, pp 313–323
3. Bialynicki-Birula I (1996) Photon wave function. In: Wolf E (ed) Progress in optics, vol. XXXVI, pp 248–294. Elsevier, Amsterdam
4. Bohm D (1954) Quantum theory. Constable, London
5. Chiao RY, Kwiat PG, Steinberg AM (1995) Quantum non-locality in two-photon experiments at Berkeley. Quantum Semiclassical Optics J Eur Optical Soc Part B 7(3):259–278
6. Cugnon J (2011) The photon wave function. Open J Microphysics 1(3):41–52. https://doi.org/10.4236/ojm.2011.13008
7. Davydov AP, Zlydneva TP (2019) One-photon light interference in terms of the photon wave function in coordinate representation. Actual Probl Mod Sci Technol Educ 10(1):156–162
8. Davydov AP, Zlydneva TP (2019) On the wave-particle duality within the framework of modeling single-photon interference. J Phys Conf Ser 1399:02219. https://doi.org/10.1088/1742-6596/1399/2/022019.
9. Grangier P, Roger G, Aspect A (1986) Experimental evidence for a photon anti-correlation effect on a beamsplitter. Europhys Lett 1(4):173–179
10. Hawton M (1999) Photon wave functions in a localized coordinate space basis. Phys Rev A 59(5):3223–3227
11. Jacques V, Wu E, Grosshans F, Treussart F, Grangier P, Aspect A, Roch J-F (2007) Experimental realization of Wheeler's delayed choice experiment. Science 315:966–968. https://doi.org/10.1126/science.1136303
12. Kramers HA (1958) Quantum mechanics (original ed. 1937). North-Holland, Amsterdam
13. Landau L, Peierls R (1930) Quantenelectrodynamik im Konfigurationsraum. Zeit F Phys 62:188–198
14. Mandel M, Wolf E (1995) Optical coherence and quantum optics. Cambridge University Press, Cambridge
15. Newton TD, Wigner EP (1949) Localized states for elementary particles. Rev Mod Phys 21:400–406
16. Peruzzo A, Shadbolt P, Brunner N, Popescu S, O'Brien JL (2012) A quantum delayed choice experiment. Science 338(6107):634–637. https://doi.org/10.1126/science.1226719
17. Power EA (1964) Introductory quantum electrodynamics. Longmans Press Ltd., London
18. Saari P (2012) Photon localization revisited. In: Lyagushyn S (ed) Quantum optics and laser experiments, pp 49–66. InTech – Open Access Publisher, Croatia
19. Scully MO, Zubairy MS (1997) Quantum optics. Cambridge University Press, Cambridge
20. Sipe JE (1995) Photon wave functions. Phys Rev A 52:1875–1883

# Employing the TIMSS International Study Tools to Evaluate Mathematical Competencies in Future Teachers of Mathematics

Olga Rozhenko⊙, Anna Darzhanya⊙, Viktoria Bondar⊙,
Marine Mirzoyan⊙, and Olga Skvortsova⊙

**Abstract** The aim of this study is to carry out a comparative assessment of the quality and progress trends in mathematical and natural science education based on a comparative analysis of the corresponding indicators for schoolchildren (grades 4 and 8). In 2019, over 50 countries joined the TIMSS monitoring study. In 2015, an extended study (Advanced TIMSS) was carried out, which implies a focus on the performance of students on their final year in school. This paper examines the results of Advanced TIMSS testing held involving undergraduate students of the North-Caucasus Federal University majoring in mathematical areas and planning to work in the field of education. The purpose of this pedagogical research is to identify the level of mathematical literacy among future teachers of mathematics, in view of international educational trends.

**Keywords** Advanced TIMSS · Mathematical education · International studies · Student's t-criterion · Mathematical competence

## 1 Introduction

One of the leading goals and driving forces behind progress in any developed country is coming to gain advance in education quality [1, 2]. The level of students' knowledge and skills, seen as a result of the education system progress, is evaluated globally by independent international monitoring studies [3]. Since the 1990-s, the Russian Federation has been actively involved in a number of international educational quality research programs. Getting better lines in such ratings has become one of the goals of the Russian national *Education* project. The data obtained through an external independent assessment of students' academic performance is one of the real tools

O. Rozhenko (✉) · V. Bondar
North Caucasus Center for Mathematical Research, North-Caucasus Federal University,
355017 Stavropol, Russia
e-mail: r.o.d@mail.ru

A. Darzhanya · M. Mirzoyan · O. Skvortsova
North-Caucasus Federal University, Stavropol, Russian Federation

© The Author(s), under exclusive license to Springer Nature Switzerland AG 2022    337
A. Tchernykh et al. (eds.), *Mathematics and its Applications in New Computer Systems*,
Lecture Notes in Networks and Systems 424,
https://doi.org/10.1007/978-3-030-97020-8_31

that can help improve the content and the quality of education. Since 1995, Russia has been part of the TIMSS International Monitoring Study (Trends in International Mathematics and Science Study), which is held every four years by the International Association for the Evaluation of Educational Achievement (IEA).

In 2019, Russia once again joined the TIMSS study of mathematical and natural science literacy. In our country, the TIMSS study is coordinated by the Center for Education Quality Assessment at the Institute for Education Development Strategy of the Russian Academy of Education jointly with the Federal Institute for Education Quality Assessment. The main purpose of the TIMSS study is to carry out a comparative analysis of the mathematical and natural science education quality in primary and secondary schools. In 2019, Russian schoolers traditionally demonstrated high mathematical achievements in the TIMSS rating (grade-4 schoolers took the 6th place out of 58; 8-graders took the 6th place out of 39).

The level of schoolers' mathematical literacy, demonstrated, among other events, through international monitoring studies, has an obvious direct link with the level of shaping mathematical competence in teachers in charge of mathematical disciplines [4–6]. In order to properly investigate the possible connection between the mathematical competence of future mathematics teachers, and the mathematical secondary school students' literacy, we held a pedagogical study aimed at examining the level of developing mathematical competence in those studying to be mathematics teachers - graduates of the Bachelor's course studying at the North-Caucasus Federal University (NCFU), majoring in mathematical areas.

## 2 Materials and Methods

The study implied students taking the Advanced TIMSS math test, which was designed to assess the level of mathematical literacy in the final (11th) grade students [7]. They had to complete 25 math tasks, including: 19 multiple-choice tasks; 1 task with a short answer, and 5 tasks with the answer to be construed independently. Table 1 shows the topics of the tasks in question.

The test involved 35 fourth-year NCFU Bachelor degree students majoring in mathematical fields. To conduct the study, the authors made themselves aware of the testing rules and prepared respective tasks relying on the standard Advanced TIMSS toolkit [7]. The time allowed to students to complete the entire test was 105 min. The countdown began once the instructions were read and questions related to organizational issues and the correct design of the completed tasks were answered. When completing the test, the NCFU students, unlike secondary school students, enjoyed no right to use calculators or any reference materials.

**Table 1** Assignment topics, Advanced TIMSS for students

| Number of assignment | Topic |
| --- | --- |
| 1 | Identical transformations of logarithmic expressions |
| 2 | Calculations by formulas |
| 3 | Calculating the area using a define integral |
| 4 | Coordinate plane |
| 5 | Movement on the plane |
| 6 | The equation of the circle |
| 7 | Geometric and physical meaning of the derivative |
| 8 | Properties of function graphs |
| 9 | Calculation of define integrals |
| 10 | Movement of a point along a circle |
| 11 | Exponential law |
| 12 | Applied Geometry |
| 13 | Complex numbers |
| 14 | Applied Geometry |
| 15 | Integrals |
| 16 | Mathematical logic |
| 17 | Elements of combinatorics |
| 18 | Elements of combinatorics |
| 19 | Properties of equalities |
| 20 | Limits of functions |
| 21 | The method of coordinates on the plane |
| 22 | Equations of the second order curves |
| 23 | Elements of probability theory |
| 24 | The method of coordinates on the plane |
| 25 | Elements of analytical geometry in space |

Table 2 offers a look at the test outcome check.

Neither spelling nor punctuation errors made by students were taken into account since they had no impact on the point.

**Table 2** Evaluation criteria

| Type of task | Maximum score possible as per task | Evaluation outcome |
|---|---|---|
| Multiple choice | 1 | 1—correct<br>0—incorrect |
| Brief answer | 1 | 1—correct<br>0—incorrect |
| Freely designed answer | 2 | 2—completely correct<br>1—partly correct<br>0—incorrect |

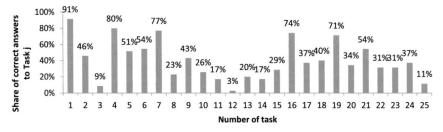

**Fig. 1** Share of correct answers to $j^{th}$ task

## 3　Results

The degree of learning of the educational material was checked through various types of educational and cognitive activities.

Figure 1 shows that Tasks 1, 4 and 7 were completed by more than 70% of the students involved in the test. Tasks 25, 3, 12 were completed by less than 11% of the students.

Figure 2 shows that only one person completed Task 12; Task 3 was done correctly by 3 persons only, while the number of students who managed Task 1 was the highest—32.

As the test results shown in Fig. 3 reveal, the worst outcome in the experimental group was 2 points, whereas the best one was 17 points scored. None of the students scored the maximum number of points possible.

As the study shows, the most difficult ones proved Tasks 3 (completion rate—8.57) and 12 (completion rate—2.86).

A comparative analysis of the test outcomes demonstrated by university students and schoolers attending advanced math classes can be seen in Table 3. The extremely low variance value for Tasks 3 and 12 points at the fact that the tasks in question proved equally hard for the vast majority of the students. It is to be noted that the completion rate for Tasks 3 and 12 in case of both Russian students and students coming from other countries is equal on average—43% - 35% and 13% - 10%,

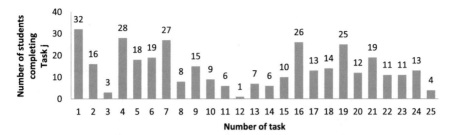

**Fig. 2** Number of students who managed $j$ th task

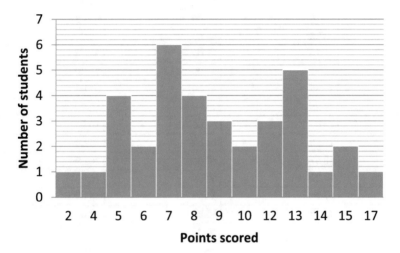

**Fig. 3** Frequency diagram

respectively. This suggests, then that Task 3 was at the average level of difficulty, whereas Task 12 turned truly hard for all the respondents.

The highest results were gained by students doing Tasks 1, 4 and 7 (completion rate—over 75%). Just to compare: 83% of Russian schoolers and 63% of these from other countries involved completed Task 1; 55% of Russian schoolers and 50% of foreign schoolers completed Task 4; 74% of Russian students and 65% of students from other countries completed Task 7.

It is interesting that there were certain tasks that schoolers managed better compared to university students. Here we will try to see the outcome revealed by the university students was so bad, and well employ statistical methods to carry out a comparative evaluation.

The Student's t-test will be used to compare the sample averages belonging to two sets of data, as well as to check whether the average values reveal any statistically significant difference.

**Table 3** Comparative presentation of Advanced TIMSS test completion

| Number of task | Average completion rate in group (university students) | Average completion rate in group (schoolers, Russia) | Average completion rate in group (schoolers, other countries) |
|---|---|---|---|
| 1 | 91.43 | 83.00 | 63.00 |
| 2 | 45.71 | 48.00 | 44.00 |
| 3 | 8.57 | 43.00 | 35.00 |
| 4 | 80.00 | 55.00 | 50.00 |
| 5 | 51.43 | 64.00 | 56.00 |
| 6 | 50.00 | 34.00 | 20.00 |
| 7 | 77.14 | 74.00 | 65.00 |
| 8 | 22.86 | 48.00 | 45.00 |
| 9 | 42.86 | 71.00 | 58.00 |
| 10 | 25.71 | 22.00 | 21.00 |
| 11 | 12.86 | 23.00 | 27.00 |
| 12 | 2.86 | 13.00 | 10.00 |
| 13 | 20.00 | 21.00 | 19.00 |
| 14 | 12.86 | 26.00 | 33.00 |
| 15 | 18.57 | 30.00 | 28.00 |
| 16 | 74.29 | 68.00 | 76.00 |
| 17 | 37.14 | 32.00 | 48.00 |
| 18 | 40.00 | 15.00 | 27.00 |
| 19 | 71.43 | 70.00 | 68.00 |
| 20 | 34.29 | 40.00 | 30.00 |
| 21 | 54.29 | 50.00 | 60.00 |
| 22 | 31.43 | 20.00 | 40.00 |
| 23 | 31.43 | 20.00 | 46.00 |
| 24 | 37.14 | 15.00 | 20.00 |
| 25 | 11.43 | 10.00 | 25.00 |

Here we are going to consider general aggregates: students of classical universities, students; graduates of advanced mathematical classes from Russia, and graduates of schools of other countries involved in the study. We will need to test the hypothesis of general average equality. For this purpose, a sample of n = 35 graduates of classical universities (NCFU students participating) was extracted from the general population. As for the test results for 11$^{th}$ grade Russian schoolers and graduates of secondary schools from the other countries involved, we used the TIMSS outcomes reported officially [7].

The study examined the test result ratio between the NCFU graduates, graduates of Russian schools, as well as graduates of schools representing the countries that joined the study.

Let us have a look at the ratio between the test results matching the NCFU graduates against graduates of Russian schools. There were two independent samples extracted from normal general populations used to find sample averages as per each group. For university students, the value is equal to $\overline{x} = 40.27$, which is the average test task completion rate, i.e. 496.42 on the international scale. The sample average demonstrated by Russian $11^{th}$ graders (on the international scale) was 561. The variance for the NCFU graduates was $D(X) = 19.09$ (6120 on the international scale). The variance for Russian school students, as the data [7] suggests (on the international scale) was 9000. Now, at a significance level set at 0.01, we will test the null hypothesis $H_0 : M(X) = M(Y)$ at the competing hypothesis of $H_1 : M(X) \neq M(Y)$.

Let us find the observed value of the criterion by the following formula.

$$t_{obs} = \frac{\overline{x} - \overline{y}}{\sqrt{\frac{D(X)}{n} - \frac{D(Y)}{m}}}, \tag{1}$$

where $t_{obs}$—is the observed value of the criterion,
$\overline{x}$ is the average rate of test completion for students,
$\overline{y}$ is the average rate of test completion for Russian $11^{th}$ graders,
$D(X)$ is the variance for students,
$n$ is the sample body for students,
$D(Y)$ is the variance for schoolers of Russia,
$m$ is the sample body for Russian schoolers.

The number of Russian school graduates involved in the Advanced TIMSS was 3500 ($m = 3500$), which, therefore, produces the following

$$t_{obs} = \frac{\overline{x} - \overline{y}}{\sqrt{\frac{D(X)}{n} - \frac{D(Y)}{m}}} = -4.92. \tag{2}$$

Given the initial instruction, the competing hypothesis is $H_1 : M(X) \neq M(Y)$, which means the critical area will be two-sided. Here we will find the right critical point by the following equality.

$$\Phi(t_{crit}) = \frac{1 - 0.01}{2} = 0.495, \tag{3}$$

where $\Phi(t_{crit})$ is Laplace's function.

Using the Laplace's function we get tcr $= 2.58$.

Since $|t_{obs}| < t_{cr}$, there is no reason to reject the null hypothesis. In other words, the sample averages show no significant difference. This allows arguing that the there was statistically insignificant mismatch between the test results demonstrated by the NCFU graduates and the results shown by schoolers of the 11th grades in specialized classes of Russian schools.

Further, we will offer a comparison of the test outcomes for students and the results gained by graduates of schools representing other countries involved in the experiment. Here we are going to test the hypothesis concerning the equality of general averages.

Two independent samples extracted from normal general populations were employed to identify sample averages as per each group—for university students, this value was $\bar{x} = 40.27$—the average completion rate (496.42 according to the international scale). The sample average for school graduates from the countries involved in the test was $\bar{z} = 40.56$ (500 by the international scale). The variance rates, following the international scale, were $D(X) = 6120$ and $D(Z) = 2500$, respectively. Taking the significance level at 0.01, we will test the null hypothesis $H_0 : M(X) = M(Z)$ with a competing hypothesis of $H_1 : M(X) \neq M(Z)$.

Here we find the observed value of the criterion:

$$t_{obs} = \frac{\bar{x} - \bar{z}}{\sqrt{\frac{D(X)}{n} - \frac{D(Z)}{m}}} = -0.27, \tag{4}$$

where $t_{obs}$ is the observed value of the criterion,

$n$ is the sample body for the students,

$m$ is the sample body for the schoolers of countries involved in the study.

The condition holds that the competing hypothesis appears as $H_1 : M(X) \neq M(Z)$ so the critical area is two-sided. The right critical point is already available $\Phi(t_{crit}) = 0,495$, tcr = 2.58.

Since $|t_{obs}| < t_{cr}$, then, subject to the rule, there is no reason to reject the null hypothesis. In other words, the average samples had no significant difference, which means we could state that the test results demonstrated by university students featured no significant difference from the results to be seen in school graduates of the countries involved in the study.

Both null hypotheses, therefore, appear confirmed, i.e. the results of students and the results of schoolchildren do not differ insignificantly from the statistical stance.

The study raised the issue of comparing the variance rate between the NCFU graduates' results and the results of school graduates coming from the countries involved the TIMSS studies. The point of interest was "How homogeneous was the sample of students compared to that of school graduates?" For this purpose, we will compare the students' corrected sample variance with the variance of school graduates.

A sample of $n = 35$ volume was extracted from the normal general population of students, with the corrected sample variance of $s^2 = 6300$ identified based on it. Given the significance level of 0.01, the point is to check the null hypothesis of $H_0 : \sigma^2 = \sigma_0^2 = 10000$, while $H_1 : \sigma^2 < 10000$ is to be taken as the competing hypothesis.

Here we find the observed value of the criterion:

$$\chi_{obs}^2 = \frac{(n-1)s^2}{\sigma_0^2} = 21.42, \tag{5}$$

where $\chi_{obs}^2$ is the observed value.

The condition holds that the competing hypothesis is expressed as $H_1 : \sigma^2 < 10000$ so the critical area is to be one-sided. Based on the table of critical distribution points of $\chi^2$, on the set significance level of $\alpha = 0.01$ as well as based on the number of degrees of freedom $k = n - 1 = 34$ we find the critical point:

$$\chi_{cr}^2\left(1 - \frac{\alpha}{2}; k\right) = 17.6. \tag{6}$$

Since $\chi_{obs}^2 > \chi_{cr}^2$, then the null hypothesis is to be rejected, i.e. there is a statistically significant variance difference. That said, the spread of values among students is significantly below that among school graduates coming from the countries involved in the study.

## 4 Discussion and Conclusion

The pedagogical comparative study relying on the Advanced TIMSS toolkit, aimed at examining the level of mathematical competence in future teachers of Mathematics—university graduates who earned their Bachelor's degree at the NCFU, and majoring in mathematical areas, allows drawing the following conclusions:

1. The test results obtained from university students and schoolers advanced mathematical classes in Russia and foreign countries differ, yet the difference was statistically insignificant, which points at certain issues in the mathematical training delivered to students. The level of the mathematical competence observed in holders of the respective Bachelor's degree is sufficient to teach Mathematics in general secondary schools; it is definitely insufficient, though, to carry out professional activities in specialized mathematical classes [8]. It is obvious that shaping an appropriate level of mathematical competence required for this specific subject teacher can be done only within the Master's degree training [9].

2. A number of applied mathematical problems proved equally difficult both for university students and for schoolers. This is indicative of insufficient coverage of applied mathematical issues through the Mathematics training course, both in secondary schools and within mathematical disciplines embraced at the university [10].

3. NCFU graduates, of whom many have plans to start teaching in schools, revealed results that are comparable to those observed among graduates of schools in other countries offering advanced Physics & Mathematics; however, the variance of the examined values among university students is now as wide.

In general, a conclusion can be made, indicating that the training of highly qualified teachers Mathematics is one of the priorities in the system higher education. The level of mathematical education in our schoolers is largely determined by the level of their teachers' respective competence, which, in turn, might allow solving numerous relevant issues faced by both our country and humanity in general, at a deeper and more serious level, which requires fundamental knowledge of Mathematics.

**Acknowledgements** The work is supported by North-Caucasus Center for Mathematical Research under agreement №. 075-02-2021-1749 with the Ministry of Science and Higher Education of the Russian Federation

# References

1. Strokova TA (2009) Kachestvo obrazovaniya: suschnost i kriterii monitoringovoi otsenki Obrazovannei nauka. Izvestiya URO RAO 4(61):36–47
2. Rastorgueva NF (2009) Kachestvo obrazovaniya – zalog konkurentosposobnosti vypusknika. Vysshee obrazovanie v Rossii 1:87–90
3. Malinetskii GG, Sirenko SN (2020) Globalizatsiya obrazovaniya v sistemnom kontekste. Mir Rossii 29(2):92–107. https://doi.org/10.17323/1811-038X-2020-29-2-92-107
4. Moroz VV (2005) Matematicheskoe obrazovanie: dukhovnoe izmerenie. Vysshee obrazovanie v Rossii 7:131–135
5. Pokornaya IYu, Titorenko SA, Ovsyannikova AN (2019) Nekotorye voprosy sovershenstvovaniya podgotovki magistrov napravleniya 44.04.01 Pedagogicheskoe obrazovanie po programme «Matematicheskoe obrazovanie» Perspektivy nauki i obrazovaniya 3(39):184–195. https://doi.org/10.32744/pse.2019.3.14
6. Testov VA (2013) Matematicheskoe obrazovanie v usloviyakh setevogo prostranstva. Obrazovannei nauka 2(101):111–120
7. Ministerstvo prosvescheniya Rossiiskoi Federatsii FGBNU «Institut strategii razvitiya obrazovaniya Rossiiskoi akademii obrazovaniya» TSentr otsenki kachestva obrazovaniya http://www.centeroko.ru/projects.html
8. Petrova AI, Petrov VV (2011) Fiziko-matematicheskoe obrazovanie: poisk nauchnoi smeny. Nauka i obrazovanie, pp 94–100
9. Lenskaya EA (2008) Kachestvo obrazovaniya i kachestvo podgotovki uchitelya. Voprosy obrazovaniya 4:81–96
10. Zainiev RM (2008) Nepreryvnoe matematicheskoe obrazovanie: shkola – vuz. Vysshee obrazovanie v Rossii 2:169–171

# Elements of Fuzzy Logic in Solving Clustering Problems

**Dzhannet A. Tambieva**⬤, **Sergey G. Shmatko**⬤, **and Dmitry V. Shlaev**⬤

**Abstract** To structure data, developers of decision support systems are increasingly using "Data Mining" methods and models, including data clustering algorithms. In this paper, the author proposes a clustering algorithm based on the graph theory and fuzzy logic methodology. Unlike the well-known clustering algorithms, where the division of a set of input vectors into groups (clusters) subject to the object similarity principle is determined by measuring the distance to a certain center(s), the formation of clusters is proposed to be done following the principle of pairwise distance of objects from each other by a value not exceeding that set by the decision-maker. The clustering problem key parameters include the distance between the objects and the number of objects in one cluster. The clear and fuzzy approaches to data cluster formation are implemented. In case of the fuzzy approach, the measure of the sample objects' similarity is determined by the decision-maker based on the fuzzy logic tools. Intended input parameters of this measure depend on the objective function of the problem. The construction of clusters of the required configuration in the fuzzy interpretation of the data clustering issue relies on the decision maker's (DM) empirical choice of the $\alpha$-slice in a fuzzy set of the distance between the objects.

**Keywords** Data Mining · Clustering · Algorithm · Fuzzy set · $\alpha$-slice

## 1 Introduction

Given total digitalization, the society has to face the issue of processing huge amounts of data. As a result, there arises a need for developing effective algorithms for structuring data and extracting knowledge. In order to structure data in decision support information systems, software developers are increasingly proposing Data Mining methods and models [1–3] that allow identifying non-obvious patterns in the available data. The best-known approaches to Data Mining include data clustering algorithms [1–4].

D. A. Tambieva (✉) · S. G. Shmatko · D. V. Shlaev
Stavropol State Agrarian University, Stavropol, Russia
e-mail: tamjannet@mail.ru

© The Author(s), under exclusive license to Springer Nature Switzerland AG 2022    347
A. Tchernykh et al. (eds.), *Mathematics and its Applications in New Computer Systems*,
Lecture Notes in Networks and Systems 424,
https://doi.org/10.1007/978-3-030-97020-8_32

Solving clustering problem implies:

— building a sampling of clustering objects
— defining the criterion for the sampling objects evaluation (key)
— identifying the degree of similarity among the objects
— selecting the data clustering method
— presentation of the clustering problem solution.

Building a sampling of clustering objects implies the selection of a vector of features for each object, quantitative and/or qualitative (categorical) ones.

As a measure revealing the similarity between objects, one of the well-known metrics is typically used, such as the Euclidean distance, the square of the Euclidean distance, the city-block distance (Manhattan distance), Chebyshev distance, power distance, etc. [4]

Besides, two main groups are identified of algorithms in cluster analysis:

— hierarchical and flat
— clear and fuzzy.

Hierarchical algorithms imply nesting of partitions, the so-called partition tree, whereas flat ones are a simple partitioning of a set of objects into clusters.

Clear and fuzzy algorithms differ by the degree of the object-cluster relation.

As noted above, clustering is viewed as the division of a set of input vectors into groups (clusters) based on the degree of similarity. Most methods employed for data clustering involve identifying objects' similarity by measuring the distance to a certain center (centers) [4]. Such centers can be represented by either really selected objects from the set of available ones or the reference samples. However, the center (centers) of clusters is not always possible to be identified clearly. We consider a case where the development of clusters is carried out based on the principle of the objects pairwise distance away from each other by a value not exceeding the one set by the decision-maker (DM). In addition, the DM can set the main evaluation parameters for the clustering problem: the permissible ratios of the number of clusters and/or the number of objects to be combined in a cluster. As for a tool to solve the issue of data clustering, the mathematical apparatus of graph theory is proposed.

Figure 1 offers examples of graphical interpretation of the data clustering problem.

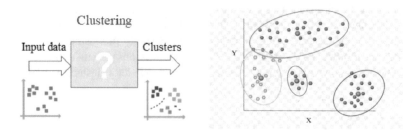

**Fig. 1** Graphic presentation of a data clustering problem

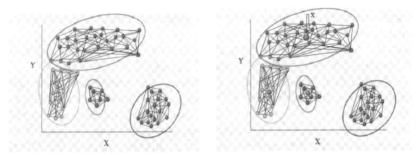

**Fig. 2** Samples of data clustering problem graphic representation (graph-theoretic interpretation)

This paper focuses on a graph-theoretic interpretation of the data clustering problem, which, following mathematical formalization, boils down to the well-known problem of covering a graph with cliques [5, 6]. The problem of covering a graph with cliques belongs to the NP-complete (NP-hard) class of problems [5]. In this regard, its solutions in research publications are limited to algorithms that are statistically efficient and asymptotically accurate (Fig. 2).

## 2 Materials and Methods

Well-known approaches to studying the issue of covering graphs with cliques include polynomial solvability, heuristic algorithms, the relationship of this problem with other ones within the graph theory, etc., which can be found in [7–10]. A paper [11] focuses on clique coverage of graph edges, as well as it dwells on methods simplifying the procedure for identifying the minimal clique covering of edges.

The mathematical apparatus of graph and hypergraph theory employed through this work incorporates the basic concepts and definitions, as they are understood within the context of [12, 13].

This work is a natural generalization and continuation of works [14–16].

The terms *clique*, in the graph-theoretic interpretation, and *cluster*, in the context of a class of related statistical aggregate elements, are viewed here as equivalent, since the data clustering problem (DCP) in the graph-theoretic interpretation can be considered relying on a certain $M$ set of objects. If the objects themselves are represented as renumbered vertices of the set $V = \{v_1, v_2, \ldots, v_n\}$, and if the links (if available) among objects are to be identified as edges $e_{ij}$, connecting the respective vertices $v_i$ and $v_j (e_{ij} = (v_i, v_j) \in E)$, then, the thus built graph $G = (V, E)$, where a set of disjoint subgraphs of a given configuration is to be detected, shall be equated here as a cluster. The set of edges $E = \{e_{ij}\}$.

As far as a given configuration is concerned, then, within in the DCT context, the similarity feature is considered, which—in the graph-theoretic formulation of the problem—can be interpreted through weighing edges or otherwise, via determining the degree of the objects' similarity by the $R$ parameter, their conditional proximity to one other.

The mathematical model of the data clustering problem in the graph-theoretic interpretation, therefore, shall look the following way.

We assume there is a certain $M$ set of objects, renumbered from 1 through $n$. Based on the $M$ set, we build the $G = (V, E)$ graph, where is a set of vertices, each vertex $v_i$, $i = 1, 2, ..., n$ mutually unambiguously corresponds to the $i$ th object of the $M$ set, while $E = \{e_{ij}\}$ is a set of edges where edge $e_{ij} = (v_i, v_j) \in E$ corresponds to the link between the $i$-th and the $j-$ th objects of the $M$ set. All edges are weighted, following the introduced criterion for evaluating the relative position of objects regarding one another. The weights of the edges shell be denoted respectively by $w(e_{ij})$. The permissible *distance* between the vertices (objects) shall be expressed through the $R$ value. The latter means that in each selected cluster the edge weights must not exceed the $R$ value. Besides, one or two evaluation parameters are also set the proposed number of clusters $K$ and/or the number of objects in one cluster $L$.

It is required to construct the $M$ set partitioning into cluster of previous set dimension (configuration). It is to be noted immediately that $L = [n/K]$ (integer division).

## 3  Results

To solve the above-mentioned problem, let us consider the *Cluster* algorithm.

1.  At the first step of the *Cluster* algorithm, we use – as the input data – the matrix equivalent of the $G = (V, E)$ graph, which we will express through $A = \{a_{ij}\}$, where $a_{ij} = w(e_{ij})$, $i, j = 1, ..., n$ (Table 1).

2.  We introduce the value of the maximum distance between objects in one cluster $R$.

3.  We introduce the cycle parameter $p$ and assign it the value of $p := 1$.

4.  We build the $B(0)$ matrix.

     The $B(0)$ matrix is built based on the $A$ matrix subject to the following principle: if $R - a_{ij} < 0$, then $b_{ij}^0 := 0$, otherwise - $b_{ij}^0 := 1$ (Table 2).

     The condition means that if the distance between the $i-$ th and the $j-$ th objects exceeds the value of $R$, then the edge $e_{ij}$ in the initial graph $G$ is not perspective, and the respective vertices (objects) cannot be part of the same cluster together.

5.  The parameter of the iterative cycle $p$ shall be expanded by the step value, which is 1 ($p := p + 1$).

     From this step, a cyclic elimination of non-perspective edges starts.

6.  Building the $B(p)$ matrix.

     The $B(p)$ matrix is built in the following way:

**Table 1** Matrix $A$, equivalent to $G = (V,\ E)$ graph

$$A =$$

|  | $v_1$ | $v_2$ | $\ldots$ | $v_k$ | $\ldots$ | $v_n$ |
|---|---|---|---|---|---|---|
| $v_1$ | $a_{11}$ | $a_{12}$ | $\ldots$ | $a_{1k}$ | $\ldots$ | $a_{1n}$ |
| $v_2$ | $a_{21}$ | $a_{22}$ | $\ldots$ | $a_{2k}$ | $\ldots$ | $a_{2n}$ |
| $\ldots$ | $\ldots$ | $\ldots$ | $\ldots$ | $\ldots$ | $\ldots$ | $\ldots$ |
| $v_k$ | $a_{k1}$ | $a_{k2}$ | $\ldots$ | $a_{kk}$ | $\ldots$ | $a_{kn}$ |
| $\ldots$ | $\ldots$ | $\ldots$ | $\ldots$ | $\ldots$ | $\ldots$ | $\ldots$ |
| $v_n$ | $a_{n1}$ | $a_{n2}$ | $\ldots$ | $a_{nk}$ | $\ldots$ | $a_{nn}$ |

**Table 2** Transformation matrix $B(0)$, based on $A$ matrix (if $R - a_{ij} < 0$, then $b_{ij}^0 := 0$, otherwise - $b_{ij}^0 := 1$)

$$B(0) =$$

|  | $v_1$ | $v_2$ | $\ldots$ | $v_k$ | $\ldots$ | $v_n$ |
|---|---|---|---|---|---|---|
| $v_1$ | $b_{11}^0$ | $b_{12}^0$ | $\ldots$ | $b_{1k}^0$ | $\ldots$ | $b_{1n}^0$ |
| $v_2$ | $b_{21}^0$ | $b_{22}^0$ | $\ldots$ | $b_{2k}^0$ | $\ldots$ | $b_{2n}^0$ |
| $\ldots$ | $\ldots$ | $\ldots$ | $\ldots$ | $\ldots$ | $\ldots$ | $\ldots$ |
| $v_k$ | $b_{k1}^0$ | $b_{k2}^0$ | $\ldots$ | $b_{kk}^0$ | $\ldots$ | $b_{kn}^0$ |
| $\ldots$ | $\ldots$ | $\ldots$ | $\ldots$ | $\ldots$ | $\ldots$ | $\ldots$ |
| $v_n$ | $b_{n1}^0$ | $b_{n2}^0$ | $\ldots$ | $b_{n2}^0$ | $\ldots$ | $b_{nn}^0$ |

if $b_{st}^{p-1} = 1$ and $\sum_{j=1}^{n} b_{sj}^{p-1} \cdot b_{tj}^{p-1} > L - 2$, then $b_{st}^p := 1$, otherwise $b_{st}^p := 0$, $s,\ t = 1,\ \ldots,\ n$.

Here we check, and in case the objects under the numbers $s$ and $t$ are both *close*, i.e. no more away than the $R$ from $(n/K - 2)$ objects, then such objects can potentially belong to one cluster; otherwise, i.e. if this value is lower than $(n/K - 2)$, then such objects can never belong to the same cluster, so this *link* is not *perspective* and has to be eliminated. If $B(p) = B(p - 1)$, then we leave the cycle (reached the fixed point), otherwise we go (return) to Step 5.

Based on the $B(p)$, we will build the graph $G(p) = (V, E(p))$. Here $E(p) = \left\{e_{ij}^p\right\}$, $i, j = 1, 2, ..., n$. The edges $e_{ij}^p$ are such that the corresponding value of the edge weight is $b_{ij}^p = 1$.

At this step, only the *perspective* edges of the original $G$ graph remain. It becomes obvious whether it is possible to build a permissible solution for the clustering problem in view of the value set for the maximum $R$ distance between objects within one cluster.

7.    Calculating the values $S_i = \sum\limits_{j=1}^{n} b_{ij}^p$.

8.    If $S_v = \min S_i < L - 1$, then there is no exact solution for the clustering problem with the parameters set for the $R$ distance. End of algorithm.

      In this case, the DM can amend the problem requirements and possible approximations in its solutions.

9.    If $S_v = \min S_i = L - 1$, then we select he vertices adjacent to vertex $v$ into a single cluster (subset) $M(v) \in M$, eliminate the respective lines and columns from the $B(p)$ matrix. In the event $M \backslash M(v) \neq \emptyset$, then we pass to Step 4, while otherwise it is end of the algorithm.

10.   If $S_v = \min S_i > L - 1$, then we will select all possible combinations of the $G(p)$ graph's full subgraphs of $L$ dimensions, and for further decision making it is possible to use algorithm of the graph coverage with *alpha L–* cliques, as shown in [14, 15].

**Theorem 1.** If the $G(p)$ graph is complete then the equality $B(p) = B(p + 1)$ is always true (fixed point).

**Proof is** straightforward. □

**Theorem 2.** If the graph contains a permissible solution, then the *Cluster* algorithm shall guarantee the identification of the fixed point, which will unequivocally determine the solution.

**Proof:** If the $G = (V, E)$ graph contains a permissible solution to the clustering problem, then $S_v = \min S_i \geq L - 1$, and this means that for each $v_s$ vertex, at each iteration step while building the $B(p)$ matrix, there will be an $L - 1$ vertex $v_t$, such one that $b_{st}^{p-1} = 1$ and $\sum\limits_{j=1}^{n} b_{sj}^{p-1} \cdot b_{tj}^{p-1} > L - 2$. This latter ensures assigning of the value 1 $\left(b_{st}^p := 1\right)$ to the respective $b_{st}^p$ element of the $B(p)$ matrix. Therefore, the potentially perspective link between the vertices of the initial $G = (V, E)$ graph can be eliminated through performing the *Cluster* algorithm. □

**Theorem 3.** The complexity of *Cluster* does not exceed the fifth-degree polynomial.

**Proof:** Here we will show that the complexity of the *Cluster* does not exceed $\tau(Cluster) \approx O(n^5)$.

1. Building the $B(0)$ matrix will take $n^2/2$ operations of subtraction and $R - a_{ij} < 0$ comparison in view of the fact that the $A$ and $B$ matrices are symmetrical regarding the main diagonal.

2. Building the $B(p)$ matrix will take, at each iteration step, $n^2/2$ comparisons of $b_{st}^{p-1}$ and 1, as well as $3n^2/2 + n$ calculations of $\sum_{j=1}^{n} b_{sj}^{p-1} \cdot b_{tj}^{p-1} > L - 2$, which will require $n$ multiplication operations $b_{sj}^{p-1} \cdot b_{tj}^{p-1}$, plus $n$ addition operations for $\sum_{j=1}^{n} b_{sj}^{p-1} \cdot b_{tj}^{p-1}$ and 1 comparison operation of $\sum_{j=1}^{n} b_{sj}^{p-1} \cdot b_{tj}^{p-1} > L - 2$ for each pair of s and t lines. Total: $- 2n + 1$. There are no more than $n^2/2 + 1$ of such pairs. Therefore, the total number of elementary actions will not exceed $(2n + 1) \cdot (n^2/2 + 1) = n^3 + n^2/2 + 2n + 1$.

The number of possible p iterations is limited, in the general case, by the $n^2/2$ value.

In view of this, we get $n^2 + \frac{n^2}{2}(n^3 + \frac{n^2}{2} + 2n + 1) = \frac{n^5}{2} + \frac{n^4}{4} + n^3 + \frac{3n^2}{2}$. The latter proves the polynomial solvability of the *Cluster* algorithm $\tau(Cluster) \approx O(n^5)$. $\square$

Given the above, the *Cluster* algorithm allows identifying a set of permissible solutions for clustering problems within the polynomial time. During that, the permissible *distance* between the vertices (objects), which is estimated by the $R$ value, is determined as a *clear* number. The latter means that in each selected cluster the edge weights must not exceed the $R$ value. Besides, there also a supposed possibility of combining two extra parameters – the number of clusters $K$ and the number of objects in one cluster $L$ ($L = n/K$).

In reality, the degree of the similarity shared by objects is a relative value. Therefore, we believe that the mathematical apparatus of fuzzy logic most properly reflects the weight indicators of the data clustering problem.

Let us consider a case where the $R$, $K$ and $L$ parameters for the above-described DCT are not set clearly, namely as
$R = \{\langle x, \ \mu_R(x)\rangle | x \in U \}$
$K = \{\langle x, \ \mu_K(x)\rangle | x \in U \}$
$L = \{\langle x, \ \mu_L(x)\rangle | x \in U \}$, where $U$ is a universal set.
During that, the DM sets the priorities within the required partitioning of the data into clusters:

a)  by the similarity degree within the cluster;
b)  subject to the required number of clusters;
c)  through setting the required set of elements within the cluster;
d)  based on the need to embrace all the objects (the $M$ set elements).

Depending on the selected priority, the decisive rule is identified, which will offer the basis for partitioning the object sets into clusters.

The computational scheme of the *Cluster* algorithm [16] relies on the procedure of eliminating *non-perspective* links among the vertices.

**Table 3** *Fuzzy* matrix equivalent of the *G* graph

|        | $v_1$         | $v_2$         | ...  | $v_n$         |
|--------|---------------|---------------|------|---------------|
| $v_1$  | $\mu_R(a_{11})$ | $\mu_R(a_{12})$ | ...  | $\mu_R(a_{1n})$ |
| $v_2$  | $\mu_R(a_{21})$ | $\mu_R(a_{22})$ | ...  | $\mu_R(a_{2n})$ |
| ...    | ...           | ...           | ...  | ...           |
| $v_n$  | $\mu_R(a_{n1})$ | $\mu_R(a_{n2})$ | ...  | $\mu_R(a_{nn})$ |

At the *Cluster* algorithm entry we build the $A = \{a_{ij}\}$ matrix, which unambiguously corresponds to the *G* graph, where $a_{ij} = w(e_{ij})$ are the edge weights $e_{ij}$, $i, j = 1, 2, ..., n$.

It is obvious that in the *fuzzy* interpretation of this problem, the *G* graph in question is complete, and the corresponding *fuzzy* matrix equivalent will appear as follows (Table 3).

Initially, it is recommended to apply the *Cluster* algorithm, with strictly-set DCT parameters. The purpose of this stage is to identify a valid solution for the DCT, if it exists. In case there is no exact permissible solution of the DCT relying on the *G* graph, yet it was possible to detect a certain approximation for solving the problem, then, based on the *fuzzy* equivalent of the *G* graph, the remaining vertices, which have not been included into any cluster, can be employed by the DM to identify the permissible value of the $\alpha$ slice [17, 18] of the *R* fuzzy set.

($R\alpha = \{x | \mu_R(x) \geq \alpha\}$, where $\alpha \in [0, 1]$).

Following the obtained value of $R\alpha$, we move to a new interpretation of the *G* graph as $G^{(1)}$ and a corresponding $A^{(1)}$ matrix. At this stage, the *Cluster* algorithm can be used repeatedly, which will allow building a permissible partitioning of the *M* set of objects into the required number of equally powerful clusters with the maximum possible indicator of object similarity.

## 4   Discussion

Clustering is known to be a mandatory procedure, it does not make any statistical conclusions, yet it allows carrying out an exploratory analysis and study the *data structure*.

The most popular fuzzy clustering algorithm is known to be the iterative algorithm of c-means, which was developed (for the $m = 2$ case) by J.C. Dunn in 1973, and was improved (for the $m > 1$ case) by J.C. Bezdek in 1981. Conceptually, the c-means offers the following:

1. Initial unclear partitioning of the set of objects *(n)* into *k* clusters by selecting the membership matrix of the $n \times k$ dimension;
2. Identification, based on the *M* matrix, the fuzzy error criterion in the clustering problem;
3. Regrouping the objects in order to reduce the value of the fuzzy error criterion;

4.  Comparing the *M matrix values with the required indicators. In case the target values fail to be met*—return to p. 2. The process is to go on until the changes in the *M* matrix become insignificant.

During that, unlike J.C. Dunn and J.C. Bezdek randomly setting the cluster centers at the initial stage, the graph theory methodology of this work allows interpreting the clustering problem as a problem of covering the graph with cliques. Further building of clusters following the set configuration is based on the DM's empirical choice of the *R* fuzzy set α-slice.


# 5  Conclusion

This work is a natural summarization and continuation of the works [14–16].

It contains a clear and a fuzzy implementation for the method of solving data clustering problems. Each implementation is the results of the DM's choice:

− the distance between the objects of a certain set;
− the number of objects in the cluster;
− the total number of clusters;
− the condition of the sets being equivalent.

The completion of the condition for partitioning sets of objects into clusters is determined by the respective decisive rule. The decision-maker's choice is made following the requirements of a specific problem.

Theorems are proofs are provided.


# References

1.  Classification, clustering, and data mining applications (2021) Proceedings of the meeting of the international federation of classification societies (IFCS), ... Data analysis, and knowledge organization). Gostekhizdat, Moskva
2.  Dan B (2021) Data mining for design and manufacturing: methods and applications (Massive Computing). RGGU, Moskva
3.  Giacomo DR (2021) Computational intelligence in data mining. RGGU, Moskva
4.  Romanov VP (2007) Intellektualnye informatsionnye sistemy v ekonomike: uchebnoe posobie dlya studentov vuzov, obuchayuschikhsya po spetsialnosti "Prikladnaya informatika" i drugim mezhdistsiplinarnym spetsialnostyam. Ekzamen, Moskva
5.  Gary M, Johnson D (1982) Vychislitelnye mashiny i trudnoreshaemye zadachi. Mir, Moskva
6.  Tambieva DA (1999) Matematicheskie modeli i metody dlya vektornoi zadachi o klikakh: dis. ... kandidata fiziko-matematicheskikh nauk: 05.13.16 - Cherkessk
7.  Zykov AA (1987) Osnovy teorii grafov. Moskva
8.  Emelichev VA, Melnikov OI, Sarvanov VI, Tyshkevich RI (1990) Lektsii po teorii grafov. Nauka, Moskva
9.  Badekha IA (2013) Issledovanie klikovykh pokrytii reber grafa. Prikladnaya diskretnaya matematika 1(19):69–83

10. Kou LT, Stockmeyer LJ, Wong CK (1978) Cliques with regard to keyword conflflicts and intersection graphs. Commun ACM 21(2):135–139. https://doi.org/10.1145/359340.359346
11. Orlin J (1977) Contentment in graph theory: covering graphs with cliques. Indagat Math 39:406–424. https://doi.org/10.1016/1385-7258(77)90055-5
12. Gramm J, Guo J, Huffner F, Niedermeier R (2006) Data reduction, exact, and heuristic algorithms for clique cover. In: Proceedings of 8th workshop on algorithm engineering and experiments, Miami, Fl, pp 86–94, 21 January 2006
13. Cavers MS (2005) Masters thesis, University of Waterloo. http://www.math.uwaterloo.ca/co/graduate-students/fifiles/mmath/MikeCavers.pdf
14. Perepelitsa VA, Tambieva DA (2009) Sistemy s ierarkhicheskoi strukturoi upravleniya: razrabotka ekonomiko–matematicheskikh i instrumentalnykh metodov, p 240. Finansy i statistika, Moskva
15. Perepelitsa VA, Tambieva DA (2016) Ob odnom teoretiko-gipergrafovom podkhode resheniya zadachi o klikakh. Materialy XII Mezhdunarodnogo seminara «Diskretnaya matematika i ee prilozheniya», imeni akademika O.B. Lupanova (Moskva, MGU, 20–25 iyunya 2016g.)/Pod redaktsiei O.M. Kasim-Zade, pp 305–308. MGU. Izd-vo mekhaniko-matematicheskogo fakulteta, Moskva
16. Tambieva DA (2018) Ob odnom podkhode klasterizatsii dannykh v kontekste metodologii "Data Mining". In: Alekseev VB Romanov, DS Danilov BR (eds) V sbornike: Diskretnye modeli v teorii upravlyayuschikh sistem. Trudy X Mezhdunarodnoi konferentsii. Otvetstvennye redaktory, pp 255–257. MAKS Press, Moskva
17. Kofman A (1982) Vvedenie v teoriyu nechetkikh mnozhestv. Moskva: Radio i svyaz, p 432
18. Orlovskii SA (1981) Problemy prinyatiya reshenii pri nechetkoi iskhodnoi informatsii. Nauka, Moskva

# Difference Methods of Solving Non-local Boundary Value Problems for a Loaded Generalized Diffusion Equation with Bessel Operator

**Murat Beshtokov**⊙, **Zariana Beshtokova**⊙, **Elbrus Olisaev**⊙, **and Marat Khudalov**⊙

**Abstract** Non-local boundary value problems for a loaded generalized diffusion equation are investigated. By the method of energy inequalities the a-priori estimates in difference-differential interpretation are obtained, whence it follows the solution uniqueness and stability based on the initial data and the right side, as well as the convergence of the solution of the differential problem to the solution of the corresponding differential problem with a speed of $O(h^2 + \tau^2)$.

**Keywords** Bessel operator · Non-local problem · Integral condition · A-priori estimate · Difference scheme · Diffusion equation · Fractional order equation · Caputo fraction derivative

## 1 Introduction

Mathematical modeling applied to a number of processes in mechanics, physics, biology often leads to the problems recently called non-local. Non-local problems are such problems where instead of usual point ("local") boundary conditions, such conditions are assigned that could connect the values of desired solution and (or) its derivatives at various points of the boundary, or at the border points and at some internal points.

Boundary value problems for parabolic equations with non-local conditions used to appear, for example, when studying particle diffusion in turbulent plasma, moisture transfer in soils, heat propagation in a thin heated rod with a specified law of variation

M. Beshtokov · Z. Beshtokova
North-Caucasus Federal University, North-Caucasus Center for Mathematical Research, Stavropol, Russia

E. Olisaev · M. Khudalov (✉)
North-Ossetian State University named after Kh.L. Khetagurov, Vladikavkaz, Russia
e-mail: hmz@mail.ru

© The Author(s), under exclusive license to Springer Nature Switzerland AG 2022　　357
A. Tchernykh et al. (eds.), *Mathematics and its Applications in New Computer Systems*,
Lecture Notes in Networks and Systems 424,
https://doi.org/10.1007/978-3-030-97020-8_33

in the total heat amount in the rod. The first works for parabolic equations with non-classical boundary conditions most likely relate to the works of Steklov V.A. [1], Canon J.R. [2], Kamynin L.A. [3], and Chudnovskii L.F. [4]. Non-local problems of mathematical physics increasingly began attracting the mathematicians' attention when the work of Bitsadze A.V. and Samarskii A.A. [5] had appeared. Various classes of non-local boundary value problems were investigated in the works [6–10].

Mathematical models creation, considering the fractal properties of various media, is of great theoretical and practical importance. It became obvious that when solving a number of physical and biological problems, the media and systems, well described as fractals, often occur, for example, highly porous media like soil. It is required to study boundary value problems for differential equations with a fractional derivative when solving such problems [11–15].

The papers [16–21] deal with numerical methods for solving boundary value problems for various fractional order equations. The results obtained in [19], allow us to apply the method of energy inequalities to obtain a-priori estimates of boundary value problems for fractional order equations in difference-differential interpretations, as in the classical case ($\alpha = 1$).

The papers [22–24] are focused on numerical methods for solving local and non-local boundary value problems for various Adler equations with Bessel operator. Uniqueness, stability and convergence between the differential problem solution and the corresponding differential problem solution are proved by the method of energy inequalities.

The paper [25] considers the problem of controlling the process of metal cylinder induction heating implying heat exchange with the environment occurring on its surface, described by the boundary value problem for an inhomogeneous heat equation containing a singular Bessel operator.

## 2  Non-local Boundary Value Problem A

In a closed rectangle $\overline{Q}_T = \{(x, t) : 0 \le x \le l, 0 \le t \le T\}$ we consider the following non-local boundary value problem for a loaded generalized diffusion equation with Bessel operator

$$\partial_{0t}^{\alpha} u = \frac{1}{x^m} \frac{\partial}{\partial x} \left( x^m k(x, t) \frac{\partial u}{\partial x} \right) + r(x, t) \frac{\partial u}{\partial x} - q(x, t) u(x_0, t) + f(x, t),$$

$$0 < x < l, 0 < t \le T, \tag{1.1}$$

$$\lim_{x \to 0} x^m k(x, t) u_x(x, t) = 0, \quad 0 \le t \le T, \tag{1.2}$$

$$-k(l, t) u_x(l, t) = \beta(t) \int_0^l x^m u(x, t) dx - \mu(t), \quad 0 \le t \le T, \tag{1.3}$$

$$u(x, 0) = u_0(x), \quad 0 \le x \le l, \tag{1.4}$$

where

$$0 < c_0 \le k(x, t) \le c_1, \ |r(x, t), r_x(x, t), k_x(x, t), q(x, t), \beta(t)| \le c_2, \ 0 \le m \le 2, \tag{1.5}$$

$\partial_{0t}^\alpha u = \frac{1}{\Gamma(1-\alpha)} \int\limits_0^t \frac{u_\tau(x,\tau)}{(t-\tau)^\alpha} d\tau$, – Caputo fraction derivative of the order $\alpha$, where $0 < \alpha < 1, c_i, i = 0, 1, 2$ – positive numbers.

When $x = 0$ the limited solution is required $|u(0, t)| < \infty$, which is equivalent to the condition (1.2), matching, in its turn, with the identical equation $k(0, t)u_x(0, t) = 0$ [26, c.173], if the functions $r(0, t), q(0, t), f(0, t)$ are finite.

Hereinafter, we will assume that the problem (1.1)–(1.4) has a unique solution possessing the necessary derivatives. We will also consider that the coefficients of the equation and the boundary conditions satisfy the necessary smoothness conditions providing the necessary approximation order of the difference scheme.

In the course of presentation we will also use constant positive numbers $M_i, i = 1, 2, \ldots$, depending only on the input data of considered problem.

## 3 The A-Priori Estimate in the Differential Form

To obtain the a-priori estimate of the problem solution (1.1)–(1.4) in the differential form we multiply the Eq. (1.1) scalarly by $x^m u$:

$$(\partial_{0t}^\alpha u, x^m u) = ((x^m k u_x)_x, u) + (r u_x, x^m u) - (q u(x_0, t), x^m u) + (f, x^m u), \tag{2.1}$$

where $(u, v) = \int_0^l uv dx$, $(u, u) = \|u\|_0^2$, where $u, v$ – specified $[0, l]$ functions.
True the following:

**Lemma 1** [18]. For every absolutely continuous function $[0, T]$ of the function $v(t)$ the following inequality is true:

$$v(t)\partial_{0t}^\alpha v(t) \ge \frac{1}{2}\partial_{0t}^\alpha v^2(t), \quad 0 < \alpha < 1.$$

**Lemma 2** [18]. The non-negative absolutely continuous function $y(t)$ satisfies for almost all t out of $[0, T]$ in the inequality

$$\partial_{0t}^\alpha y(t) \le c_1 y(t) + c_2(t), \quad 0 \le \alpha \le 1,$$

where $c_1 > 0, c_2(t)$ – summarized with $[0, T]$ is non-negative function. Then

$$y(t) \le y(0) E_\alpha(c_1 t^\alpha) + \Gamma(\alpha) E_{\alpha,\alpha}(c_1 t^\alpha) D_{0t}^{-\alpha} c_2(t),$$

where $E_\alpha(z) = \sum_{n=0}^{\infty} \frac{z^n}{\Gamma(\alpha n+1)}$, $E_{\alpha,\mu}(z) = \sum_{n=0}^{\infty} \frac{z^n}{\Gamma(\alpha n+\mu)}$ — Mittag-Leffler function.

We transform the integrals which are part of the identical equation (2.1), using Cauchy inequality with $\varepsilon$ [26, p. 100] and the Lemma 1, then from (2.1) we find

$$\partial_{0t}^\alpha \|x^{\frac{m}{2}} u\|_0^2 + \|x^{\frac{m}{2}} u_x\|_0^2 \le M_1 \|x^{\frac{m}{2}} u\|_0^2 + M_2(\|x^{\frac{m}{2}} f\|_0^2 + \mu^2(t)). \qquad (2.2)$$

Then, applying to the both parts (2.2) operator of fractional integration $D_{0t}^{-\alpha}$, we will obtain

$$\begin{aligned}
&\|x^{\frac{m}{2}} u\|_0^2 + D_{0t}^{-\alpha} \|x^{\frac{m}{2}} u_x\|_0^2 \\
&\le M_1 D_{0t}^{-\alpha} \|x^{\frac{m}{2}} u\|_0^2 + M_2(D_{0t}^{-\alpha}(\|x^{\frac{m}{2}} f\|_0^2 + \mu^2(t)) + \|x^{\frac{m}{2}} u_0(x)\|_0^2).
\end{aligned} \qquad (2.3)$$

With the help of the Lemma 2 from (2.3) we obtain the following required a-priori estimate

$$\|x^{\frac{m}{2}} u\|_0^2 + D_{0t}^{-\alpha} \|x^{\frac{m}{2}} u_x\|_0^2 \le M(D_{0t}^{-\alpha}(\|x^{\frac{m}{2}} f\|_0^2 + \mu^2(t)) + \|x^{\frac{m}{2}} u_0(x)\|_0^2), \qquad (2.4)$$

where $M - const > 0$, depending only on the input data of the problem (1.1)–(1.4), $D_{0t}^{-\alpha} u = \frac{1}{\Gamma(\alpha)} \int_0^t \frac{u d\tau}{(t-\tau)^{1-\alpha}}$ — Riemann-Liouville fractional integral of the order $\alpha$, $0 < \alpha < 1$.

**Theorem 1.** Let the conditions (1.5) be met, then for the solution $u(x, t)$ of the problem (1.1)–(1.4) the a-priori estimate (2.4) is true, whence it follows the solution uniqueness and stability based on the initial data and the right side.

## 4   Stability and Convergence of the Difference Scheme

To solve the problem (1.1)–(1.4) we apply the method of finite differences. On the uniform quid $\overline{\omega}_{h\tau}$ the differential problem (1.1)–(1.4) we assign the following difference scheme:

$$\begin{aligned}
\varkappa \Delta_{0t_{j+\sigma}}^\alpha y_i &= \frac{\Theta x}{x_i^m} (x_{i-0.5}^m a_i^j y_{\bar{x},i}^{(\sigma)})_x + \frac{b-j}{x_i^m} \left( x_{i-0.5}^m a_i^j y_{\bar{x},i}^{(\sigma)} \right) \\
&+ \frac{b+j}{x_i^m} \left( x_{i+0.5}^m a_{i+1}^j y_{x,i}^{(\sigma)} \right) - d_i^j \left( y_{i_0}^{(\sigma)} x_{i_0}^- + y_{i_0+1}^{(\sigma)} x_{i_0}^+ \right) + \varphi_i^j, \ (x, t) \in \omega_{h,\tau},
\end{aligned} \qquad (3.1)$$

$$\varkappa_0 a_1 y_{x,0}^{(\sigma)} = \frac{0.5h}{m+1} (\Delta_{0t_{j+\sigma}}^\alpha y_0 + d_0(y_{i_0}^{(\sigma)} x_{i_0}^- + y_{i_0+1}^{(\sigma)} x_{i_0}^+)) - \mu_1, \in \overline{\omega}_\tau, \ x = 0, \qquad (3.2)$$

$$-\varkappa_N a_N y_{\bar{x},N}^{(\sigma)} = \tilde{\beta} \sum_{i=0}^{N} x_i^m y_i^{(\sigma)} \hbar + 0.5 h d_N \left( y_{i_0}^{(\sigma)} x_{i_0}^- + y_{i_0+1}^{(\sigma)} x_{i_0}^+ \right) + 0.5 h \Delta_{0t_{j+\sigma}}^{\alpha} y_N - \mu_2,$$

$$(3.3)$$

$$y(x,0) = u_0(x), \ x \in \bar{\omega}_h, \ t = 0,$$

$$(3.4)$$

where $\Delta_{0t_{j+\sigma}}^{\alpha} y = \frac{\tau^{1-\alpha}}{\Gamma(2-\alpha)} \sum_{s=0}^{j} c_{j-s}^{(\alpha,\sigma)} y_t^s$ – the discrete analogue of Caputo fraction derivative of the order $\alpha$, $0 < \alpha < 1$, approximating to the order of accuracy $O(\tau^{3-\alpha})$ [19].

$$a_0^{(\alpha,\sigma)} = \sigma^{1-\alpha}, a_l^{(\alpha,\sigma)} = (l+\sigma)^{1-\alpha} - (l-1+\sigma)^{1-\alpha}, \ l \geq 1, \ \sigma = 1 - \frac{\alpha}{2},$$

$$b_l^{(\alpha,\sigma)} = \frac{1}{2-\alpha}[(l+\sigma)^{2-\alpha} - (l-1+\sigma)^{2-\alpha}] - \frac{1}{2}[(l+\sigma)^{1-\alpha} + (l-1+\sigma)^{1-\alpha}], l \geq 1,$$

$$\text{at } j = 0, \ c_0^{(\alpha,\sigma)} = a_0^{(\alpha,\sigma)};$$

$$\text{at } j > 0, \ c_s^{(\alpha,\sigma)} = \begin{cases} a_0^{(\alpha,\sigma)} + b_1^{(\alpha,\sigma)}, & s = 0, \\ a_s^{(\alpha,\sigma)} + b_{s+1}^{(\alpha,\sigma)} - b_s^{(\alpha,\sigma)}, & 1 \leq s \leq j-1, \\ a_j^{(\alpha,\sigma)} - b_j^{(\alpha,\sigma)}, & s = j, \end{cases}$$

$$c_s^{(\alpha,\sigma)} > \frac{1-\alpha}{2}(s+\sigma)^{-\alpha} > 0, \ a_i^j = k(x_{i-0.5}, t^{j+\sigma}), \eta_{,i}^{\pm j} = \frac{\varkappa_i r_i^{\pm j+\sigma}}{k_i^{j+\sigma}},$$

$$\tilde{\beta} = \tilde{\varkappa} \beta^{j+\sigma}, \mu_1 = \frac{0.5h}{m+1} \varphi_0^j, \ \mu_2 = \tilde{\varkappa} \mu^{j+\sigma} + 0.5 h \varphi_N^j,$$

$$y^{(\sigma)} = \sigma y^{j+1} + (1-\sigma) y^j, r_N = r(l,t) = r_N^{j+\sigma} \geq 0,$$

$$r = r^+ + r^-, r_0 = r(0,t) = r_0^{j+\sigma} \leq 0, \ |r| = r^+ - r^-, \ r^- = 0.5(r - |r|) \leq 0,$$

$$\bar{\varkappa}_i = 1 + \frac{m(m-1)h^2}{24x_i^2}, = \overline{1, N-1}, \ r^+ = 0.5(r + |r|) \geq 0,$$

$$d_i^j = \begin{cases} \bar{\varkappa}_i q_i^{j+\sigma}, i \neq 0, N, \\ q_i^{j+\sigma}, i = 0, N. \end{cases} \varphi_i^j = \begin{cases} \bar{\varkappa}_i f_i^{j+\sigma}, \neq 0, N, \\ f_i^{j+\sigma}, i = 0, N. \end{cases} \hbar = \begin{cases} 0.5h, i = 0, \\ h, i \neq 0, N, \end{cases}$$

$$\varkappa_i = \frac{1}{1+R_i}, \ R_i = \frac{0.5h|r_i|\bar{\varkappa}_i}{k_{i-0.5}}, \ \varkappa_0 = \frac{1}{1+\frac{0.5h|r_0|}{(m+1)a_1}}, \ r_0 \leq 0,$$

$$Y = \hat{y} + y, \ \hat{y} = y^{j+1}, \ y_t = \frac{\hat{y}-y}{\tau}, = y_i^j = y(x_i, t_j),$$

$$\bar{\varkappa} = 1 + \frac{0.5hm}{l} = \frac{1}{1 - \frac{0.5hm}{l}}, \varkappa_N = \frac{1}{1 + 0.5h\frac{|r_N^{j+\sigma}|}{k_{N-0.5}}}, \text{если } r_N^{j+\sigma} \geq 0.$$

$$x_{i_0}^- = \frac{x_{i_0+1} - x_0}{h}, \quad x_{i_0}^+ = \frac{x_0 - x_{i_0}}{h}, \quad x_{i_0} \leq x_0 \leq x_{i_0+1}.$$

We rewright the problem (3.1)–(3.4) in the operator form

$$\overline{\overline{\varkappa}} \Delta_{0t_{j+\sigma}}^\alpha y = \overline{\Lambda}(t^{j+\sigma}) y^{(\sigma)} + \overline{\Phi}, \tag{3.5}$$

$$y(x, 0) = u_0(x), \tag{3.6}$$

where

$$\overline{\overline{\varkappa}} = \begin{cases} \overline{\varkappa}_i, x \in \omega_h, \\ 1, x = 0, l, \end{cases} \quad \overline{\varkappa}_i = 1 + \frac{m(m-1)h^2}{24x_i^2}, \quad \overline{\Phi} = \begin{cases} \varphi = \varphi_i, (x, t) \in \omega_{h\tau}, \\ \varphi^- = \frac{(m+1)}{0.5h}\mu_1, x = 0, \\ \varphi^+ = \frac{1}{0.5h}\mu_2, x = l. \end{cases}$$

$$\overline{\Lambda}(t^{j+\sigma}) y^{(\sigma)} = \begin{cases} \widetilde{\Lambda}(t^{j+\sigma}) y_i^{(\sigma)} = \frac{\varkappa_i}{x^m} (x_{i-0.5}^m a_i^j y_{\bar{x},i}^{(\sigma)})_x + \frac{b^{-j}}{x_{im}} (x_{i-0.5}^m a_i^j y_{\bar{x},i}^{(\sigma)}) + \frac{b^{+j}}{x_i^m} (x_{i+0.5}^m a_{i+1}^j y_{x,i}^{(\sigma)}) \\ \qquad - d_i^j (y_{i_0}^{(\sigma)} x_{i_0}^- + y_{i_0+1}^{(\sigma)} x_{i_0}^+), \qquad\qquad (x, t) \in \omega_{h\tau}, \\ \Lambda^- y_0^{(\sigma)} = \frac{(m+1)(\varkappa_0 a_1 y_{x,0}^{(\sigma)} - \frac{0.5h}{m+1} d_0(y_{i_0}^{(\sigma)} x_{i_0}^- + y_{i_0+1}^{(\sigma)} x_{i_0}^+))}{0.5h}, \qquad x = 0, \\ \Lambda^+ y_N^{(\sigma)} = -\frac{\varkappa_N a_N y_{\bar{x},N}^{(\sigma)} + \beta \sum_{i=0}^N y_i^{(\sigma)} h + 0.5hd_N \left(y_{i_0}^{(\sigma)} x_{i_0}^- + y_{i_0+1}^{(\sigma)} x_{i_0}^+\right)}{0.5h}, \qquad x = l, \end{cases}$$

We input the scalar product and the norm as follows

$$(u, v) = \sum_{i=1}^{N-1} u_i v_i h, \quad (u, u) = \|u\|_0^2,$$

$$(u, v] = \sum_{i=1}^{N} u_i v_i \hbar, \quad \|u\|_0^2 = \sum_{i=1}^{N} u_i^2 \hbar,$$

For the solution of the problem (3.5) we obtain the a-priori estimate by multiplying (3.5) scalarly by $x^m y^{(\sigma)}$

$$(\overline{\overline{\varkappa}} \Delta_{0t_{j+\sigma}}^\alpha y, x^m y^{(\sigma)}] = (\overline{\Lambda}(t_{j+\sigma}) y^{(\sigma)}, x^m y^{(\sigma)}] + (\overline{\Phi}, x^m y^{(\sigma)}]. \tag{3.7}$$

True the following

**Lemma 3** [19]. For any function $y(t)$, specified on the quid $\overline{\omega}_\tau$, the inequality $y^{(\sigma)} \Delta_{0t_{j+\sigma}}^\alpha y \geq \frac{1}{2} \Delta_{0t_{j+\sigma}}^\alpha (y^2)$ is true.

**Lemma 4** [24]. We assume that non-negative sequences $y^j, \varphi^j, j = 0, 1, 2, \ldots$ satisfy the inequality

$$\Delta^{\alpha}_{0t_{j+\sigma}} y^j \le \lambda_1 y^{j+1} + \lambda_2 y^j + \varphi^j, \quad j \ge 1,$$

where $\lambda_1 \ge 0, \lambda_2 \ge 0-$ the constants, then there is such $\tau_0$, that if $\tau \le \tau_0$,

$$y^{j+1} \le 2\left(y^0 + \frac{t_j^{\alpha}}{\Gamma(1+\alpha)} \max_{0 \le j' \le j} \varphi^{j'}\right) E_{\alpha}\left(2\lambda t_j^{\alpha}\right), \quad 1 \le j \le j_0,$$

where $E_{\alpha}(z) = \sum_{k=0}^{\infty} \frac{z^k}{\Gamma(1+k\alpha)} -$ Mittag-Leffler funcion, $\lambda = \lambda_1 + \frac{\lambda_2}{2+2^{1-\alpha}}$.

Assessing the sums forming (3.7) and considering the Lemma 3 and the conditions (3.2), (3.3) we obtain:

$$\left(\frac{\overline{\overline{\varkappa}}}{2}, \Delta^{\alpha}_{0t_{j+\sigma}}\left(x^{\frac{m}{2}} y\right)^2\right] + \frac{1}{1+hM_1}\left(\bar{x}^m a_i \varkappa, \left(y_{\bar{x}}^{(\sigma)}\right)^2\right] + \frac{h}{4(m+1)} x_{0.5}^m \Delta^{\alpha}_{0t_{j+\sigma}} y_0^2$$

$$+\frac{h}{4}(\bar{x}_N^m - x_N^m)\Delta^{\alpha}_{0t_{j+\sigma}} y_N^2 \le \varepsilon \left\|\bar{x}^{\frac{m}{2}} y_{\bar{x}}^{(\sigma)}\right\|\Big|_0^2 + M_2^{\varepsilon} \left\|x^{\frac{m}{2}} y^{(\sigma)}\right\|\Big|_0^2$$

$$-\left(d\left(y_{i_0}^{(\sigma)} x_{i_0}^- + y_{i_0+1}^{(\sigma)} x_{i_0}^+\right), x^m y^{(\sigma)}\right)$$

$$-\bar{x}_N^m \tilde{\beta} y_N^{(\sigma)} \sum_{i=0}^{N} x_i^m y_i^{(\sigma)} \hbar - 0.5 h d_N \bar{x}_N^m y_N^{(\sigma)}\left(y_{i_0}^{(\sigma)} x_{i_0}^- + y_{i_0+1}^{(\sigma)} x_{i_0}^+\right)$$

$$-x_{0.5}^m \frac{0.5h}{m+1} d_0 y_0^{(\sigma)}(y_{i_0}^{(\sigma)} x_{i_0}^- + y_{i_0+1}^{(\sigma)} x_{i_0}^+) + (\varphi, x^m y^{(\sigma)}) + \bar{x}_N^m \mu_2 y_N^{(\sigma)} + x_{0.5}^m \mu_1 y_0^{(\sigma)}.$$

$$(3.8)$$

Considering that $x_{N-0.5}^m \ge \frac{1}{6} x_N^m$, we transform some summands into (3.8)

$$\left(\frac{\overline{\overline{\varkappa}}}{2}, \Delta^{\alpha}_{0t_{j+\sigma}}\left(x^{\frac{m}{2}} y\right)^2\right] + \frac{h}{4}(\bar{x}_N^m - x_N^m)\Delta^{\alpha}_{0t_{j+\sigma}} y_N^2 \ge \left(\frac{\overline{\overline{\varkappa}}}{2}, \Delta^{\alpha}_{0t_{j+\sigma}}(x^{\frac{m}{2}} y)^2\right)$$

$$+\frac{h}{4} x_{N-0.5}^m \Delta^{\alpha}_{0t_{j+\sigma}} y_N^2 \ge \frac{M_3}{2}\left(1, \Delta^{\alpha}_{0t_{j+\sigma}}\left(x^{\frac{m}{2}} y\right)^2\right)$$

$$+\frac{0.5h}{12} \Delta^{\alpha}_{0t_{j+\sigma}}\left(x_N^{\frac{m}{2}} y_N\right)^2 \ge \frac{1}{12}\left(1, \Delta^{\alpha}_{0t_{j+\sigma}}(x^{\frac{m}{2}} y)^2\right)$$

$$(3.9)$$

$$+\frac{0.5h}{12} \Delta^{\alpha}_{0t_{j+\sigma}}(x^{\frac{m}{2}} y_N)^2 \ge \frac{1}{12}(1, \Delta^{\alpha}_{0t_{j+\sigma}}(x^{\frac{m}{2}} y)^2] \ge \frac{1}{12}\Delta^{\alpha}_{0t_{j+\sigma}} \|x^{\frac{m}{2}} y\|_0^2,$$

where $M_3 = \begin{cases} 1, \text{если } m = 0, m \ge 1, \\ \frac{1}{2}, \text{если } m \in (0,1), h \le h_0 = \sqrt{\frac{12x^2}{m(1-m)}}, \end{cases}$

$$-\left(d\left(y_{i_0}^{(\sigma)}x_{i_0}^{-} + y_{i_0+1}^{(\sigma)}x_{i_0}^{+}\right), x^m y^{(\sigma)}\right) - \bar{x}_N^m \tilde{\beta} y_N^{(\sigma)} \sum_{i=0}^{N} x_i^m y_i^{(\sigma)} \hbar$$

$$-0.5 h d_N \bar{x}_N^m y_N^{(\sigma)} (y_{i_0}^{(\sigma)} x_{i_0}^{-} + y_{i_0+1}^{(\sigma)} x_{i_0}^{+})$$

$$-x_{0.5}^m \frac{0.5h}{m+1} d_0 y_0^{(\sigma)} (y_{i_0}^{(\sigma)} x_{i_0}^{-} + y_{i_0+1}^{(\sigma)} x_{i_0}^{+}) + (\varphi, x^m y^{(\sigma)}) + \bar{x}_N^m \mu_2 y_N^{(\sigma)} + x_{0.5}^m \mu_1 y_0^{(\sigma)}$$

$$\leq \varepsilon \|\bar{x}^{\frac{m}{2}} y_{\bar{x}}^{(\sigma)}\|_0^2 + M_4^\varepsilon (\|x^{\frac{m}{2}} y^{(\sigma)}\|_0^2 + (x_{0.5}^{\frac{m}{2}} y_0^{(\sigma)})^2)$$

$$+ M_5(\|x^{\frac{m}{2}} \varphi\|_0^2 + \mu_1^2 + \mu_2^2).$$

(3.10)

We rewrite (3.8) considering (3.9) and (3.10)

$$\Delta_{0t_{j+\sigma}}^\alpha \|x^{\frac{m}{2}} y\|_1^2 + \|\bar{x}^{\frac{m}{2}} y_{\bar{x}}^{(\sigma)}\|_0^2 \leq \varepsilon M_6 \|\bar{x}^{\frac{m}{2}} y_{\bar{x}}^{(\sigma)}\|_0^2 + M_7 \|x^{\frac{m}{2}} y^{(\sigma)}\|_1^2$$

$$+ M_8(\|x^{\frac{m}{2}} \varphi\|_0^2 + \mu_1^2 + \mu_2^2),$$

(3.11)

where $\|x^{\frac{m}{2}} y\|_1^2 = \|x^{\frac{m}{2}} y\|_0^2 + (x_{0.5}^{\frac{m}{2}} y_0)^2$.
Opting $\varepsilon = \frac{1}{2}$, from (3.11) we find

$$\Delta_{0t_{j+\sigma}}^\alpha \|x^{\frac{m}{2}} y\|_1^2 \leq M_9^\sigma \|x^{\frac{m}{2}} y^{j+1}\|_1^2 + M_{10}^\sigma \|x^{\frac{m}{2}} y^j\|_1^2 + M_{11}(\|x^{\frac{m}{2}} \varphi\|_0^2 + \mu_1^2 + \mu_2^2).$$

(3.12)

On the basis of the Lemma 4 from (3.12) we obtain

$$\| x^{\frac{m}{2}} y^{j+1}\|_1^2 \leq M(\| x^{\frac{m}{2}} y^0\|_1^2 + \max_{0 \leq j' \leq j} (\| x^{\frac{m}{2}} \varphi \|_0^2 + \mu_1^2 + \mu_2^2)). \quad (3.13)$$

where $M - const > 0$, independent on $h$ and $\tau$.
True the following

**Theorem 2.** Let the conditions (1.5) be met, then there are such $h_0, \tau_0$, that if $h \leq h_0, \tau \leq \tau_0$, for the solution of the differential problem (3.1)–(3.4) the a-priori estimate (3.13) is true.

From (3.13) it follows the solution uniqueness and stability of the difference scheme solution (3.1)–(3.4) based on the initial data and the right side and also convergence of differential problem solution (3.1)–(3.4) to the differential problem solution (1.1)–(1.4) in terms of norm $\|x^{\frac{m}{2}} z^{j+1}\|_1^2$ in each layer, that if there are such $h_0, \tau_0$, then at $h \leq h_0, \tau \leq \tau_0$ the a-priori estimate is true

$$\|x^{\frac{m}{2}} (y^{j+1} - u^{j+1})\|_1 \leq M \|x^{\frac{m}{2}-1}\|_1 (h^2 + \tau^2) \leq \overline{M}(h^2 + \tau^2).$$

where $\overline{M} - const > 0$, independent on $h$ and $\tau$.

## 5 Non-local Boundary Value Problem B and the A-Priori Estimate in the Differential Form

Let us consider the following non-local boundary value problem for the Eq. (1.1). We change the condition (1.3) to the condition

$$-k(l,t)u_x(l,t) = \beta(t)u(0,t) + \int_0^t \rho(t,\tau)u(0,\tau)d\tau - \mu(t), |\beta| \le c_2. \quad (4.1)$$

To obtain the a-priori estimate of the solution we multiply (1.1) scalarly by $x^m u$. Then, considering the transformations (2.2)–(2.4), from (2.1) we obtain

$$\frac{1}{2}\partial_{0t}^\alpha \|x^{\frac{m}{2}}u\|_0^2 + c_0\|x^{\frac{m}{2}}u_x\|_0^2 \le x^m uku_x|_0^l + \varepsilon\|x^{\frac{m}{2}}u_x\|_0^2$$
$$+M_1^\varepsilon\|x^{\frac{m}{2}}u\|_0^2 + \frac{1}{2}\|x^{\frac{m}{2}}f\|_0^2. \quad (4.2)$$

After some simple transformations we obtain from (4.2)

$$\partial_{0t}^\alpha \|x^{\frac{m}{2}}u\|_0^2 + \|x^{\frac{m}{2}}u_x\|_0^2$$
$$\le M_2\|x^{\frac{m}{2}}u\|_0^2 + \int_0^t \left(\varepsilon_1 M_3\|x^{\frac{m}{2}}u_x\|_0^2 + M_4^{\varepsilon_1}\|x^{\frac{m}{2}}u\|_0^2\right)d\tau \quad (4.3)$$
$$+M_5(\|x^{\frac{m}{2}}f\|_0^2 + \mu^2(t)).$$

Applying to the both sides (4.3) the operator of the fractional integration $D_{0t}^{-\alpha}$, from (4.3) we find

$$\|x^{\frac{m}{2}}u\|_0^2 + D_{0t}^{-\alpha}\|x^{\frac{m}{2}}u_x\|_0^2 \le M_2 D_{0t}^{-\alpha}\|x^{\frac{m}{2}}u\|_0^2 + D_{0t}^{-\alpha}\int_0^t (\varepsilon_1 M_3\|x^{\frac{m}{2}}u_x\|_0^2$$
$$+M_4^{\varepsilon_1}\|x^{\frac{m}{2}}u\|_0^2)d\tau + M_5(D_{0t}^{-\alpha}(\|x^{\frac{m}{2}}f\|_0^2 + \mu^2(t)) + \|x^{\frac{m}{2}}u_0(x)\|_0^2). \quad (4.4)$$

We transform the second summand in the right side (4.4) as follows

$$D_{0t}^{-\alpha}\int_0^t \|x^{\frac{m}{2}}u\|_0^2 d\tau = \frac{1}{\Gamma(\alpha)}\int_0^t \frac{d\tau}{(t-\tau)^{1-\alpha}}\int_0^\tau \|x^{\frac{m}{2}}u\|_0^2 ds$$
$$= \frac{1}{\Gamma(\alpha)}\int_0^t \|x^{\frac{m}{2}}u\|_0^2 \int_s^t \frac{d\tau}{(t-\tau)^{1-\alpha}} = \frac{1}{\Gamma(\alpha)}\int_0^t \|x^{\frac{m}{2}}u\|_0^2(-\frac{(t-\tau)^\alpha}{\alpha}|_s^t)ds$$
$$= \frac{1}{\alpha\Gamma(\alpha)}\int_0^t (t-s)^\alpha \|x^{\frac{m}{2}}u\|_0^2 ds = \frac{1}{\Gamma(\alpha+1)}\int_0^t (t-\tau)^\alpha \|x^{\frac{m}{2}}u\|_0^2 d\tau \quad (4.5)$$
$$\le \frac{1}{\alpha\Gamma(\alpha)}\int_0^t \frac{(t-\tau)\|x^{\frac{m}{2}}u\|_0^2 d\tau}{(t-\tau)^{1-\alpha}} \le \frac{T}{\alpha}D_{0t}^{-\alpha}\|x^{\frac{m}{2}}u\|_0^2.$$

With the help of (4.5) from (4.4) at $\varepsilon_1 = \frac{\alpha}{2TM_6}$ we find

$$\|x^{\frac{m}{2}}u\|_0^2 + D_{0t}^{-\alpha}\|x^{\frac{m}{2}}u_x\|_0^2 \le M_6 D_{0t}^{-\alpha}\|x^{\frac{m}{2}}u\|_0^2$$
$$+M_7\left(D_{0t}^{-\alpha}\left(\|x^{\frac{m}{2}}f\|_0^2 + \mu^2(t)\right) + \|x^{\frac{m}{2}}u_0(x)\|_0^2\right). \tag{4.6}$$

On the basis of the Lemma 2 from (4.6) we obtain the required a-priori estimate

$$\|x^{\frac{m}{2}}u\|_0^2 + D_{0t}^{-\alpha}\|x^{\frac{m}{2}}u_x\|_0^2 \le M(D_{0t}^{-\alpha}(\|x^{\frac{m}{2}}f\|_0^2 + \mu^2(t)) + \|x^{\frac{m}{2}}u_0(x)\|_0^2), \tag{4.7}$$

where $M - const > 0$, depending only on the input data of the problem (1.1), (1.2), (4.1), (1.4).

True the following

**Theorem 3.** If the conditions (1.5), (4.1) be met, then for the solution $u(x, t)$ of the problem (1.1), (1.2), (4.1), (1.4) the a-priori estimate (4.7) is true, whence it follows the solution uniqueness and stability based on the initial data and the right side.

## 6 Stability and Convergence of the Difference Scheme

On the uniform quid $\overline{\omega}_{h\tau}$ the differencial problem (1.1), (1.2), (4.1), (1.4) we assign the following difference scheme:

$$\begin{cases} \overline{\overline{\varkappa}}\Delta_{0t_{j+\sigma}}^{\alpha} y = \overline{\Lambda}(t^{j+\sigma})y^{(\sigma)} + \overline{\Phi}, \\ y(x, 0) = u_0(x), \end{cases} \tag{5.1}$$

where

$$\overline{\overline{\varkappa}} = \begin{cases} \varkappa_i, x \in \omega_h, \\ 1, x = 0, l, \end{cases} \quad \varkappa_i = 1 + \frac{m(m-1)h^2}{24x_i^2}, \quad \overline{\Phi} = \begin{cases} \varphi = \varphi_i, (x, t) \in \omega_{h\tau}, \\ \varphi^- = \frac{(m+1)}{0.5h}\mu_1, x = 0, \\ \varphi^+ = \frac{1}{0.5h}\mu_2, x = l. \end{cases}$$

$$\overline{\Lambda}(t^{j+\sigma})y^{(\sigma)} = \begin{cases} \tilde{\Lambda}(t^{j+\sigma})y_i^{(\sigma)} = \frac{\varkappa_i}{x_i^m}(x_{i-0.5}^m a_i^j y_{\overline{x},i}^{(\sigma)})_x + \frac{b^{-j}}{x_{im}}(x_{i-0.5}^m a_i^j y_{\overline{x},i}^{(\sigma)}) + \\ +\frac{b^{+j}}{x_i^m}(x_{i+0.5}^m a_{i+1}^j y_{x,i}^{(\sigma)}) - d_i^j(y_{i_0}^{(\sigma)}x_{i_0}^- + y_{i_0+1}^{(\sigma)}x_{i_0}^+), \\ \Lambda^- y_0^{(\sigma)} = \frac{(m+1)(\varkappa_0 a_1 y_{x,0}^{(\sigma)} - d_0(y_{i_0}^{(\sigma)}x_{i_0}^- + y_{i_0+1}^{(\sigma)}x_{i_0}^+))}{0.5h}, x = 0, \\ \Lambda^+ y_N^{(\sigma)} = -\frac{\varkappa_N a_N y_{\overline{x},N}^{(\sigma)} + \tilde{\beta}y_0^{(\sigma)} + \sum_{s=0}^j \rho_s^j y_s^s \overline{\tau} + 0.5hd_N\left(y_{i_0}^{(\sigma)}x_{i_0}^- + y_{i_0+1}^{(\sigma)}x_{i_0}^+\right)}{0.5h}, x = l, \end{cases}$$

We input the scalar product and the norm as follows

$$(u, v] = \sum_{i=1}^{N} u_i v_i \hbar, \quad \|u]\|_0^2 = \sum_{i=1}^{N} u_i^2 \hbar, \quad \hbar = \begin{cases} 0.5h, = N, \\ h, i \neq N. \end{cases}$$

Then we multiply (5.1) scalarly by $x^m y^{(\sigma)}$, and obtain

$$(\overline{\varkappa} \Delta_{0t_{j+\sigma}}^{\alpha} y, x^m y^{(\sigma)}] = (\overline{\Lambda}(t_{j+\sigma}) y^{(\sigma)}, x^m y^{(\sigma)}] + (\overline{\Phi}, x^m y^{(\sigma)}]. \tag{5.2}$$

After some simple transformations we obtain from (5.2)

$$\Delta_{0t_{j+\sigma}}^{\alpha} \|x^{\frac{m}{2}} y]\|_1^2 + \|\bar{x}^{\frac{m}{2}} y_{\bar{x}}^{(\sigma)}]\|_0^2 \leq \sum_{s=0}^{j} \|\bar{x}^{\frac{m}{2}} y_{\bar{x}}]\|_0^2 \bar{\tau}$$

$$+ M_1 \|x^{\frac{m}{2}} y^{(\sigma)}]\|_1^2 + M_2 \sum_{s=0}^{j} \|x^{\frac{m}{2}} y]\|_0^2 \bar{\tau} + M_3 (\|x^{\frac{m}{2}} \varphi\|_0^2 + \mu_1^2 + \mu_2^2), \tag{5.3}$$

where $\|x^{\frac{m}{2}} y]\|_1^2 = \|x^{\frac{m}{2}} y]\|_0^2 + (x_{0.5}^{\frac{m}{2}} y_0)^2$.

We assess the first summand in the right side (5.3) and rewrite (5.3) in another form

$$\|\bar{x}^{\frac{m}{2}} y_{\bar{x}}^{j+1}]\|_0^2 \leq \sum_{s=0}^{j} \|\bar{x}^{\frac{m}{2}} y_{\bar{x}}^s]\|_0^2 \bar{\tau} + F, \tag{5.4}$$

where $F = M_1 \|x^{\frac{m}{2}} y^{(\sigma)}]\|_1^2 + M_2 \sum_{s=0}^{j} \|x^{\frac{m}{2}} y]\|_0^2 \bar{\tau} + M_3 (\|x^{\frac{m}{2}} \varphi\|_0^2 + \mu_1^2 + \mu_2^2)$.
Applying the Lemma 4 (see [27, p. 171]) to (5.4), we find

$$\|\bar{x}^{\frac{m}{2}} y_{\bar{x}}^{j+1}]\|_0^2 \leq M_4 F, \tag{5.5}$$

Considering (5.5), after some simple transformations we obtain from (5.3)

$$\Delta_{0t_{j+\sigma}}^{\alpha} \|x^{\frac{m}{2}} y]\|_1^2 + \|\bar{x}^{\frac{m}{2}} y_{\bar{x}}^{(\sigma)}]\|_0^2 \leq M_5 \|x^{\frac{m}{2}} y^{(\sigma)}]\|_1^2 + F, \tag{5.6}$$

where $F = M_6 \sum_{s=0}^{j} \|x^{\frac{m}{2}} y]\|_0^2 \bar{\tau} + M_7 \sum_{s=0}^{j} (\|x^{\frac{m}{2}} \varphi\|_0^2 + \mu_1^2 + \mu_2^2) \tau$.
On the basis of the Lemma 4 we obtain from (5.6)

$$\|x^{\frac{m}{2}} y^{j+1}]\|_1^2 \leq M_8 \left( \|x^{\frac{m}{2}} y^0]\|_1^2 + \max_{0 \leq j' \leq j} (\sum_{s=0}^{j'} \|x^{\frac{m}{2}} y]\|_0^2 \bar{\tau} + \sum_{s=0}^{j'} (\|x^{\frac{m}{2}} \varphi^s\|_0^2 + \mu_1^2 + \mu_2^2) \tau) \right). \tag{5.7}$$

Considering that

$$\max_{0 \leq j' \leq j} \sum_{s=0}^{j'} \|x^{\frac{m}{2}} y^s]\|_1^2 \bar{\tau} \leq \sum_{j'=0}^{j} \max_{0 \leq s \leq j'} \|x^{\frac{m}{2}} y^s]\|_1^2 \bar{\tau} \leq \sum_{j'=0}^{j} \max_{0 \leq s \leq j'} \|x^{\frac{m}{2}} y^s]\|_1^2 \tau$$

and introducing another designation $g^j = \max_{0 \le j' \le j} \| x^{\frac{m}{2}} y]\|_1^2$ we obtain from (5.7)

$$g^{j+1} \le M_5 \sum_{s=0}^{j} g^s \tau + M_9 F_1^j, \tag{5.8}$$

where $F_1^j = \| x^{\frac{m}{2}} y^0]\|_1^2 + \max_{0 \le j' \le j} \Sigma_{s=0}^{j'} (\| x^{\frac{m}{2}} \varphi^s \|_0^2 + \mu_1^2 + \mu_2^2)\tau.$

On the basis of the Lemma 4 (see [27, p.171]) from (5.8) we obtain the required a-priori estimate

$$\| x^{\frac{m}{2}} y^{j+1}]\|_0^2 \le M(\| x^{\frac{m}{2}} y^0]\|_1^2 + \max_{0 \le j' \le j} \Sigma_{s=0}^{j'} (\| x^{\frac{m}{2}} \varphi^s \|_0^2 + \mu_1^2 + \mu_2^2)\tau), \tag{5.9}$$

where $M - const > 0$, independent on $h$ и $\tau$.

True the following

**Theorem 4.** Let the conditions (1.5), (4.1) be met, then there are such $h_0, \tau_0$, that if $h \le h_0, \tau \le \tau_0$, then for the differencial problem solution (5.1) the a-priori estimate (5.9) is true.

From (5.9) it follows the solution uniqueness and stability of the difference scheme solution (5.1) based on the initial data and the right side and also convergence of differencial problem solution (5.1) to the differential problem solution (1.1), (1.2), (4.1), (1.4) in terms of norm $\|x^{\frac{m}{2}} z^{j+1}]\|_0^2$ in each layer, that if there are such $h_0, \tau_0$, then at $\tau \le \tau_0, h \le h_0$ the a-priori estimate is true

$$\|x^{\frac{m}{2}} (y^{j+1} - u^{j+1})]\|_0^2 \le \overline{M}(h^2 + \tau^2).$$

where $\overline{M} - const > 0$, independent on $h$ и $\tau$.

**Remark.** The results obtained in this work are true in case if:

1) The condition (1.3) is changed to the condition:

$$\partial_{0t}^\alpha \int_0^l x^m (x, t)dx = \mu(t), 0 \le t \le T.$$

2) The Eq. (4.1) is represented as follows:

$$\partial_{0t}^\alpha u = \frac{1}{x^m} \frac{\partial}{\partial x} \left( x^m k(x, t) \frac{\partial u}{\partial x} \right) + r(x, t) \frac{\partial u}{\partial x} - \sum_{s=1}^{p} q_s(x, t) u(x_s, t) + f(x, t),$$

$$0 < x < l, 0 < t \le T,$$

If it is required to meet the conditions $|q_s| \le c_2$.

# References

1. Steklov VA (1983) Basic problems of mathematical physics. Nauka, Moscow
2. Canon JR (1963) The solution of the heat equation subject to the specification of energy. Quart Appl Math 21(2):155–160
3. Kamynin LI (1964) On a boundary value problem of the theory of thermal conductivity with nonclassical boundary conditions. Comput Math Math Phys 4(6):1006–1024
4. Chudnovskii LF (1969) Some corrections in the formulation and solution of problems of heat and moisture transfer in the soil. In: Collection of proceedings of the AFI, vol 23, pp 41–54
5. Bitsadze AV, Samarskii AA (1969) On some simplest generalizations of linear elliptic boundary value problems. Dokl USSR Acad Sci 185(4):739–740
6. Ionkin NI, Morozova VA (2000) Two-dimensional equation of thermal conductivity with non-local boundary conditions. Differ Eq 36(7):884–888
7. Kozhanov AI (2004) On a non-local boundary value problem with variable coefficients for the equations of thermal conductivity and Aller. Differ Eq 40(6):763–774
8. Gulin AV, Ionkin NI, Morozova VA (2005) Difference schemes for non-local problems. Izv vuzov Math 1:40–51
9. Khudalov MZ (2002) A non-local boundary value problem for a loaded equation of parabolic type. Vladikavkaz Math J 4(4):59–64
10. Olisaev EG, Lafisheva MM (2002) On the convergence of a difference scheme for a parabolic equation with a non-local condition in cylindrical coordinates. Vladikavkaz Math J 4(2):50–56
11. Nakhushev AM (2003) Fractional calculus and its application. Fizmatlit, Moscow
12. Podliubnyi I (1999) Fractional differential equations. Academic Press, San Diego
13. Samko SG, Kilbas AA, Marichev OI (1987) Integrals and derivatives of fractional order and some of their applications. Nauka i Tekhnika, Minsk
14. Kochubei AN (1990) Diffusion of fractional order. Differ Eq 26(4):485–492
15. Nigmatullin RR (1992) Fractional integral and its physical interpretation. Theor Math Phys 90(3):354–368
16. Diethelm K, Walz G (1997) Numerical solution of fractional order differential equations by extrapolation. Numer Algorithms 16:231–253
17. Taukenova FI, Shkhanukov-Lafishev MH (2006) Difference methods for solving boundary value problems for fractional differential equations. Comput Math Math Phys 46(10):1785–1795
18. Alikhanov AA (2010) A-priori estimates for solutions of boundary value problems for fractional-order equations. Differ Eq 46(5):658–664
19. Alikhanov AA (2015) A new difference scheme for the time fractional diffusion equation. J Comput Phys 280:424–438
20. Chen C, Liu F, Anh V, Turner I (2010) Numerical schemes with high spatial accuracy for a variable-order anomalous subdiffusion equations. SIAM J Sci Comput 32(4):1740–1760
21. Lin Y, Li X, Xu C (2011) Finite difference/spectral approximations for the fractional cable equation. Math Comput 80:1369–1396
22. Beshtokov MK (2016) Difference method for solving a nonlocal boundary value problem for a degenerating third-order pseudo-parabolic equation with variable coefficients. Comput Math Math Phys 56(10):1763–1777. https://doi.org/10.1134/S0965542516100043
23. Beshtokov MK (2018) Boundary value problems for degenerate and nondegenerate Sobolev type equations with a nonlocal source in differential and difference interpretations. Differ Eq 54(2):250–267
24. Beshtokov MK (2019) Nonlocal boundary value problems for Sobolev-type fractional equations and grid methods for solving them. Sib Adv Math 29(1):1–21
25. Kipriianov IA, Kulikov AA (1994) Optimal control of processes described by singular equations of parabolic type. Diff Eq 30(11):1982–1987
26. Samarskii AA (1983) Teoriya raznostnykh skhem (Theory of difference schemes). Nauka, Moscow
27. Samarskii AA, Gulin AV (1983) Stability of difference schemes. Nauka, Moscow

# Discrete Implementation of Korteweg de Vries Equation Based on a Modified "Cabaret" Scheme

**Elena Timofeeva**⊚, **Aleksandr Sukhinov**⊚, **Aleksandr Chistiakov**⊚, and **Nadezhda Timofeeva**⊚

**Abstract** The work is focused on creation and investigation of the scheme applied to solving the problems related to non-linear wave processes described by Korteweg de Vries model equation. The behavior of the numerical solution in this case depends mostly on the difference scheme choice. The article proposes to implement a problem numerical solution on the basis of the improved "cabaret" scheme. Its difference operator is a linear combination of difference schemes operators called "cross" and "cabaret", while the modified scheme is obtained from schemes with optimal weight coefficients. Subject to certain values of the weighting coefficients, the combination leads to mutual compensation of approximation error, and the resulting scheme acquires better properties than the original schemes. The scheme called "left corner" was considered as a test finite-difference scheme for the results assessment in a model problem numerical solution. The results of stability and accuracy investigation showed that the proposed difference scheme has the same limitations as the "left corner" scheme with an approximation error equal $O\left(\tau^2 + h^3\right)$.

**Keywords** Mathematical model · Korteweg de Vries equation · Numerical modeling · Difference "cabaret" scheme · Difference scheme "cross" · Parallel computations

## 1 Introduction

Non-linear wave processes are intensively investigated in various scientific and technological fields, including optics, physics, radiation physics and hydrodynamics [1].

E. Timofeeva (✉)
North Caucasus Center for Mathematical Research, North-Caucasus Federal University, 355017 Stavropol, Russia
e-mail: teflena@mail.ru

A. Sukhinov · A. Chistiakov
Don State Technical University, Rostov-on-Don, Russia

N. Timofeeva
North-Caucasus Federal University, Stavropol, Russia

© The Author(s), under exclusive license to Springer Nature Switzerland AG 2022
A. Tchernykh et al. (eds.), *Mathematics and its Applications in New Computer Systems*,
Lecture Notes in Networks and Systems 424,
https://doi.org/10.1007/978-3-030-97020-8_34

When considering issues related to mathematical modeling of wave processes on heavy liquid free surface, it is possible to obtain the hydrodynamic models including both non-linear and dispersion effects [2]. Korteweg de Vries equation is the simplest model in the class of non-linear dispersion waves [3]. The equation possesses space-localized soliton solutions, represented by a classical solitary wave with a single hill and infinite period, dying out monotonically at infinity [4]. There are also generalized-solitary waves and solitary wave packages among Korteweg de Vries equation solutions.

The solution of non-linear partial differential equations is one of the most urgent and complex problems. Unlike linear differential equations, for which general solution methods have already been developed (for example, Fourier, Laplace's methods, etc.), there are no common solution methods for non-linear differential equations in partial derivatives. Each non-linear equation or a small group of equations of the same type requires its own, specific solution methods development.

The materials [5–10] comprise a rather wide range of approaches for numerical integration of equations of long non-linear waves. The methods based on the use of explicit and implicit finite-difference schemes as well as pseudospectral approach should be highlighted among the major methods. In this paper, we consider the solution of the problem of non-linear wave dynamics of solitons, including Korteweg de Vries equation on the example of a modified "cabaret" scheme. This scheme detailed analysis was implemented in [11, 12].

## 2 Materials and Methods

### 2.1 Problem Statement

Korteweg de Vries equation is considered to be a test problem in order to study waves of small but finite amplitude in dispersion media. In a rectangle $\overline{Q_T} = \{(x, t) : 0 \leq x \leq l, \ 0 \leq t \leq T\}$ Korteweg de Vries equation

$$\frac{\partial u}{\partial t} + u \frac{\partial u}{\partial x} + \beta \frac{\partial^3 u}{\partial x^3} = 0 \tag{1}$$

or

$$u'_t + u u'_x + \beta u'''_{xxx} = 0, \beta = const > 0.$$

It is required to find the solution $u(x, t)$ of Cauchy problem under the following conditions:

– at initial time point $t = 0$

$$u(x, \ 0) = u_0(x), 0 \leq x \leq l, \tag{2}$$

- at the boundaries

$$u(0, t) = u(l, t) = u''(0, t) = 0, t \geq 0, \tag{3}$$

- or space-periodic conditions

$$u(0, t) = u(l, t), \ u'(0, t) = u'(l, t), \ u''(0, t) = u''(l, t), \tag{3*}$$

where $u(x, t) = v + \delta v$, $v$ – average speed, $\delta v$ – oscillating speed, $\beta$ – dispersion coefficient.

## 2.2 Difference Scheme "Left Corner" for Korteweg de Vries Equation

We find a numerical solution of the model problem (1)–(3) using the finite-difference scheme "left corner" [13]. This problem will be considered as a as a test one.

We insert in the problem definition domain the uniform computation grid $\omega = \omega_\tau \times \overline{\omega}_h$, где $\omega_\tau = \{t^n \mid n = 0, 1, \ldots\}$, with the constant time step $\tau = t_{n+1} - t_n$, $\overline{\omega}_h = \{x_i \mid x_i = ih; \ i = 0, 1, \ldots, N; \ Nh = l\}$ with the step $h$ by spatial variable, $N$ – a number of spatial steps.

The finite-difference scheme "left corner" for Korteweg de Vries equation is the following:

$$\frac{u_i^{n+1} - u_i^n}{\tau} + u_{i-1/2} \frac{u_i^n - u_{i-1}^n}{h} + \beta \frac{u_{i+1}^n - 3u_i^n + 3u_{i-1}^n - u_{i-2}^n}{h^3} = 0, \tag{4}$$

where $u_{i-1/2} = (u_i + u_{i-1})/2$.

## 2.3 Stability Investigation of the Scheme "Left Corner"

We investigate the scheme "left corner" (4) stability for Korteweg de Vries equation. Let $u_i^n = \varphi^n \cdot e^{jki}$, где $j = \sqrt{-1}$, we insert this expression into the equality (4), divide the equality by $\varphi^n \cdot e^{jki}$, and obtain:

$$\frac{\varphi - 1}{\tau} + u \cdot \frac{1 - e^{-jk}}{h} + \beta \cdot \frac{e^{jk} - 3 + 3 \cdot e^{-jk} - e^{-2jk}}{h^3} = 0,$$

After the transformations this expression is shown as:

$$\varphi = 1 - \frac{u\tau}{h}(1 - \cos k) - \frac{u\tau j}{h} \sin k - \beta\tau \frac{4\cos k - 3 - \cos 2k - 2j\sin k + j\sin 2k}{h^3} = 0.$$

We introduce the designations $a = \frac{u\tau}{h}$, $b = \frac{\beta\tau}{h^3}$, $c = a + 2b(\cos k - 1)$, highlight actual and supposed parts and obtain:

$$\varphi = 1 + (\cos k - 1)c - jc \sin k.$$

To keep the system stable, it is necessary and sufficient that inequality be respected $|\varphi| \leq 1$. It is obvious that:

$$|\varphi|^2 = (1 - c + c \cos k)^2 + c^2 \sin^2 k = (1 - c)^2 + 2c \cos k(1 - c) + c^2$$

$$\leq (1 - c)^2 + 2c(1 - c) + c^2 = 1, \text{ at } 0 \leq c \leq 1 \text{ or } 0 \leq \frac{u\tau}{h} + 2\frac{\beta\tau}{h^3}(\cos k - 1) \leq 1.$$

In consequence, the scheme is stable under the condition $\beta \leq uh^2/4$, $\frac{u\tau}{h} \leq 1$.

## 2.4 Difference Schemes "Cabaret" and "Cross" for Korteweg de Vries Equation

The major functional points in construction of efficient grid schemes like "cabaret" with improved dispersion properties [11, 12] for non-linear Korteweg de Vries equations are illustrated by solving the test problem (1)–(3).

We insert in the problem definition domain the uniform computation grid $\omega = \omega_\tau \times \overline{\omega}_h$, где $\omega_\tau = \{t^n | n = 0, 1, ...\}$, with the constant time step $\tau = t_{n+1} - t_n$, $\overline{\omega}_h = \{x_i | x_i = ih; i = 0, 1, ..., N; Nh = l\}$ with the step $h$ by spatial variable, $N$ – a number of spatial steps.

For the problem numerical solution (1)–(3) it is possible to use the finite-difference schemes [14, 15]:

– "cabaret"

$$\frac{u_i^{n+1} - u_i^n}{2\tau} + \frac{u_{i-1}^n - u_{i-1}^{n-1}}{2\tau} + u_{i-1/2}^n \frac{u_i^n - u_{i-1}^n}{h} + \beta \frac{u_{i+1}^n - 3u_i^n + 3u_{i-1}^n - u_{i-2}^n}{h^3} = 0, \quad (5)$$

– "cross"

$$\frac{u_i^{n+1} - u_i^{n-1}}{2\tau} + u_{i+1/2}^n \frac{u_{i+1}^n - u_i^n}{2h} + u_{i-1/2}^n \frac{u_i^n - u_{i-1}^n}{2h} + \beta \frac{u_{i+2}^n - 2u_{i+1}^n + 2u_{i-1}^n - u_{i-2}^n}{2h^3} = 0. \quad (6)$$

In the "cross" scheme, for the finite-difference approximation of the third-order derivative, a symmetric difference derivative is used, obtained as the half-sum of the right and left difference derivatives, where the approximation of the second derivative acts as a function.

These schemes extended templates considering the approximation of the third derivative with respect to the spatial variable are shown in the Figs. 1 and 2.

Let us consider the application of a modified scheme, representing the linear combination of the difference schemes "cabaret" and "cross" with weight coefficients

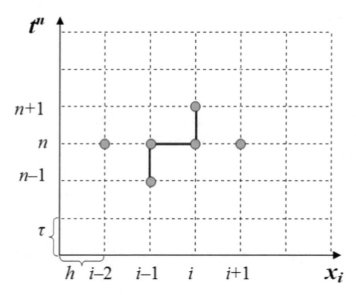

**Fig. 1** The extended template of the scheme "cabaret" for Korteweg de Vries equation

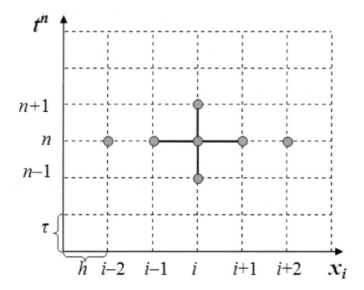

**Fig. 2** The extended template of the scheme "cross" for Korteweg de Vries equation

2/3 and 1/3, respectively [12], to solve Korteweg de Vries model equation. The approximation of Eq. (1) will be shown as:

$$\frac{u_i^{n+1} - u_i^n}{\tau} + 2\frac{u_{i-1}^n - u_{i-1}^{n-1}}{3\tau} + \frac{u_i^n - u_i^{n-1}}{3\tau} + u_{i+1/2}^n\frac{u_{i+1}^n - u_i^n}{3h} + 5u_{i-1/2}^n\frac{u_i^n - u_{i-1}^n}{3h}$$
$$+ \frac{1}{3}\beta\frac{u_{i+2}^n + 2u_{i+1}^n - 12u_i^n + 14u_{i-1}^n - 5u_{i-2}^n}{h^3} = 0.$$

$$(7)$$

## 2.5 Stability Investigation of the Scheme Obtained from a Linear Combination of "Cabaret" and "Cross" Schemes

We investigate the scheme (7) stability with the help of harmonics methods [16, 17]. Let $u_i^n = \varphi^n \cdot e^{jki}$, где $j = \sqrt{-1}$, we insert this expression into the equality (7) and obtain:

$$\frac{\varphi^{n+1} \cdot e^{jki} - \varphi^n \cdot e^{jki}}{\tau} + 2\frac{\varphi^n \cdot e^{jk(i-1)} - \varphi^{n-1} \cdot e^{jk(i-1)}}{3\tau} + \frac{\varphi^n \cdot e^{jki} - \varphi^{n-1} \cdot e^{jki}}{3\tau}$$

$$+ u\frac{\varphi^n \cdot e^{jk(i+1)} + 4\varphi^n \cdot e^{jki} - 5\varphi^n \cdot e^{jk(i-1)}}{3h}$$

$$+ \frac{1}{3}\beta\frac{\varphi^n \cdot e^{jk(i+2)} + 2\varphi^n \cdot e^{jk(i+1)} - 12\varphi^n \cdot e^{jki} + 14\varphi^n \cdot e^{jk(i-1)} - 5\varphi^n \cdot e^{jk(i-2)}}{h^3} = 0.$$

We divide the equality by $\varphi^n \cdot e^{jki}$, and then multiply the equation by $\varphi \cdot \tau$

$$\varphi^2 + \left(\frac{2}{3} \cdot e^{-jk} - \frac{2}{3} + u\tau \cdot \frac{e^{jk} + 4 - 5 \cdot e^{-jk}}{3h} + \beta\tau \cdot \frac{e^{2jk} + 2e^{jk} - 12 + 14 \cdot e^{-jk} - 5e^{-2jk}}{3h^3}\right)\varphi$$

$$- \frac{2e^{-jk} + 1}{3} = 0.$$

With introduced designations $a = u\tau/h$, $b = \beta\tau/h^3$, $c = a + 2b(\cos k - 1)$ and Euler's formula applied $e^{jk} = \cos k + j\sin k$, the equation is shown as following:

$$\varphi^2 + \left(\frac{2}{3} \cdot (\cos k - 1) - j\frac{2}{3} \cdot \sin k + c \cdot \frac{-4(\cos k - 1) + 6j\sin k}{3}\right)\varphi - \frac{2e^{-jk} + 1}{3} = 0,$$

$$\varphi^2 + \left(\frac{2}{3} \cdot (\cos k - 1)(1 - 2c) - j \cdot \sin k\left(\frac{2}{3} - 2c\right)\right)\varphi - \frac{2e^{-jk} + 1}{3} = 0.$$

The quadratic equation solution in reference to $\varphi$:

$$\varphi_{1,2} = -\left(\frac{1}{3} \cdot (\cos k - 1)(1 - 2c) - j \cdot \sin k\left(\frac{1}{3} - c\right)\right)$$

$$\pm\sqrt{\left(\frac{1}{3} \cdot (\cos k - 1)(1 - 2c) - j \cdot \sin k\left(\frac{1}{3} - c\right)\right)^2 + \frac{2e^{-jk} + 1}{3}}.$$

Let us consider the case $c = 0$:

$$\varphi_{1,2} = -\left(\frac{1}{3} \cdot (\cos k - 1) - \frac{1}{3} j \cdot \sin k\right) \pm \sqrt{\left(\frac{1}{3} \cdot (\cos k - 1) - j \cdot \frac{1}{3}\sin k\right)^2 + \frac{2e^{-jk} + 1}{3}}$$

$$\text{or } \varphi_{1,2} = -\frac{1}{3} \cdot e^{-jk} + \frac{1}{3} \pm \left(\frac{1}{3} \cdot e^{-jk} + \frac{2}{3}\right).$$

As a result we obtain $\varphi_1 = 1$, $\varphi_2 = -\frac{2}{3} \cdot e^{-jk} - \frac{1}{3}$, therefore, $\varphi_2$ is not a solution.
Let us consider the case $c = 1$:

$$\varphi_{1,2} = -\left(-\frac{1}{3} \cdot (\cos k - 1) + \frac{2}{3} j \cdot \sin k\right) \pm$$

$$\pm\sqrt{\left(-\frac{1}{3} \cdot (\cos k - 1) + \frac{2}{3} j \cdot \sin k\right)^2 + \frac{2e^{-jk} + 1}{3}} \quad \text{или}$$

$$\varphi_{1,2} = \frac{1}{3} \cdot (\cos k - 1) - \frac{2}{3} j \cdot \sin k \pm \left(\frac{2}{3}\cos k - \frac{1}{3} j\sin k + \frac{1}{3}\right).$$

As a result we obtain $\varphi_1 = e^{-jk}$, т.е. $\varphi_1 = \cos k - j \sin k$, then, $|\varphi_1|^2 = \cos^2 k + \sin^2 k = 1$, $|\varphi_1| = 1$, $\varphi_2 = -\frac{1}{3}e^{-jk} - \frac{2}{3}$, therefore, $\varphi_2$ is not a solution.
We designate $\psi(k, c)$ the function absolute values $\varphi_{1,2}(k, c)$:

$$\psi = \left|-\left(\frac{1}{3} \cdot (\cos k - 1)(1 - 2c) - j \cdot \sin k\left(\frac{1}{3} - c\right)\right)\right.$$

$$\pm\sqrt{\left(\frac{1}{3} \cdot (\cos k - 1)(1 - 2c) - j \cdot \sin k\left(\frac{1}{3} - c\right)\right)^2 + \frac{2e^{-jk} + 1}{3}}\left.\right|.$$

The function values $\psi(k, c)$ response could be checked by a numerical method. As a result, at the values $k \in [0; 2\pi]$ and $c \in [0, 1]$ it could be shown, that the equality $\psi \leq 1$ is ensured, i.e. $|\varphi| \leq 1$, indicating the difference scheme stability.

## 2.6 Accuracy Investigation of a Modified Scheme

We represent the function $u(x, t)$ in the form of a finite trigonometric Fourier series [18] in complex form:

$$u(x, t) = \sum_{m=-N}^{N} C_m(t)e^{j\omega mx}, \quad u^2(x, t)/2 = \sum_{m=-N}^{N} D_m(t)e^{j\omega mx}, \qquad (8)$$

where $\omega = \frac{\pi}{l}$, $m$ – harmonic number, $C_m(t) = \frac{2}{l}\int_0^l u(x, t)e^{-j\omega mx}dx$ – complex amplitude of $m$ harmonic, $j = \sqrt{-1}$. After inserting (8) into (1) we obtain:

$$\left(\sum_{m=-N}^{N} C_m(t)e^{j\omega mx}\right)_t' + \left(\sum_{m=-N}^{N} D_m(t)e^{j\omega mx}\right)_x' + \beta\left(\sum_{m=-N}^{N} C_m(t)e^{j\omega mx}\right)_{xxx}''' = 0.$$

Let us change the sequence of operations of differentiation and summation in the series partial sum, calculate the spatial derivative, and bearing in mind that the functions $e^{j\omega mx}$ are linearly independent for different values of $m$, we find:

$$(C_m(t))_t' = -j\omega m\, D_m(t) + j\beta\omega^3 m^3 C_m(t). \qquad (9)$$

Let us consider a modified scheme application of, representing a linear combination of the difference schemes "cabaret" and "cross" (7), for the solution of Korteweg de Vries model equation. With respect to (8) and that $x_i = ih$, we represent (7) as follows

$$\sum_{m=-N}^{N} \frac{C_m^{n+1}e^{j\omega mhi} - C_m^n e^{j\omega mhi}}{\tau} + 2\sum_{m=-N}^{N} \frac{C_m^n e^{j\omega mh(i-1)} - C_m^{n-1}e^{j\omega mh(i-1)}}{3\tau}$$

$$+ \sum_{m=-N}^{N} \frac{C_m^n e^{j\omega mhi} - C_m^{n-1}e^{j\omega mhi}}{3\tau} + \sum_{m=-N}^{N} \frac{D_m^n e^{j\omega mh(i+1)} + 4D_m^n e^{j\omega mhi} - 5D_m^n e^{j\omega mh(i-1)}}{3h}$$

$$+ \frac{1}{3}\beta\sum_{m=-N}^{N} \frac{C_m^n e^{j\omega mh(i+2)} + 2C_m^n e^{j\omega mh(i+1)} - 12C_m^n e^{j\omega mhi} + 14C_m^n e^{j\omega mh(i-1)} - 5C_m^n e^{j\omega mh(i-2)}}{h^3} = 0.$$

After the transformations we obtain:

$$\sum_{m=-N}^{N} \frac{C_m^{n+1} - C_m^n}{\tau}e^{j\omega mhi} + 2\sum_{m=-N}^{N} \frac{C_m^n - C_m^{n-1}}{3\tau}e^{-j\omega mh}e^{j\omega mhi}$$

$$+ \sum_{m=-N}^{N} \frac{C_m^n - C_m^{n-1}}{3\tau}e^{j\omega mhi} + \sum_{m=-N}^{N} \frac{D_m^n e^{j\omega mh} + 4D_m^n - 5D_m^n e^{-j\omega mh}}{3h}e^{j\omega mhi}$$

$$+\frac{1}{3}\beta\sum_{m=-N}^{N}\frac{C_m^n e^{2j\omega mh} + 2C_m^n e^{j\omega mh} - 12C_m^n + 14C_m^n e^{-j\omega mh} - 5C_m^n e^{-2j\omega mh}}{h^3}e^{j\omega mhi} = 0.$$

Due to linear independence of $e^{j\omega mi}$, the last expression we rewrite as follows:

$$\frac{C_m^{n+1} - C_m^n}{\tau} + 2\frac{C_m^n - C_m^{n-1}}{3\tau}e^{-j\omega mh} + \frac{C_m^n - C_m^{n-1}}{3\tau} + D_m^n\frac{e^{j\omega mh} + 4 - 5e^{-j\omega mh}}{3h}$$
$$+\frac{1}{3}\beta C_m^n\frac{e^{2j\omega mh} + 2e^{j\omega mh} - 12 + 14e^{-j\omega mh} - 5e^{-2j\omega mh}}{h^3} = 0.$$

$$(10)$$

At $\tau \to 0$ from (10) it follows:

$$(C_m(t))'_t = -\frac{e^{j\omega mh} + 4 - 5e^{-j\omega mh}}{2h(2 + e^{-j\omega mh})}D_m^n - \beta\frac{e^{2j\omega mh} + 2e^{j\omega mh} - 12 + 14e^{-j\omega mh} - 5e^{-2j\omega mh}}{2h^3(2 + e^{-j\omega mh})}C_m^n.$$

$$(11)$$

**Lemma 1.** *When approximating the problem (1)–(3) by the difference scheme (7), for each harmonic the solutions of the problem of wave propagation speed and dispersion term are less than real values and differ, respectively, by values:*

$$\alpha_1 = 1 - \frac{e^{j\omega mh} + 4 - 5e^{-j\omega mh}}{2j\omega mh(2 + e^{-j\omega mh})}, \alpha_2 = 1 + \frac{e^{2j\omega mh} + 2e^{j\omega mh} - 12 + 14e^{-j\omega mh} - 5e^{-2j\omega mh}}{2j\omega^3 m^3 h^3(2 + e^{-j\omega mh})}.$$

**Proof:** From (11) at $\tau \to 0$ it follows:

$$(C_m(t))'_t = -j\omega m D_m^n\frac{e^{j\omega mh} + 4 - 5e^{-j\omega mh}}{2j\omega mh(2 + e^{-j\omega mh})}$$

$$-j\omega^3 m^3 \beta C_m^n\frac{e^{2j\omega mh} + 2e^{j\omega mh} - 12 + 14e^{-j\omega mh} - 5e^{-2j\omega mh}}{2j\omega^3 m^3 h^3(2 + e^{-j\omega mh})}.$$

Due to (9) the solution, obtained on the basis of scheme (7), the following corresponds to the equation solution.

$$u'_t = -(u^2/2)_x(1 - \alpha_1) - \beta u'''_{xxx}(1 - \alpha_2),$$
$$\alpha_1 = 1 - \frac{e^{j\omega mh} + 4 - 5e^{-j\omega mh}}{2j\omega mh(2 + e^{-j\omega mh})}, \alpha_2 = 1 + \frac{e^{2j\omega mh} + 2e^{j\omega mh} - 12 + 14e^{-j\omega mh} - 5e^{-2j\omega mh}}{2j\omega^3 m^3 h^3(2 + e^{-j\omega mh})}.$$

The lemma is proved.

**Lemma 2.** *Modified difference scheme for the problem (1)–(3) has an approximation error equal to $O(\tau^2 + h^2)$.*

**Proof:** From (10) at $h \to 0$ it follows:

$$\frac{C_m^{n+1} - C_m^n}{\tau} + 2\frac{C_m^n - C_m^{n-1}}{3\tau}e^{-j\omega mh} + \frac{C_m^n - C_m^{n-1}}{3\tau} + D_m^n\frac{e^{j\omega mh} + 4 - 5e^{-j\omega mh}}{3h}$$

$$+\frac{1}{3}\beta C_m^n\frac{e^{2j\omega mh} + 2e^{j\omega mh} - 12 + 14e^{-j\omega mh} - 5e^{-2j\omega mh}}{h^3} = \frac{C_m^{n+1} - C_m^{n-1}}{\tau} + 2j\omega m D_m^n$$

$$-2j\beta\omega^3 m^3 C_m^n = 2(C_m(t))_t' + 2j\omega m D_m^n - 2j\beta\omega^3 m^3 C_m^n + O(\tau^2).$$

Let us investigate the order of approximation error of the convective special term. We make a replacement $j\omega mh = s$:

$$\alpha_1 = 1 - \frac{e^s + 4 - 5e^{-s}}{2s(2 + e^{-s})} = -\frac{s^3}{36} + O(s^4) \text{ или } \alpha_1 = j\frac{(\omega mh)^3}{36} + O(h^4).$$

It has become obvious that the modified difference scheme approximates the convective term with the third order of spatial accuracy. Let us consider the order of approximation error of the spatial dispersion term:

$$\alpha_2 = 1 - \frac{e^{2s} + 2e^s - 12 + 14e^{-s} - 5e^{-2s}}{2s^3(2 + e^{-s})} = -\frac{s^2}{12} + O(s^3).$$

The lemma is proved.

The investigation of the approximation error order of the modified difference scheme for the solution of Korteweg de Vries equation has demonstrated that the difference scheme (9) has an approximation error equal to $O(\tau^2 + h^3)$.

## 3 Results

There has been carried out a numerical implementation of a non-linear dispersion model for the solution of Korteweg de Vries equation represented by a software module out. The numerical solution of Korteweg de Vries problem based on the scheme (9) with the initial conditions $\delta v_0(x) = 0.1 \sin(\pi(x - 10)/20)(h(10 - x) - h(30 - x))$, where $h(x)$ – Heaviside function, is represented in the Fig. 3.

In the Figs. 3a) and b) are represented: by a solid line – the numerical solution of the linearized Korteweg de Vries problem; by an intermittent line – its analytical (exact) solution. Figures 3c) and d) demonstrate: by an intermittent line – the linearized problem solution; by a solid line – the nonlinear Korteweg de Vries problem. The numerical solution of the problem obtained on the basis of the "left corner" is represented in the Figs. 3a) and c). The results of the numerical experiment with the use of the proposed modified scheme (linear combination of the schemes "cabaret" and "cross") are presented in the Figs. 3b) and d).

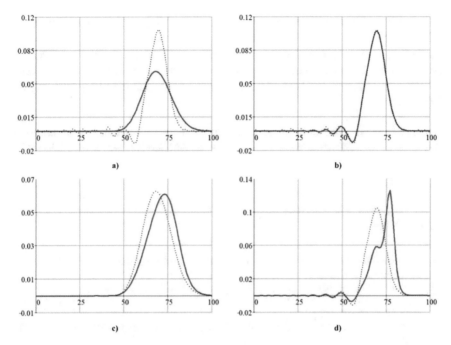

**Fig. 3** The test problem solution on the basis of considered schemes

The results of numerical calculations of the linearized Korteweg de Vries problem based on the difference scheme obtained from a linear combination of "cabaret" and "cross" schemes coincide with the analytical solution almost completely. The use of the scheme "left corner" is unreasonable for solving the problem, because this scheme shows high inaccuracy caused by dissipation, which is evident from the calculation results. The obtained numerical solutions of the non-linear and linearized Korteweg de Vries problem do not differ much from each other. The numerical solution response in this case depends more on the choice of the difference scheme.

# 4 Conclusion

A difference scheme with improved dispersion properties is proposed and investigated for Korteweg de Vries equation numerical solution. It is based on a linear combination of difference operators of the "cabaret" and "cross" schemes with weight coefficients obtained as a result of minimizing the approximation error order. The disadvantage of the scheme "left corner" is a considerable dissipation. To eliminate large net viscosity, it becomes necessary to reduce the step $h$ by the spatial variable, which leads to stricter limitations on the parameter $\beta$. As a result, the class of problems for which the scheme (4) is applicable becomes narrower. The results of stability and accuracy investigation have showed that the proposed difference scheme has the same limitations as the scheme "left corner" with an approximation error equal to $O\left(\tau^2 + h^3\right)$. The results of numerical calculations prove that it makes sense to use a difference scheme obtained on the basis of the linear combination of the schemes "cabaret" and "cross" with optimal weight coefficients for Korteweg de Vries problem.

**Acknowledgements** The work was implemented in North-Caucasus center of mathematical research under the agreement with the Ministry of science and higher education No 075-02-2021-1749.

# References

1. Whitham GB (1977) Linear and non-linear waves. In: Shabat AB (ed) Mir, 622 p. Translated from English Zharinova VV
2. Karpman VI (1973) Non-linear waves in dispersive media, Nauka, 176 p
3. Samarskii AA, Vabishchevich PN, Matus PP (1998) Difference schemes with operator multipliers, Minsk, 442 p
4. Berezin IL (1977) Numerical research of non-linear waves in rarefied plasma. Nauka, Novosibirsk, p 109
5. Samarskii AA, Mazhukin VI, Matus PP, Mikhailiuk IA (1997) L2-conservative schemes for Korteweg de Vries equation. Reports of Russian Academy of Sciences, vol 357, no 4, pp 458–461
6. Mazhukin VI, Matus PP, Mikhailiuk IA (2000) Difference schemes for Korteweg de Vries equation. Diff Eq 36(5):709–716
7. Bykovskaia EN, Shapranov AV, Mazhukin VI (2021) Error analysis of approximation of two-layer difference schemes for Korteweg de Vries equation. Preprints of the Institute of Applied Mathematics named after M.V. Keldysh, vol 1, 17 p
8. Golovizin VM, Karabasov SA, Sukhodulov DA (2000) Variational approach to obtaining a difference scheme with a spatially-split time derivative for Kortweg de Vries equation. Math Model 12(4):105–116
9. Sukhinov AI, Chistiakov AE, Protsenko EA, Sidoriakina VV, Protsenko SV (2020) A complex of integrated models of sediment and suspension transport with respect to three-dimensional hydrodynamic processes in the coastal zone. Math Model 32(2):3–23. https://doi.org/10.20948/mm-2020-02-01
10. Sukhinov AI, Chistiakov AE, Protsenko EA, Sidoriakina VV, Protsenko SV (2019) An accounting method of cells fullness applied to solution of hydrodynamic problems with complex geometry of computational domain. Math Model 31(8):79–100. https://doi.org/10.1134/S0234087919080057
11. Sukhinov AI, Chistiakov AE (2019) The difference scheme "cabaret" with improved dispersion properties. Math Model 31(3):83–96. https://doi.org/10.1134/S0234087919030067
12. Sukhinov AI, Chistiakov AE, Protsenko EA (2019) About difference schemes "cabaret" and "cross." Comput Meth Program 20(2):170–181. https://doi.org/10.26089/NumMet.v20r216
13. Samarskii AA (1977) Теория разностных схем: Study guide, Nauka. Chief Editorial Board of Physical And Mathematical Literature, 656 p
14. Golovizin VM, Samarskii AA (1998) Some features of "cabaret" difference scheme. Math Model 10(1):101–116
15. Gushchin VA (2016) About one class of quasi-monotone difference schemes of the second approximation order. Math Model 28(2):6–18. https://doi.org/10.1134/S2070048216050094

16. Samarskii AA, Popov IP (1992) Difference methods for solving problems of gas dynamics: study guide: for HEI, 2nd edn. Chief Editorial Board of Physical and Mathematical Literature, Nauka, 424 p
17. Samarskii AA (1967) Classes of stable schemes. Comput Math Math Phys 7(5):1096–1133. English version: https://doi.org/10.1007/s42452-021-04889-7
18. Sukhinov AI, Kuznetsova II, Chistiakov AE, Protsenko EA, Belova IuV (2020) Investigation of a difference scheme accuracy and applicability for solving the diffusion-convection problem at large Peclet's grid numbers. Comput Mech Contin Media 13(4):437–448. https://doi.org/10.7242/1999-6691/2020.13.4.34

# Revealing Chaos-Based Steganographic Transmission by the Recurrence Quantification Analysis

Timur Karimov⬭, Vyacheslav Rybin⬭, Olga Druzhina⬭,
Valerii Ostrovskii⬭, and Daria Protasova⬭

**Abstract** Among modern approaches to cryptography and steganography, a technique based on the use of chaos theory is rapidly developing. Using noise-like chaotic signals and special modulation methods, it is possible to reliably conceal both the message content and the fact of its transmission. To attack this encryption scheme, various techniques such as spectral or envelope analysis can be applied. In this paper, we explore the possibilities of recurrent quantitative analysis to recover the bitstream encrypted with the chaotic parameter modulation. The results show its higher performance compared to the method based on the spectrogram analysis.

**Keywords** Chaos-based communication · Encryption · Message · Recovery · Steganography · Nonlinear signals analysis

## 1 Introduction

Covert and secure data transmission methods are widely used in current communications. One of the promising technology in this field is chaos-based data transmission, which provides a very high possible level of secrecy due to the fact that a chaotic signal has a noise-like waveform and spectrum [1]. Using various methods, a message can be embedded into this signal, but any interceptor will not suspect that data is being transferred.

T. Karimov (✉) · V. Rybin · O. Druzhina · V. Ostrovskii · D. Protasova
St. Petersburg State Electrotechnical University, St. Petersburg, Russia
e-mail: tikarimov@etu.ru

V. Rybin
e-mail: vgrybin@etu.ru

O. Druzhina
e-mail: osdruzhina@etu.ru

V. Ostrovskii
e-mail: vyostrovskii@etu.ru

D. Protasova
e-mail: daprotasova@etu.ru

© The Author(s), under exclusive license to Springer Nature Switzerland AG 2022      385
A. Tchernykh et al. (eds.), *Mathematics and its Applications in New Computer Systems*,
Lecture Notes in Networks and Systems 424,
https://doi.org/10.1007/978-3-030-97020-8_35

Over the last three decades, various scientific groups have proposed different chaotic communication systems based on chaotic masking [2], chaotic shift keying [3], nonlinear mixing of the carrier signal with a chaotic signal [4], modulation of the control parameters [5]. With the development of technologies for chaotic data transmission, hacking methods were also proposed. According to [6], all possible attacks on chaotic communication systems can be divided into three groups: chaotic component extraction from the transmitted signal, direct message extraction from the transmitted chaotic signal, chaotic communication system parameters estimation. All of these methods are based on transmitted signal processing. Therefore, when studying attack methods, it is sufficient to consider only the side of the transmitter, without going into the details of the implementation of the receiving system.

In this paper, we investigate ways to decrypt message transmission based on the transmitting chaotic system parameter modulation. As a reference method, we use the spectra analysis method. Despite it is quite effective, other methods can have better performance, for example, recurrent quantification analysis (RQA), which has found application in the field of physical processes analysis, that are characterized by stochasticity and nonlinearity [7]. The results obtained with RQA for the feature extraction from chaotic sensing system signals [8] suggest that this method will also be effective for attacking the chaotic communication systems.

The paper is organized as follows. In Sect. 2, we present a brief overview of the chaos-based communication scheme and its hacking methods, including the RQA. In Sect. 3, we present results. Section 4 concludes the paper. Web of Science.

## 2 Materials and Methods

### 2.1 Chaos-Based Communication System

Consider the parameter modulation method for chaotic communication systems [9]. We used the procedure described in [10]. First, the message is encoded into a binary symbol alphabet. Each symbol of the binary alphabet corresponds to a specific value of the parameter. Obtained message forms a signal $m(t)$ which affects the corresponding parameter on the transmitter side. The receiver consists of two slave systems. Each has a specific parameter corresponding to a particular symbol. The channel signal $x(t)$ is applied to all receiver systems, and the timing error $\Delta x$ between $x(t)$ and each receiver response is calculated. When the value of the synchronization error $\Delta x$ at one of the receivers falls below a certain threshold, it is considered receiving a symbol corresponding to the symmetry factor in that receiver system. Thus, the received message $m^*(t)$ is decoded symbol by symbol [10]. Figure 1 illustrates the architecture of the proposed chaotic communication system.

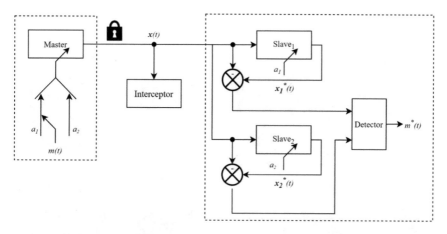

**Fig. 1** Block-diagram of the peer-to-peer chaotic communication system with binary alphabet

Let us consider the Pecora-Carrol synchronization of the well-known Rössler system. A transceiver system:

$$\dot{x}_1 = -y_1 - z_1$$
$$\dot{y}_1 = x_1 + ay_1 \qquad . \qquad (1)$$
$$\dot{z}_1 = b + z_1(x_1 - c)$$

Here, $a$, $b$, and $c$ are the system parameters. In our study, we choose a nonlinear parameter as $a = 0.2$, $b = 0.2$, $c = 5.7$. The receiver system:

$$\dot{x}_2 = -y_2 - z_2 + k(x_1 - x_2)$$
$$\dot{y}_2 = x_2 + ay_2 \qquad . \qquad (2)$$
$$\dot{z}_2 = b + z_2(x_2 - c)$$

Here $k = 1.4$ is synchronization coefficient selected according to [11]. Waveforms of the transmitter signals when logical 0 and 1 are encoded with parameter pairs $a = \{0.2, 0.3\}$ ($\Delta a = 0.1$) and $a = \{0.27, 0.3\}$ ($\Delta a = 0.03$) are presented in Fig. 2a and Fig. 2b, respectively. Figure 2c and Fig. 2d present corresponding spectrograms.

One may see that in the case of $\Delta a = 0.1$, spectrogram demasks 0 and 1 being transmitted, but in case of $\Delta a = 0.03$ there is no visual difference.

With the Rössler system having a mean spectral frequency of 0.2 Hz, each symbol was transmitted for 50 s. Integration step for simulation was chosen $h = 0.01$ s. In real communication systems, the transmission rate can be increased many times by scaling the chaotic system and choosing a small integration step. However, the methods of encryption and attacks will not change.

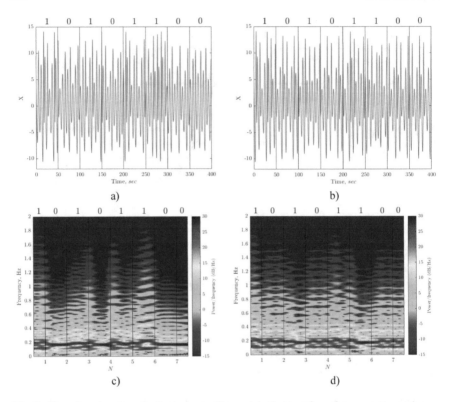

**Fig. 2** Chaos-based communication scheme. Here: **a**) is the waveform for parameters pair $a =$ $\{0.2, 0.3\}$ ($\Delta a = 0.1$) to transfer 0 and 1, respectively; **b**) is the waveform for the case $a = \{0.27, 0.3\}$ ($\Delta a = 0.03$); **c**) is the spectrogram for case $a = \{0.2, 0.3\}$, **d**) is the spectrogram for case $a = \{0.27, 0.3\}$

## 2.2   Chaotic Transmission Decryption Method

a)   *Chaotic component extraction from the transmitted signal*
      This approach is one of the first proposed methods to attack chaotic commu-
      nication systems. The principle of the method is as follows: the chaotic carrier
      signal is extracted from the transmitted signal, which is then removed from the
      transmitted signal to reveal the message. Unfortunately, despite the relatively
      simple implementation of this method, this approach cannot accurately extract
      the transmitted message and cannot be used in the case of some modulation
      methods [12, 13].

b)   *Power spectral analysis and filtering*
      In their standard normalized form, popular chaotic systems such as Lorentz,
      Rössler, Chua, etc., do not produce a genuinely wideband signal but produce
      a narrowband signal instead [14, 15]. Therefore not every chaotic signal can
      successfully hide the spectrum of the transmitted message in the case of a high

message frequency (for chaotic masking). For example, the power spectrum of the Lorenz generator output signal shows that this spectrum hardly exceeds 4–5 Hz, which is certainly not a wideband signal. Therefore, a message signal such as a 5 Hz sine wave after masking with this chaotic signal will be easily detected using a simple high-pass filter [16]. In the case of chaotic signal parameter modulation or CSK, changes in the signal dynamics can be so significant that it will easily recover the hidden message [17–19].

c) *Generalized synchronization technique*

The generalized synchronization (GS) method was first proposed in [20]. This approach is also suitable for attacking chaotic communication systems, in particular the CSK method. A key feature of this method is the measurement of synchronization error over time. When the transmitted symbol changes, the signal dynamics changes, and as a result, the synchronization error also changes. However, significant changes in the dynamics by switching transmitted symbols are necessary for the successful working of the GS method [21].

d) *Return Map analysis*

Return Map (RM) is one of the popular methods of dealing with chaotic communication methods. The RM method is based on constructing a particular graph based on a simple transformation of the local minima and maxima of the signal. For example, in [16], RM was successfully used to hack CSK-based techniques. This method is described in more detail in [22], where methods of chaotic communication, based on chaotic masking and modulation of the parameter of a chaotic signal, have also shown their consistency in attacks.

## 2.3 Recurrence Quantification Analysis

The idea of recurrence plots [7] is the visualization of the repeated states of the system $x(t)$ in the $m$-dimensional phase space in neighborhood $\varepsilon$. The repetition of the state of the system that existed in time $i$, at some other time $j$ is marked in a square binary matrix, in which 1 (black dot) corresponds to repetition, and both coordinate axes (X and Y) are time axes. The state of some dynamical systems (including chaotic ones) cannot be repeated completely equivalent to the initial state, therefore, the states that appear into an $m$-dimensional neighborhood with radius $\varepsilon$ and center $i$ are considered to be recurrent. Mathematically, a recurrence plot is a matrix $\mathbf{r}$ with elements

$$r_{i,j} = \theta\left(\varepsilon - \|\mathbf{x}_j - \mathbf{x}_i\|\right), \quad x_i \in \Re^m, \ i, j \in [1; N]. \tag{3}$$

Here $\theta$ is the Heaviside step function, $N$ is a signal length and $\|\cdot\|$ denotes Euclidean distance.

In practical application, researchers start from the distance matrix $\mathbf{d}$ with elements

$$d_{i,j} = \|\mathbf{x}_j - \mathbf{x}_i\|, \quad i, j \in [1; N], \tag{4}$$

which may be presented in pseudo-colors and than, selecting a threshold value, obtain informative binary matrices. To recognize the obtained distance matrices and binary matrices, both simple metrics can be used, for example, the number of single vertical columns or horizontal rows, and more complex methods, for example, multilayer feed-forward artificial neural networks [23].

In our work, we use the following simple metrics: trapping time (TT), entropy (ENT) and recurrence rate (RR). Trapping time is the mean time of the recurrent states, entropy represents an average level of uncertainty according to Shannon, recurrence rate is the number of repeating states per certain time unit.

## 3 Results

We assume that the length of the symbol is known; if it is not so, than it may be established by an iterative procedure if the hacking method is sensitive enough to differ symbol 0 from 1. The intercepted waveform is divided into 8 segments corresponding to each symbol. For every segment («window»), the distance matrix (pseudo-color recurrence plot) is calculated. Then, using a selectable threshold value, distance matrix $d$ is converted to binary recurrence matrix $r$. Having it, we calculate RR, ENT, and TT values.

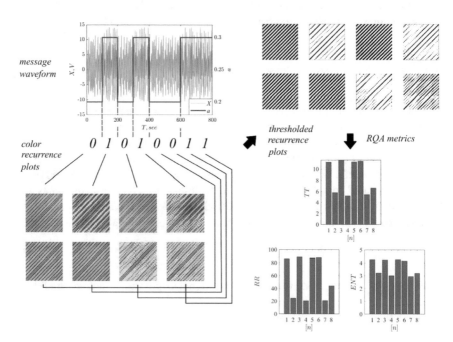

**Fig. 3** Decryption of the chaotic communication system message with the windowed recurrence quantification analysis

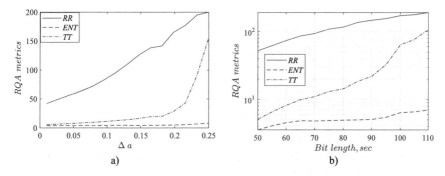

**Fig. 4** Relationship of the metrics of recurrent analysis on: a) the difference in the value of the modulation parameter and b) the duration of the single bit transfer. The list of calculated metrics includes recurrence rate (RR), entropy (ENT) and trapping time (TT)

Figure 3 presents the developed algorithm and results of its application for the case when logical 0 and 1 are encoded with parameter pairs $a = \{0.2, 0.3\}$. The difference between symbols is very apparent. Moreover, while spectrograms for pair $a = \{0.27, 0.3\}$ are almost unrecognizable (Fig. 2d), using RQA it is possible to distinguish symbols even in case of difference $\Delta a = 0.01$, see Fig. 4a. This plot shows the difference between the averaged metric value for symbol 0 and symbol 1. With an increase in the length of the transmitted symbol, this difference increases (Fig. 4b), which means that the recognition reliability also increases.

## 4 Conclusion

This paper proposes to use recurrence quantification analysis (RQA) to decrypt binary messages transmitted by the chaotic communication system. We briefly overview the chaos-based messaging techniques and how to attack them. There is a lack of information on the application of RQA for this purpose in the scientific literature, so the current study is intended to fill this gap. We have obtained reliable results in recognition of bit messages encrypted by the dissipative chaotic Rössler system parameter modulation. In further studies, we will use RQA to decrypt bit messages transmitted by other methods and systems, and numerically evaluate its performance in comparison with other hacking methods, such as return maps, in the presence of non-zero noise and interference.

**Acknowledgements** This study was supported by the grant of the Russian Science Foundation (RSF), project 20-79-10334.

# References

1. Riaz A, Ali M (2008) 2008 6th international symposium on communication systems, networks and digital signal processing
2. Cuomo KM, Oppenheim AV, Strogatz SH (1993) Synchronization of Lorenz-based chaotic circuits with applications to communications. IEEE Trans Circ Syst II Analog Digital Signal Process 40(10):626–633
3. Dedieu H, Kennedy MP, Hasler M (1993) Chaos shift keying: modulation and demodulation of a chaotic carrier using self-synchronizing Chua's circuits. IEEE Trans Circ Syst II Analog Digital Signal Process 40(10):634–642
4. Dmitriev AS, Panas AI, Starkov SO (1995) Experiments on speech and music signals transmission using chaos. Int J Bifurcat Chaos 5(04):1249–1254
5. Yang T, Chua LO (1996) Secure communication via chaotic parameter modulation. IEEE Trans Circ Syst I Fundam Theory Appl 43(9):817–819
6. Alvarez G, Li S (2006) Some Basic Cryptographic requirements for chaos-based cryptosystems. Int J Bifurcat Chaos 16:2129–2151
7. Marwan N, Kurths J, Saparin P (2007) Generalised recurrence plot analysis for spatial data. Phys Lett A 360:545–551
8. Karimov TI, Druzhina OS, Ostrovskii VY, Karimov AI, Butusov DN (2020) 2020 IEEE conference of Russian young researchers in electrical and electronic engineering (EIConRus)
9. Yang T, Chua LO (1996) Secure communication via chaotic parameter modulation. IEEE Trans Circ Syst I Fundam Theory Appl 43:817–819
10. Karimov T, Rybin V, Kolev G, Rodionova E, Butusov D (2021) Chaotic communication system with symmetry-based modulation. Appl Sci 11:3698
11. Rybin V, Tutueva A, Karimov T, Kolev G, Butusov D, Rodionova E (2021) 10th Mediterranean conference on embedded computing (MECO)
12. Short KM (1994) Steps toward unmasking secure communications. Int J Bifurcat Chaos 4:959–977
13. Short KM (1996) Unmasking a modulated chaotic communications scheme. Int J Bifurcat Chaos 6:367–375
14. Alvarez G, Li S (2004) Breaking network security based on synchronized chaos. Comput Commun 27:1679–1681
15. Alvarez G, Hernández L, Muñoz J, Montoya F, Li S (2005) Security analysis of communication system based on the synchronization of different order chaotic systems. Phys Lett A 345:245–250
16. Perez G, Cerdeira HA (1995) Extracting messages masked by chaos. Phys Rev Lett 74:1970–1973
17. Karimov TI, Druzhina OS, Andreev VS, Tutueva AV, Kopets EE (2021) 2021 IEEE conference of Russian young researchers in electrical and electronic engineering (ElConRus)
18. Alvarez G, Montoya F, Romera M, Pastor G (2004) Breaking parameter modulated chaotic secure communication systems. Chaos Solitons Fract 21:783–787
19. Alvarez G, Montoya F, Romera M, Pastor G (2004) Breaking two secure communication systems based on chaotic masking. IEEE Trans Circ Syst II Express Briefs 51:505–506
20. Yang T, Yang LB, Yang CM (1998) Breaking chaotic switching using generalized synchronization: examples. IEEE Trans Circ Syst I Fundam Thoery Appl 45:1062–1067
21. Alvarez G, Li S, Montoya F, Pastor G, Romera M (2005) Breaking projective chaos synchronization secure communication using filtering and generalized synchronization. Chaos Solitons Fract 24(3):775–783
22. Yang T, Yang LB, Yang CM (1998) Cryptanalyzing chaotic secure communication using return maps. Phys Lett A 245:495–510
23. Chen Y, Yang H (2012) Multiscale recurrence analysis of long-term nonlinear and nonstationary time series. Chaos Solitons Fract 45(7):978–987

# Using Artificial Neural Networks and Wavelet Transform for Image Denoising

**Dmitry Kaplun, Alexander Voznesensky, Aleksandr Sinitca, Alexander Veligosha, and Nikolay Malyshko**

**Abstract** The paper presents an approach for image denoising that reduces the noise components level such as Gaussian and salt-and-pepper when performing a discrete wavelet transform (DWT) of images. Denoising is achieved by using artificial neural networks (ANN) as an element that allows restoring information distorted due to noise exposure. The structure of the ANN and data for its training are given in the paper. It is shown that using ANN with DWT increases the image restoration quality for Gaussian and salt-and-pepper noise types by an average of 10 dB in the peak signal-to-noise ratio (PSNR). The obtained results were confirmed through the simulation of the proposed approach for the test images like "Lena" and "Forest". The comparison with the non-local means (NLM) algorithm and CNN shows the advantage of our approach in PSNR.

**Keywords** Digital image processing · Image denoising · Discrete wavelet transform · Daubechies wavelets · Gaussian noise · Salt-and-pepper noise · Artificial neural networks · Non-local means · CNN

## 1 Introduction

Digital image processing has a wide range of applications in various fields of science and technology, such as machine vision, medicine, security systems, and others.

D. Kaplun (✉) · A. Voznesensky · A. Sinitca
Department of Automation and Control Processes, Saint Petersburg Electrotechnical University "LETI", 5 Professor Popova Street, 197376 Saint Petersburg, Russia
e-mail: dikaplun@etu.ru

A. Voznesensky
e-mail: asvoznesenskiy@etu.ru

A. Sinitca
e-mail: amsinitca@etu.ru

A. Veligosha · N. Malyshko
Military Academe of Strategic Missile Forces named after Peter the Great, 17 Brigadnaya Street, 142210 Serpukhov, Russia

© The Author(s), under exclusive license to Springer Nature Switzerland AG 2022      393
A. Tchernykh et al. (eds.), *Mathematics and its Applications in New Computer Systems*,
Lecture Notes in Networks and Systems 424,
https://doi.org/10.1007/978-3-030-97020-8_36

Digital image processing applications that are used in the above areas should have, on the one hand, simplicity of hardware implementation, low power consumption of the system, high speed of information processing, and, on the other hand, mathematical processing methods that will lead to improved image quality (visual comparison, peak signal-to-noise ratio), which is obtained during processing [1].

It is known, that images are exposed to noise during video processing. Therefore, their recovery operations with a given quality are associated with the elimination of various noise types [2]. Digital images are exposed to noise when received or transmitted, which leads to a deterioration in visual quality and loss of image areas. Sensors' (digital photocells) operation depends on various factors, such as external conditions in image acquisition and the quality of the sensors themselves [3]. Repairing damaged pixels before the main processing is an important task of digital image processing [4].

Currently, there are various filtering methods to remove noise components. The image distorted by impulse noise looks with white and black pixels randomly scattered across the frame. It is known that the use of linear filters for this purpose causes a strong blurring of the image parts, leading to the loss of details and image contours [5]. To overcome this drawback, median filtering was proposed, which replaces the image pixels with the corresponding median values of a certain neighborhood of the image area [6]. Median filters also lead to image blurring, which is much less noticeable than the results of processing with linear filters [7]. This becomes especially noticeable when processing images distorted by noise with high intensity. Approaches based on adaptive filtering have been proposed in [8, 9] to reduce this negative effect. But this still does not solve all the above problems. For this reason, it should be considered the possibility of using artificial neural networks (ANN), especially convolutional neural networks (CNN), that are widely used in image processing [10–14].

An artificial Neural Network (ANN) is a group of multiple perceptrons or neurons at each layer. ANN is also known as a Feed-Forward Neural network because inputs are processed only in the forward direction. This type of neural network is one of the simplest variants of neural networks. They pass information through various input nodes in one direction until it makes it to the output node. The network may or may not have hidden node layers, making its functioning more interpretable [15, 16].

Convolutional neural networks (CNN) are one of the most popular models used today. This neural network computational model uses a variation of multilayer perceptrons and contains one or more convolutional layers that can be either entirely connected or pooled. These convolutional layers create feature maps that record a region of the image, which is ultimately broken into rectangles and sent out for nonlinear processing [15, 16].

CNN-based denoising methods are also popular recently. Jain et al. [17] proposed the first CNN-based denoising method. Compared with other traditional methods, this method achieves similar or even better results. DnCNN is presented by Zhang et al. [18], which combines batch normalization with residual learning and obtains the latest developments. Then, an automatic encoder with a symmetric jump connection network is proposed by Mao et al. [19]. The method realizes ten pairs of symmetric and deconvolution layers, and the first five layers are the coding layer while the last five are the decoding layer. The idea of subband denoising with CNN is presented in [20], but CNN has high computational complexity. Therefore, the structures of image denoising networks based on CNN become more and more complicated. However, such networks are rather cumbersome and require significant computing resources.

An approach that reduces the level of noise components such as "Gaussian" and "salt-and-pepper" is proposed in the paper. Reducing the level of noise components is achieved through the use of ANN (considered ANN has only one hidden layer) with discrete wavelet transform (DWT).

The main content of the paper is organized as follows. The mathematical image processing model based on discrete wavelet transform (DWT) is given in Sect. 2.1. In Sect. 2.2, the architecture of the proposed ANN for image denoising is presented. The simulation results of the proposed method are presented in Sect. 3. The results are discussed in Sect. 4. Conclusions are given in Sect. 5.

## 2 Materials and Methods

### 2.1 Mathematical Model of Image Processing Based on Discrete Wavelet Transform

To study the use of ANN for denoising, a one-level DWT of image was performed [21]. An image $I$, consisting of $X$ rows and $Y$ columns is represented as a function $I(x, y)$, where $0 \leq x \leq X - 1$ and $0 \leq y \leq Y - 1$ are the spatial coordinates of $I$. The pixel values are dependent on the kind of image (binary, grayscale, or color). In this paper, we focus primarily on color images. Thus, the values of the pixels are referred to as $I(x, y, z)$ for color images, where $z = 1, 2, 3$ – color number (red, green, and blue for example). DWT of image is implemented by sequentially using filter banks [22] (wavelet filters). The scheme of a one-level two-dimensional DWT of images is shown in Fig. 1. Daubechies wavelets are used. The wavelets can be orthogonal, when the scaling functions have the same number of coefficients as the wavelet functions, or biorthogonal, when the number of coefficients differ. The JPEG 2000 compression standard uses the biorthogonal Daubechies 5/3 wavelet (also called the LeGall 5/3 wavelet) for lossless compression and the Daubechies 9/7 (also known as the Cohen-Daubechies-Fauraue 9/7 or the "CDF 9/7") for lossy compression. In general, Daubechies wavelet has extremal phase and highest number of vanishing moments for defined support width. The wavelet is also easy to put into practice with minimum-phase filters. Daubechies wavelet is widely used in solving a broad range of problems, e.g., self-likely properties of a signal or fractal problem, signal discontinuities, etc. [23].

1.  Row analysis is performed by decomposing the image along the rows with lowpass $LD$ and highpass $HD$ wavelet filters and downsampling $\downarrow 2$.

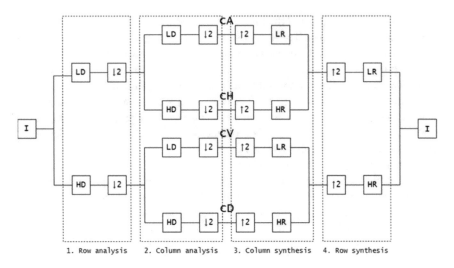

1. Row analysis    2. Column analysis    3. Column synthesis    4. Row synthesis

**Fig. 1** The scheme of a one-level two-dimensional DWT of images

2. Column analysis is performed by decomposing the coefficients obtained at stage 1, by columns with the technique similar to the row analysis.

We get four sets of coefficients $cA, cH, cV, cD$ of image decomposition, called approximating and detailing (horizontal, vertical, and diagonal), respectively, as a result of direct DWT of the original image $I$.

3. Column synthesis is performed by upsampling $\uparrow 2$ the coefficients $cA, cH, cV, cD$, restoration with lowpass $LR$ and highpass $HR$ filters and summation of results.

4. Row synthesis is performed for coefficients obtained at stage 3, by rows with the technique similar to the column synthesis.

The original image $I$ is restored as a result of the synthesis (inverse DWT) from the coefficients $cA, cH, cV, cD$. The original image should be completely restored.

We will assume that the wavelet filters $F$ consist of the coefficients $f_{F,i}$, where $i = 0, ..., k - 1$ is the coefficient number, where $k$ is the number of the filter coefficients. The next operation is called a convolution and is performed as follows:

$$I'(x, y) = \sum_{i=0}^{k-1} I(x, y + i) \cdot f_{F,k-1-i} \quad I''(x, y) = \sum_{i=0}^{k-1} I(x + i, y) \cdot f_{F,k-1-i} \quad (1)$$

where $I'$ – result of row convolution, $I''$ – result of column convolution. We shall consider only wavelets with compact support.

The coefficients of the wavelet filters are related by the equation:

$$f_{HD,i} = (-1)^{i+1} f_{LD,k-1-i}, \quad f_{LR,i} = f_{LD,k-1-i}, \quad f_{HR,i} = (-1)^i f_{LD,i} \quad (2)$$

We used peak signal-to-noise ratio between two images to quantify the image processing quality. This characteristic is measured in decibels (dB) and is calculated by the following formula [24]:

$$PSNR = 10 \log_{10}\left(\frac{M^2}{MSE}\right) \qquad (3)$$

where $M$ is the maximum amplitude of the input image and $MSE$ is the mean square error of image:

$$MSE_{color} = \frac{1}{3}\sum_{z=1}^{3}\sum_{x=0}^{X-1}\sum_{y=0}^{Y-1}\frac{[I_1(x, y, z) - I_2(x, y, z)]^2}{XY} \qquad (4)$$

where $I_1$ and $I_2$ are input and restored images.

Larger the $PSNR$ value indicates the better image quality and for identical images $PSNR = \infty$.

## 2.2 Architecture of the Proposed ANN for Image Denoising

This section provides a concise and precise description of the experimental results, their interpretation as well as the experimental conclusions that can be drawn.

To study the application of ANN, a one-level DWT was performed on the images shown in Figs. 2, 3, 4 and 5.

(a)          (b)          (c)          (d)          (e)

**Fig. 2** **a** the original image, **b** the salt-and-pepper noise intensity 0.02, **c** the salt-and-pepper noise intensity 0.04, **d** the salt-and-pepper noise intensity 0.2, **e** the salt-and-pepper noise intensity 0.4

(a)          (b)          (c)          (d)          (e)

**Fig. 3** **a** the original image, **b** the Gaussian noise intensity 0.02, **c** the Gaussian noise intensity 0.04, **d** the Gaussian noise intensity 0.2, **e** the Gaussian noise intensity 0.4

398 D. Kaplun et al.

**Fig. 4** **a** the original image, **b** the salt-and-pepper noise intensity 0.02, **c** the salt-and-pepper noise intensity 0.04, **d** the salt-and-pepper noise intensity 0.2, **e** the salt-and-pepper noise intensity 0.4

**Fig. 5** **a** the original image, **b** the Gaussian noise intensity 0.02, **c** the Gaussian noise intensity 0.04, **d** the Gaussian noise intensity 0.2, **e** the Gaussian noise intensity 0.4

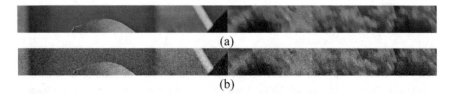

**Fig. 6** **a** An example of the composite image for training the ANN, **b** An example of the composite image for training ANN with noise "Gaussian" intensity 0.02

The dimension of the images is 512 by 512 pixels, the color depth is 24 bits. The DWT coefficients obtained by decomposing the original composite image for training the ANN (Fig. 6) were used as the target response, and the DWT coefficients obtained by processing noisy composite images were used as input.

For image processing, the ANN of the "multilayer perceptron" architecture was developed, training was carried out by the method of error back propagation, the activation function is exponential. The proposed approach is shown in Fig. 7.

Processing is not performed for the entire image at one time, but sequentially after applying the DWT coefficient vectors, regardless of the image size ($I = K = M \cdot N$). In one processing cycle, the ANN input receives from 6 to 96 DWT coefficients, and so on until the end of the coefficient vector is reached.

When using the ANN, according to the scheme shown in Fig. 7, the input image is subjected to DWT, the coefficients of which (approximating and detailing) are fed to the input of the ANN for each of the three colors channels. Further, after processing in the input layer, the samples arrive at the input of a hidden layer of dimension in neurons. Then, after processing in a hidden layer, samples are sent to the output layer

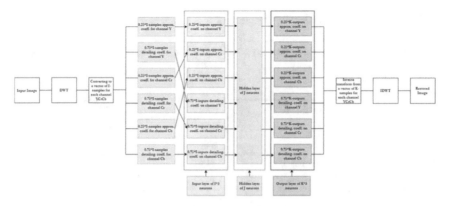

**Fig. 7** The scheme of proposed approach

and processed by color channel for each of the three color channels. The resulting ANN responses are converted into a sequence of coefficients, and the inverse DWT with image restoration is performed. Following the scheme shown in Fig. 7, ANN was developed that made it possible to implement DWT with a decrease in noise components in the images.

The estimation of the signal recovery accuracy was determined using the correlation coefficient between the target response and the response received after processing by ANN according to the expression:

$$r = \frac{\sum (x - \overline{x})(y - \overline{y})}{\sqrt{\sum (x - \overline{x})^2 (y - \overline{y})^2}} \tag{5}$$

where $x$ – is the target response value, $\overline{x}$ – is the mean value vector of the target responses, $y$ – is the response value obtained with ANN, $\overline{y}$ – is the mean value vector of the responses obtained with ANN.

Loss function is determined as following:

$$\lambda(y) = \sum (y - y')^2 \tag{6}$$

where $y$ – true model output (which should be obtained in the ideal case), $y'$ – actual model output.

The ANN architecture, the type and intensity of noise, the number of training cycles, and the accuracy of signal recovery are shown in Tables 1, 2 and 3.

From the data shown in Table 1, it can be concluded that the use of architecture with a dimension of input and output layers equal to six allows us to achieve a relatively low accuracy of signal recovery ~50 ... 60%. In this regard, it is necessary to consider the construction of ANN, with the dimension of layers with many elements. The results of constructing the ANN are given in Table 2, which shows that an increase in the

**Table 1** ANN architecture, type and intensity of noise, the number of training cycles and the accuracy of signal recovery

| ANN architecture (input layer, hidden layer, output layer), number of elements | Type and intensity of noise | Number of training cycles | Correlation coefficient |
|---|---|---|---|
| 6-6-6 | Gaussian 0.02 | 353 | 0.533 |
| 6-32-6 | | 444 | 0.58 |
| 6-48-6 | | 638 | 0.60 |
| 6-6-6 | Gaussian 0.2 | 1245 | 0.52 |
| 6-32-6 | | 1292 | 0.603 |
| 6-48-6 | | 1375 | 0.616 |

**Table 2** ANN architecture, type and intensity of noise, the number of training cycles and the accuracy of signal recovery (composite images)

| ANN architecture (input layer, hidden layer, output layer), number of elements | Type and intensity of noise | Number of training cycles | Correlation coefficient |
|---|---|---|---|
| 48-48-48 | Gaussian 0.02 | 567 | 0.67 |
| 48-64-48 | | 473 | 0.66 |
| 96-96-96 | | 651 | 0.64 |
| 96-128-96 | | 1987 | max → 0.673 |
| 48-48-48 | Gaussian 0.04 | 552 | 0.653 |
| 48-64-48 | | 447 | 0.653 |
| 96-96-96 | | 595 | 0.626 |
| 96-128-96 | | 887 | max → 0.656 |
| 48-48-48 | Gaussian 0.2 | 540 | max → 0.66 |
| 48-64-48 | | 382 | 0.65 |
| 96-96-96 | | 759 | 0.64 |
| 96-128-96 | | 805 | 0.643 |
| 48-48-48 | Gaussian 0.4 | 396 | 0.613 |
| 48-64-48 | | 644 | max → 0.626 |
| 96-96-96 | | 1322 | 0.61 |
| 96-128-96 | | 1294 | 0.61 |

number of elements in the layers of the ANN can improve the accuracy of signal recovery after noise exposure.

The data presented in Table 2 allow us to determine the most effective ANN according to the accuracy of signal prediction, depending on the intensity of the Gaussian noise. Moreover, an increase in the dimension of the hidden layer does not

**Table 3** ANN architecture, type and intensity of noise, the number of training cycles and the accuracy of signal recovery (composite images)

| ANN architecture (input layer, hidden layer, output layer), number of elements | Type and intensity of noise | Number of training cycles | Correlation coefficient |
|---|---|---|---|
| 48-48-48 | Salt-and-pepper 0.02 | 629 | 0.68 |
| 48-64-48 | | 339 | 0.666 |
| 96-96-96 | | 838 | 0.66 |
| 96-128-96 | | 1256 | max → 0.683 |
| 48-48-48 | Salt-and-pepper 0.04 | 365 | 0.63 |
| 48-64-48 | | 455 | 0.63 |
| 96-96-96 | | 891 | 0.63 |
| 96-128-96 | | 1461 | max → 0.646 |
| 48-48-48 | Salt-and-pepper 0.2 | 311 | max → 0.536 |
| 48-64-48 | | 222 | 0.49 |
| 96-96-96 | | 298 | 0.49 |
| 96-128-96 | | 394 | 0.496 |
| 48-48-48 | Salt-and-pepper 0.4 | 222 | 0.393 |
| 48-64-48 | | 186 | 0.396 |
| 96-96-96 | | 202 | 0.426 |
| 96-128-96 | | 303 | max → 0.43 |

always allow one to achieve an increase in the accuracy of restoration of the signal distorted by the influence of noise.

The results of determining the accuracy of signal recovery with a decrease in the level of noise components of the salt-and-pepper noise are shown in Table 3.

# 3 Results

The simulation was carried out in the MATLAB software version R2020b for different test images. Gaussian noise and salt-and-pepper noise of different intensities were applied to the original images using the "imnoise" command. The application results of ANN, Non-Local Means (NLM) and CNN for test images, based on the maximum accuracy presented in Tables 2, 3, are shown in Table 4.

**Table 4** ANN architecture, type and intensity of noise, PSNR value before and after application of ANN, NLM and CNN ("Lena" - first row, "Forest" - second row)

| ANN architecture (input layer, hidden layer, output layer), number of elements | Type and intensity of noise | PSNR (dB) | PSNR using ANN/NLM/CNN (dB) | Δ |
|---|---|---|---|---|
| 96-128-96 | Gaussian 0,02 | 19.96 | 26.26/22.42/26.06 | 6.30/2.46/6.10 |
|  |  | 20.03 | 29.73/22.59/29.50 | 9.70/2.56/9.47 |
| 96-128-96 | Gaussian 0.04 | 19.46 | 26.27/21.59/25.96 | 6.81/2.13/6.50 |
|  |  | 19.48 | 29.42/21.68/26.75 | 9.94/2.20/7.27 |
| 48-48-48 | Gaussian 0.2 | 13.20 | 26.08/13.58/14.04 | 12.88/0.38/0.84 |
|  |  | 13.04 | 29.51/13.44/13.94 | 16.47/0.40/0.90 |
| 48-64-48 | Gaussian 0.4 | 8.25 | 24.81/8.29/8.42 | 16.56/0.04/0.17 |
|  |  | 7.76 | 29.14/7.84/7.98 | 21.38/0.08/0.22 |
| 96-128-96 | Salt-and-pepper 0,02 | 22.11 | 27.01/25.39/24.35 | 4.90/3.28/2.24 |
|  |  | 21.95 | 30.72/25.88/23.87 | 8.77/3.93/1.92 |
| 96-128-96 | Salt-and-pepper 0.04 | 19.07 | 25.61/22.16/23.23 | 6.54/3.09/4.16 |
|  |  | 18.98 | 28.94/22.26/22.72 | 9.96/3.28/3.74 |
| 48-48-48 | Salt-and-pepper 0,2 | 12.05 | 20.38/14.57/19.82 | 8.33/2.52/7.77 |
|  |  | 11.98 | 20.64/14.55/20.02 | 8.66/2.57/8.04 |
| 96-128-96 | Salt-and-pepper 0.4 | 9.06 | 19.56/11.12/17.56 | 10.5/2.06/8.50 |
|  |  | 8.96 | 21.18/11.01/17.29 | 12.22/2.05/8.33 |
| Mean value |  | 15.33 | 25.95/17.40/20.09 | 10.62/2.06/4.76 |

Non-local means is an algorithm of spatiotemporal image processing with one of the best denoising qualities [25, 26]. Unlike "local mean" filters, which take the mean value of a group of pixels surrounding a target pixel to smooth the image, NLM filtering takes a mean of all pixels in the image, weighted by how similar these pixels are to the target pixel. This results in much greater post-filtering clarity and less detail loss in the image than local mean algorithms.

The following parameters were used for NLM: $r_{sim} = 3$; (radius of similarity window);$r_{win} = 5$; (radius of searching window), $sigma\_range = 0.1$ (filter strength).

The pretrained built-in CNN DnCNN with default parameters (59 layers) is used [27, 28].

The visual quality of image processing using the "Lena" image as an example is shown in Fig. 8.

**Fig. 8** The result of using
ANN for image processing
"Lena": **a** salt-and-pepper
0.2, PSNR = 12.05 dB, **b**
ANN, PSNR = 20.38 dB

(a)                                        (b)

## 4 Discussion

This paper proposes an ANN-based denoising method implemented in the wavelet domain, which could solve the denoising problem by a unified network structure. ANN handle image denoising problem in the wavelet transform domain. It combines the advantages of both DWT and ANN-based methods to get a high PSNR result. ANN is implemented in the wavelet domain instead of in the spatial domain directly for the reason that different high-frequency corresponding different detailed information should be processed differently, which cannot be distinguished well in the spatial domain. The experimental results (Table 4) show that our method not only could remove typical images noise (Gaussian noise and salt-and-pepper noise) but also could enhance the spatial resolution (Fig. 8).

In comparison with NLM, the proposed method provides significantly better results in most of the experiments conducted. This is due to the NLM's inability to learn, the strict set of parameters when using NLM, and the inability to adjust to the noise level.

In comparison with CNN, the proposed method also provides significantly better results in most experiments conducted. This is due to CNN's inability to detect local features in the spatial domain.

Note that NLM is very expensive to apply directly. The image denoising networks based on CNN can also provide good results but require much more computing resources than the proposed approach.

## 5 Conclusion

The paper proposes an approach that reduces noise components such as Gaussian and salt-and-pepper when performing images denoising. The results are shown in Table 4 allow us to conclude that using ANN with DWT increases the image restoration quality for various noise types by an average of 10.62 dB (NLM – 2.06 dB, CNN – 4.76 dB).

To improve the images denoising quality, the possibility of training ANN on a wide range of composite images when using ANN with a more complex architecture should be considered in the future.

**Acknowledgements** Dmitry Kaplun and Alexander Voznesensky were supported by the grant from the Ministry of Science and Technology of Israel and RFBR according to the research project no. 19-57-06007. Aleksandr Sinitca was supported by the Ministry of Science and Higher Education of the Russian Federation 'Goszadanie' № 075-01024-21-02 from 29.09.2021 (project FSEE-2021-0014).

# References

1. Al B (2005) Handbook of image and video processing. Elsevier, Texas, p 1372
2. Gonzalez RC, Woods RE, Eddins SL (2003) Digital image processing using MATLAB. Pearson Prentice Hall, Hoboken, 609p
3. Schirrmacher F et al (2018) Temporal and volumetric denoising via quantile sparse image prior. Med Image Anal 48:131–146
4. Gonzalez RC, Woods RE (2007) Digital image processing, 3rd edn. Pearson, London
5. Lyakhov PA, Nagornov NN, Chervyakov NI, Kaplun DI (2019) Analysis of the quantization noise of linear time-invariant filters for image processing. In: 2019 IEEE conference of russian young researchers in electrical and electronic engineering (EIConRus), Saint Petersburg and Moscow, Russia, pp 1192–1196
6. Tukey JW (1977) Exploratory data analysis. Pearson, London
7. Chervyakov NI, Lyakhov PA, Orazaev AR (2018) Two methods of adaptive median filtering of impulse noise in images. Comput Opt 42(4):667–678. https://doi.org/10.18287/2412-6179-2018-42-4-667-678
8. Lyakhov PA, Orazaev AR, Chervyakov NI, Kaplun DI (2019) A new method for adaptive median filtering of images. In: IEEE conference of russian young researchers in electrical and electronic engineering (EIConRus), Saint Petersburg and Moscow, Russia, 2019, pp 1197–1201
9. Hwang H, Haddad RA (1995) Adaptive median filters: new algorithms and results. IEEE Trans Image Process 4:499–502
10. Litjens G et al (2017) A survey on deep learning in medical image analysis. Med Image Anal 42:60–88
11. Brill A, Feng Q, Humensky TB, Kim B, Nieto D, Miener T (2019) Investigating a deep learning method to analyze images from multiple gamma-ray telescopes. In: 2019 New York scientific data summit (NYSDS), New York, NY, USA, pp 1–4
12. Fang W et al (2019) Recognizing global reservoirs from landsat 8 images: a deep learning approach. IEEE J See Top Appl Earth Observ Remote Sens 12(9):3168–3177
13. Sze V, Chen Y, Yang T, Emer JS (2017) Efficient processing of deep neural networks: a tutorial and survey. Proc IEEE 105(12):2295–2329
14. Krizhevsky A, Sutskever I, Hinton GE (2012) ImageNet classification with deep convolutional neural networks. In: Advances in neural information processing systems, vol 25, no 2
15. https://www.geeksforgeeks.org/difference-between-ann-cnn-and-rnn/
16. https://www.analyticsvidhya.com/blog/2020/02/cnn-vs-rnn-vs-mlp-analyzing-3-types-of-neural-networks-in-deep-learning/
17. Jain V, Seung H (2008) Natural image denoising with convolutional networks. In: Proceedings of the advances in neural information processing systems, Vancouver, BC, Canada, pp 769–776
18. Zhang K, Zuo W, Chen Y, Meng D, Zhang L (2017) Beyond a Gaussian denoiser: residual learning of deep CNN for image denoising. IEEE Trans Image Process 26:3142–3155
19. Mao X, Shen C, Yang Y (2016) Image restoration using convolutional auto-encoders with symmetric skip connections. arXiv:1606.08921

20. Zhao R, Lam K, Lun DPK (2019) Enhancement of a CNN-based denoiser based on spatial and spectral analysis. In: 2019 IEEE international conference on image processing (ICIP), Taipei, Taiwan, pp 1124–1128. https://doi.org/10.1109/ICIP.2019.8804295
21. Chervyakov N, Lyakhov P, Kaplun D, Butusov D, Nagornov N (2018) Analysis of the quantization noise in discrete wavelet transform filters for image processing. Electronics 7(8):135
22. Kaplun DI, Klionskiy DM, Voznesenskiy AS, Gulvanskiy VV (2014) Application of polyphase filter banks to wideband monitoring tasks. In: Proceedings of the 2014 IEEE NW Russia young researchers in electrical and electronic engineering conference, pp 95–98. https://doi.org/10.1109/ElConRusNW.2014.6839211
23. Daubechies I(1992) Ten lectures on wavelets. SIAM
24. Rao KR, Yip PC (2001) The transform and data compression handbook. CRC Press, Boca Raton, p 399
25. Erdem E (2007) The slides are adapted from the course "A Gentle Introduction to Bilateral Filtering and its Applications". In: Paris S, Kornprobst P, Tumblin J, Durand F. http://people.csail.mit.edu/sparis/bf_course/, http://web.cs.hacettepe.edu.tr/~erkut/bil717.s12/w09-bilateral-nlmeans.pdf
26. Goossens B, Luong H, Pizurica A, Philips W (2008) An improved non-local denoising algorithm. In: Proceedings of the International Workshop on Local and Non-local Approximation in Image Process, Tuusalu, Finland, pp 143–156
27. ImageNet. http://www.image-net.org
28. Russakovsky O, Deng J, Su H et al (2015) ImageNet large scale visual recognition challenge. Int J Comput Vis (IJCV) 115(3):211–252

# Computational Techniques for Investigating Low-Potential Gupta Clusters of Extremely Large Dimensions

**Pavel Sorokovikov** and **Anton Anikin**

**Abstract** The paper considers the problem of finding the geometric structure of an atomic cluster, for which the interaction energy of the atoms included in it is minimal. It is the problem of finding low-energy metallic Gupta clusters, which reduces to the global minimization of the multiextremal objective function. We developed algorithms and computational techniques for the investigation of these atomic clusters. Problems with a different number of particles were solved to check the performance of the developed methods. System computational experiments were performed to search for a global minimum in the Gupta model of extra-large dimensions (from 126 to 156 atoms). The authors are not aware of other works in which the results of computations for clusters of the specified dimensions are issued.

**Keywords** Atomic clusters · Gupta low-potential clusters · Computational technology · Global optimization · Multiextremal function

## 1 Introduction

Contemporary methods of molecular physics [1] permit scientists to get a substantial amount of information about the numerous materials, which makes it attainable to construct mathematical models that considerably well ponder their inner composition. Understanding the features of the formation of atomic cluster structures can improve the efficiency of nanocluster production and impart new optical, magnetic, and electrical properties to materials. Recently, many mathematical models have been developed focused on different microstates and types of materials (metal, ionic, gas clusters, biomolecules, supercooled liquids, etc. [4, 14]). The amount of developed models has stretched nearly a few hundred, according to specialists.

P. Sorokovikov (✉) · A. Anikin
Matrosov Institute for System Dynamics and Control Theory of SB RAS, Irkutsk, Russia
e-mail: sorokovikov.p.s@gmail.com

A. Anikin
e-mail: anikin@icc.ru

© The Author(s), under exclusive license to Springer Nature Switzerland AG 2022     407
A. Tchernykh et al. (eds.), *Mathematics and its Applications in New Computer Systems*,
Lecture Notes in Networks and Systems 424,
https://doi.org/10.1007/978-3-030-97020-8_37

A specific trait of atomic clusters is the nonmonotonic dependence of their features on the number of particles. Thus, the numerical tasks for investigating models of stable states of the materials are very hard in most cases. We consider these tasks as the problems for nonlocal minimization of multiextremal potential functions. The amount of local optima is increased quickly in these models, depending on the number of atoms in the cluster. Nonetheless, modern computational technologies for global optimization can obtain the "best of known" solutions that may be absolute extremes.

## 2 Materials and Methods

### 2.1 Gupta Model

We consider the problem of finding low-potential Gupta clusters [6] of extra-large dimensions. The Gupta potential is often used in computations of the nanoclusters properties of metals such as zinc, cadmium, lead, sodium, cobalt, and others [2–4, 11].

The Gupta problem is the following box-constrained global optimization task:

$$f(x) \to \min, \ x \in B \subseteq R^n, \tag{1}$$

$$B = \left\{ x \mid x = (x_1, x_2, ..., x_n), \ \alpha_i \leq x_i \leq \beta_i, \ i = \overline{1, n} \right\}, \tag{2}$$

$$f(x) = \frac{1}{2} \sum_{i=1}^{N} \left[ A \sum_{j \neq i} \exp\left[-p\left(\frac{r_{ij}}{r_0} - 1\right)\right] - \sqrt{\sum_{j \neq i} \exp\left[-2q\left(\frac{r_{ij}}{r_0} - 1\right)\right]} \right]. \tag{3}$$

Here $f(x)$ is the smooth and nonconvex potential function; $r_{ij}$ is the distance between atoms $i$ and $j$; $A, p, q, r_0$ are special parameters ($A = 0.1477$, $p = 9.689$, $q = 4.602$, $r_0 = 1.0$); $N$ is the number of atoms, $n = 3N$ is the number of variables, $\alpha$, $\beta$ are vectors of box constraints.

### 2.2 Description of Computational Technology

We implemented three-phase computational technology to investigate the problems of optimizing the conformation of Gupta clusters.

The first phase is based on the random approximation of a neighborhood of a known record point and depends on a collection of algorithms-generators of different starting points [13]:

1. "level" generator (LG);
2. "averaged" generator (AG);

3.  "gradient" generator (GG);
4.  "level-averaged" generator (LAG);
5.  "level-gradient" generator (LGG);
6.  "averaged-gradient" generator (AGG);
7.  "level-averaged-gradient" generator (LAGG).

In the second phase, we solved the problem of fast descent of the approximations obtained at the first stage [15] applying one of the "starter algorithms". These
algorithms permit finding low-energy states of atomic clusters. We developed three
modifications of "starter algorithms": the variant of B.T. Polyak's algorithm [12, 13],
the "raider algorithm" [13], and the proposed globalized modification of the Powell
method [7, 8] that is described below in the article.

For local search performed in the third phase, we implemented the limited-
memory Broyden-Fletcher-Goldfarb-Shanno (L-BFGS) algorithm [9, 10, 17].

Powell method [7] is a classical search method, on iterations of which, the conjugate descent directions are built, due to the solution of multiple one-dimensional
tasks. We developed a modification of this method, which consists in applying the
modified Evtushenko's algorithm of nonlocal univariate search as a subsidiary one.

Algorithm:

1.  Set the starting point $x^0 \in B$.
2.  Perform a univariate search of the minimum of the function $f$ along the direc-
    tion $p^n = e^n = (0, \ldots, 0, 1)^T$. Find the step size $h_0$ from the condition
    $f(x^0 + h_0 p^n) = \min_h f(x^0 + h p^n)$. Assume $x^0 = x^0 + h_0 p^n$. Set iteration
    number $k = 1$.
3.  Take the directions of the coordinate axes as the initial directions of the search,
    i.e. $p^i = e^i$, $i = \overline{1, n}$, where $e^1 = (1, \ldots, 0, 0)^T$, $\ldots$, $e^n = (0, \ldots, 0, 1)^T$.
4.  Perform $n$ one-dimensional searches from the point $x^0$ along directions $p^i = e^i$, $i = \overline{1, n}$. Moreover, each subsequent search is performed from the minimum
    point obtained in the previous step. Find the step size $h_i$ from the condition
    $f(x^{i-1} + h_i p^i) = \min_h f(x^{i-1} + h p^i)$. Assume $x^i = x^{i-1} + h_i p^i$.
5.  Choose a new direction $p = x^n - x^0$. Replace directions $p^1, \ldots, p^n$ by
    $p^2, \ldots, p^n, p$, respectively.
6.  Perform a univariate search from the point $x^n$ along the direction $p = p^n = x^n - x^0$. Find the step size $h_{n+1}$ from the condition $f(x^n + h_{n+1} p) = \min_h f(x^n + h p)$. Assume $x^{n+1} = x^n + h_{n+1} p$, $k = k + 1$.
7.  If the condition $k \leq n$ is met, then go to step 8. Otherwise, go to step 9.
8.  If the stopping criterion for the algorithm is satisfied, then terminate the search,
    set the resulting point $x^* = x^{n+1}$. Otherwise, go to step 9.
9.  Assume $x^0 = x^{n+1}$, take this point as the starting point $x^0$ for the next iteration,
    go to step 4.

One of the methods of nonlocal one-dimensional search, in which the objective
function must satisfy the Lipschitz condition, is the algorithm proposed by Yu.G.

Evtushenko [5, 16]. In the specified algorithm, a sequence of points $x_1, x_2, \dots$ of univariate function $f$ on the interval $[a, b]$ is constructed as follows:

$$x_1 = a, \quad x_k = x_{k-1} + \frac{f(x_{k-1}) - f_k^* + 2\varepsilon}{L}. \tag{4}$$

Here $f_k^* = \min\{f(x_1), \dots, f(x_k)\}$ is the record value of a function, $\varepsilon$ is the function accuracy, $L$ is the Lipschitz constant. Minimization stops when $x_k \geq b$.

A feature of this algorithm is that, for its operation, it is necessary to know a priori a reliable estimate of the Lipschitz constant. This problem is critical when we use the algorithm as an auxiliary in solving multivariate problems. In the multilevel hierarchy of algorithms, it will function at the lowest level and be executed many times in the computation process.

The paper proposes a modification of Evtushenko's algorithm with automatic estimation of the growth constant of a derivative based on the Hölder condition:

$$x_1 = a, \quad x_k = x_{k-1} + \frac{f(x_{k-1}) - f_k^* + 2\varepsilon}{G_j}. \tag{5}$$

Here $G_j$ is the estimation of the Hölder constant at the $j$-th iteration of the algorithm.

The base of the proposed algorithm is the idea of a multi-stage scanning of the search area and the use of information about the growth rate of the function at subsequent stages. All the samples performed in the previous steps are saved and affect the further course of the computations.

At the start of the algorithm, the estimate of the Hölder constant is chosen to be

$$G_0 = \frac{|f(b) - f(a)|}{|b - a|^{\frac{1}{N}}}, \tag{6}$$

calculated at each iteration of the algorithm by the following formula:

$$G_j = K_c \cdot \max \frac{|f(x_{k+1}) - f(x_k)|}{|x_{k+1} - x_k|^{\frac{1}{N}}}. \tag{7}$$

Here $K_c \geq 1$ is the caution ratio, $N \geq 1$ is the Hölder index (algorithm parameters). The cycle of passage along the segment is repeated several times, taking into account the resulting assessment and the performed samples. The stopping criterion of the algorithm is the condition $G_{j+1} = G_j, j$ is the segment scan cycle number.

# 3 Results

We performed computational experiments to search for low-potential states of the Gupta clusters applying the proposed computational techniques. We conducted calculations using a computer having the following characteristics: $2 \times$ Intel Xeon E5-2680 v2 2.8 GHz (20 cores, 40 threads); 128 Gb DDR3 1866 MHz.

We made an investigation of the efficiency of three-phase computational technology for clusters of 3–125 atoms. We compared the resulting values with the best of the known ones: $\delta = \left| \tilde{f} - \overline{f} \right|$, where $\tilde{f}$ is the value obtained by the optimization algorithm, $\overline{f}$ is the probable value of global optimum. The values of potential energy received under the investigation coincided with the known solutions issued in the Cambridge Cluster Database [14].

We used three-phase computational technology to investigate models with an extra-large number of particles because we successfully solved problems of smaller dimensions with its help. We investigated problems for clusters with the number of particles from 126 to 156 (solutions with found atoms conformations are accessible only for clusters up to 125 atoms in the Cambridge Cluster Database [14] and other public sources [2–4]).

We executed computational experiments for clusters with the number of particles from 126 to 156 (for each initial points generator and starter algorithm). We chose an excess of the given time limit (24 h) as a stop criterion. Table 1 demonstrates the results of computational experiments for the task with 126 atoms in the cluster.

As can be seen from Table 2, among all the investigated variants, there is a leading computational technology with a combination of the AGG generator and the raider algorithm.

Figure 1 demonstrates the numerical comparison of "starter algorithms" and the dependence of the best-found values of potential energy on the number of particles in the form of a graph. We plot the number of atoms on the horizontal axis, the found values – on the vertical axis. The best result was shown by the raider algorithm, quite a bit ahead of the modification of the Powell method.

**Table 1** The results of computational experiments (126 atoms)

| Generator | Starter | Value | Starter | Value | Starter | Value |
|-----------|---------|-------|---------|-------|---------|-------|
| 1 (LG)    | Polyak  | −164.4158 | Raider | −168.8011 | Powell | −168.8447 |
| 2 (AG)    | Polyak  | −164.3241 | Raider | −168.6678 | Powell | −168.7077 |
| 3 (GG)    | Polyak  | −164.4237 | Raider | −168.8129 | Powell | −168.8012 |
| 4 (LAG)   | Polyak  | −164.3586 | Raider | −168.7932 | Powell | −168.7784 |
| 5 (LGG)   | Polyak  | −164.3833 | Raider | −168.8275 | Powell | −168.7975 |
| 6 (AGG)   | Polyak  | −164.3715 | Raider | −168.8758 | Powell | −168.7861 |
| 7 (LAGG)  | Polyak  | −164.4423 | Raider | −168.7786 | Powell | −168.8501 |

**Table 2** The best-found solutions (number of atoms: from 126 to 156)

| N | Value | N | Value |
|---|---|---|---|
| 126 | −168.8758040477 | 141 | −189.0125295014 |
| 127 | −170.1994387092 | 142 | −190.4253739388 |
| 128 | −171.5821078823 | 143 | −191.7931293144 |
| 129 | −172.8529959138 | 144 | −193.1261879660 |
| 130 | −174.2539299691 | 145 | −194.4009431875 |
| 131 | −175.5918622806 | 146 | −196.1142359317 |
| 132 | −176.9870755985 | 147 | −197.1433042187 |
| 133 | −178.2694908282 | 148 | −198.4914310785 |
| 134 | −179.6825456752 | 149 | −199.7856001350 |
| 135 | −180.9472907460 | 150 | −201.2787305527 |
| 136 | −182.3723327456 | 151 | −202.5000313936 |
| 137 | −183.6309031715 | 152 | −203.8983896360 |
| 138 | −185.0976794276 | 153 | −205.2144669158 |
| 139 | −186.3657423488 | 154 | −206.6136278650 |
| 140 | −187.7640497248 | 155 | −207.9296264180 |
| | | 156 | −209.2856372769 |

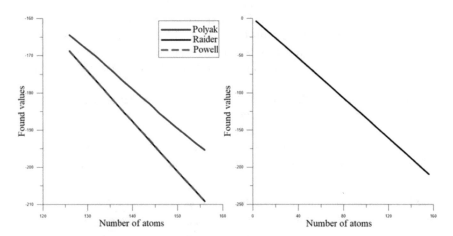

**Fig. 1** Numerical comparison of "starters" (left) and dependence of the best-found values of potential energy on the number of particles (right)

Table 2 demonstrates the best-found solutions. We do not know of other efforts to perform computational experiments of optimization the Gupta clusters potentials of the extra-large dimensions.

Established on the conducted numerical experiments, we can express the following findings:

1.  The raider algorithm demonstrated the best results on average, and the modification of the Powell method is quite a bit worse than the raider method.
2.  The modified Polyak's algorithm is worse than the modification of the Powell method algorithm and raider algorithm.
3.  The best-found solutions were achieved, for the most part, applying the raider algorithm and modification of the Powell method.
4.  The "averaged-gradient" generator (AGG) is leading between the implemented initial points generators.

## 4   Conclusion

In the article we proposed computational technologies for investigating low-potential Gupta clusters. The performance of the implemented algorithms for problems with small, medium, large, and extra-large number of particles was tested. Numerical comparison of the developed optimization algorithms was performed.

Probable low-energy states of Gupta clusters of extremely large dimensions (from 126 to 156 atoms) were obtained in the course of numerical study. Comparative analysis of the experimental results did not reveal sharp deviations from the observed regularity of the found values of potential energies, which describes their growth depending on the number of atoms.

**Acknowledgements**  This work was partially supported by the organization Ministry of Education and Science of Russia, project No 121041300060-4.

## References

1.  Brooks RL (2013) The fundamentals of atomic and molecular physics. Springer, New York
2.  Doye JP (2003) Identifying structural patterns in disordered metal clusters. Phys Rev B 68(19):195418
3.  Doye JP (2006) Lead clusters: different potentials, different structures. Comput Mater Sci 35(3):227–231
4.  Doye JP, Wales DJ (1997) Structural consequences of the range of the interatomic potential a menagerie of clusters. J Chem Soc Faraday Trans 93(24):4233–4243
5.  Evtushenko YG, Potapov MA, Korotkich VV (2014) Numerical methods for global optimization. In: Recent advances in global optimization. Princeton University Press, pp 274–297
6.  Gupta RP (1981) Lattice relaxation at a metal surface. Phys Rev B 23(12):6265
7.  Himmelblau DM (2018) Applied nonlinear programming. McGraw-Hill, New York
8.  Levin AY (1965) A minimization algorithm for convex functions. In: Dokl. Akad. Nauk SSSR, vol 160, no 6, pp 1244–1247
9.  Liu DC, Nocedal J (1989) On the limited memory BFGS method for large scale optimization. Math Program 45(1):503–528
10. Mokhtari A, Ribeiro A (2015) Global convergence of online limited memory BFGS. J Mach Learn Res 16(1):3151–3181

11. Noya EG, Doye JP, Wales DJ, Aguado A (2007) Geometric magic numbers of sodium clusters: interpretation of the melting behaviour. Eur Phys J D 43(1):57–60
12. Polyak BT (1969) Minimization of unsmooth functionals. USSR Comput Math Math Phys 9(3):14–29
13. Sorokovikov P, Gornov A, Anikin A (2020) Computational technology for the study of atomic-molecular Morse clusters of extremely large dimensions. In: IOP Conference Series: Materials Science and Engineering, vol 734, no 1. IOP Publishing, p 012092
14. Wales DJ et al (2012) Cambridge Cluster Database. http://www-wales.ch.cam.ac.uk/CCD.html
15. Wales DJ, Doye JP (1997) Global optimization by basin-hopping and the lowest energy structures of Lennard-Jones clusters containing up to 110 atoms. J Phys Chem A 101(28):5111–5116
16. Zhigljavsky AA, Zilinskas AG (1991) The methods of global extremum searching. Nauka, Moscow
17. Zhu C, Byrd RH, Lu P, Nocedal J (1997) Algorithm 778: L-BFGS-B: Fortran subroutines for large-scale bound-constrained optimization. ACM Trans Math Softw (TOMS) 23(4):550–560

# Software Engineering for Optimal Control Problems

**Alexander Gornov**⬤, **Tatiana Zarodnyuk**⬤, **Anton Anikin**⬤,
**Pavel Sorokovikov**⬤, **and Alexander Tyatyushkin**⬤

**Abstract** This article discusses the features and difficulties of the constructing soft-ware for optimal control problems. A brief overview of Russian and foreign optimiza-tion software for nonlinear controlled dynamic systems is presented. We describe the statement of the optimal control problem and the general approach to application software development. The authors used the presented technique to create software systems for solving applied optimal control problems.

**Keywords** Software · Numerical algorithms · Optimal control · Nonlinear dynamical systems · Applied problems

## 1 Introduction

The development of algorithms for solving optimal control problems (OCP) began in the face of numerous orders from practical applications in the 60 s and was carried out in many organizations. However, most of the proposed methods were implemented in the form of the scientific research prototypes of programs and did not pass the necessary steps to achieve the level of the software product.

A. Gornov · T. Zarodnyuk (✉) · A. Anikin (✉) · P. Sorokovikov (✉) · A. Tyatyushkin
Matrosov Institute for System Dynamics and Control Theory SB RAS, Irkutsk, Russia
e-mail: tzarodnyuk@gmail.com

A. Anikin
e-mail: anton.anikin@gmail.com

P. Sorokovikov
e-mail: pavel2301s@gmail.com

A. Tyatyushkin
e-mail: tjat@icc.ru

A. Gornov · A. Anikin
Moscow Institute of Physics and Technology, National Research University, Moscow, Russia

© The Author(s), under exclusive license to Springer Nature Switzerland AG 2022     415
A. Tchernykh et al. (eds.), *Mathematics and its Applications in New Computer Systems*,
Lecture Notes in Networks and Systems 424,
https://doi.org/10.1007/978-3-030-97020-8_38

## 1.1   Russian Software for the OCP

The first technologically designed software tool for optimal control problems can be considered the CONTROL software (R. Fedorenko, V. Popov, Keldysh Institute of Applied Mathematics of Academy of Sciences of the USSR, electronic computer BESM-6). The implemented technology is described in detail in the monograph of R. Fedorenko, which for a long time served as a unique training tool for practical problem solving.

Within the framework of the DISO project ("DIalogue Optimization System") by Yu. Evtushenko, N. Grachev (Computing Center of the Academy of Sciences of the USSR, BESM-6, IBM PC), the "Optimal control" block was created that implements methods of reduction the OCP to the mathematical programming problem. The software included a developed dialogue system, which allowed the use of a variety of interactive solution technologies (Yu. Evtushenko). Somewhat later, the capabilities of DISO were expanded to the class of discontinuous problems (N. Grachev).

Within the framework of the MAPR program project – "Mathematical Programming in Multidimensional Problems" (A. Tyatyushkin, A. Gornov, Irkutsk Computer Center of the SB RAS, BESM-6, Elbrus computer) for nonlinear OCPs, interactive solutions technology was implemented and about ten years developed. The "Nonlinear Optimal Control" block of the MAPR software was used to solve a significant number of applied optimization problems (A. Gornov).

At the same time, the development of the "PPP for ZOU" (V. Gurman, V. Baturin, BESM-6) being conducted in the same organization included methods based on sufficient optimality conditions and was actively applied to solve navigation and environmental-economic problems (V. Gurman).

The KONUS software – "Comprehensive Optimization of Nonlinear Controlled Systems" (A. Zholudev, Irkutsk Computer Center of the SB RAS, ES computer) was based on the reduction method to mathematical programming and algorithmic components similar to the MINOS software (A. Zholudev).

The software OPTCON (A. Gornov, Irkutsk Computer Center of the SB RAS, IBM PC), was aimed at creating a compact, but multifunctional technology for solving OCP. The multimethod technology implemented on its basis is applicable to a wide class of problems and has been used in practical applications for more than twenty years (A. Gornov).

## 1.2   Foreign Software for the OCP

Abroad, according to the site of the Control Engineering Virtual Library, about 400 organizations are working on the optimization of dynamic systems. However, as in Russia, only a few of them pay attention to the creation and development of software technologies for optimal control problems.

As part of the MATLAB project, the RIOTS application – "Recursive Integration Optimal Trajectory Solver" (A. Schwartz, E. Polak, Y. Chen) was implemented. According to the authors, RIOTS is "the most powerful optimal control problem solver", but "distributed without any performance or accuracy guarantees". The accompanying documents indicate that the user must know the "theory of optimal control, optimization, and numerical approximation methods." The results of serious testing or application of the technology to solve practical problems are not given in the literature.

SOCS – "Sparse Optimal Control Software" (J. Betts) is a software implemented by Boeing Computer Services from Boeing. The optimization core is composed of SPRNLP and BARNLP (sequential quadratic programming) algorithms. Available publications describe a fairly developed software technology that is successfully applied to solving practical problems of navigation, robotics, and chemical kinetics.

PDECON (K. Schittkowski, University of Bayreuth, Germany, 1997) is the software technology for the wide class of optimal control problems, including, in addition to classical formulations, systems of partial differential equations and algebraic-differential systems. The algorithmic basis of the technique consists of the original methods of sequential quadratic programming, developed by this scientific group for more than 40 years. According to the authors, this is the leading development in this class of software products.

With the support of the US Department of Energy, the Argonne National Laboratory have been developed and tested technologies for solving optimal control problems (J. More, A. Bondarenko, D. Bortz) based on advanced commercial mathematical programming packages (DONLP2, LANCELOT, MINOS, SNOPT, LOQO).

MISER3 (K.L. Teo, C.J. Goh, Hong Kong Polytechnic University) is the software for the optimization of the dynamic systems, developed for about 20 years. The proposed technology based on both optimal control theory and classical finite-dimensional optimization. There are a large number of publications on solving practical problems using MISER3.

DIRCOL (O. von Stryk, Darmstadt University, Germany) is a software package based on collocation methods and reduction to the mathematical programming problem. The well-known NPSOL and SNOPT packages (M. Saunders, Systems Optimization Laboratory, Stanford University) are used as the basic optimization technology. It is claimed that DIRCOL is operated in 60 organizations. There are publications on technology applications for solving applied problems from robotics, navigation, biology and economics.

Of course, there are other technologies focused on the study of optimization problems. For example, the MINOT software, which includes optimization techniques for dynamic systems (C. Schweiger, C. Floudas, Princeton University) and is an advanced interface between the problem and a set of finite-dimensional optimization packages (CPLEX, LPSOLVE, MINOS, NPSOL, SNOPT, DASOLV, DAESSA). But information on their testing and application results in solving applied optimal control problems is not always presented in the open press.

## 2   The Statement of the Optimal Control Problem

The classic optimal control problem. Let there be a controlled process described by an ordinary system of differential equations with initial conditions

$$\dot{x} = f(x, u, t), \; x(t_0) = x^0,$$

defined on the interval $T = [t_0, t_1]$.

Here t is an independent variable, $x(t)$ is the phase coordinate vector, $u(t)$ is the vector of control functions; $f(x(t),u(t),t)$ is the vector function assumed to be continuously differentiable with respect to all arguments except t. The initial phase vector $x(t0) = x0$ is given. We call the admissible control functions $u(t) \in U \subset E^r$ for $\forall t$, where U is a convex closed set from the corresponding space.

The optimal control problem with the free right end of the trajectory consists in finding the vector function $u(t)$ that satisfies the constraints and allows to achieve the minimum value for the objective functional $I_0(u) = 0_0(x(t_1))$. In the optimal control problem with terminal constraints, there are also restrictions of the form $I_j(u) = 0_j(x(t_1)) = (\leq)0, j = \overline{1, m}$. In the problem with phase constraints, phase restrictions of the inequality type $I_j(u) = g_j(x(t), u(t), t) \leq 0, j = \overline{m+1, mt}$ are also added to the OCP.

All functions $0_j(x(t_1)), j = \overline{0, m}$ and $g_j(x(t), u(t), t), j = \overline{m+1, mt}$ are assumed to be continuously differentiable with respect to all arguments.

## 3   General Approach to the Application Software Development

### 3.1   The Task of Creating High-Quality Software for Computing Problems

The problem of creating high-quality software for computational problems was formulated in the 60 s [1]. For 10 years, the psychological barrier [2] was overcome, associated with new research methods, and this area turned into an independent scientific discipline. At the first stage, the efforts of specialists from various scientific centers were combined in the implementation of projects to create the algorithm libraries: IMSL [3], NAG [4], NATS [5].

## 3.2 Fundamentals of Designing Software Systems for Optimal Control Problems

The practice of using software to optimize dynamic systems confirms the view that the current software implementations, despite the considerable efforts of a large number of specialists, are excessively difficult to operate and not convenient enough for the mass user. The main factor aggravating the problem is the objective complexity and multivariance of the optimal control problem, which impedes standardization of the approaches and finding typical computing schemes with a wide area of effective application. Other significant factors in this situation include the inconvenience of implementing "continuous" algorithms for optimizing dynamic systems on existing discrete computing platforms and the lack of implementation of the numerical methods to the level of the robust algorithms. The multilevelness of the optimal control methods developed so far, the attributive presence in their constructions of not always trivial auxiliary Cauchy problems, either direct or conjugate, and the lack of theoretical methods for estimating the informational complexity of solving and matching errors play an important role. A significant factor is also the underdevelopment of testing methods for optimization programs and the lack of a universally recognized set of test problems for optimal control.

These factors, acting multiplicatively, significantly increase the requirements for the mathematical qualifications of the user and, as a result, prevent the widespread introduction of the developed software systems in the practice of the scientific research and applied work.

With the increase in the power of the available computers and the desire of the specialists in the various fields to investigate more and more complex controlled systems, the problem of automation of software for optimal control problems becomes more urgent.

The creation of new optimization algorithms and tools has already required a lot of effort, these problems can and should be study with further. According to the authors, a qualitatively new level of the software for the optimal control problems is achieved mainly through the development of the metacomponents, the use of fundamentally new approaches and technologies, as well as by creating and replenishing the testing base.

The basic principles roposed by the author for the design of software systems for the optimal control problems are as follows:

- multimethod;

    The multivariance of the optimal control problem and the absence of a single optimization method capable of satisfactorily solving various problems naturally leads to the need to have a set of algorithms with significantly different computational characteristics. The functional redundancy provided by the multimethod approach is also necessary to take into account the user's qualifications and effectively solve auxiliary instrumental problems. The authors also proposes to associate the notion of "multimethod" with the presence in the software package of the developed tools for studying OCP and different algorithms for solving auxiliary problems.

- specialization in user types;

    The user's qualifications, his ability to spend time solving the problem, the real goals of the computational experiment are taken into account by: a) implementing various operating modes of the software, b) applying a flexible technological formulation of the problem.

- completeness of the coverage of the problem area;

    The variety of the formulations of the optimal control problems leads to the practical impossibility of creating specialized software for each type of the problems. In order not to lose the functional sufficiency of the developed programs, it is necessary to carefully select the canonical formulation of the problem and have the large collection of the methods for reducing the heterogeneous problems to canonical one.

- two-level structure;

    Ensuring the reliability of the functioning of the software and monitoring the overall result of the calculations require coordination of all basic software components and the creation of a special meta-level component. This is the supervisor of the computing process.

- multilingualism;

    The diverse components of the software should be implemented in the programming languages oriented towards the creation of the heterogeneous components. The C language is used to construct the computing components, and specialized languages (JAVA, etc.) are used to implement the graphical interfaces. The interaction of the components of the programs is done through the built-in file system or using special technologies for integrating heterogeneous components.

## 3.3 The Basic Elements of the Computing Technology for the OCP

The authors propose to classify the software modules that underlie software tools for optimizing dynamic systems as follows:

- algorithmic components;

    The algorithmic components consist of a set of the optimization algorithms and the algorithms for solving Cauchy problems, a set of the algorithms for solving auxiliary problems of one-dimensional search, a set of the algorithms for accounting for terminal and phase constraints, a library of the algorithms for solving lower level computing problems (calculating integrals, solving systems of linear equations, generating pseudorandom numbers, etc.).

- tool and service components;

    The instrumental and service components contain the tools for constructing the suboptimal sampling grid; the algorithms for estimating the integration errors and discretization, the criteria for evaluating the convergence of a sequence of improved controls, the differential derivative estimation algorithms, and others.

- metacomponents;

It is proposed to include metacomponents a dialogue system that implements the user interface; the automation and control problems; the design and fixation tools for new computing schemes; mechanisms for tuning the algorithmic parameters and others.

- testing base.

The testing base includes a library of test problems with well-known solutions and a set of testing methods for the optimal control search algorithms.

## 3.4  Functional Subsystems of Software Technology

The software technology for implementing the software for solving the OCP, according to the authors, should include the following functional subsystems:

- auditor of the program statement of the problem;

The program statement of the problem, defined in a programming language, for example C, is the translation of the technological statement of the problem into a form consistent with the input formats of the software.

- start state constructor;

A fairly common problem in the initial research of the controlled systems is the lack of the acceptable starting point, i.e. the initial approximation, from which you can start the optimization process. As a rule, this is due to the nonphysicality or rigidity of the system at random initial approximations chosen or to the unsatisfactory discretization of the system. The main approach to overcoming this problem is the formulation and investigation of the simplified auxiliary problems, the solutions of which give the initial approximation in the original problem. For each approximation under study, the discretization customizer should fulfill its functions.

- discretization customizer;

The speed of the software depends on the adequate tuning of the interconnected discretization and integration algorithms of the system of differential equations (about 99% of the processor time is spent on solving Cauchy problems). A satisfactory solution to this problem is a compromise between the integration accuracy and the computational speed. The parameters available for adjustment are the sampling grid, the integration method, and the local integration accuracy. The manipulation by discretization accuracy is also used in the post-optimization analysis manager.

- resident of emergency handling;

Working with the wide class of the optimal control problems leads to the regular appearance of the problems that have various features (nonphysicality, rigid systems of differential equations, etc.) or have no solutions at all with some controls. In such situations, the software should be protected from "avosts" (automatic shutdowns) and fatal errors that entail the forced termination of calculations, and the user should receive the information necessary to correct the problem.

- designer of computing schemes;

The computational schemes underlying the software can be classified according to four independent design criteria ("the coordinate axes"): the optimization method, the discretization algorithm, the technique for decomposing the variable set; the method for parameterizing the elements of the problem statement.

Each point (element) of the set in the considerable discrete four-dimensional space corresponds to a specific computational algorithm. The properties of this set of the algorithms are studied rather poorly. The software has a set of fixed circuits (the points in the considered set), tested in practice and offered to the user to choose from. To solve the complex problems, this set may not be enough, therefore it is also advisable to implement interactive tools that allow a qualified user, after carrying out trial calculations, to select a unique scheme that is most effective for the particular problem in manual control mode.

- manager for setting control parameters;

Tuning algorithmic parameters is not the most important, but still an essential way to increase the efficiency of the algorithms. In the software for solving the OCP, it is proposed to maintain compromise values of the "default" parameters and to develop the adaptive mechanisms for adjusting parameters to the specific situation.

- the post-optimization analysis manager;

The use of the computational methods is always associated with the risk of getting the wrong solution. For the optimal control problems, this risk multiplicatively increases due to the multilevelness of the applied algorithms. The general approach to checking the calculation results, implemented when creating the software, is characterized by a simplification of the structure of the applied algorithms with a simultaneous sharp increase in the accuracy of solving auxiliary problems. Unfortunately, this approach cannot guarantee the adequacy of the result in the original problem, but significantly reduces the risks of an incorrect answer.

The problem of assessing the quality of the resulting optimal control can be divided into two main parts:

1) the quality of the approximation of the initial continuous problem approximating discrete;

The quality of the discretization of the problem is estimated according to the Runge rule: a denser sampling grid is selected and the system is integrated by a more accurate method with reduced local accuracy. The comparison of the scaled values of the current and refined integration give an estimate of the errors of integration and discretization at the same time.

2) the quality of the solution of the discrete optimal control problem;

To assess the optimality of the solution of the discrete problem, in addition to checking the residuals of the necessary optimality conditions, an attempt is made to further improve control using the search methods (the coordinate descent, Powell's method) that are independent of the accuracy of the gradients calculation (the integration of the conjugate system, the difference estimates of the gradients). In the absence of the significant improvement in the functionals, the current control is considered approximately optimal.

• supervisor of the computing process.

It is unrealistic to foresee all possible situations during the operation of the software. Nevertheless, it is possible to fix empirically and try to formalize the internal constraints, the implementation of which will avoid existing ambiguities and errors, and thereby increasing the probability of ultimate success. The task of this subsystem is general monitoring of the internal situation during the operation of the software in an automated mode.

## 3.5   Algorithm Selection Technique

The problem of searching from variants of algorithms for one that has the best characteristics is a main problem for all areas of computational mathematics. The traditional descriptions of the algorithms in books cannot be transferred without any preliminary research to any computer language. As a rule, they are inaccurate, contain many omissions, allow for ambiguity in interpretation, etc. Therefore, in reality, the book descriptions contain a whole set of the algorithms on which the scatter of the individual properties can be very large. As a result of laborious studies, the best performing algorithm is selected from this set and is programmed [7]. On the other hand, there is an opinion that "writing two good programs is an order of magnitude easier than deciding which one is better" [8].

It is natural to consider the main characteristics of the algorithm: 1) Efficiency (E); 2) Accuracy (A); 3) Reliability (R).

According to the test results, you can determine a point in the three-dimensional space of criteria (E, A, R), reflecting its basic properties and allowing to predict its behavior.

The proposed methodology for constructing the algorithm includes iteratively applied stages of its development and selection: 1) the choice of mathematical construction; 2) the implementation of the algorithms for the auxiliary problems; 3) the assembly of the first version; the comparative testing; the efficiency estimation; 4) the construction of the adaptive mechanisms; 5) the comparative testing; 6) the statistical testing; the stress testing; 7) the integration with the software package; 8) the trial operation.

After equipping the algorithm with the necessary adaptive mechanisms, its size (in the form of text in the algorithmic language) can grow 3–10 times.

The adaptation mechanisms include:

1) the stopping criteria;
2) the mechanisms for controlling the operation of the algorithm (the algorithmic parameters; the limiters in the iterations number, the processor time, the number of solved Cauchy problems, etc.);

3)  the ways to coordinate the accuracy of solving all subproblems;
4)  the algorithms for taking into account the history of the iterative process;
5)  the protection mechanisms from computational features (the automatic stops, the destructive machine zeros, loss of the significant digits), the normalization of intermediate data;
6)  the self-control mechanisms of the algorithm (the looping, the number of the inefficient iterations, the excessive smallness of the variations);
7)  the means of the interaction with the user (the default parameter values, the multilevel output of the intermediate results, the output diagnostics).

The quality of the algorithmic scheme can be fundamentally evaluated by the following formula

$$K = K_1 \cdot K_2 \cdot K_3 \cdot K_4 \cdot K_5$$

Here, $K_1$ is the coefficient reflecting the quality of the mathematical method underlying the algorithm, $K_2$ depends on the quality of the algorithms for solving auxiliary problems, $K_3$ reflects the power of the adaptation mechanisms built into the algorithm, $K_4$ depends on the amount of effort to research and refine the algorithm, the number of options studied, $K_5$ reflects the duration of its operation, the experience its practical application, the number of the errors found over time.

## 4 Conclusion

Applying the principles presented in the article, the authors developed a series of software for the study of the optimization problems: AS OPTCON-F, focused on the numerical study of functional-differential equations of pointwise type [9, 10]; AS OPTCON-M, constructed to solve the optimization problems of the atomic-molecular clusters [11]; Internet-technology for remote user support OPTCON [12, 13]; AS OPTCON-III based on the stochastic algorithms and used for solving optimal control problems [14, 15].

The authors created a unique collection of the non-convex optimal control problems, which allows us to study the properties of the specialized algorithms and compare new ones with existing methods [16], as well as testing methods focused on estimating the quality of the developed approaches.

Using the developed software, a number of applied problems from different areas were solved, namely from nanophysics: the problem of minimizing the potential Keating function in the germanium-silicon heterostructures [17], the problem of studying quantum logical operations in the nanostructures with the quantum dots (Rzhanov Institute of Semiconductor Physics SB RAS, Novosibirsk) [18, 19]; robotics: the problem of finding control for transferring the robot from its original state to the required state (Kyushu University, Japan) [20, 21]; technical ecology:

the problem of estimating the influence of the natural and social environment factors on the public health [22], the problem of assessing the atmospheric air pollution caused by the forest fires (Scientific Research Institute of Occupational Medicine and Human Ecology, State Scientific Center for Scientific and Technical Research, VSNS SB RAMS, Angarsk) [23]; energetics: the problem of an indicative analysis of energy security factors (Melentyev Energy Systems Institute SB RAS, Irkutsk) [24, 25]; earthquake resistant constructing: study of the strength resource of the building structures under the seismic effects based on the single-mass model of the elasto-plastic deformation with two degrees of freedom (Institute of the Earth's Crust SB RAS, Irkutsk) [26]; medicine: the problem of optimizing the investment programs of the Republic of Buryatia (Institute of Mathematics, Economics and Informatics of Irkutsk State University) [27], the problem of estimating the concentration of the individual fractions of the cholesterol in the blood of patients (Irkutsk Diagnostic Center) [28] and others.

Based on the experience of creating optimization software, it can be argued that the general approach to the software development proposed in the article is useful and effective for solving different applied problems from various fields.

**Acknowledgements** This work was carried out within the state assignment of the Ministry of Education and Science of Russia under the project "Theory and Methods of Research of Evolutionary Equations and Controlled Systems with their Applications" (State Registration No. 121041300060-4).

# References

1. Rice JR (1971) Mathematical software. Academic Press, New York
2. Gear CW (1979) Numerical software: science or alchemy. University of Illinois, Department of Computer Science Report UIUCDCS-R-79-969
3. Aird TJ (1977) The IMSL Fortran converter: an approach to solving portability problems. Lect Notes Comput Sci 57:368–388
4. Du Croz JJ, Hague SJ, Siemieniuch JL (1977) Aids to portability within the NAG project. Lect Notes Comput Sci 57:390–404
5. Boyle JM et al (1972) NATS, a collaborative effort to certify and disseminate mathematical software. In: Proceedings of the national ACM conference, New York, vol II, pp 630–635
6. Bychkov IV, Oparin GA, Feoktistov AG, Bogdanova VG, Pashinin AA (2014) Multi-agent methods and control tools in a service-oriented distributed computing environment. Proc Inst Syst Program RAS 26(5):65–82 (In Russian)
7. Voevodin VV, Voevodin VlB (2002) Parallel computing. SPb: BHV-Petersburg
8. Forsythe J, Malcolm M, Mowler C (1980) Computer methods of mathematical calculations. Mir, Moscow
9. Zarodnyuk TS, Anikin AS, Finkelshtein EA, Beklaryan AL, Belousov FA (2016) The technology for solving the boundary value problems for nonlinear systems of functional differential equations of pointwise type. Mod. Technol. Syst. Anal. Simul. 1(49):19–26
10. Beklaryan A, Beklaryan L, Gornov A (2019) Solutions of Traneling Wave Type for Korteweg-de Vries-type system with polynomial potential. In: Proceedings of the 9th international conference on communications in computer and information science (OPTIMA 2018), vol 974, pp 291–305

11. Sorokovikov P, Gornov A, Anikin A (2020) Computational technology for the study of atomic-molecular Morse clusters of extremely large dimensions. IOP Conf. Ser. Mater. Sci. Eng. 734:012092
12. Massel L, Gornov A, Zarodnyuk T (2018) Internet-technology for remote user support OPTCON. In: 5th international workshop "critical infrastructures: IWCI 2018", Advances in intelligent systems research, vol 158, pp 124–128
13. Massel LV, Gornov AY, Podkamennii DV (2002) Creation of computing resources on the Internet based on legacy software. Comput Technol 7:247–252
14. Gornov AY, Zarodnyuk TS (2013) Tunneling algorithm for solving nonconvex optimal control problems. In: Optimization, simulation, and control. Springer optimization and its applications, vol 76, pp 289–299
15. Gornov AY, Zarodnyuk TS, Finkelshtein EA, Anikin AS (2016) The method of uniform monotonous approximation of the reachable set border for a controllable system. J Glob Optim 66(1):53–64
16. Gornov AY, Zarodnyuk TS, Madzhara TI, Daneyeva AV, Veyalko IA (2013) A collection of test multiextremal optimal control problems. In: Springer optimization and its applications, pp 257–274
17. Anikin AS, Gornov AY (2012) Software implementation of Newton's method for Keating's potential optimization problems with use of sparse matrix technology. J Studia Informatica Universalis 9(3):11–19
18. Gornov AY, Dvurechenskii AV, Zarodnyuk TS, Zinov'eva AF, Nenashev AV (2011) Problem of optimal control in the system of semiconductor quantum points. Autom Remote Control 72(6):1242–1247
19. Nenashev AV, Zinovieva AF, Dvurechenskii AV, Gornov AY, Zarodnyuk TS (2015) Quantum logic gates from time-dependent global magnetic field in a system with constant exchange. J Appl Phys 117(11):113905
20. Finkelstein E, Svinin M, Gornov A (2016) Numerical study of the problem of spherical mobile robot optimal control. In: Proceedings of the 7th international conference on OPTIMA-2016, pp 51–52
21. Gornov AY, Tyatyushkin AI, Finkelstein EA (2016) Numerical methods for solving terminal optimal control problems. Computat Math Math Phys 56(2):221–234
22. Efimova NV et al (2010) Environmental factors: the experience of integrated assessment. In: Rukavishnikov VS (ed) NTSRVH SB RAMS, Irkutsk
23. Gornov AY, Zarodnyuk TS, Efimova NV (2018) Air pollution and population morbidity forecasting with artificial neural networks. IOP Conf Ser Earth Environ Sci 211:012053
24. Edelev AV, Rudenko EM, Gornov AY, Zarodnyuk TS, Anikin AS (2009) The assessment of the adequacy of neural-type models in the problem of the indicative analysis of energy security. In: Proceedings of the XIV Baikal all-Russian conference on "information and mathematical technologies in science, Technology and Education", Part I, ISEM SB RAS, Irkutsk, pp 47–52
25. Gornov A, Zarodnyuk T (2018) Computational technology for solving nonconvex optimal control problems for power systems. In: IWCI 2018: 5th International workshop on "critical infrastructures". Advances in intelligent systems research, vol 158, pp 68–72
26. Berzhinsky YA, Ordynskaya AP, Berzhinskaya LP, Gornov AY, Finkelstein EA (2019) Analysis of the mechanism of transition to the limit state of series 111 residential buildings during the 1988 Spitak earthquake. Geodyn. Tectonophys. 10(3):715–730
27. Bokmelder EP, Dyakovich MP, Efimova NV, Gornov AY, Zarodnyuk TS (2010) The experience of using models of dynamic systems in solving medical-social and medical-environmental problems. Inform Contr Syst 2(24):161–164 (in Russian)
28. Kuz'menko VV, Gornov AY, Anikin AS (2018) Estimation of mathematical models accuracy for calculation of LDL-cholesterol concentration. In: IWCI 2018: 5th International workshop on "critical infrastructures". Advances in intelligent systems research, vol. 158, pp 111–116

# Numerical Methods for Solving Boundary Value Problems for the Generalized Integro-differential Modified Moisture Transfer Equation with the Bessel Operator

**M. KH. Beshtokov**⊙

**Abstract** Initial-boundary value problems for the generalized integro-differential modified moisture transfer equation with the Bessel operator are investigated. Difference schemes are constructed that approximate these problems on uniform grids. The case of an equation is considered in which the coefficients of the highest derivatives are separated from zero by a positive constant. For the solution of the problems under consideration, a priori estimates are obtained in differential and difference interpretations. The obtained estimates imply the uniqueness and stability of the solution with respect to the right-hand side and initial data, as well as the convergence of the solution of the difference problem to the solution of the corresponding differential problem. An algorithm for finding approximate solutions of the problems under consideration is constructed and numerical calculations of test examples are carried out, illustrating the theoretical calculations obtained in the work.

**Keywords** Boundary value problems · A priori estimate · Moisture transfer equation · Integro-differential equation · Fractional-order differential equation · Fractional Caputo derivative

## 1 Introduction

One-dimensional movement of moisture in capillary-porous media, which include soils with a fractal structure, can occur under the influence of a wide variety of driving forces. Based on the analysis of the diffusion mechanism in a porous mass, when the occurrence of moisture flows under the action of a capillary pressure gradient is taken into account, a nonlinear equation is obtained

M. KH. Beshtokov (✉)
North Caucasian Center for Mathematical Research, North Caucasian Federal University, Stavropol, Russia
e-mail: beshtokov-murat@yandex.ru

© The Author(s), under exclusive license to Springer Nature Switzerland AG 2022    427
A. Tchernykh et al. (eds.), *Mathematics and its Applications in New Computer Systems*,
Lecture Notes in Networks and Systems 424,
https://doi.org/10.1007/978-3-030-97020-8_39

$$\frac{\partial u}{\partial t} = \frac{\partial}{\partial x}\left(D(u)\frac{\partial u}{\partial x}\right),$$

where $D(u)$ is the diffusivity coefficient, $u$ is the humidity, $t$ is the time, $x$ is the depth.

The diffusion model, which assumes that if at the initial moment an uneven humidity is set, then a moisture flow from the more humid to the less humid layers should appear, is often not justified. Direct, sufficiently convincing and repeated experiments sometimes demonstrate the opposite sign of the flow from layers with low to layers with high moisture content [1, 2]. A real explanation of the experimental facts and the correct interpretation of when and under what conditions the movement of moisture in the forward and reverse directions takes place, is possible using the modified diffusion equation, or, as it is called, the Aller equation [3]:

$$\frac{\partial u}{\partial t} = \frac{\partial}{\partial x}\left(D(u)\frac{\partial u}{\partial x} + A\frac{\partial^2}{\partial t\partial x}\right)$$

where $A$ is a variable parameter.

The theory of fractals is widely used to describe the structure of disordered media and the processes occurring in them [4, 5]. Porous bodies are examples of disordered media. In this case, fractals can be pore space, rock skeleton, rock skeleton surface, etc. In the case when cracks and solid porous blocks are represented by homogeneous interpenetrating continua, the Barenblat-Zheltov model is usually used to describe fluid filtration [1]. In the case when the space is a fractal with Hausdorff-Bezikovich dimension $d_f$, immersed in a continuous medium with dimension $d$, $(d \geq d_f$, $d = 2, 3)$, a differential equation of fractional order is used to describe the motion of an impurity in a flow of a homogeneous fluid [6].

In this paper, we numerically study the solution of a three-dimensional integro-differential equation with a fractional Caputo derivative containing an integral over the time variable

$$\partial_{0t}^\alpha u = Lu + f(x, t), \quad (x, t) \in Q_T, \tag{*}$$

where $Lu = \sum_{s=1}^3 L_s u$, $x = (x_1, x_2, x_3)$, $L_s u = \frac{\partial}{\partial x_s}(k_s(x, t)\frac{\partial u}{\partial x_s}) + \partial_{0t}^\alpha \frac{\partial}{\partial x_s}(\eta_s(x)\frac{\partial u}{\partial x_s}) + h_s(x, t)\frac{\partial u}{\partial x_s} + \int_0^t p_s(x, t, \tau)u(x, \tau)d\tau$.

When passing from a three-dimensional equation to a cylindrical coordinate system $(r, \varphi, z)$ in the case when the solution $u = u(r)$ does not depend on either $z$ or $\varphi$ (axial symmetry takes place), (*) takes the form (we denote $x = r$) [7, p. 170]:

$$\partial_{0t}^\alpha u = \frac{1}{r}(rk(r, t)u_r)_r + \frac{1}{r}\partial_{0t}^\alpha(r\eta(r)u_r)_r + h(r, t)u_r$$

$$+ \int_0^t p_s(r, t, \tau)u(r, \tau)d\tau + f(r, t),$$

and in the case of spherical symmetry, Eq. (*) takes the form:

$$\partial_{0t}^{\alpha} u = \frac{1}{r^2}(r^2 k(r, t)u_r)_r + \frac{1}{r^2}\partial_{0t}^{\alpha}(r^2\eta(r)u_r)_r + h(r, t)u_r$$

$$+ \int_0^t p_s(r, t, \tau)u(r, \tau)d\tau + f(r, t),$$

where $k(r, t) = k_1(x, t) = k_2(x, t) = k_3(x, t)$, $\eta(r) = \eta_1(x) = \eta_2(x) = \eta_3(x)$, $h(r, t) = h_1(x, t) = h_2(x, t) = h_3(x, t)$, $p(r, t, \tau) = p_1(x, t, \tau) = p_2(x, t, \tau) = p_3(x, t, \tau)$ - are the symmetry conditions for the coefficients due to the symmetry of $r$ with respect to the variables $x_1, x_2, x_3$.

The appearance of the integral term in the equation is associated with the need to take into account the dependence of the instantaneous values of the characteristics of the described object on their previous values, i.e. influence on the current state of the system of its prehistory.

Note that this study is a continuation of the author's works, in which difference methods were proposed for solving local and nonlocal boundary value problems for a degenerate equation of pseudoparabolic type.

## 2   Materials and Methods

### 2.1   Problem Statement

In a closed cylinder $\overline{Q}_T = \{(x, t) : 0 \le x \le l, 0 \le t \le T\}$ consider the boundary value problem

$$\partial_{0t}^{\alpha} u = \frac{1}{x^m}\frac{\partial}{\partial x}\left(x^m k(x, t)\frac{\partial u}{\partial x}\right) + \frac{1}{x^m}\partial_{0t}^{\alpha}\frac{\partial}{\partial x}\left(x^m\eta(x)\frac{\partial u}{\partial x}\right) + r(x, t)\frac{\partial u}{\partial x}$$

$$+ \int_0^t p(x, t, \tau)u(x, \tau)d\tau + f(x, t), \quad 0 < x < l, 0 < t \le T, \quad (1)$$

$$\lim_{x \to 0} x^m \Pi(x, t) = 0, 0 \le t \le T, \quad (2)$$

$$u(l, t) = 0, 0 \le t \le T, \quad (3)$$

$$u(x, 0) = u_0(x), 0 \le x \le l, \quad (4)$$

where $0 < c_0 \leq k(x, t)$, $\eta(x) \leq c_1$, $|r(x, t), r_x, k_x, p(x, t, \tau)| \leq c_2$, $0 \leq m \leq 2$, (5)

$\partial_{0t}^\alpha u = \frac{1}{\Gamma(1-\alpha)} \int_0^t \frac{u_\tau(x,\tau)}{(t-\tau)^\alpha} d\tau$, —is a fractional derivative in the sense of Caputo of order $\alpha$, $0 < \alpha < 1$, $c_i$, $i = 0, 1, 2$—are positive numbers, $\Pi(x, t) = ku_x + \partial_{0t}^\alpha(\eta(x)u_x)$.

Note that, for $x = 0$, the condition of boundedness of the solution $|u(0, t)| < \infty$ is imposed, which is equivalent to condition (2), which in turn is equivalent to the identity $\Pi(x, t) = 0$ [7], if the functions $r(0, t)$, $k(0, t)$, $q(0, t)$, $f(0, t)$ are finite.

We will assume that problem (1)–(4) has a unique solution possessing the derivatives required in the course of the presentation, the coefficients of the equation and boundary conditions satisfy the smoothness conditions necessary in the course of the presentation, which ensure the required order of approximation of the difference scheme.

In the course of the presentation, we will also use positive constants $M_i$, $i = 1, 2$, ..., depending only on the input data of the problem under consideration.

## 2.2   A Priori Estimate in Differential Form

To obtain an a priori estimate for the solution of problem (1)–(4) in differential form, we introduce the scalar product and the norm in the following form: $(u, v) = \int_0^l uv dx$, $(u, u) = \|u\|_0^2$, where, $v$ are functions given on $[0, l]$.

We multiply Eq. (1) scalarly by $x^m u$:

$$\left(\partial_{0t}^\alpha u, x^m u\right) = \left((x^m ku_x)_x, u\right) + \left(\partial_{0t}^\alpha(x^m \eta u_x)_x, u\right) + \left(ru_x, x^m u\right)$$
$$+ \left(\int_0^t pud\tau, x^m u\right) + \left(f, x^m u\right).$$
(6)

Using the Cauchy inequality with $\varepsilon$ and Lemma 1 [8], from identity (6), after simple transformations taking into account (2), we obtain:

$$\partial_{0t}^\alpha \left\|x^{\frac{m}{2}} u\right\|_0^2 + \int_0^l \eta \partial_{0t}^\alpha (x^{\frac{m}{2}} u_x)^2 dx + \left\|x^{\frac{m}{2}} u_x\right\|_0^2$$
$$\leq M_1 \left\|x^{\frac{m}{2}} u\right\|_{w_2^1(0,l)}^2 + M_2 \int_0^t \left\|x^{\frac{m}{2}} u\right\|_0^2 d\tau + M_3 \left\|x^{\frac{m}{2}} f\right\|_0^2.$$
(7)

where $\|x^{\frac{m}{2}}\|_{W_2^1(0,l)}^2 = \|x^{\frac{m}{2}} u\|_0^2 + \|x^{\frac{m}{2}} u_x\|_0^2$.

Applying the fractional integration operator $D_{0t}^{-\alpha}$ to both sides of (7), we obtain

$$\|x^{\frac{m}{2}} u\|^2_{W^1_2(0,l)} + D^{-\alpha}_{0t} \|x^{\frac{m}{2}} u_x\|^2_0 \leq M_4 D^{-\alpha}_{0t} \|x^{\frac{m}{2}} u\|^2_{W^1_2(0,l)}$$

$$+ M_5 D^{-\alpha}_{0t} \int_0^t \|x^{\frac{m}{2}} u\|^2_0 d\tau + M_6 \left( D^{-\alpha}_{0t} \|x^{\frac{m}{2}} f\|^2_0 + \|x^{\frac{m}{2}} u_0(x)\|^2_{W^1_2(0,l)} \right) \qquad (8)$$

The second term on the right-hand side of (8) is estimated as

$$D^{-\alpha}_{0t} \int_0^t \|x^{\frac{m}{2}} u\|^2_0 d\tau = \frac{1}{\Gamma(\alpha)} \int_0^t \frac{d\tau}{(t-\tau)^{1-\alpha}} \int_0^\tau \|x^{\frac{m}{2}} u\|^2_0 ds$$

$$= \frac{1}{\Gamma(\alpha)} \int_0^t \|x^{\frac{m}{2}} u\|^2_0 ds \int_s^t \frac{d\tau}{(t-\tau)^{1-\alpha}} = \frac{1}{\Gamma(\alpha)} \int_0^t \|x^{\frac{m}{2}} u\|^2_0 \left( -\frac{(t-\tau)^\alpha}{\alpha} \right) \Big|_s^t ds$$

$$= \frac{1}{\alpha \Gamma(\alpha)} \int_0^t (t-s)^\alpha \|x^{\frac{m}{2}} u\|^2_0 ds = \frac{1}{\Gamma(\alpha+1)} \int_0^t (t-\tau)^\alpha \|x^{\frac{m}{2}} u\|^2_0 d\tau$$

$$\leq \frac{1}{\alpha \Gamma(\alpha)} \int_0^t \frac{\|x^{\frac{m}{2}} u\|^2_0 (t-\tau)}{(t-\tau)^{1-\alpha}} d\tau \leq \frac{T}{\alpha} D^{-\alpha}_{0t} \|x^{\frac{m}{2}} u\|^2_0. \qquad (9)$$

In view of (9), from (8) we find

$$\|x^{\frac{m}{2}} u\|^2_{W^1_2(0,l)} + D^{-\alpha}_{0t} \|x^{\frac{m}{2}} u_x\|^2_0 \leq M_7 D^{-\alpha}_{0t} \|x^{\frac{m}{2}} u\|^2_{W^1_2(0,l)}$$

$$+ M_6 \left( D^{-\alpha}_{0t} \|x^{\frac{m}{2}} f\|^2_0 + \|x^{\frac{m}{2}} u_0(x)\|^2_{W^1_2(0,l)} \right). \qquad (10)$$

Using Lemma 2 [8], from (10), we obtain the following a priori estimate

$$\|x^{\frac{m}{2}} u\|^2_{W^1_2(0,l)} + D^{-\alpha}_{0t} \|x^{\frac{m}{2}} u_x\|^2_0 \leq M \left( D^{-\alpha}_{0t} \|x^{\frac{m}{2}} f\|^2_0 + \|x^{\frac{m}{2}} u_0(x)\|^2_{W^1_2(0,l)} \right), \qquad (11)$$

where $M = const > 0$, depending only on the input data of problem (1)–(4).
$D^{-\alpha}_{0t} u = \frac{1}{\Gamma(\alpha)} \int_0^t \frac{u d\tau}{(t-\tau)^{1-\alpha}}$—is a fractional derivative in the sense of Riemann-Liouville of order $\alpha$, $0 < \alpha < 1$.

## 2.3 Stability and Convergence of the Difference Scheme

On the uniform grid $\overline{\omega}_{h\tau}$, we associate the differential problem (1)–(4) with the difference scheme:

$$\varkappa\Delta_{0t_{j+\sigma}}^{\alpha}y = \frac{\varkappa}{x_i^m}(x_{i-0.5}^m a_i^j y_{\bar{x}}^{(\sigma)})_x + \frac{1}{x_i^m}\Delta_{0t_{j+\sigma}}^{\alpha}(x_{i-0.5}^m \gamma_i y_{\bar{x},i})_x + \frac{b^{-j}}{x_i^m}(x_{i-0.5}^m a_i^j y_{\bar{x},i}^{(\sigma)})$$

$$+\frac{b^{+j}}{x_i^m}\left(x_{i+0.5}^m a_{i+1}^j y_{x,i}^{(\sigma)}\right) + \sum_{s=0}^{j+\frac{1}{2}}\rho_{s,i}^j y_i^s \bar{\tau} + \varphi_i^j, \ (x,t) \in \omega_{h,\tau}, \tag{12}$$

$$\varkappa_0 a_1 y_{(x,0)}^{(\sigma)} + \Delta_{0t_{j+\sigma}}^{\alpha}(\gamma_1 y_{x,0}) = \frac{0.5h}{m+1}(\Delta_{0t_{j+\sigma}}^{\alpha} y_0 - \sum_{s=0}^{j+\frac{1}{2}}\rho_{0,s}^j y_0^s \bar{\tau}) - \mu, \ t \in \bar{\omega}_\tau, = 0,$$
$$\tag{13}$$

$$y_N^{(\sigma)} = 0, \ t \in \bar{\omega}_\tau, \ x = l, \tag{14}$$

$$y(x,0) = u_0(x), \ x \in \bar{\omega}_h, \ t = 0, \tag{15}$$

where $\Delta_{0t_{j+\sigma}}^{\alpha}y = \frac{\tau^{1-\alpha}}{\Gamma(2-\alpha)}\sum_{s=0}^{j}c_{j-s}^{(\alpha,\sigma)}y_t^s$—is a discrete analogue of the fractional Caputo derivative of order $\alpha$, $0 < \alpha < 1$[9].

$$a_0^{(\alpha,\sigma)} = \sigma^{1-\alpha}, \ a_l^{(\alpha,\sigma)} = (l+\sigma)^{1-\alpha} - (l-1+\sigma)^{1-\alpha}, \ l \geq 1, \ \sigma = 1 - \frac{\alpha}{2},$$

$$b_l^{(\alpha,\sigma)} = \frac{1}{2-\alpha}[(l+\sigma)^{2-\alpha} - (l-1+\sigma)^{2-\alpha}] - \frac{1}{2}[(l+\sigma)^{1-\alpha} + (l-1+\sigma)^{1-\alpha}], \ l \geq 1,$$

$$\text{при } j = 0, \quad c_0^{(\alpha,\sigma)} = a_0^{(\alpha,\sigma)};$$

$$\text{при } j > 0, \quad c_s^{(\alpha,\sigma)} = \begin{cases} a_0^{(\alpha,\sigma)} + b_1^{(\alpha,\sigma)}, & s = 0, \\ a_s^{(\alpha,\sigma)} + b_{s+1}^{(\alpha,\sigma)} - b_s^{(\alpha,\sigma)}, & 1 \leq s \leq j-1, \\ a_j^{(\alpha,\sigma)} - b_j^{(\alpha,\sigma)}, & s = j, \end{cases}$$

$c_s^{(\alpha,\sigma)} > \frac{1-\alpha}{2}(s+\sigma)^{-\alpha} > 0$, $a_i^j = k(x_{i-0.5}, t^{j+\sigma})$, $\gamma_i = \eta(x_{i-0.5})$, $b_i^{\pm j} = \frac{\varkappa_i r_i^{j+\sigma}}{k_i^{j+\sigma}}$, $y^{(\sigma)} = \sigma y^{j+1} + (1-\sigma)y^j$, $r_N = r(l,t) = r_N^{j+\sigma} \geq 0$, $r = r^+ + r^-$, $r(0,t) = r_0^{j+\sigma} \leq 0$, $\varkappa_i = 1 + \frac{m(m-1)h^2}{24x_i^2}$, $i = \overline{1, N-1}$, $r^+ = 0.5(r + |r|) \geq 0$, $r^- = 0.5(r - |r|) \leq 0$,

$$\varkappa_i = \frac{1}{1+R_i}, \ \rho_{i,s}^j = p_{i,s}^{j+\sigma}, \ \varphi_i^j = \begin{cases} \varkappa_i f_i^{j+\sigma}, & \neq 0, N, \\ f_i^{j+\sigma}, & i = 0, N. \end{cases} \ \hbar = \begin{cases} 0.5h, & i = 0, \\ h, & i \neq 0, N, \end{cases}$$

$$R_i = \frac{0.5h|r_i|\varkappa_i}{k_{i-0.5}}, \ \varkappa_0 = \frac{1}{1 + \frac{0.5h|r_0|}{(m+1)a_1}}, \ r_0 \leq 0, \ |r| = r^+ - r^-, \ \bar{x}^m = x_{i-0.5}^m,$$

$$\mu = \frac{0.5h}{m+1}\varphi_0, \ Y = \hat{y} + y, \ \hat{y} = y^{j+1}, \ y_t = \frac{\hat{y} - y}{\tau}, \ y = y_i^j = y(x_i, t_j),$$

$$\sum_{s=0}^{j+\frac{1}{2}} \vartheta^s \overline{\tau} = \sum_{s=1}^{j-1} \vartheta^s \tau + 0.5\tau\left(\vartheta^0 + \vartheta^j + \vartheta^{j+0.5}\right), \ \overline{\tau} = \begin{cases} 0.5\tau, \ s = 0, \ j, \ j+0.5, \\ \tau, \ s \neq 0, \ j, \ j+0.5. \end{cases}$$

We introduce the scalar product and the norm in the following form: $(u, v) = \sum_{i=1}^{N-1} u_i v_i h$, $(u, u) = \|u\|_0^2$, $(1, u_x^2] = \|u_{\overline{x}}]\|_0^2$, $\|u\|_0^2 = \sum_{i=1}^{N} u_i^2 \hbar = (1, u^2]$.

Now let us find an a priori estimate; for this purpose, we multiply (12) scalarly by $x^m y^{(\sigma)}$ :

$$(\overline{\varkappa}\Delta^{\alpha}_{0t_{j+\sigma}} y, \ x^m y^{(\sigma)}) = (\varkappa(x^m_{i-0.5} a^j_i y^{(\sigma)}_{\overline{x}})_x, \ y^{(\sigma)}) + (\Delta^{\alpha}_{0t_{j+\sigma}} (x^m_{i-0.5} \gamma_i y_{\overline{x},i})_x, \ y^{(\sigma)})$$

$$+ \left(b^{-j} x^m_{i-0.5} a^j_i y^{(\sigma)}_{\overline{x},i}, \ y^{(\sigma)}\right) + \left(b^{+j} x^m_{i+0.5} a^j_{i+1} y^{(\sigma)}_{x,i}, \ y^{(\sigma)}\right)$$

$$+ \left(\sum_{s=0}^{j+\frac{1}{2}} \rho^j_{s,i} y^s_i \overline{\tau}, \ x^m y^{(\sigma)}\right) + \left(\varphi, \ x^m y^{(\sigma)}\right). \tag{16}$$

Estimating the sums in (16), using the $\varepsilon$-Cauchy inequality and Lemma 1 [9], we obtain:

$$\frac{1}{2}\Delta^{\alpha}_{0t_{j+\sigma}} \|x^{\frac{m}{2}} y\|_0^2 + M_1 \|x^{\frac{m}{2}} y^{(\sigma)}_{\overline{x}}]\|_0^2 + \frac{c_0}{2}\Delta^{\alpha}_{0t_{j+\sigma}} \|\overline{x}^m y_{\overline{x}}]\|_0^2 \leq \overline{x}^m y^{(\sigma)} (\varkappa a y^{(\sigma)}_{\overline{x}}$$

$$+ \Delta^{\alpha}_{0t_{j+\sigma}} (\gamma_i y_{\overline{x}}))|_0^N + M_2(\|x^{\frac{m}{2}} y^{(\sigma)} + \|\overline{x}^{\frac{m}{2}} y^{(\sigma)}_{\overline{x}}]\|_0^2) + M_3 \sum_{s=0}^{j+\frac{1}{2}} \|x^{\frac{m}{2}} y^s\|_0^2 \overline{\tau}$$

$$+ \frac{1}{2}\|x^{\frac{m}{2}} \varphi\|_0^2. \tag{17}$$

We transform the first expression on the right-hand side of (17), then we get

$$\overline{x}^m y^{(\sigma)} \left[\varkappa a y^{(\sigma)}_{\overline{x}} + \Delta^{\alpha}_{0t_{j+\sigma}} (\gamma_i y_{\overline{x}})\right]|_0^N = -x^m_0 y^{(\sigma)}_0 \left(\varkappa_0 a_1 y^{(\sigma)}_{x,0} + \Delta^{\alpha}_{0t_{j+\sigma}} \gamma_1 y_{x,0}\right)$$

$$= -x^m_{0.5} y^{(\sigma)}_0 [\frac{0.5h}{m+1} (\Delta^{\alpha}_{0t_{j+\sigma}} y_0 - \sum_{s=0}^{j+\frac{1}{2}} \rho^j_{0,s} y^s_0 \overline{\tau}) - \mu] = x^m_{0.5} y^{(\sigma)}_0 \mu - \frac{0.5h}{m+1} x^m_{0.5} y^{(\sigma)}_0 \Delta^{\alpha}_{0t_{j+\sigma}} y_0$$

$$+ \frac{hx^m_{0.5}}{2(m+1)} y^{(\sigma)}_0 \sum_{s=0}^{j+\frac{1}{2}} \rho^j_{0,s} y^s_0 \overline{\tau} \leq \mu^2 + M_4 (x^{\frac{m}{2}}_{0.5} y^{(\sigma)}_0)^2$$

$$- \frac{0.5h}{2(m+1)} \Delta^{\alpha}_{0t_{j+\sigma}} (x^{\frac{m}{2}}_{0.5} y_0)^2 + M_5 \sum_{s=0}^{j+\frac{1}{2}} (x^{\frac{m}{2}}_{0.5} y^s_0)^2 \overline{\tau}. \tag{18}$$

Taking into account (18), from (17) we obtain

$$\Delta_{0t_{j+\sigma}}^{\alpha} \|x^{\frac{m}{2}} y\|_1^2 + \|\overline{x}^{\frac{m}{2}} y_{\overline{x}}^{(\sigma)}]\|_0^2 \le M_6 \|x^{\frac{m}{2}} y^{(\sigma)}\|_1^2$$

$$+M_7 \sum_{s=0}^{j+\frac{1}{2}} (\|x^{\frac{m}{2}} y^s\|_0^2 + (x_{0.5}^{\frac{m}{2}} y_0^s)^2)\overline{\tau} + M_8\left(\|x^{\frac{m}{2}}\varphi\|_0^2 + \mu^2\right), \tag{19}$$

where $\|x^{\frac{m}{2}} y\|_1^2 = \|x^{\frac{m}{2}} y\|_0^2 + \|\overline{x}^{\frac{m}{2}} y_{\overline{x}}]\|_0^2 + (x_{0.5}^{\frac{m}{2}} y_0)^2$.

Considering that

$$\sum_{s=0}^{j+\frac{1}{2}} (\|x^{\frac{m}{2}} y^s\|_0^2 + (x_{0.5}^{\frac{m}{2}} y_0^s)^2)\overline{\tau}$$

$$= \sum_{s=0}^{j} (\|x^{\frac{m}{2}} y^s\|_0^2 + (x_{0.5}^{\frac{m}{2}} y_0^s)^2)\overline{\tau} + 0.5\tau(\|x^{\frac{m}{2}} y^j\|_0^2 + (x_{0.5}^{\frac{m}{2}} y_0^j)^2)$$

We rewrite (19) in another form

$$\Delta_{0t_{j+\sigma}}^{\alpha} \|x^{\frac{m}{2}} y\|_1^2 \le M_9^{\sigma} \|x^{\frac{m}{2}} y^{j+1}\|_1^2 + M_{10}^{\sigma} \|x^{\frac{m}{2}} y^j\|_1^2 + M_{11} F^j. \tag{20}$$

where $F^j = \sum_{s=0}^{j}(\|x^{\frac{m}{2}} y^s\|_0^2 + (x_{0.5}^{\frac{m}{2}} y_0^s)^2)\overline{\tau} + \|x^{\frac{m}{2}}\varphi\|_0^2 + \mu^2$.

Based on Lemma 7 [10], from (20) we obtain

$$\|x^{\frac{m}{2}} y^{j+1}\|_1^2 \le M_{12}\left(\|x^{\frac{m}{2}} y^0\|_1^2 + \frac{t_j^{\alpha}}{\Gamma(1+\alpha)} \max_{0\le j'\le j} F^{j'}\right), \tag{21}$$

where $M_{12} = const > 0$, independent of $h$ and $\tau$.

From (21), we get

$$\|x^{\frac{m}{2}} y^{j+1}\|_1^2 \le M_{12}\Big(\|x^{\frac{m}{2}} y^0\|_1^2$$

$$+\frac{t_j^{\alpha}}{\Gamma(1+\alpha)} \max_{0\le j'\le j} \left(\sum_{s=0}^{j}\left(\|x^{\frac{m}{2}} y^s\|_0^2 + \left(x_{0.5}^{\frac{m}{2}} y_0^s\right)^2\right)\overline{\tau} + \|x^{\frac{m}{2}}\varphi\|_0^2 + \mu^2\right)\Big)\Big). \tag{22}$$

Introducing the notation $g^j = \max_{0\le j'\le j} \|x^{\frac{m}{2}} y^{j'}\|_2^2$, in view of

$$\|x^{\frac{m}{2}} y\|_1^2 = \|x^{\frac{m}{2}} y\|_0^2 + \|\overline{x}^{\frac{m}{2}} y_{\overline{x}}]\|_0^2 + (x_{0.5}^{\frac{m}{2}} y_0)^2,$$

$$\max_{0\le j'\le j} \sum_{s=0}^{j'} (\|x^{\frac{m}{2}} y^s\|_0^2 + (x_{0.5}^{\frac{m}{2}} y_0^s)^2)\overline{\tau} \le \sum_{j'=0}^{j} \max_{0\le s\le j'} (\|x^{\frac{m}{2}} y^s\|_0^2 + (x_{0.5}^{\frac{m}{2}} y_0^s)^2)\overline{\tau}$$

$$\leq \sum_{j'=0}^{j} \max_{0 \leq s \leq j'} (\|x^{\frac{m}{2}} y^s\|_0^2 + (x_{0.5}^{\frac{m}{2}} y_0^s)^2) \tau$$

from (22), we obtain

$$g^{j+1} \leq M_{13} \sum_{s=0}^{j} g^s \tau + M_{14} F_1^j, \tag{23}$$

where $F_1^j = \|x^{\frac{m}{2}} y^0\|_1^2 + \frac{t_j^{\alpha}}{\Gamma(1+\alpha)} \max_{0 \leq j' \leq j} \left( \|x^{\frac{m}{2}} \varphi\|_0^2 + \mu^2 \right),$

$$\|x^{\frac{m}{2}} y\|_2^2 = \|x^{\frac{m}{2}} y\|_0^2 + (x_{0.5}^{\frac{m}{2}} y_0)^2.$$

Based on Lemma 4 (see [11, p. 171]), from (23) we obtain

$$\|x^{\frac{m}{2}} y^{j+1}\|_2^2 \leq M \left( \|x^{\frac{m}{2}} y^0\|_1^2 + \max_{0 \leq j' \leq j} \left( \|x^{\frac{m}{2}} \varphi^{j'}\|_0^2 + \mu^2 \right) \right). \tag{24}$$

where $M = const > 0$, independent of h and $\tau$.

## 2.4 Statement of the Third Boundary Value Problem and a Priori Estimate in Differential Form

Consider the third boundary value problem for Eq. (1). For this, we replace condition (3) by a condition of the form

$$-\Pi(l, t) = \beta(t) u(l, t) - \mu(t), \ |\beta| \leq c_2. \tag{25}$$

To obtain an a priori estimate for the solution, we multiply (1) scalarly by $x^m u$. After simple transformations from (6) we obtain

$$\partial_{0t}^{\alpha} \|x^{\frac{m}{2}} u\|_0^2 + \int_0^l \eta \partial_{0t}^{\alpha} (x^{\frac{m}{2}} u_x)^2 dx + \|x^{\frac{m}{2}} u_x\|_0^2$$

$$\leq M_1 \|x^{\frac{m}{2}} u\|_{W_2^1(0, l)}^2 + M_2 \int_0^t \|x^{\frac{m}{2}} u\|_0^2 d\tau + M_3 \left( \|x^{\frac{m}{2}} f\|_0^2 + \mu^2(t) \right), \tag{26}$$

where $\|x^{\frac{m}{2}} u\|_{W_2^1(0, l)}^2 = \|x^{\frac{m}{2}} u\|_0^2 + \|x^{\frac{m}{2}} u_x\|_0^2.$

Applying the fractional integration operator $D_{0t}^{-\alpha}$ to both sides of inequality (26) and repeating arguments (7)–(26), from (26) we find the a priori estimate

$$\|x^{\frac{m}{2}}u\|^2_{W_2^1(0,l)} + D_{0t}^{-\alpha}\|x^{\frac{m}{2}}u_x\|^2_0 \leq M\left(D_{0t}^{-\alpha}(\|x^{\frac{m}{2}}f\|^2_0 + \mu^2(t))\right.$$

$$\left. + \|x^{\frac{m}{2}}u_0(x)\|^2_{W_2^1(0,l)}\right), \tag{27}$$

where $M = const > 0$, depending only on the input data of problem (1), (2), (25), (4), $D_{0t}^{-\alpha}u = \frac{1}{\Gamma(\alpha)}\int\limits_0^t \frac{u d\tau}{(t-\tau)^{1-\alpha}}$—is the fractional Riemann-Liouville integral of order $\alpha$, $0 < \alpha < 1$.

## 2.5  Stability and Convergence of the Difference Scheme

On the uniform grid $\overline{\omega}_{h\tau}$, we associate the differential problem (1), (2), (25), (4) with the difference scheme:

$$\begin{cases} \overline{\overline{\varkappa}}\Delta_{0t_{j+\sigma}}^\alpha y = \overline{\Lambda}(t^{j+\sigma})y^{(\sigma)} + \overline{\delta}y + \overline{\Phi}, \\ y(x,0) = u_0(x), \end{cases} \tag{28}$$

where

$$\overline{\overline{\varkappa}} = \begin{cases} \varkappa_i, & x \in \omega_h, \\ 1, & x = 0, l, \end{cases} \quad \varkappa_i = 1 + \frac{m(m-1)h^2}{24x_i^2}, \quad t^* = t^{j+1/2},$$

$$\overline{\Lambda}y^{(\sigma)} = \begin{cases} \tilde{\Lambda}y_i^{(\sigma)} = \frac{\varkappa_i}{x_i^m}(x_{i-0.5}^m a_i^j y_{\overline{x},i}^{(\sigma)})_x + \frac{b-j}{x_i^m}(x_{i-0.5}^m a_i^j y_{\overline{x},i}^{(\sigma)}) + \frac{b+j}{x_i^m}(x_{i+0.5}^m a_{i+1}^j y_{x,i}^{(\sigma)}) + \sum\limits_{s=0}^{j+\frac{1}{2}} \rho_{i,s}^j y_i^s \overline{\tau}, \\ \Lambda^- y_0^{(\sigma)} = \frac{m+1}{0.5h}(\varkappa_0 a_1 y_{x,0}^{(\sigma)}) + \frac{h}{2}\sum\limits_{s=0}^{j+\frac{1}{2}} \rho_{0,s}^j y_0^s \overline{\tau}), \ i = 0 \\ \Lambda^+ y_N^{(\sigma)} = -\frac{1}{0.5h}(\varkappa_N a_N y_{\overline{x},N}^{(\sigma)}) + \tilde{\varkappa}\beta y_N^{(\sigma)} - 0.5h\sum\limits_{s=0}^{j+\frac{1}{2}} \rho_{N,s}^j y_N^s \overline{\tau}), \ x = l, \end{cases}$$

$$\overline{\delta}y = \begin{cases} \delta y_i = \frac{1}{x_i^m}\Delta_{0t_{j+\sigma}}^\alpha (x_{i-0.5}^m \gamma_i y_{\overline{x},i})_x, & (x,t) \in \omega_{h\tau}, \\ \delta^- y_0 = \frac{m+1}{0.5h}\Delta_{0t_{j+\sigma}}^\alpha (\gamma_1 y_{x,0}), & x = 0, \\ \delta^+ y_N = -\frac{1}{0.5h}\Delta_{0t_{j+\sigma}}^\alpha (\gamma_N y_{\overline{x},N}), & x = l, \end{cases}$$

$$\overline{\Phi} = \begin{cases} \varphi = \varphi_i, & (x,t) \in \omega_{h\tau} \\ \varphi^- = \frac{m+1}{0.5h}\mu_1, & x = 0 \\ \varphi^+ = \frac{1}{0.5h}\mu_2, & x = l. \end{cases}$$

$$\mu_1 = 0.5h\varphi_0^j, \quad \mu_2 = \tilde{\varkappa}\mu^{j+\sigma} + 0.5h\varphi_N^j, \quad \tilde{\varkappa} = 1 + \frac{0.5hm}{l} = \frac{1}{1 - \frac{0.5hm}{l}},$$

$$\varkappa_0 = \frac{1}{1 + \frac{0.5h|r_0|}{(m+1)k_{0\,5}^{j+\sigma}}}, \quad r_0^{j+\sigma} \leq 0, \quad \varkappa_N = \frac{1}{1 + 0.5h\frac{|r_N^{j+\sigma}|}{k_{N-0\,5}}}, \quad \text{если} \quad r_N^{j+\sigma} \geq 0.$$

We introduce the scalar product and the norm in the following form

$$(u, v] = \sum_{i=1}^{N} u_i v_i \hbar, \quad \|u\|_0^2 = \sum_{i=1}^{N} u_i^2 \hbar, \quad \hbar = \begin{cases} 0.5h, & = N, \\ h, & i \neq N. \end{cases}$$

We now multiply (28) scalarly by $x^m y^{(\sigma)}$, then we obtain

$$\left(\overline{\overline{\varkappa}}\Delta_{0t_{j+\sigma}}^{\alpha} y, \, x^m y^{(\sigma)}\right] = \left(\overline{\Lambda}(t_{j+\sigma}) y^{(\sigma)}, \, x^m y^{(\sigma)}\right] + \left(\overline{\delta} y, \, x^m y^{(\sigma)}\right] + \left(\overline{\Phi}, \, x^m y^{(\sigma)}\right]. \tag{29}$$

After some transformations from (29) we obtain

$$\Delta_{0t_{j+\sigma}}^{\alpha} \|x^{\frac{m}{2}} y]|_1^2 + \|\overline{x}^{\frac{m}{2}} y_{\overline{x}}^{\sigma}]|_0^2 \leq M_1 \|x^{\frac{m}{2}} y^{(\sigma)}]|_1^2 + M_2 \sum_{s=0}^{j+\frac{1}{2}} \|x^{\frac{m}{2}} y^s]|_1^2 \overline{\tau}$$

$$+ M_3 \left(\|x^{\frac{m}{2}} \varphi^j\|_0^2 + \mu_1^2 + \mu_2^2\right). \tag{30}$$

where $\|x^{\frac{m}{2}} y]|_1^2 = \|x^{\frac{m}{2}} y]|_0^2 + \|\overline{x}^{\frac{m}{2}} y_{\overline{x}}]|_0^2 + (x_{0.5}^{\frac{m}{2}} y_0)^2$.

Repeating the reasoning (19)–(24), from (30) we obtain the a priori estimate

$$\|x^{\frac{m}{2}} y^{j+1}]|_1^2 \leq M(\|x^{\frac{m}{2}} y^0]|_1^2 + \max_{0 \leq j' \leq j} (\|x^{\frac{m}{2}} \varphi^{j'}\|_0^2 + \mu_1^2 + \mu_2^2)). \tag{31}$$

where $M = const > 0$, не зависящее от $h$ и $\tau$.

## 3 Results

The following theorem are true.

**Theorem 1.** If $k(x, t) \in C^{1,0}(\overline{Q}_T)$, $\eta(x) \in C^1[0, l]$, $r(x, t)$, $q(x, t)$, $f(x, t)$, $\rho(x, t, \tau) \in C(\overline{Q}_T)$, $u(x, t) \in C^{2,0}(Q_T) \cap C^{1,0}(\overline{Q}_T)$, $\partial_{0t}^{\alpha} u(x, t) \in C(\overline{Q}_T)$,

$\partial_{0t}^{\alpha} u_{xx}(x, t) \in C(\overline{Q}_T)$ and conditions (5) be satisfied, then estimate (11) is valid for the solution $u(x, t)$ of problem (1)–(4).

**Theorem 2.** Let conditions (5) be satisfied, then there exist $h_0$, $\tau_0$, such that if $h \leq h_0$, $\tau \leq \tau_0$, then estimate (24) is valid for the solution of the difference problem (12)–(15).

The a priori estimate (24) implies the uniqueness and stability of the solution to the difference scheme (12)–(15) with respect to the right-hand side and the initial data, as well as the convergence of the solution of the difference problem (12)–(15) to the solution of the differential problem (1)–(4) in the sense of the norm $\|xz^{j+1}\|_2^2$ on each layer so that if there exist such $\tau_0$, $h_0$, then for $\tau \leq \tau_0$, $h \leq h_0$, the a priori estimate $\|x(y^{j+1} - u^{j+1})\|_2 \leq \overline{M}(h^2 + \tau^2)$ is valid, where $\overline{M} = const > 0$.

**Theorem 3.** If $k(x, t) \in C^{1,0}(\overline{Q}_T)$, $\eta(x) \in C^1[0, l]$, $r(x, t)$, $q(x, t)$, $f(x, t)$, $\rho(x, t, \tau) \in C(\overline{Q}_T)$, $u(x, t) \in C^{2,0}(Q_T) \cap C^{1,0}(\overline{Q}_T)$, $\partial_{0t}^{\alpha} u(x, t) \in C(\overline{Q}_T)$, $\partial_{0t}^{\alpha} u_{xx}(x, t) \in C(\overline{Q}_T)$ and conditions (5) are satisfied, then for the solution $u(x, t)$ of problem (1), (2), (25), (4) estimate (27) is valid.

**Theorem 4.** Let conditions (5) be satisfied, then there exist $h_0$, $\tau_0$, such that if $h \leq h_0$, $\tau \leq \tau_0$, then estimate (31) is valid for the solution of the difference problem (28).

The a priori estimate (31) implies the uniqueness and stability, as well as the convergence of the solution of the difference problem (28) to the solution of the differential problem (1), (2), (25), (4) in the sense of the norm $\|xz^{j+1}\|_1^2$ on each layer so that if there exist such $\tau_0$, $h_0$, then for $\tau \leq \tau_0$, $h \leq h_0$, the a priori estimate $\|x(y^{j+1} - u^{j+1})]\|_1 \leq \overline{M}(h^2 + \tau^2)$. is valid.

**Numerical Experiment**

The coefficients of the equation and boundary conditions of problem (28) are selected so that the exact solution of the problem is the function $u(x, t) = t^3 x^4$.

Below in table for different values of the parameters $\alpha = 0.01; 0.5; 0.99$, $m = 0; 1; 2$ we present the maximum value of the error $(z = y - u)$ and the computational order of convergence (CO) in the norms $\|[\cdot]\|_0$ and $\|\cdot\|_{C(\overline{w}_{h\tau})}$, where $\|y\|_{C(\overline{w}_{h\tau})} = \max_{(x_i, t_j) \in \overline{w}_{h\tau}} |y|$, when $h = \tau$, while the mesh size is decreasing. The error is being reduced in accordance with the order of approximation $O(h^2 + \tau^2)$. The order of convergence is determined by the following formula: $CO = \log_{\frac{h_1}{h_2}} \frac{\|[z_1]\|_0}{\|[z_2]\|_0}$, where $z_i$ is the error corresponding $h_i$ (Table 1).

**Table 1** We present the maximum value of the error ($z = y - u$) and the computational order of convergence (CO) in the norms $\|z^j\|_0$ and $\|z\|_{C(\overline{w}_{h\tau})}$ when the grid size is reduced by $t = 1$, if $h = \tau$

| $\alpha$ | $m$ | $h$ | $\max\limits_{0<j<m} |[z^j]|_0$ | CO in $\|[\cdot]\|_0$ | $\|z\|_{C(\overline{w}_{h\tau})}$ | CO in $\|\cdot\|_{C(\overline{w}_{h\tau})}$ |
|---|---|---|---|---|---|---|
| 0.01 | 1 | 1/10 | 0.059822153 | | 0.024798721 | |
| | | 1/20 | 0.014884147 | 2.0069 | 0.005956540 | 2.0461 |
| | | 1/40 | 0.003708226 | 2.0050 | 0.001459095 | 2.0235 |
| | | 1/80 | 0.000925207 | 2.0029 | 0.000361044 | 2.0119 |
| | | 1/160 | 0.000231055 | 2.0015 | 0.000089796 | 2.0060 |
| | 2 | 1/10 | 0.042407707 | | 0.053990105 | |
| | | 1/20 | 0.010296342 | 2.0422 | 0.013145040 | 2.0865 |
| | | 1/40 | 0.002530004 | 2.0249 | 0.003240011 | 2.0448 |
| | | 1/80 | 0.000626611 | 2.0135 | 0.000804130 | 2.0228 |
| | | 1/160 | 0.000155892 | 2.0070 | 0.000200299 | 2.0115 |
| 0.5 | 1 | 1/10 | 0.079856635 | | 0.041714318 | |
| | | 1/20 | 0.019946185 | 2.0013 | 0.010150116 | 2.0390 |
| | | 1/40 | 0.004976000 | 2.0031 | 0.002501595 | 2.0206 |
| | | 1/80 | 0.001242057 | 2.0023 | 0.000620892 | 2.0104 |
| | | 1/160 | 0.000310214 | 2.0014 | 0.000154667 | 2.0052 |
| | 2 | 1/10 | 0.057977170 | | 0.027315531 | |
| | | 1/20 | 0.014161201 | 2.0335 | 0.006487950 | 2.0739 |
| | | 1/40 | 0.003489152 | 2.0210 | 0.001579365 | 2.0384 |
| | | 1/80 | 0.000865205 | 2.0118 | 0.000389568 | 2.0194 |
| | | 1/160 | 0.000215357 | 2.0063 | 0.000096745 | 2.0096 |
| .99 | 1 | 1/10 | 0.107653172 | | 0.057770899 | |
| | | 1/20 | 0.026922002 | 1.9995 | 0.014051048 | 2.0397 |
| | | 1/40 | 0.006722651 | 2.0017 | 0.003461966 | 2.0210 |
| | | 1/80 | 0.001679095 | 2.0013 | 0.000859057 | 2.0108 |
| | | 1/160 | 0.000419534 | 2.0008 | 0.000213958 | 2.0054 |
| | 2 | 1/10 | 0.078270796 | | 0.038907159 | |
| | | 1/20 | 0.019141843 | 2.0317 | 0.009238178 | 2.0744 |
| | | 1/40 | 0.004721898 | 2.0193 | 0.002248148 | 2.0389 |
| | | 1/80 | 0.001171855 | 2.0106 | 0.000554370 | 2.0198 |
| | | 1/160 | 0.000291838 | 2.0056 | 0.000137638 | 2.0100 |

# 4 Conclusion

This work is devoted to the study of boundary value problems for the generalized integro-differential modified moisture transfer equation with the Bessel operator, as well as difference schemes that approximate these problems on uniform grids. The case of an equation is considered in which the coefficients of the highest derivatives are separated from zero by a positive constant. For the solution of boundary value problems, a priori estimates are obtained in the differential and difference interpretations, which implies the uniqueness and stability of the solution with respect to the initial data and the right-hand side, as well as the convergence of the solution of the difference problem to the solution of the differential problem with a rate of $O\left(h^2 + \tau^2\right)$.

**Remark.** The results obtained in this paper are also valid in the case when we consider an equation with a nonlocal linear source of the form

$$
\partial_{0t}^{\alpha} u = \frac{1}{x^m} \frac{\partial}{\partial x} \left( x^m k(x, t) \frac{\partial u}{\partial x} \right) + \frac{1}{x^m} \partial_{0t}^{\alpha} \frac{\partial}{\partial x} \left( x^m \eta(x) \frac{\partial u}{\partial x} \right)
$$
$$
+ r(x, t) \frac{\partial u}{\partial x} - \int_0^x p(s, t) u(s, t) ds + f(x, t).
$$

# References

1. Barenblatt GI, Zheltov YuP, Kochina IN (1960) Basic concepts in the theory of seepage of homogeneous fluids in fissurized rocks. J Appl Math Mech 24(5):1286–1303
2. Trubacheva GA, Nerpin SV, Rymshin OV (1988) Theory and mathematical modeling of evaporation and condensation of moisture in the soil. Soviet Soil Sci 20(4):118–128
3. Nerpin SV (1975) Thermodynamic and rheological peculiarities of soil water and their role in energy - and mass transfer. In: Seminar on heat and mass transfer in the environment of vegetation, heat and mass transfer in the biosphere, pp 49–65
4. Dinariev OY (1990) Filtration in fractured medium with fractal geometry of fractures. Doklady AN, MZhG 5:66–70
5. Kochubei AN (1990) Diffusion of fractional order. Differ Eq 26(4):485–492
6. Nigmatullin RR (1986) The realization of the generalized transfer equation in a medium with fractal geometry. Phys Status Solidi 133(1):425–430
7. Samarskiy AA (1983) Teoriya raznostnykh skhem. [Theory of difference schemes]. Nauka, Moscow
8. Alikhanov AA (2010) A priori estimates for solutions of boundary value problems for fractional-order equations. Differ Eq 46(5):658–664
9. Alikhanov AA (2015) A new difference scheme for the time fractional diffusion equation. J Comput Phys 280:424–438
10. Beshtokov MK (2019) Nonlocal boundary value problems for Sobolev-type fractional equations and grid methods for solving them. Sib Adv Math 29(1):1–21
11. Samarskii AA, Gulin AV (1983) Stability of Difference Schemes. Nauka, Moscow

# Modeling Hyperchaotic Datasets for Neural Networks

**Egor Shiriaev**⏺, **Ekaterina Bezuglova**⏺, **Nikolai Kucherov**⏺, **and Georgii Valuev**⏺

**Abstract** This work is aimed at the studies related to neurocryptography. The paper represents the studies of hyperchaotic mappings and their construction based on the attractors and the research of image noise characteristics using the attractors and their performance. The conducted experiments have demonstrated that Liapunov hyperchaos generator possesses the best performance ratio and noise characteristics. In prospect we are going to conduct the experiments with a compiled data set and neural networks focused on the work with chaotic models and cryptographic algorithms.

**Keywords** Chaos theory · Hyperchaos · Neurocryptography · Lorentz attractor · Ressler attractor · Saito generator · Liapunov generator

## 1 Introduction

There are a number of areas in cryptography which are practically applied and rather credible in the scientific community. However, there are also many relatively new and promising research areas which do not still have practical application. One of these areas is neurocryptography [1]. This field uses various stochastic methods and algorithms, for example, neural networks used either as encoders and decoders, or in cryptanalysis. The most suitable application area is continuous key generation.

The first theoretical result in this field was obtained by Sebastian Dorlens in 1995 [2]. His experiment implied the neural networks application to cryptanalysis, namely,

E. Shiriaev (✉)
North-Caucasus Federal University, Stavropol, Russia
e-mail: eshiriaev@ncfu.ru

E. Bezuglova · N. Kucherov · G. Valuev
North Caucasus Center for Mathematical Research, North-Caucasus Federal University, 355017 Stavropol, Russia
e-mail: nkucherov@ncfu.ru

G. Valuev
e-mail: mail@gvvaluev.ru

© The Author(s), under exclusive license to Springer Nature Switzerland AG 2022
A. Tchernykh et al. (eds.), *Mathematics and its Applications in New Computer Systems*,
Lecture Notes in Networks and Systems 424,
https://doi.org/10.1007/978-3-030-97020-8_40

in training on inversion of S-permutation to the Data Encryption Standard (DES). As a result, 50% of the key was obtained. Khalil Shabab also received a public key encryption protocol based on a multi-level neural network and learning by a back propagation algorithm [3].

Many encryption algorithms additionally "noise" information, preventing the usage of methods of brute force excess, etc. There is also a so-called "chaotic communication channel" where data transmission security is ensured by generating a chaotic signal overlaid on the information, and decoding consists in signal filtering. There is a work [4] in this field where the authors use a convolutional neural network as a decoder and transmit a signal and a key to it enabling to filter the signal. The use of neural networks with chaotic data has some general traits with cryptography methods. From there, it can be concluded that a neural network capable to process chaotic information with a key can be applied cryptography tasks. Thus, this work can be attributed to neurocryptography and it proves that a neural network can be used as an encoder and decoder. In fact, the encrypted information is a random set of data for an observer, which can be decrypted if there is a key and algorithm knowledge. So, in this case, the neural network accepts a random set of input data, determines the encrypted information and decrypts it. However it is necessary to consider the fact that a large data set is required for training and testing in order to implement large-scale research in this field. Signal generation is a quite primitive type of data creation and transmission. A big number of modern neural network models are aimed at image processing, for example as the dataset as the Modified National Institute of Standards and Technology (MNIST) [22], i.e., the neural networks need the large sets of image data.

Such data sets creation means large expenditure of resources, it should be noted that a truly random number generator creation is one of the major problems of modern computer mathematics. The task is to determine the most effective method of creation of data sets with a large noise component, which filtering requires the same or more time compared with modern ciphers decryption. Thus, this paper implies the study of the methods and algorithms for modeling hyperhaotic functions and superimposing them on images.

## 2    Materials and Methods

### 2.1    Related Works

Researchers have been creating mathematical models with chaos for several decades. In this section we will consider the most interesting works devoted to hyperchaos investigation.

In the previous section, we have mentioned the work [4] devoted to a convolutional neural network construction based on a hyperchaotic signal. According to the authors, such a neural network can be part of a secure communication channel, which security is based on hyperchaos use, which is a random signal for a third party (a user). However, if there is a hyperchaos key, it is possible to filter the signal and obtain its true value. The authors propose to use a convolutional neural network as a filter. They provide hyperchaos by using Saito hysteresis chaos generator. The authors claim that the restored signal will be true by additional filtering. In general, this method can be used to generate a dataset, but no performance studies have been found.

The authors investigate a neural network in [5], used to build a 4-D hyperchaos. The authors justify the research by the fact that their developed neural network allows calculating and visualizing a four-dimensional hyperchaos in real time. The authors consider their model as a contribution to cryptography research, since continuous chaotic models can help ensuring data security and confidentiality. In the frameworks of the research implemented in this paper we interested in the four-dimensional hyperchaos taken as the basis of our research.

The studies of possibility to ensure security in a communication channel by encoding information with hyperchaotic attractors have provided the major fame for chaos theory. Our study considers this method, but we investigate the image noise to create a dataset. Pham V. T. [7] investigates the images encoding for transmission in a communication channel by using hyperchaos. He developed a five-dimensional neural network based on a five-dimensional hyperchaos. Their research has shown that by knowing the parameters for a system of equations and using them as a secret key, it is possible to encrypt and decrypt images with high accuracy.

Taking into consideration that the four-dimensional chaos was used in [6] and the information was subjected to additional filtering, we can conclude that the research in this field have made a reasonable progress in the last 20 years. Considering also such works as [7–9], we do not see scientific and practical interest in additional research on the obtained data set decryption. The hyperchaos attractors considered in this paper in the works above have the possibility of full compensation of the noise superimposed on the images in the presence of a secret key (the parameters of equation systems). Based on the situation described above, our work will investigate the models performance and image noise degree depending on the source data.

## 2.2 Hyperchaotic Models

The founder of chaos theory is a meteorologist Edward Norton Lorenz. He introduced the basic concepts of chaos theory in [10]. From a mathematical perspective, this theory describes the non-linear dynamical systems as absolute determinism, i.e. modeling and prediction of real processes. Lorenz justified his theory from the side of meteorology and weather prediction. He considered atmospheric flows as a deterministic system, but during his research he found that the slightest errors in the input data had a great influence on the output ones, i.e. with an error of 0.001% the result differed significantly. Such systems are called chaotic and the attractors are used to describe them. Now we are going to consider Lorentz attractor in details in the next section.

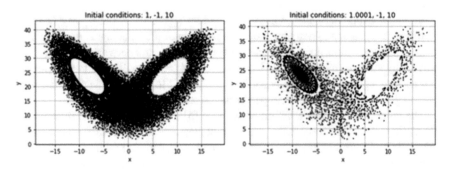

**Fig. 1** Lorentz attractor

## 2.3 Lorenz Attractor

Lorenz used Navier-Stokes system of equations to obtain a non-linear autonomous system of ordinary differential equations of the third order applied to dynamical systems description. With the use of this equation, he described air flows movement in a flat liquid layer of constant thickness when the flow velocity and temperature are decomposed into double Fourier series with subsequent truncation to the first and second harmonics. He obtained the following expression:

$$\begin{cases} \dot{x} = s(y - x), \\ \dot{y} = rx - y - xz \\ \dot{z} = xy - bz \end{cases} \tag{1}$$

where $s, r$ and $b$ – the system parameters.

Then, taking the values $s = 10, r = 28$ and $b = \frac{8}{3}$ it is possible to construct Lorentz attractor (Fig. 1) describing the pendulum movement from side to side. Based on the Lorentz attractor, we can consider the main consequence of chaos theory: with the slightest change in the input data, the output data undergo significant changes.

As we can observe, an error of 0.1% makes big changes to the data on the attractor. This makes it possible to "noise" the information more efficiently and complicate the ordering of this system.

Despite a rather simple mapping, this attractor is interesting because it represents a real oscillatory system simulation. It can be used as noise for an image required for the study.

## 2.4 Ressler Attractor

Ressler attractor is the first hyperchaotic attractor [11]. Such attractor can be constructed on the basis of a system of Ressler differential equations:

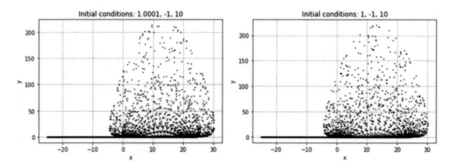

**Fig. 2** Ressler attractor

$$
\begin{cases}
\dot{x} = -y - z \\
\dot{y} = x + \alpha y \\
\dot{z} = b + z(x - c)
\end{cases}
, \tag{2}
$$

where $\alpha, b, c$ are the constants greater than zero. The value of these constants determines the attractor behavior. The system of equations has a stable limit cycle at $\alpha = 0.2, b = 0.2$ and $2.6 \le c \le 4.2$. However, when $c > 4.2$, a chaotic attractor arises (Fig. 2), upon its analysis it can be found that the phase space is filled with limit cycles lines with many trajectories possessing fractal properties.

Further, we will examine the hyperchaos models of greater dimensionality.

## 2.5  Saito Attractor

The following hyperchaos model can be considered in the light of Chua generator:

$$
\begin{cases}
\tau\dot{x} = -x_j + \sum_{k} a_{jk} y_{jk} + \sum_{l} s_{jl} x_l + i_j \\
y_j = f(x_j) = 0.5 * \left( \left| x_j + 1 \right| - \left| x_j - 1 \right| \right)
\end{cases} \tag{3}
$$

where $\tau$ is a time constant.

There is an explanation in [4]; on its basis this generator can be transformed into a system of Saito equations describing a hyperchaotic model:

$$
\begin{cases}
\dot{x} = -z - w \\
\dot{y} = \gamma(2\delta y + z) \\
\dot{z} = \rho(x - y) \\
\dot{w} = \frac{x - h(w)}{\varepsilon}
\end{cases} \tag{4}
$$

where $h(w) = w - (|w + 1| - |w - 1|)$, $\dot{x}, \dot{y}, \dot{z}$ – state variables, a $\gamma, \delta, \rho$ and $\varepsilon$ is the system parameters.

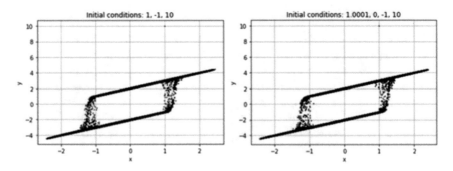

**Fig. 3** Saito attractor

Let us consider the constructed attractor based on this hyperchaotic system of equations (Fig. 3).

This attractor can be viewed from different dimensions, but the most informative pattern in the phase space is based on 1 and 4 spaces. This attractor has 2 static elements, namely, symmetrical segments. Chaotic points appear between the segments. Next, we are going to examine a five-dimensional attractor.

## 2.6 Liapunov Attractor

Liapunov attractor [12] is a 4-D fourteen-membered hyperchaotic system with one real equilibrium point. This attractor is the most hyperchaotic and complex compared to the other 4-D hyperchaotic attractors. Its range of parameters is quite large. A detailed theoretical and numerical system analysis is presented in [21] with the use of the hyperchaotic attractor, Poincare mapping and Liapunov frequency spectrum. In addition, the following features associated with this attractor can be distinguished:

1. This attractor has only one real equilibrium point, however, it is possible to consider such works as [14–16, 18–20] where the interactions with more than one equilibrium point are presented.
2. If we examine the following works [14–21], Liapunov attractor is more hyperchaotic.

It is possible to investigate this system other properties in the work [21]. Liapunov attractor is based on the following system:

$$
\begin{cases}
\dot{x}_1 = -dx_1 + (d-1)x_2 - x_3 + x_4 \\
\dot{x}_2 = x_1(r - 10x_3) + ax_2 + x_4 \\
\dot{x}_3 = x_1(10x_2 + x_3 - c) - bx_3 \\
\dot{x}_4 = px_1 + px_2
\end{cases}
\tag{5}
$$

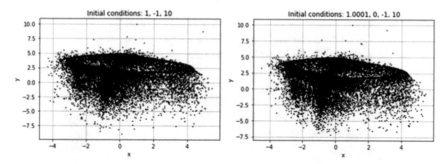

**Fig. 4** Liapunov attractor

where $\dot{x}_1, \dot{x}_2, \dot{x}_3, \dot{x}_4$ – state variables, a $a, b, c, d, r$ – positive parameters of the system (5). The attractor reaches the hyperchaos condition at $p < 0$.

Considering the range of parameters and the work results [17], we conclude, that, with Liapunov values $d = 55, r = 46.6, a = 12, b = 6, c = 11, p = -4$ the attractor reaches the highest possible hyperchaos. Let us pay attention to the attractor picture constructed with theses parameters (Fig. 4).

Considering the attractor in 1 and 4 dimensions, we notice that the majority of the points are constructed in the form of a ring. There are random points around the ring, representing hyperchaos. The next section is devoted to these attractors superposition on the image and the performance of hyperchaos construction and its superposition on information.

## 3 Results

We have used a color (four-channel) image of 128 by 128 pixels (Fig. 5) for our research.

By superimposing attractors on images with standard parameters, the following images can be obtained (Fig. 6). This noise is performed with standard parameters with the time characteristic equal to 1000. When observing the illustrations, it can be noticed that the greatest granularity is achieved with Liapunov attractor. Saito attractor also distorts the image, however, with less intensity. While considering Lorentz and Ressler attractors, it was found out that their equations systems are 3-D, i.e., we can only expose 3 of 4 image channels with their help, so 1 channel will remain clean, what simplifies image filtering.

Let us implement a more detailed study by constructing an attractor with different time characteristics (Fig. 7). Considering the study results, we can conclude the following. The major image grain size is achieved during an attractor construction based on 10,000 time points. However, the white spectrum fills almost the whole image with such a time characteristic. If we examine the Fig. 7 a–d.a, we can observe that noise occurs only in the upper part, and the rest of the image is smoothed into

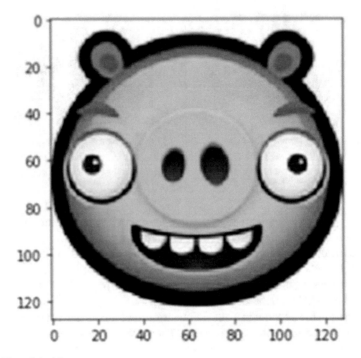

**Fig. 5** The original image

the white spectrum, which prevents further processing. With the time characteristic equal to 10, the image is noised insufficiently and its shape can be distinguished. With the characteristics equal to 100 and 1000, the image also reaches granularity. The characteristic equal to 1000 provides the best image noising, because, on the one hand, it gives high granularity, and, on the other hand, the white spectrum does not prevail over the rest ones, what allows filtering the image out. With the characteristic of 10,000, the image becomes almost completely white, i.e. indistinguishable.

When comparing the attractors, it can be concluded that Lorentz and Ressler attractors provide the worst noise for color four-channel images. When constructing attractors with the characteristic of 10,000, too much impact occurs, what makes an image almost completely indistinguishable. Then, we will examine the performance of attractor and data set calculation (Table 1).

Let us examine an illustration in the form of a graph with the results obtained (Fig. 8). When observing the result, it can be noticed that Liapunov attractor has the lowest performance, what is explained by the fact that the operations on 14 terms are necessary to calculate the attractor. Despite the fact that this attractor gives the best noise characteristic, its calculations performance is reasonably low. Saito attractor possesses the performance improved more than twice compared to Liapunov attractor, but it has a high noise characteristic and granularity.

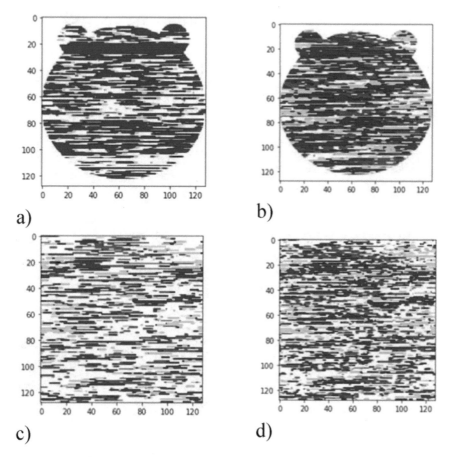

**Fig. 6** Image noising with the use of: **a** Lorentz attractor **b** Ressler attractor **c** Saito attractor **d** Liapunov attractor

Let us consider the performance of noise superimposition operation (Table 2, Fig. 9).

Examining a dataset calculation performance, the following situation is highlighted: one can quickly build a dataset consisting of 1000 elements with any time characteristic with the help of Liapunov hyperchaos attractor. It is reasonably equal in performance comparing to Saito attractor.

In summary, the study results have demonstrated that the hyperchaos generator, based on Liapunov equation system, is the most suitable for constructing datasets used in training and testing neural networks in neurocryptography with a time characteristic equal to 1000. But it is to take into account that the attractor will be calculated at the pre-calculation stage.

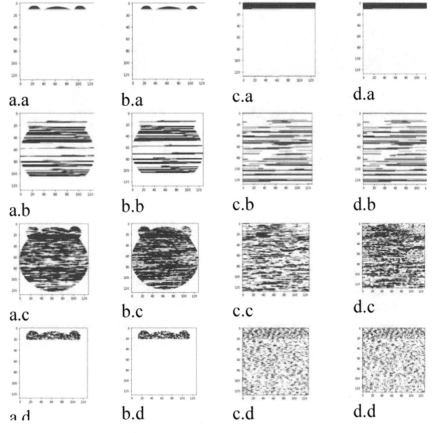

**Fig. 7** Image noising with the help of a.x – Lorentz attractor, b.x – Ressler attractor, c.x – Saito attractor, d.x – Liapunov attractor. At time characteristics x.a – from 0 to 10, x.b – from 0 to 100, x.c – from 0 to 1000, x.d – from 0 to 10,000

**Table 1** The results of modeling the attractors construction with different time characteristic, sec

| Attractors | Time characteristic | | | |
|------------|------|------|------|--------|
|            | 10   | 100  | 1000 | 10,000 |
| Lorenz     | 0.013 | 0.127 | 1.305 | 4.460 |
| Rossler    | 0.003 | 0.036 | 0.349 | 3.630 |
| Saito      | 0.023 | 0.256 | 2.617 | 26.957 |
| Lyapunov   | 0.062 | 0.602 | 5.957 | 60.387 |

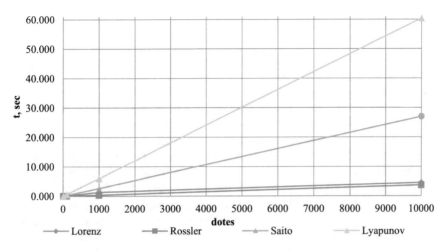

**Fig. 8** The graph of construction performance of attractors

**Table 2** The results of modeling datasets for constructing attractors with different time characteristics, sec

| Time characteristic | | | | |
| --- | --- | --- | --- | --- |
| | 10 | 100 | 1000 | 10,000 |
| Saito | 0.29136 | 0.29122 | 0.29061 | 0.29118 |
| Liapunov | 0.29139 | 0.29119 | 0.29117 | 0.29126 |

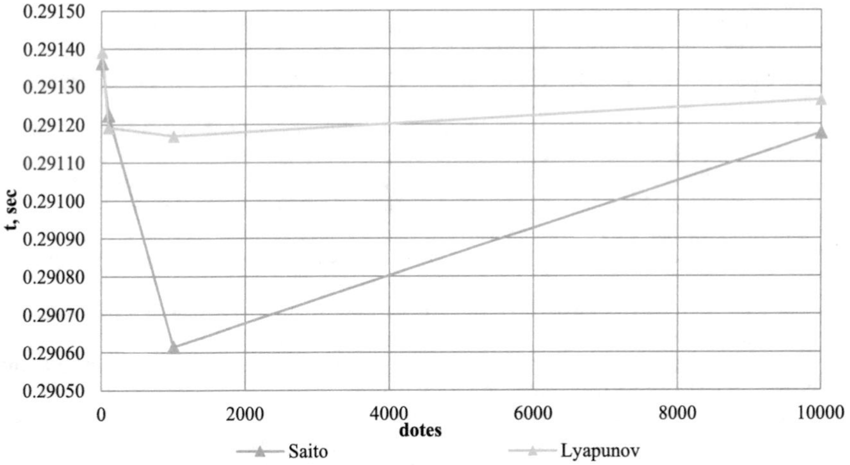

**Fig. 9** Graph of construction performance of data set

# 4 Conclusion

A number of studies related to chaos theory have been implemented in this paper, in particular, the studies with chaotic and hyperchaotic attractors. We have examined the attractors constructed in 3 and 4 dimensions. The use of these attractors for datasets creation allows creating the datasets for training the neural networks used in the field of neurocryptography. Our work comprises a comparative study evaluating the performance of attractor constructions and image noise. It was found in the course of the study that Liapunov attractor has the best noise characteristic, but low performance. When comparing this hyperchaos generator with the other ones, it was determined that Liapunov attractor has the best ratio of performance and noise characteristics. In this regard, this attractor is recommended for a dataset creation.

In the future, we are planning to conduct experiments with a dataset made on the basis of Liapunov attractor and neural networks used for the work with hyperchaotic models and encryption systems.

**Acknowledgements** The work is supported by North-Caucasus Center for Mathematical Research under the agreement No 075-02-2021-1749 with the Ministry of Science and Higher Education of the Russian Federation and the part study was funded by RFBR, the project number is 20-37-70023.

# References

1. Tirumala SS, Narayanan A (2017) Transpositional neurocryptography using deep learning. In:Proceedings of the 2017 international conference on information technology
2. Dourlens S (1995) The first definition of the Neuro-Cryptography (AI Neural-Cryptography) applied to DES cryptanalysis – 1995, France
3. Shihab K (2006) A backpropagation neural network for computer network security. J Comput Sci 2(9):710–715
4. Caponetto R et al (1998) Hyperchaotic dynamic generation via SC-CNNs for secure transmission applications. In: 1998 IEEE international joint conference on neural networks proceedings. IEEE world congress on computational intelligence (Cat. No. 98CH36227), vol 1. IEEE
5. Koyuncu I et al (2020) Artificial neural network-based 4-d hyper-chaotic system on field programmable gate array. Int J Intell Syst Appl Eng 8(2):102–108
6. Pham VT et al (2014) Hidden hyperchaotic attractor in a novel simple memristive neural network. Optoelectron Adv Mater Rapid Commun 8(11–12):1157–1163
7. Yang F et al (2020) An image encryption algorithm based on BP neural network and hyperchaotic system. China Commun 17(5):21–28
8. Mohammadzadeh A, Ghaemi S (2017) Synchronization of uncertain fractional-order hyperchaotic systems by using a new self-evolving non-singleton type-2 fuzzy neural network and its application to secure communication. Nonlinear Dyn 88(1):1–19
9. Locquet A (2006) Chaotic optical communications using delayed feedback systems. Georgia Institute of Technology
10. Lorenz EN (1963) Deterministic nonperiodic flow. J Atmos Sci 20:130–141
11. Rossler OE (1979) (1979) Continuous chaos - four prototype equations. Ann N Y Acad Sci 316(1):376–392
12. Fang J, Deng W, Wu Y, Ding G (2014) A novel hyperchaotic system and its circuit implementation. Optik 125:6305–6311

13. Feng C, Cai L, Kang Q, Wang S, Zhang H (2015) Novel hyperchaotic system and its circuit implementation. J Comput Nonlinear Dyn 10:061012
14. Wang FQ, Liu CX (2006) Hyperchaos evolved from the Liu chaotic system. Chin Phys 15:963–968
15. Li Y, Tang WKS, Chen G (2005) Generating hyperchaos via state feedback control. Int J Bifurcat Chaos 5(10):3367–3375
16. Bao BC, Liu Z (2008) A hyperchaotic attractor coined from chaotic Lu system. Chin Phys Lett 25(7):2396–2401
17. Qiang J (2007) Hyperchaos generated from the Lorenz chaotic system and its control. Phys Lett A 366:217–222
18. Chen A, Lu J, Lu J, Yu S (2006) Generating hyperchaotic Lu attractor via state feedback control. Phys A 364:103–110
19. Chen Y, Yang Q (2015) A new Lorenz-type hyperchaotic system with a curve of equilibria. Math Comput Simul 112:40–55
20. Li C-L, Xiong J-B, Li W (2014) A new hyperchaotic system and its generalized synchronization. Optik 125:575–579
21. Singh JP, Roy BK (2015) Analysis of an one equilibrium novel hyperchaotic system and its circuit validation. Int J Control Theor Appl 8(3):1015–1023
22. LeCun Y (1998) The MNIST database of handwritten digits. http://yann.lecun.com/exdb/mnist/

# Neural Network Analysis for Image Classification

**Vershkov Nikolay**📷, **Babenko Mikhail**📷, **Kuchukov Viktor**📷,
and **Kuchukova Natalia**📷

**Abstract** The article considers the possibility of modeling artificial neural networks using the mathematical apparatus of information theory. The issues of pattern recognition, classification and clustering of images using neural networks are represented by two main architectures: a direct distribution network and convolutional networks. The possibility of using orthogonal transformations to increase the efficiency of neural networks, the use of wavelet transformations in convolutional networks is investigated. Based on the theoretical studies carried out, the directions on practical application of the obtained results are proposed.

**Keywords** Neural networks · Convolutional neural networks · Sub-band coding · Sub-band filtering · Wave model

## 1  Introduction

Artificial neural networks (ANN) are built following their biological analogues' arrangement and functioning principles [1]. The work, which is rightfully considered fundamental in this field, was written in 1943 by W. McCulloch and Y. Pitts [2], and it focuses on the brain functioning theory. In more than 80 years of progress, the ANN theory has gained significant potential both in terms of architecture and teaching methods. However, it is to be noted that the evolution of ANN is more of intuitive & algorithmic nature rather than of mathematical. Many ANN architectures have been transferred from the biological field of research [3]. A significant breakthrough in

V. Nikolay (✉) · B. Mikhail · K. Viktor
North Caucasus Center for Mathematical Research NCFU, Stavropol, Russia
e-mail: vernick61@yandex.ru

K. Viktor
e-mail: vkuchukov@ncfu.ru

K. Natalia
North-Caucasus Federal University, Stavropol, Russia
e-mail: nkuchukova@ncfu.ru

© The Author(s), under exclusive license to Springer Nature Switzerland AG 2022     455
A. Tchernykh et al. (eds.), *Mathematics and its Applications in New Computer Systems*,
Lecture Notes in Networks and Systems 424,
https://doi.org/10.1007/978-3-030-97020-8_41

mathematical description of ANN were respective works by Kolmogorov-Arnold [4, 5] and Hecht-Nielsen [6, 7].

During that, if viewed from the point of view of information theory, ANN research has been done quite rarely. In his classical work, Claude Shannon [8] outlined the basics of information theory. Shannon's significant contribution was the possibility of information quantitative measurement, the possibility of matching the signal source with the communication channel through information optimal encoding, as well as the effect that coding redundancy has on the ability to detect errors and correct them. ANN studies, which are basically information processing systems, rely on the mathematical apparatus of information theory to study self-organizing models [3].

A number of works completed by the authors of this article, while focusing on information processes in direct distribution ANN [9–11] have shown possible advantage gained in terms of learning speed, reducing redundancy and energy intensity when taking into account the specifics of information processing in neural networks. The unique ability of the neuron mathematical model that allows performing various transformations in the ANN layers suggests that input information processing could be viewed from the angle of information theory. The classical approach (the McCulloch-Pitts model [2]) views the mathematical model of a neuron as

$$y_{k,l} = f\left(\sum_{i=1}^{n} w_i^{k,l} x_i^{k,l}\right), \tag{1}$$

where $k$, $l$ is the number of the layer and the number of the neuron in the layer; $y_{k,l}$ is the neuron output, $x_i^{k,l}$ are neuron inputs, $w_i^{k,l}$ are the weights (synapses) of incoming signals, $f$ is the output function of the neuron, which can be either linear or nonlinear.

A number of linear transformations in information theory feature a similar structure: orthogonal transformations, convolution, correlation, filtering in the frequency domain, etc. Studies [9–11] focused on solving specific problems: finding the optimal loss function, nonlinear features of a neuron, optimizing the volume of a neural network, etc.

The purpose of this article is to study neural networks in order to carry out image processing from the information theory stance, as well as to develop general rules and principles for building an ANN for solving specific problems. The study is completely theoretical, and this article pursues no task of experimental confirmation of certain provisions that the authors obtained through employing the mathematical apparatus of information theory.

## 2  Materials and Methods

### 2.1  Wave Model of Feed-Forward ANN

In general, the information model of a feed-forward ANN [9] may look as follows: the ANN input gets multidimensional vector values of $X_i = \{x_1^i, x_2^i, \ldots, x_n^i\}$, which, within the proposed model, can represent – discretized in time and level – values of some input function $x(t)$. The ANN processes the input value of $X_i$ subject to the expression (1) in each neuron of each layer, and at the output offers discrete values $Y_i = \{y_1^i, y_2^i, \ldots, y_m^i\}$, obtained through discretizing a certain function of $y(t)$, which is completely known at ANN supervised learning. Discretizing of the functions $x(t)$ and $y(t)$ is carried out following the requirements of the sampling theorem [12], which is more commonly known abroad as Nyquist criterion. It is also to be noted that the $\{X_i\}_{i=1,2,\ldots,n}$ set is not complete, i.e. when solving an applied problem, classification, for instance, there may values at the input, which are not included in the ANN training alphabet. This is the fundamental difference between decoding in the ANN and the receiving path of the data reception-transmission channel as described by Shennon [8]. In Shennon's works, the alphabet of the transmitted discrete messages is finite and set in advance. It will also be assumed that the set of weights in all the ANN neurons is filled randomly prior to the start of learning.

ANN may contain an input layer (which is typically not involved data processing) of the $X_i$ capacity, an output layer (where the number of neurons is equal to the $Y_i$ capacity), as well as one or more hidden layers. Since ANN can perform various functions (e.g. classification of images, clustering, approximation of functions, etc.), then we will view the operation of the network in a sphere that is the most convenient to understand the functions in question.

The ANN output layer performs a crucial function in classification, clustering, etc., which means that it is in the output layer where the problem of the input signal assigning to the appropriate class (cluster) is solved. Drawing an analogy with the receiving path of the information transmission system, it can be said that here is the task of the $X_i$ observation to see which if the $Y_i$ was transmitted [13]. Just like in an information transmission system, an ANN has interferences. Their nature, however, is determined not by the connection channel yet by the set of input information, meant for classification (clustering). Given that, a mathematical description of an ANN can be performed by the $p[x(t)|y(t)]$ transition probability, which describes the probability of the obtained representation being transformed by the ANN into a respective class or cluster. Just like in the communication theory, a data model with additive white Gaussian noise can be employed as a mathematical model [14]. This model is appropriate for a significant amount of data in the network (in the MNIST database [15], for instance, the training set contains 60,000 records). The transition probability, then, decreases exponentially along with the growth of the Euclidean distance square $d^2(x, y)$ between the obtained $X_i$ value and the "ideal representation" of the $Y_i$ class:

$$p[x(t)|y(t)] = k\exp\left(-\frac{1}{N_0}d^2(x, y)\right),\tag{2}$$

where k is the coefficient not depending of $x(t)$ and $y(t)$, $N_0$ is the spectral density of the noise, while

$$d^2(x, y) = \int_0^T [x(t) - y(t)]^2 dt\tag{3}$$

If, in problems of approximation, forecasting, etc. signals $x(t)$ and $y(t)$ are similar by their period, then within the studied classes of problems they are different. When solving the problem of the MNIST database image classification, the input vector contains 784 values (pixels), while the number of image classes is 10. To solve such problems, an unambiguous correspondence $Y_i \leftrightarrow \tilde{X}_i$ must be established. In other words, the $X_i$ observation is compared to the "ideal representation" of the $\tilde{X}_i$ class, and in case the match, the decision is made that the $X_i$ observation belongs to the $Y_i$ class, etc.

$$(X_i \in Y_i) = \min_j d^2(X_i, \tilde{X}_j)\tag{4}$$

Here we will open the brackets in expression (3) and use the representation $\tilde{x}(t)$ instead of $y$:

$$d^2(x, \tilde{x}) = \int_0^T x(t)^2 dt - 2\int_0^T x(t)\tilde{x}(t)dt + \int_0^T \tilde{x}(t)^2 dt = x^2 - 2z + \tilde{x}^2\tag{5}$$

The values $\|x\|^2$ and $\|\tilde{x}\|^2$ in the expression (5) are the energy of the input implementation and representation of the $y$ cluster, so the values are stable (in case of $x(t)$) normalizing, whereas $z$ is the following correlation

$$z = \int_0^T x(t)\tilde{x}(t)dt,\tag{6}$$

which is called mutual energy of two signals. In this case, the expression (4) in view of (5) and (6) may be represented as

$$(X_i \in Y_i) = \max_j z_j^i\tag{7}$$

Here $z_j^i$ is mutual correlation of the $X_i$ implementation and a representation of the $j$-th cluster of $\tilde{X}_j$. During that, cluster representations should be normalized so that there is no gain through an energetically more powerful signal (which is obvious from expression (5)). Should the input data has different lengths of signal vectors, such an arrangement shall be called volumetric packing. In this case, the average energy $\overline{E} = \frac{1}{n} \sum\limits_{i=1}^{n} E_i = const$ is fixed. However, in the event all signal vectors are of the same length, i.e. the ends of the vectors are located on a spherical surface, such an arrangement is called a spherical packing.

Returning to expression (1), on which all neurons function, it is to be noted that if random values are set as the weights of the output layer $W^{k,l}$, then the $W^{k,l}$ vector appears as a multiplicative interference, which increases the volume of packing. However, through learning, due to calculating the error function and further transforming it into the $W^{k,l}$ vector gradient, the weight values adopt some meaningful value defined by the expression (7). This type of operation is known in information theory as "matched filtering", whereas the ANN output layer, while getting optimized through the process of learning, comes to appear as a matched filter. The theory of information has the condition for obtaining the maximum response from a device having an impulse response [16]:

$$h(t) = kx(-t) \tag{8}$$

This means that the weights of the neuron defining the $Y_i$ class are to have a feature, which is conjugated by Hilbert with an ideal representation of the $\tilde{X}_i$ class. Therefore, if we set weights in each neuron of the output layer subject to expression (8) for each class, and use the output layer function as a $\max\limits_{i} Y_i$, function, then we will get a matched filter of $m$ dimension. However, given all the clarity of the proposed solution, there is a certain issue. The correlation integral (6) is known to as able to be represented both in a temporal format and in that of frequency:

$$z_i = \int x(t)\tilde{x}_i(t - \tau)d\tau = X(j\omega)\tilde{X}_i(j\omega) \tag{9}$$

Model (1), therefore, is no good for solving the problem of calculating the correlation function in a temporal format. To be more precise, if the $X$ and $\tilde{X}$ vectors are represented as a decomposition in the orthogonal basis (e.g., Fourier), then all products whose indices do not match, will turn into zero, the expression (1) to be obtained finally. Should the orthogonality condition be not met, then, in case operation (1) is performed, the correlation value (9) will be obtained with an error. This means that the classification error value will grow and may not correspond to the expected outcome.

Expression (9) makes it clear that the most successful solution will be the case where the inputs of the output layer are orthogonal vectors.

To find the conversion coefficients, a set of orthogonal $\{u_n(t)\} = \{u_1(t), u_2(t), \ldots, u_n(t)\}$ functions is used, where the following condition is met for each pair:

$$\int\limits_{0}^{T} u_i(t)u_j(t)dt = \begin{cases} a, & \forall i = j \\ 0, & \forall i \neq j \end{cases} \tag{10}$$

Conversion coefficients are not hard to identify as

$$c_j = \frac{1}{a} \int_0^T x(t) u_j(t) dt, \quad j = 1, 2, \ldots, m \tag{11}$$

Initially, expression (11) symbolized the transition from the $x(t)$ space of continuous images to a discrete space of classes (clusters). Further on, when passing to the digital processing of the images, the integral in expression (11) was converted into the sum of:

$$c_j = \frac{1}{a} \sum_{k=0}^{n-1} x_k u_j^k \tag{12}$$

Various types of orthogonal transformations for pattern recognition have been presented in detail in [17]. Orthogonal transformations are linear and offer a one-to-one correspondence of the input vector X and the output vector of the C coefficients. The output vector of the coefficients, then, will also be $n$-dimensional. Matching expressions (12) and (1) will make it easy to see that they are identical, so if we take the $u_j^k$ values as the $w_j^k$ weights, then the ANN layer can be seen as an orthogonal transformation, i.e., at the layer output there will be the $\{c_j\}$ values. Presenting the $\tilde{X}$ vector as a $\tilde{C}_x$ orthogonal transformation, we will get expression (9) as follows:

$$Z_i = \sum_{j=0}^{n-1} X_i \tilde{X}_i = \sum_{j=0}^{n-1} x_j \tilde{x}_j$$

Therefore, an orthogonal transformation will allow implementing an image recognition system based on a direct distribution ANN.

While studying the ANN wave model [9–11], the authors could see repeatedly that during the training process both models (standard and wave) revealed about the same classification error, yet within different training times. This can be accounted for by the fact that the standard learning algorithm (most commonly, the algorithm of error back propagation based on the gradient method) relies on the learning principle, which includes the modification of weights from the last layer to the first (error back propagation). In other words, the decomposition functions in the former layer are selected in view of the classification errors in the latter. The gradient's most important property used in ANN training is that it indicates the direction where a certain $f(x)$ function increases the most:

$$\nabla f(x) = \frac{df}{dx_1} e_1 + \frac{df}{dx_2} e_2 + \cdots + \frac{df}{dx_n} e_n \tag{13}$$

Here $E = (e_1, e_2, \ldots, e_n)$ is the error function vector. This means that the direction where the $f(x)$ function does not increase could be defined as opposite to the gradient. Therefore, the correction vector of the last layer weights is calculated from the error function, while the correction vector of the previous layer will be calculated based on the last layer errors. The result of this algorithm will be the selected values of the decomposition functions in the first hidden layer. Taking into account the nonlinear nature of the neuron's transfer function, however, we will get decomposition functions of a rather complex form, as predicted by V.I. Arnold [5].

The above examples show that an orthogonal transformation if used in an ANN will allow more efficient information processing as all operations of correlation, convolution, etc. are performed in the respective planes, while multiplication operations of elements featuring mismatched indexes (i.e., located in different planes) will be reset automatically due to the orthogonal transformation properties.

## 2.2 Convolutional ANN Wave Model

Classification (clustering) rarely uses direct distribution ANN. In this case, the architecture of convolutional networks is much more suitable. Convolutional Neural Network (CNN) is an architecture proposed by Yan LeCun in 1988, which was designed for efficient image recognition. A CNN contains one or more convolutional layers that perform a specific operation – convolution, while using a small kernel. The convolution operation, if applied, will reduce the image.

This is especially valuable for color images, when a 3D core allows obtaining one image at the output of the layer instead of 3. The convolutional layer is usually the first in the ANN structure, following which, pooling (subdiscretization) operations can be performed. However, this is not to be the focus now. Applying convolution with one core, we obtain a feature map that can be classified using the last layer of the direct propagation ANN.

Since the convolution integral shares a lot with the correlation integral (9), then for this operation, too, all the arguments will be good, which were listed in the previous section. The convolution integral, them, is much easier and more efficient to calculate if the input signal and the kernel are represented as an orthogonal transformation. And in turn, it appears reasonable to place a layer performing an orthogonal transformation as the first layer of the standard ANN.

In information theory, signal processing is commonly performed through a linear transformation. Sub-band encoding (known as sub-band filtering) is an individual case of linear transformation and features a significant number of useful properties for their application in the ANN theory. Encoders based on linear transformation are divided into 2 types – encoders with transformation and sub-band encoders [18]. The classical transformation of the first type is the Fourier transform, which decomposes the signal into sinusoidal components. The discrete cosine transform (DCT) and the Karhunen-Loeve transform (KLT) can serve as examples of the second type.

These transformations are found through calculating the convolution of a finite-length signal with a family of basis functions. This results in a number of coefficients, which are processed further. In practice, many of such transformations have efficient computational algorithms. A specific feature of the transformations is that they are typically applied to non-overlapping signal blocks [18].

Sub-band encoding is implemented via the signal convolution with several band-pass filters and the result decimation (thinning). Each signal resulting from the conversion bears information regarding the spectral component of the original signal on a certain spatial (temporal) scale. The most critical properties at encoding images include [18]:

- scale and orientation;
- spatial localization;
- orthogonality;
- fast algorithms of calculation.

In communication theory, orthogonality is not usually viewed within the context of sub-band coding, and orthogonal transformations are used to decorrelate signal samples. Fourier bases are localized by frequency, yet not in space. For signal encoding as described by the Gaussian process, this does not constitute a disadvantage. However, the images contain contours that cannot be described by this model and will take bases localized in space, for which reason filter blocks, being local as well as in space, ensure better decorrelation on average. The correlation between pixels decreases exponentially along with increasing distance, i.e.:

$$R_l = e^{-\omega_0 |\delta|},$$

where $\delta$ is a distance variable. The respective spectral power density is

$$\Phi_l(\omega) = \frac{2\omega_0}{\omega_0^2 + (2\pi\omega)^2} \tag{14}$$

Expression (14) makes it rather clear that obtaining flat spectrum segments will take accurate division of the spectrum at low frequencies, and rough – at high ones. The sub-bands resulting through this will be described by white noise with a dispersion proportional to the power spectrum within this said range.

The Fourier transform is known to have a significant disadvantage – obtain a single conversion coefficient takes all information regarding the behavior of the signal in time. Therefore, the signal time peak will spread all over the frequency domain of the Fourier transform. To get rid of this issue, the Short Time Fourier Transform is used:

$$\Phi_x(\omega, b) = \int x(t) e^{-j\omega t} w(t - b) dt \tag{15}$$

In this case, a time window appears in the description of the $w(t - b)$ conversion, due to which the transform gets time-dependent, i.e., there appears a time-frequency matrix of the signal [19]. In case the Gaussian function is selected as the window, the inverse transform can be carried out employing the same function.

A significant downside of expression (15) is the window size constancy, which will not allow it to be adjusted to the image landscape. If the Fourier transform window is substituted with the following wavelet transform:

$$\psi_{a,b}(t) = a^{-1/2}\psi\left(\frac{t-b}{a}\right),\tag{16}$$

then it is easy to see that the basic functions are, first, real, and second – they occupy a different position near the abscissa axis. The wavelets are defined on a short time period (shorter than the signal period), whereas the basis functions can be seen as scaled and time-shifted versions (16). Parameters $b$ and $a$ can be viewed as the position in time and the zoom value, respectively. In this case, the direct wavelet transform will appear as:

$$\Phi_x(a,b) = a^{-1/2}\int x(t)\psi\left(\frac{t-b}{a}\right)dt\tag{17}$$

The functioning of the ANN convolutional layer is known to imply calculating the convolution of the $X$ input signal block with the $J$ core that is $s \times s$ of size, i.e.,

$$C_{i,j} = \sum_{k=0}^{s-1}\sum_{l=0}^{s-1} X_{i+k,j+l}J_{k,l}\tag{18}$$

Converting expression (17) for sampled signals and functions and comparing it with (18), will produce an understanding of when the wavelet transform basic function plays the role of the convolutional layer core. This means that using several basic functions is equivalent to using several filters different in their core size. Here selection of window parameters adaptive to the signal appears possible.

The idea of using wavelet transformations in an ANN is far from a new option [22]. However, the design of such ANN relies on the idea of involving the wavelet transform in the first layer of the CNN direct propagation [23]. The idea appears much more attractive when the wavelet function performs the function of a convolutional layer core. Since a convolutional layer can function with several cores simultaneously, several approximations can be obtained in one layer.

When analyzing signals, communication theory often holds that it is useful to represent a signal as a set of its successive approximations. When transmitting an image, for instance, transmit a rough version of it could be done first, and then it can be refined sequentially. This transmission strategy features its benefits, i.e., when selecting images from a certain database, when a large number of images has to be viewed quickly. A similar approach can be employed for image recognition. When an image comparison with class standards in the roughest approximation reveals that it cannot be assigned to a certain class, then there is no need to carry out a comparison in a more accurate approximation. This approach is called multiscale analysis.

The multiscale analysis is understood as the description of the $L^2(R)$ space through hierarchical nested subspaces $V_m$, which do not intersect and which, if united, will produce in the limit of $L^2(R)$, i.e. $\cdots \cup V_2 \cup V_1 \cup V_0 \cup V_{-1} V_{-2} \cup \cdots$, $\bigcap_{m \in Z} V_m = \{0\}$, $\bigcup_{m \in Z} V_m = L^2(R)$. These spaces feature the following property: for any $f(x) \in V_m$ function its compressed version will belong to the $V_{m-1}$, т.е. $f(x) \in V_m \Leftrightarrow f(2x) \in V_{m-1}$ space. And, finally, there is the $\varphi(x) \in V_0$ function available, whose shifts $\varphi_{0,m}(x) = \varphi(x - m)$ will create an orthonormal basis of the $V_0$ space. Since the $\varphi_{0,m}(x)$ functions make up an orthonormal basis of the $V_0$ space, then the $\varphi_{n,m}(x) = 2^{-m/2}\varphi(2^{-m}x - n)$ functions make up an orthonormal basis of the $V_m$ space. These basis functions are called scaling functions because they create scaled versions of functions in $L^2(R)$ [18]. Hence follows that the $f(x)$ function in $L^2(R)$ may be represented by a set of its successive $f_m(x)$ in $V_m$ approximations.

Consequently, there appears a possibility to analyze the image at different resolution or scale levels. The $m$ value is called the scale factor or the analysis level. In case of large values of $m$, the approximation is quite rough, with no details available, yet it is possible to highlight larger generalizations, and as long as the scale factor is decreased, it allows identifying details.

In reality, $f_m(x)$ is an orthogonal projection of $f(x)$ onto $V_m$ [18], i.e.,

$$f_m(x) = \sum_n \varphi_{m,n}(x), \ f(x)\varphi_{m,n}(x) = \sum_n c_{m,n}\varphi_{m,n}(x) \qquad (19)$$

Without getting down to details of the wavelet analysis here, we shall note that any $f(x)$ function, which is defined in the $L^2(R)$ space, may be described as a sum of orthogonal projections. In case the function analysis is caried out up to a certain $m$ scale, then $f(x)$ will be represented by the sum of its rough approximation and a whole number of details. Such opportunities are offered, for instance, by the Haar family of wavelets [19].

As there were sub-band transformations considered as filtering with subsequent thinning at the beginning of the chapter, it is to be noted that there is a possibility of creating filter banks [18, 20]. If we take into account a two-band filter bank, then the low-frequency part of the filter approximates a function lacking details, while the high-frequency part does contain them. Depending on the processing task, the ANN can compare low-frequency approximation to highlight large and smooth surfaces or, vice versa, to match details in order to highlight the desired feature.

Therefore, wavelets used as the core of the CNN allows separating and detailing of the required image features. This approach is not a novelty within the theory of information transmission and processing. Employing it to build an information model of a convolutional neural network, though, allows not only better understanding the processes taking place during the development of a feature map, yet also building a lifting scheme for information processing in a multilayer CNN.

# 3 Conclusion

When working with images, regardless of the ANN architecture, orthogonal transformations offer an undeniable advantage. In a direct propagation ANN, orthogonal transformations enhance the efficiency of the network's last layer where the classification (clustering) of the image is performed. Orthogonalization helps improve the accuracy of the correlation integral calculating for the signal that is being classified, as well as for the ideal representation of the class.

In a CNN, just like in a direct propagation ANN, it is direct propagation networks that are used as the last layer (or several layers), whereas such networks are require for the classification of the feature map. The use of orthogonal transformation, like in direct propagation ANN, increases the efficiency of the last layer in charge of classification. Employing the Fourier transform (or similar ones), though, does not ensure a significant gain when it comes to analyzing the image details. In this case, it is of interest to use wavelet transformations, which are localized not only in frequency (such as the Short Time Fourier Transform), yet also in time. Wavelets can feature the property of orthogonal transformation; however, apart from that, they allow creating banks of filters and performing both general image analysis following the accepted criterion, and detail analysis. This approach will allow carrying out not general classification of images only (as, for instance, within the MNIST database, where only one image occupies the entire field), yet also comparing details, and classifying certain details of complex images.

The entire approach described above requires experimental proof, which means that the next step should imply analyzing the already existing wavelet transformations for CNN, and their application in convolutional layers, as well as detailing feature maps for their effective processing in the last layers.

**Acknowledgements** This work has been supported by the North-Caucasus Center for Mathematical Research subject to Agreement №. 075-02-2021-1749 with the Ministry of Science and Higher Education of the Russian Federation, while part of the study was funded by RFBR, Project Number 20-37-70023.

# References

1. Kruglov VV, Borisov VV (2001) Artificial neural networks. Theory and practice. Hotline-telecom, p 382
2. McCulloch WS, Pitts Y (1956) A logical calculus of ideas related to nervous activity. In: Shannon CE, McCarthy J (ed) Automata. Publishing House of Foreign Literature, pp 363–384
3. Khaikin S (2008) Neural networks: a complete course, 2nd edn. Williams Publishing House
4. Kolmogorov AN (1957) On the representation of continuous functions of many variables by superposition of continuous functions of one variable and addition. Dokl Akad Nauk SSSR 114(5):953–956
5. Arnold VI (1958) On the representation of functions of several variables in the form of a superposition of functions of a smaller number of variables. Math Educ 3:41–61

6.  Hecht-Nielsen R (1990) Neurocomputing. Addison-Wesely Publishing Company
7.  Hornik K, Stinchcombe M, White H (1989) Multilayer feedforward networks are universal approximators. Neural Netw 2(5):359–366
8.  Shannon K (1963) Works on information theory and cybernetics. In: Dobrushin RL, Lupanov OB (eds) Per. with eng. Publishing House Foreign Literature
9.  Vershkov NA, Kuchukov VA, Kuchukova NN, Babenko M (January 2020). The wave model of artificial neural network. In 2020 IEEE conference of Russian young researchers in electrical and electronic engineering (EIConRus). IEEE, pp 542–547
10. Vershkov NA, Babenko MG, Kuchukov VA, Kuchukova NN (2021) Advanced supervised learning in multi-layer perceptrons to the recognition tasks based on correlation indicator. Proc Inst Syst Program RAS (Proc ISP RAS) 33(1):33–46
11. Vershkov NN, Kuchukov VA, Kuchukova NN (2019) The theoretical approach to the search for a global extremum in the training of neural networks. Proc Inst Syst Program RAS 31(2):41–52
12. Kotelnikov VA (1956) The theory of potential noise immunity. Radio and Communication, St. Petersburg, USSR
13. Kharkevich AA (1972) Selected works. Information theory. Recognition of images. T.3. Science, Moscow, USSR
14. Ipatov V (2007) Broadband systems and code division of signals. Principles and applications. Technosphere, Moscow, Russia
15. LeCun Y (1998) The MNIST database of handwritten digits. http://yann.lecun.com/exdb/mnist/
16. Cook C (2012) Radar signals: an introduction to theory and application. Elsevier
17. Ahmed N, Rao KR (2012) Orthogonal transforms for digital signal processing. Springer, Heidelberg
18. Vorobiev VI, Gribunin VG (1999) Theory and practice of wavelet transform. Military University of Communications, Saint Petersburg, Russia
19. Sikarev AA, Lebedev ON (1983) Microelectronic devices for the formation and processing of complex signals. Radio and Communication, Moscow, USSR
20. Haar A (1910) On the theory of orthogonal systems of functions. Math Ann 69:331–371
21. Genchai R, Selcuk F, Whitcher B (2001) Introduction to wavelets and other filtering techniques in finance and economics. Academic Press, New York
22. Alexandridis AK, Zapranis AD (2013) Wavelet neural networks: a practical guide. Neural Netw 42:1–27
23. Cui Z, Chen W, Chen Y (2016) Multi-scale convolutional neural networks for time series classification. arXiv preprint arXiv:1603.06995

# An Overview of the Methods Used to Recognize Garbage

**Ekaterina Bezuglova**⬦, **Egor Shiriaev**⬦, **Nikolai Kucherov**⬦, and **Georgii Valuev**⬦

**Abstract** This article offers an overview of methods employed to recognize garbage. There are methods discussed, which rely on machine vision to detect objects, as well as hardware for garbage sorting intelligence systems. There has been a comparative analysis carried out, which embraces various methods based on machine vision and optical sensors aimed at detecting metal in garbage. There have been technologies identified, which feature the best ratio of indicators. The main criteria included the cost of building a system based on the method, and accuracy. Further on, there are plans to carry out research focusing on developing an original system for the recognition of garbage patterns, its classification and sorting.

**Keywords** Machine vision · Convolutional neural networks · Deep learning · Mineral recognition · Solid household waste · Household garbage

## 1 Introduction

The modern society pays much attention to environmental issues, environment protection and reducing human influence on nature. While we are living in the times of enhanced consumption, one of the major environmental challenges still implies waste recycling and disposal. Currently, solid household waste (SHW) in Russia reveals the following ratios: paper and cardboard – 35%, food waste – 41%, plastics – 3%, glass – 8%, metals – 4%, textiles and other – 9% (Fig. 1).

E. Bezuglova (✉) · N. Kucherov · G. Valuev
North-Caucasus Center for Mathematical Research NCFU, Stavropol, Russia
e-mail: bezuglovakaterina@mail.ru

N. Kucherov
e-mail: nkucherov@ncfu.ru

G. Valuev
e-mail: mail@gvvaluev.ru

E. Shiriaev (✉)
North-Caucasus Federal University, Stavropol, Russia

© The Author(s), under exclusive license to Springer Nature Switzerland AG 2022    467
A. Tchernykh et al. (eds.), *Mathematics and its Applications in New Computer Systems*,
Lecture Notes in Networks and Systems 424,
https://doi.org/10.1007/978-3-030-97020-8_42

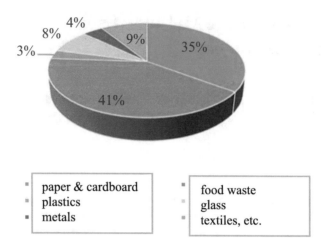

|  |  |  |  |
|---|---|---|---|
| ▪ | paper & cardboard | ▪ | food waste |
| ▪ | plastics | ▪ | glass |
| ▪ | metals | ▪ | textiles, etc. |

**Fig. 1** SHW ratios in Russia

On average, 10–15% of garbage gets recycled [1]. SHW, in turn, can be divided into waste and household garbage (non-biological solid waste of artificial or natural origin) [2].

In this paper, we will focus on household garbage recycling, or, to be exact, on the available automated recognition and sorting methods. One of the first tasks in terms of sorting household garbage is correct separation, i.e. separation of non-recyclable items and waste, following certain criteria, into qualitatively different elements. The purpose of this entire thing is to increase the percentage of recycled waste, i.e. the use of waste to produce goods (products), performance of works, provision of services, including reusing waste for its intended purpose (recycling), its return to the production cycle after respective preparation (regeneration), as well as the extraction of useful components to be reused (recovery) [3].

Household garbage sorting can be broken into the following stages:

1. Preliminary sorting, which implies separation of glass, hazardous waste, fine fraction and plastic bag shreds.
2. Main sorting, where recyclables are separated, e.g., paper, cardboard, polymeric elements.
3. Treatment of recyclables – paper, cardboard and polymers are pressed and packed, glass is loaded into special containers for further transportation and processing. The third stage may include additional monitoring to observe compliance with processing standards.
4. Recovery of waste with residual resources, which stage implies sorting the remaining garbage, separating elements that feature a high heating potential in case of burning to be processed into fuel. Besides, biodegradable waste for composting, and inert materials to be used in construction, are separated. Just like in stage 1, the waste is checked for potential hazard and harm.

5. Processing of bulky waste elements, such as furniture, wood, windows, doors, carpets, mattresses. This stage is mostly manual, and is carried out in order to separate stuff for further processing as recyclables, thus reducing the volume of materials that will be buried later.

6. Processing of fine fraction. This stage implies sifting (or other ways to obtain recyclable stuff) of a small fraction.

7. Processing of hazardous materials – here valuable elements (lithium, silver, zinc, nickel) are extracted from hazardous materials, electrolyte is removed from batteries for further reuse, as well as metal and plastic elements are released for recasting.

Each of the stages above includes additional sub-stages, which are not to be discussed within this paper. Section 2 offers a look at methods for sorting SHW through machine vision.

## 2 Materials and Methods

### 2.1 Machine Vision for Recognition

A more economical and environmentally friendly garbage sorting method relies on optical sorting systems used as an analytical device that employs machine vision to recognize objects. This area is already well studied, with a lot of contribution made into object recognition algorithms by researchers like Alfred Haar, Herbert Freeman, Chris Harris, Mike Stephens, Navnit Dalal, Bill Triggs, Yann LeCun, Yanqing Jia, Yu.B. Zubarev, V.P. Dvorkovich, A.B. Dvorkovich have made a significant contribution to the creation of [4]. The downside of this method is the performance dependence on the camera and the overall speed of data processing. Here below we will have a look at the prerequisites for employing neural networks for garbage sorting, in view of the following works.

In 2015, He et al. [5] introduced the deep residual learning method, also known as ResNet, which earned him the 1$^{st}$ place at the 2015 ILSVRC classification contest. This structure adds skipped connections to create a residual mapping that eliminates the issue of vanishing gradient when training deeper networks. In 2017, Huang et al. [5] proposed a multilayer Convolutional Neural Network (CNN), also known as DenseNet, which connects each layer of the image with every subsequent layer through direct communication. The connections between functions in various channels facilitate the reuse of functions, so the number of the required parameters is reduced significantly with no reduction in accuracy. In the same year, Xie et al. [6] introduced a highly modulated convolutional network architecture, aka ResNeXt, where a set of transformations with a similar topology is aggregated in each block. There is empirical evidence showing that increasing power can improve classification accuracy. In 2019, Efficient Net, proposed by Tan et al. [7] presented a system that uses a simple yet effective component coefficient for uniform scaling of three dimensions of the network width, depth and resolution, which will reduce further network parameters as well as enhance recognition accuracy [8]. These works served the basis for further progress in using neural networks in image processing.

Further, we will take a look at the work by Zhihong et al. [9], which proposes a robotic capture system for automatic garbage sorting based on machine vision. The system ensures identification and detection of the target objects position against a complex background to use the manipulator of sorted objects automatic capture. Identifying an object against a complex background is a key issue that the machine vision algorithm is aimed at solving. The authors use the deep learning method to identify the authenticity of the target object against a complex background. To arrive at an accurate capture of the target object, the Region Proposal Network (RPN) and the VGG-16 model are used for object recognition and for assessing location in space. The machine vision system sends information concerning the coordinates of the geometric center and the angle of the target object's long side to the manipulator, which completes the classification and capture of the target object. Let us have a more detailed look at their neural network model.

The Fast R-CNN method proposed by R. Girshick [10] ensures a speed close to real time, which is done using very deep networks. Fast R-CNN is based on previous work on object proposal efficient classification using deep convolutional networks. Compared to the previous work, Fast R-CNN involves several innovations to improve the speed of training and testing, as well as to improve the detection accuracy [11]. Fast R-CNN consists of two subnets: RPN and VGG-16. RPN highlights convolutional full image functions with a detection network, which is a fully convolutional network. This network predicts simultaneously the object boundaries and its position assessment.

The target object moves along the conveyor belt, there is a repeated fixing and processing of the target object carried out repeatedly. The data deletion method used for repeated identification can be viewed as an abstraction of useful data. The target object is divided into two groups: one group includes the objects that get into the camera view, while the other group includes all objects that are in the camera view. The authors assume that the time when the same target appears in the image is equal to Nc and is to be calculated subject to the following formula:

$$N_C = \frac{S_y - 2L_{msx}}{V_c} \times F_{ps}, \tag{1}$$

where $V_c$ is the conveyer belt speed, $F_{PS}$ is the camera frame rate, $L_{msx}$ is the maximum length of the target towards the $Y$ axis. The abstraction of the payload will be written $N_c$ times. The authors, therefore, present a robotic capture system based on machine vision, which can detect, recognize and capture objects in different positions, using the deep learning model. RPN and the VGG-16 model for object recognition and position detection are used to identify the target object against a complex background. Finally, the results of the experiment focusing on sorting bottles from garbage revealed that the review algorithm and the method of manipulator control within the proposed system is effectively in doing this.

The paper [12] presents a new system for garbage classification and localization to be captured and placed in the respective trashcan, while such system is integrated into a mobile manipulator. The authors, in particular, first introduce and train a deep neural network (GarbageNet) to identify various types of garbage that are good for recycling. Further, the capture localization method is involved to determine the appropriate position for picking up the garbage from the ground, and finally, objects are captured and sorted with a mobile robot operate by a whole-body control system.

In view of the need for a real-time garbage detection and localization using a mobile robot, the authors decided to employ the YOLACT system as presented in [13], and train it to operate garbage objects. The new trained network is called is GarbageNet. This type of network structure allows obtaining the bounding box and the type of object, as well as object images at the pixel level, which will improve the robot's awareness of the environment. Real-time performance and high accuracy make it better when compared to other types of object segmentation methods, such as Mask R-CNN [14] and TensorMask [15]. There was a source network integrated into the ROS shell, where the robot's visual sensor operates as input data, with messages generated concerning the segmentation of the garbage piece, the bounding box, the position and capture. The structure creates images and evaluates them through image coefficients. Images are combined using the Network Management System (NMS) to avoid overlap between objects, as well as to preserve useful information.

In the future, the authors have plans to test the system outside, in the wild, under various weather conditions, and go on working on the method of constructing the movement path and operation. In particular, the issues of where to find garbage in a large open space, as well as how to collect it while saving energy and time are further steps towards solving the problem set by the GarbageNet developers. There is also a work available, which has a similar name – PublicGarbageNet), which is to be considered below.

The authors propose a public garbage classification algorithm, which is based on the CNN architecture, PublicGarbageNet namely [8]. The proposed algorithm is a multitask classification, whose one task is to identify the type of household garbage, and the other implies recognition of the garbage subclass. The two classification tasks are related, and the function of joint loss helps improve the accuracy of garbage recognition. Given the fact that the existing garbage datasets are incomplete and extremely scarce, the authors developed a new garbage dataset that includes 10 subclasses and 10,624 images. Further, let us have a look at the network architecture.

In their research, the authors rely on the *four classifications* standard, which has been adopted in many Chinese cities, in order to classify garbage as food, hazardous, recyclable and other waste. Detailed garbage sorting takes breaking some basic categories into several subcategories. The main category of recyclable materials, for instance, is divided into five subcategories, i.e. recycled plastics, recyclable paper, metal, electronic products and glass. In view of specific garbage classification scenarios, there is a two-task CNN-based classification algorithm proposed, which is called PublicGarbageNet.

First, the initial garbage image is uploaded to the network, after which the hierarchical elements are extracted using an optimized backbone. Finally, the obtained features are derived as two branches of classification – one for classification into the main category, and the other for classification into subcategories.

While training a convolutional neural network, the choice of an optimizer is a decisive factor helping achieve the best minimum for the algorithm. There are many popular optimizers available, such as SGD [16], AdaGrad [17], Adam [18], etc. Of them, Adam is used widely in the real learning process of the deep learning network. It can adaptively implement step-by-step annealing and dynamically adjust the network learning rate. In the initial learning process, network gradients are typically very large. In case the learning rate is set too high, this will lead to a gradient explosion problem, so during initial training, the learning rate should be set to a low value. After a certain learning time, the learning rate can be respectively increased. At the end of the training period, the learning rate should be reduced to a small value in order to ensure better convergence for the network.

The author's technology, therefore, implies building a large-scale publicly available garbage dataset, which would contain four main categories and ten subcategories of waste – a total of 10,624 images. Then a two-task CNN-based garbage classification algorithm – PublicGarbageNet – was proposed. After employing a number of useful techniques, such as selecting the most suitable backbone, multitask training, increasing the data, adjusting the speed of dynamic warm-up training and label smoothing, PublicGarbageNet's accuracy comes to 96.35%.

Table 1 shows the classification of neural networks that are used for sorting SHW, the employed architecture, the distribution and recognition accuracy, the system openness as well as its applicability. This comparison was carried out in order to identify the most effective technology among those described, such is PublicGarbageNet.

**Table 1** Classification of neural networks used for SHW sorting

|   | Name | Architecture | Accuracy | System openness | Applicability |
|---|------|-------------|----------|-----------------|---------------|
| 1 | RPN &VGG-16 | Region Proposal Generation | 94% | – | – |
| 2 | Fast R-CNN | Region Proposal Generation + CNN | 96% | – | + |
| 3 | GarbageNet | ROS + CNN; Solo + TenzorMask | 90% | – | – |
| 4 | PublicGarbageNet | CNN + Adam | 96.35% | + | + |

## 2.2  Hardware Systems for Intelligent Systems of Household Waste Sorting

A review of the work on machine vision used for sorting garbage will show that cameras account for one of the most important components of the system. However, they can be replaced with optical sensors that will sort the garbage depending on the object properties, such as the material spectrum. Based on this, various properties of objects can be identified. Traditional sorting technologies, such as magnetic sorting and eddy current sorting, allow rough processing for only certain special types of waste mixture ingredients, e.g., ferrous and non-ferrous metals separation, since there are respective force fields between the waste particles and separators. Some other properties of solid particles, however, such as color, shape and texture, can also be accepted as sorting criteria, yet but there is no sufficient force field between these properties and separators. Feature recognition with optical sensors could offer good results, then. Let us consider the available methods based on optical technologies.

The simplest system implies sorting by transparency, and the system presented by Scott, D. M. [19] offers a perfect example of it. This method relies on a two-color near-infrared sensor with a fixed filter in combination with a simple ratio scheme for sorting recycled household plastic waste. The method has proven effective when sorting polyethylene terephthalate and polyvinyl chloride. A sensor based on this identification method is inexpensive to manufacture, yet it does not ensure the speed and performance required by the recycling industry. Figure 2 contains an example of an industrial belt using this sorting method.

One of the options is sorting solid waste, which is a mixture of various materials that has been previously crushed, classified and sent to treatment facilities. Of all these steps, sorting is the decisive one in terms of further waste recycling and reuse. Jiu Huang [20] presented the process of indirect sorting involving an optical sensor and a mechanical separation system. This system allows identifying the size and position of the particles, their color and shape, which further allows using these as sorting criteria. The mechanical sorting device embraces a compressed air nozzle,

**Fig. 2**  Industrial belt by Scott D.M.

which is controlled by computers; the target particles detected by the sensor are blown out of the main waste stream. This study has proposed a new approach to recognition and offers good results overall, not counting the speed.

Work [9] proposes a method where a 3D camera with color scanning of the BASLER avA1000-120 km/khz region operates as a visual sensor. It can measure 3D visual parameters, as well as the position and the color of objects. Its maximum measurement frequency is 120 frames/s, while the resolution is 1 megapixel. 3D images are obtained relying on the principle of triangulation scanning. The laser beam is required to shape a triangulation scanning system. The beam emitter used in this study has a wavelength of 662 nm, whereas its effective laser beam length can reach 750 mm, which is beyond the width of the conveyor belt.

The laser emitter is installed above the conveyor belt. The laser was emitted vertically onto the conveyor belt surface, while the laser beam lay on the belt surface vertically to the belt edges [21]. A 3D scanning camera was, too, installed above the conveyor belt and left for the laser beam. The camera focused on the laser radiation on the belt. All waste particles were fed to the conveyor belt separately, with no overlap, which was done to avoid failure through sorting. This type of 3D scanning got the name of the triangulation principle.

Depending on how far the laser falls on the surface, the laser beam comes with different shades of gray within the camera view area. This method is called triangulation scanning because the laser beam, the camera and the laser emitter make up a triangle. The length of the triangle's one side and the distance between the camera and the laser emitter are known values. The laser emitter's inclination angle is known, too. The camera's view angle can be identified through focusing on the laser beam location within the camera's view field. These three bits of data determine completely the shape and the size of the triangle, as well as they point at the location of the triangle's laser beam.

The lens focus and the point of laser beam emission are O and P. The laser emitter emits several parallel lines on the belt surface. From the conveyor belt surface, the height of the camera and the laser generator is equal to $h_1$ and $h_2$, and the height/depth of the $W$ surface spot is $h$. $L$ is the horizontal distance between the camera and the generator, i.e. $L = \overline{O_1 P_1}$. $\alpha_0$ is the lens axis installation angle in the vertical direction, whereas $\alpha$ is the angle between the $\overline{OW}$ line and the lens axis. $\beta_0$ is the projection angle between the laser beam and the vertical direction, while $\Delta\beta$ is the angle of the laser line beam increment from the first line. The triangles $\Delta OO_1W$ and $\Delta PP_1W$ allow drawing the following equation [22]:

$$\tan(a + a_0) = \frac{L - (h_1 - h)\tan(\beta + \Delta\beta)}{h_2 - jh}, \tag{2}$$

as well as:

$$\tan\alpha = \frac{y}{z}. \tag{3}$$

The camera focal length is $f$, while $(i, j)$ is the pixel in the image plane, which corresponds to the $W(x, y, z)$ point on the conveyer belt surface, and then the two similar triangles produce [22]:

$$\frac{y}{j} = \frac{z}{f}.$$ (4)

The coordination of the four edges of the AABB bounding rectangle makes it possible to obtain the approximate geometric center and the particle size of objects. Then, given Eq. (4), it is also possible to navigate following the actual central positions of the waste particles on the conveyor belt. The positional navigation of each waste particle is to be used for the sorting.

This, therefore, is a reliable method helping recognize a variety of features, which can be used to sort solid waste. The results of the experiments presented in the paper reveal that almost all the waste ingredients with significant color and shape features could be recognized to be further separated. Visual specifics of individual waste particles can be quickly captured and processed with 3D scanning cameras and virtual instrument software. After pretreatment and direct sorting, the sorting efficiency of conventional household waste can be improved significantly through the introduction of sorting technology based on optical sensors.

Further on, we will have a look at a method that relies on combining an electromagnetic and a two-energy X-ray transmission sensor [23]. Experts of Delft University of Technology, optimized and tested the DE-XRT system in collaboration with L3 Communications. The sensor submits information concerning the average atomic number of the tested material. The system is similar to the one used for baggage screening and for medical applications, the main difference being the software for data processing and image analysis. The lab machine has been modified so as to operate at a constant scanning speed of approximately 1 m/s. The particles used for the experiments were selected from a category featuring a low or high atomic number. The DE-XRT system ensures an image of the particles, whereas the color and the intensity determine the relationship between the atomic number and the particle tested. Metals with high density or high atomic number (atomic number of 26 and above), which have high transmission attenuation, appear in the image darker than metals with low density (atomic number of 13 and below). As is obvious, the shape and the size of the particles can also be determined depending on the resolution.

The experiments employed a prototype of electromagnetic sensors (EMS), which was developed at the Delft University of Technology, while its design and application can be found in an earlier work [20]. The EMS measures the interaction between a metal particle and the electromagnetic field generated by the transmitter coil. There is a set of receiving coils measuring this interaction, which is then analyzed by the data processing unit [24]. The EMS detects differences in the electrical conductivity ($\sigma$) of the metals. Earlier experiments using the EMS revealed that non-ferrous metals with $\sigma \ll b30\%$ (stainless steel, lead and tin (CCO)) can be easily distinguished from metals where $\sigma > 30\%$ CCO, such as copper, pure aluminum, aluminum alloys and magnesium [24].

The study focused on differentiating various metals with high density, such as copper, brass, zinc and stainless steel, and metals with low density and a high $\sigma$. Besides, the size and the shape of the particles have an impact on the sensor's output signal, which explains why it is difficult to distinguish groups of metals, such as aluminum alloys, from irregular-shaped magnesium alloys. Zinc and brass have almost the same $\sigma$ – approximately 27% of the CCO, which makes it impossible to tell apart these two metals with an EMS.

The authors also offer the following features of their DE-XRT system:

light and heavy NF metals can be divided into two different groups with rather high accuracy;
86 of 92%-pure magnesium was extracted from a mixture of light metals;
68 of 81%-pure cast aluminum was extracted;
50 of 70%-frequency deformable aluminum was extracted;

The shape, the size, the contour and the purity of the particles can be determined from the DE-XRT. 5.2 images.

The authors also make conclusions regarding the EMS specific features:

separation of SS from a mixture of NF-metals is effective. 90% of SS was removed from the NF heavy metal mixture, the purity being at 95%;
separation of copper and brass is a harder issue. 65% of copper was removed from an NF mixture of heavy metals with a purity of 80%;
separation of a mixture of light metals into aluminum cast, forged and magnesian ones is impossible;
separation of SS from a mixture of all groups, light and heavy NF-metals is effective.

The conclusions about combined sorting involving EMS and DE-XRT.
The combination of EMS and DE-XRT to sort mixtures of NF-metals allows:

dividing light and heavy NF-metals into separate groups;
separation of the cast aluminum group, which is significantly lower in terms of magnesium and deformable aluminum levels;
separation of the deformable aluminum group, which is significantly lower in terms of magnesium and cast aluminum levels;
separation of the magnesium group, which is significantly lower in terms of aluminum levels;
separation of brass fraction that is lower in copper content;
separation of copper fraction that is lower in brass content.

Table 2 contains a comparative analysis of the advantages and downsides pertaining to the system, which were discussed above. The criterion of applicability has also been introduced, and is aimed to check whether the system is being used or not.

**Table 2** Classification of recognition systems based on sensors used for garbage sorting

| | Name | Properties | Advantage | Downside | Applicability |
|---|---|---|---|---|---|
| 1 | NIR | Two-color near-infrared sensor with fixed filter + simple ratio scheme | Inexpensive production cost | Does not ensure required speed and production capacity | Not identified |
| 2 | By Jiu Huang №1 | Optical sensor + mechanical separation system | High accuracy | Low speed | Not identified |
| 3 | By Jiu Huang №2 | Laser + 3D camera | Accuracy, speed, easy-to-use | Expensive | Yes |
| 4 | By M.B. Mesina | Dual-energy X-ray sensor | Accuracy, speed | Expensive, recognized metals only | Not identified |

## 3 Conclusion

The overview of the available technologies and methods for intelligent garbage sorting, suggests that this is quite a popular issue in the modern society, with numerous research projects carried out by Chinese scientists. However, when developing an intelligent garbage sorting system, the major thing is to identify the ration of mobility and scale. In the case of personal use of such technologies, it is necessary to develop a rather simple system that would be not resource-demanding, and that could be used on mobile devices. Otherwise, it is a system involving various cameras and optical sensors, while such systems take a lot of resources and are installed at processing plants. For our research, of greater interest are methods that allow building intelligent garbage sorting systems for mobile devices. The study outcomes have revealed that the most effective technologies are GarbareNet and PublicGarbareNet as they allow sorting garbage in a real time mode.

Our further plans include research focusing on the development of an original system for recognizing garbage patterns, its classification and sorting.

**Acknowledgements** The work is supported by North-Caucasus Center for Mathematical Research under agreement №. 075-02-2021-1749 with the Ministry of Science and Higher Education of the Russian Federation and part study was funded by RFBR, project number 20-37-70023.

## References

1. Krauss J (2012) Infographics: more than words can say. Learn Lead Technol 39(5):10–14

2. Bespalov V, Paramonova O, Gurova O, Samarskaya N (2017) Selection of ecologically efficient and energetically economic engineering-ecological system for municipal solid wastes transportation. MATEC Web Conf 106:07021
3. Law F (1998) On Production and Consumption Waste dated 24.06.1998 No. 89-FZ (latest edition). Russ. Available from: http://www.consultant.ru/document/cons_doc_LAW_19109/. Accessed 26 Jun 2020
4. Yuzhakov AA Hierarchical recognition method in the subsystems of machine vision ACS sorting and disposal of household waste
5. He K, Zhang X, Ren S, Sun J (2016) Deep residual learning for image recognition. In: Proceedings of the IEEE conference on computer vision and pattern recognition, pp 770–778
6. Xie S, Girshick R, Dollár P, Tu Z, He K (2017) Aggregated residual transformations for deep neural networks. In: Proceedings of the IEEE conference on computer vision and pattern recognition, pp 1492–1500
7. Tan M, Le Q (May 2019) EfficientNet: rethinking model scaling for convolutional neural networks. In: International conference on machine learning. PMLR, pp 6105–6114
8. Zeng M, Lu X, Xu W, Zhou T, Liu Y (July 2020) PublicGarbageNet: a deep learning framework for public garbage classification. In: 2020 39th Chinese control conference (CCC). IEEE, pp 7200–7205
9. Zhihong C, Hebin Z, Yanbo W, Binyan L, Yu L (July 2017) A vision-based robotic grasping system using deep learning for garbage sorting. In: 2017 36th Chinese control conference (CCC). IEEE, pp 11223–11226
10. Girshick R (2015) Fast R-CNN. In: Proceedings of the IEEE international conference on computer vision, pp 1440–1448
11. Ren S, He K, Girshick R, Sun J (2015) Faster R-CNN: towards real-time object detection with region proposal networks. Adv Neural Inf Process Syst 28:91–99
12. Liu J, et al (July 2021) Garbage collection and sorting with a mobile manipulator using deep learning and whole-body control. In: 2020 IEEE-RAS 20th international conference on humanoid robots (humanoids). IEEE, pp 408–414
13. Bolya D, Zhou C, Xiao F, Lee YJ (2019) YOLACT: real-time instance segmentation. In: Proceedings of the IEEE/CVF international conference on computer vision, pp 9157–9166
14. He K, Gkioxari G, Dollár P, Girshick R (2017) Mask R-CNN. In: Proceedings of the IEEE international conference on computer vision, pp 2961–2969
15. Chen X, Girshick R, He K, Dollár P (2019) TensorMask: a foundation for dense object segmentation. In: Proceedings of the IEEE/CVF international conference on computer vision, pp 2061–2069
16. Taddy M (2019) Business data science: Combining machine learning and economics to optimize, automate, and accelerate business decisions. McGraw Hill Professional
17. Gupta MR, Bengio S, Weston J (2014) Training highly multiclass classifiers. J Mach Learn Res 15(1):1461–1492
18. Kingma DP, Ba J (2014) Adam: a method for stochastic optimization. arXiv preprint arXiv: 1412.6980
19. Scott DM (1995) A two-colour near-infrared sensor for sorting recycled plastic waste. Meas Sci Technol 6(2):156
20. Huang J, Pretz T, Bian Z (October 2010) Intelligent solid waste processing using optical sensor based sorting technology. In: 2010 3rd international congress on image and signal processing, vol 4. IEEE, pp 1657–1661
21. Jähne B (2005) Digital image processing: with 155 exercises and CD-ROM, Bernd Jähne
22. Su J, Xu B (1999) Fabric wrinkle evaluation using laser triangulation and neural network classifier. Opt Eng 38(10):1688–1693
23. Majid Z, Setan H, Chong A (2008) Integration of stereophotogrammetry and triangulation-based laser scanning system for precise mapping of craniofacial morphology. Int Arch Photogram Remote Sens Spat Inf Sci 37:805–811
24. Killmann D, Scharrenbach T, Pretz T (May 2007) Perspectives of sensor based sorting for processing of solid waste material. In: Proceeding of II International Symposium MBT, pp 296–307

# Implementation of Spline-Wavelet Robust Bent Code in Code-Division Multiple Access

**Alla Levina**⑩ **and Gleb Ryaskin**

**Abstract** This paper presents the application of spline wavelet robust bent codes based on bent functions in communication systems with code-division multiple access (CDMA). The scheme of a spline wavelet robust bent code provides higher robust parameters and gives better protection from side-channel attacks in comparison with existing solutions. The comparison was made with the codes used in CDMA protocols. As a result, schemes have been proposed for introducing robust codes into the CDMA communication system. Implementation of these code constructions in CDMA can present a better level of providing information security.

**Keywords** Robust codes · Boolean functions · Spline-wavelet decomposition · Bent functions · CDMA

## 1 Introduction

Nowadays the volume of circulating information grows exponentially and due attention needs to be paid to the security of this information [1]. One of the methods of protecting information is the coding theory, but hardware implementations of error correction codes, data storage systems, and cryptographic algorithms are vulnerable to malicious analyses that exploit the physical properties of the designs and work architecture. For a more successful analysis, an attacker can collect side-channel information, such as runtime, behavior in the presence of malfunctions that can be used. Side-channel attacks use information obtained from an incorrectly functioning implementation of an algorithm to derive the secret information. An incorrect operation can occur due to malfunctions in the data transmission system or the encoding device, which can be caused by natural effects or caused by an attacker [2].

By performing various actions on the hardware components of a data transfer device in order to distort information, while performing control and analysis of errors, an attacker can change the information transmitted over the channel. This

A. Levina (✉) · G. Ryaskin
Saint-Petersburg Electrotechnical University "LETI", 197376 Saint-Petersburg,
Russian Federation
e-mail: alla_levina@mail.ru

© The Author(s), under exclusive license to Springer Nature Switzerland AG 2022     479
A. Tchernykh et al. (eds.), *Mathematics and its Applications in New Computer Systems*,
Lecture Notes in Networks and Systems 424,
https://doi.org/10.1007/978-3-030-97020-8_43

type of attack is called a computational error attack [3–9]. To provide protection against attacks of this type, robust codes are used, built on nonlinear functions, since linear functions do not detect all errors in the channel due to the linear properties of the space [3–9]. The use of bent functions and spline wavelets for constructing a robust code is considered in [12, 13].

In this work, several schemes of spline wavelet robust bent codes based on bent functions and wavelet decomposition with various degrees of multiplicative elements were selected, as a result, we obtain a lower indicator for the maximum probability of masking error, but the time spent on encoding information increases. The scheme of implementation of the selected designs into the CDMA communication system has been drawn up. A comparison with the codes used in CDMA is made, followed by an analysis of the results.

## 2 Robust Codes

Linear codes, which are used in most communication protocols and standards, are not suitable for protecting against malicious attacks because you can always choose an error that will not be detected by the recipient. To solve this problem, Mark Karpovsky in his work [5] proposed the use of robust codes. At present, this class of codes is being actively investigated in order to increase the level of information protection against attacks based on algebraic manipulation errors.

**Definition 1.** Robust codes are nonlinear systematic error-detecting codes that provide uniform protection against all errors without any (or that minimize) assumptions about the error and fault distributions, capabilities and methods of an attacker.

Let $M = |C|$, this is the number of codewords in code $C$. By the definition of an $R$-robust code, there are no more than $R$ code words that cannot be detected for any fixed error $e$.

$$R = max|\{x|x \in C, x + e \in C\}| \tag{1}$$

The probability of masking any error can be defined as:

$$\max Q(e) = max\frac{|\{x|x \in C, x + e \in C\}|}{M} = \frac{R}{M} \tag{2}$$

The use of linear codes does not allow detecting all errors in the data transmission channel, as a rule, only errors with a certain Hamming weight. Robust codes are opposed to linear codes, and detect all errors in the channel with a probability not less than the maximum probability of masking errors, but do not allow to correct errors due to the insufficient value of the minimum code distance. Accordingly, the lower the value of the parameter of the probability of masking any error and, the more protected the system is from attacks based on algebraic manipulation errors. The value of this parameter is indirectly related to the nonlinearity of the function, which is used to build robust code.

**Definition 2.** The nonlinearity of a function $f$ is the distance from $f$ to a class of affine functions. Lets denote the nonlinearity of the function $f$ in terms of $N_f$: $N_f = d(f, A(n)) = \min_{g \in A(n)} d(f, g)$, where $A(n)$ is the class of linear functions.

The function $f \in P_2(n)$ is called maximally nonlinear if $N_f = 2^{n-1} - 2^{(n/2)-1}$. To construct robust codes, it is possible to use various classes of nonlinear functions. In this paper, we will consider the construction of robust codes on maximally nonlinear functions - bent functions of various degrees.

**Definition 3.** A bent function is a Boolean function with an even number of variables for which the Hamming distance from the set of affine Boolean functions with the same number of variables is maximal [10].

Bent functions were first investigated by O. Rothaus in the middle of the twentieth century. At present, the study of bent functions is widespread. However, many questions in this topic remain unexplored and require careful consideration [10, 11].

Consider robust codes based on bent functions and spline-wavelet transformation that have a minimum parameter of maximum error masking probability in comparison with the most used robust codes [9].

## 3 Spline-Wavelet Bent Robust Codes

The idea of the wavelet transform is based on the partition of the signal s(t) into two components, approximating $A_m(t)$ and detailing $D_m(t)$.

$$s(t) = A_m(t) + \sum_{i=1}^{m} D_i(t) \tag{3}$$

where m denotes the decomposition (reconstruction) level.

Using spline-wavelet decomposition, we will create a large number of different code constructs. The key feature of this method is the ability to change the grid values while executing the algorithm. Therefore, a spline-wavelet mesh can be used as an algorithm for changing the parameters of a robust code. More information on spline wavelets can be found in [13, 14]. The coding algorithm uses the spline-wavelet bent function, which depends on the information word and the spline-wavelet element $Wave_k$.

For each number of variables, bent functions were constructed on the basis of spline wavelets. In this construction, for all code, a grid is selected $x = \{x_1, x_2, \ldots, x_{n-1}, x_k\}$, based on static or on the information part of the codeword

**Table 1** Spline wavelet bent-functions for n = 8

| Number of function | Grid | Function | Deg (f) |
|---|---|---|---|
| 1 | $x_i = c_i$ | $f_i = c_{i+1}c_{i+3}c_{i+4} + c_{i+2}c_{i+3}c_{i+5} + Wave_{i+2}c_{i+6} + c_ic_{i+3} + c_ic_{i+5} + c_{i+2}c_{i+3} + c_{i+2}c_{i+4} + c_{i+2}c_{i+5} + c_{i+3}c_{i+4} + c_{i+3}c_{i+5} + c_{i+6}c_{i+7}$ | 4 |
| 2 | Static | $f_i = c_ic_{i+1}Wave_{i+2} + c_{i+1}c_{i+3}Wave_{i+4} + c_ic_{i+1} + c_ic_{i+3} + c_{i+1}c_{i+5} + c_{i+2}c_{i+4} + c_{i+3}c_{i+4} + c_{i+6}c_{i+7}$ | 3 |
| 3 | Static | $f_i = c_i * Wave_i + c_{i+1} * Wave_{i+2} + c_{i+2} * Wave_{i+4} + c_{i+4} * Wave_{i+6} + c_i * c_{i+3} + c_{i+1} * c_{i+5} + c_{i+2} * c_{i+3} + c_{i+2} * c_{i+5} + c_{i+3} * c_{i+4} + c_{i+6} * c_{i+7}$ | 2 |

$c = (c_1, c_2, \ldots, c_n)$. For all functions, the wavelet element is calculated from the same function:

$$Wave_k = c_k - c_{k+1} - (x_{k+2} - x_{k-1})(x_{k+2} - x_k)^{-1}(c_{k-1} - c_{k+1}) \qquad (4)$$

where $k$ - the number of characters in code.

Using the spline-wavelet element, spline wavelet bent functions are constructed, which are bent functions regardless of the possible value of the spline-wavelet element. Spline-wavelet functions for $n = 8$ are presented in Table 1 and are given taking into account the grid conditions and the degree of function. All functions are bent functions regardless of grid values.

**Definition 4.** Spline-wavelet bent robust code. Let $c = \{c_1, c_2, \ldots, c_{n-1}, c_n\}$ denotes the code word of some shared $(n, k)$ code. Then $\{c_1, c_2, \ldots, c_{k-1}, c_k\}$ is the information part, and $\{c_{k+1}, \ldots, c_n\}$ - additional, $n = k + 2$. Grid is selected depending on the spline wavelet function, $\$f_i(c_1, c_2, \ldots, c_{k-1}, c_m)$ is a spline-wavelet bent function from Table 1, $m = 8$. The vector c belongs to the spline wavelet bent robust code if

$$c_{k+1} = f_0(c_1, \ldots, c_m) + c_{m+1}c_{m+2} + \ldots + c_{k-1}c_k;$$

$$c_{k+2} = f_1(c_1, \ldots, c_m) + c_{m+1}c_{m+2} + \ldots + c_{k-1}c_k.$$

This construction allows better protection against side-channel attacks, because parameter $R$ and maximum probability of error concealment $Q(e)$ lower than existed solutions, but it takes more time for the coding information [9].

The use of the spline-wavelet robust code allows the most efficient protection against side-channel attacks. Let us construct an algorithm for the operation of robust codes in the existing data transmission channel, the best example for construction is a data transmission channel based on a CDMA communication system. Next, we will analyze the structure of CDMA, and the use of robust coding in it.

## 4 Implementation of Spline-Wavelet Robust Bent Code in Code Division Multiple Access

Code-division multiple access (CDMA) is a channel access method used by various radio communication technologies. CDMA is an example of multiple access, where several transmitters can send information simultaneously over a single communication channel [15]. Each transmitter is assigned a code.

The CDMA standard uses orthogonal Walsh codes for code division. Walsh codes are formed from the rows of the Walsh matrix.

The feature of the Walsh matrix is that each of its rows is orthogonal to any other row of the matrix, as a result of which the orthogonality properties can be used [2, 7]. With orthogonal signals, the system can be configured in such a way that the receiver of information when receiving a signal will always have a logical "0", except for those cases when the signal to which it is tuned is received - this is achieved by using orthogonal properties. Signals not intended for the receiver when multiplied by an element of the Walsh matrix will always equal 0, except when there is interference in the channel and the original signal was distorted. Figure 1 shows the coding process in the CDMA communication system.

To implement robust coding in CDMA, it is necessary to choose the same number of informational and additional symbols as in the existing variants of the coding architecture. In various versions of CDMA, the Hamming code or convolutional codes are used, in which the number of information symbols is equal to the number of additional symbols, the length of a block of information symbols $k$ should not exceed the value of nine - $k < 9$, respectively $n = 2\,k$.

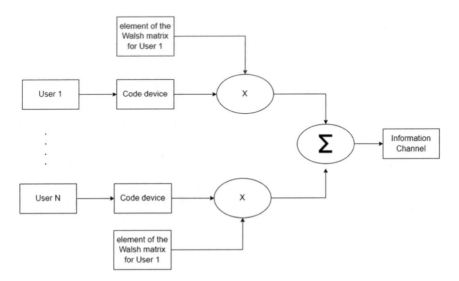

**Fig. 1** The architecture of the CDMA communication system

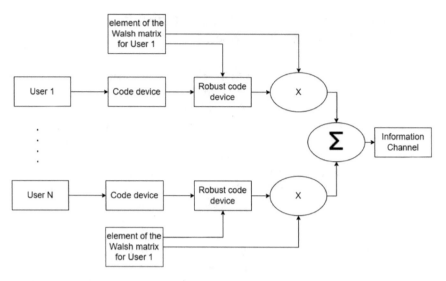

**Fig. 2** The architecture of the CDMA communication system with Robust code

Let us take the number of information symbols $k = 8$ and, accordingly, the length of the codeword $n = 16$. When introducing a robust code, we divide the encoder into 2 parts, first we encode the information word with a line code in order to correct single errors, then with a robust code to protect against side-channel attacks. As a result, when using robust coding with $k = 8$, the redundant symbols of the linear code (Hamming code) is 6, and the redundant symbols of the robust code are 2, for which the input word is information word + Hamming code.

CDMA is a very convenient scheme for spline wavelet robust bent code, because in the case of using a static grid in spline wavelet function, the question arises of transferring this value between the participants, and for CDMA you can use an element of the Walsh matrix, which will be the same size as spline-wavelet grid. An example of an implementation robust scheme is shown in Fig. 2.

As a result, additional redundant characters will not be introduced into the system, and no additional changes are required to be introduced into the system, except for the coding module. Let's compare the characteristics that we obtain when coding information with a robust code with existing solutions, take algorithms of a spline-wavelet robust bent code with different degrees of bent functions, in which we will use functions Table 1. The results of calculating the parameter R, the maximum probability of masking error, the encoding time of 6000 B of information vectors and the maximum number of bits that can be corrected are given in Table 2.

For the spline-wavelet robust bent codes, there are no undetectable errors, unlike the coding system in CDMA, it is also worth clarifying that if the Kerdock code were used as a robust code, then, according to Table 2, this code would have worse parameters of maximum probability of masking error than the spline-wavelet robust designs. In CDMA, an error-correcting code is used to correct minimum errors,

**Table 2** Comparison of codes

| Code | Maximum probability of masking error Q (e) | The encoding time, s | The maximum number of bits that can be corrected |
|---|---|---|---|
| Spline-wavelet robust bent code with function 1 | 0,375 | 0,096 | 1 |
| Spline-wavelet robust bent code with function 2 | 0,4675 | 0,082 | 1 |
| Spline-wavelet robust bent code with function 3 | 0,5 | 0,081 | 1 |
| Robust code with Kerdock code | 1 | 0,085 | 1 |
| CDMA with Hamming code | 1 | 0,035 | 1 |
| CDMA with convolutional code | 1 | 0,036 | 1 |

for this, in the proposed architecture, in addition to robust coding, linear coding is used. In the developed designs and in existing solutions, only single errors in the channel can be corrected. According to this parameter, the developed architecture is not inferior to the current solutions, thus one of the main meanings of using an error-correcting code in CDMA is not violated. The implementation of robust coding greatly improves the robustness of the CDMA system against side-channel attacks and malicious actions, but the time for encoding information takes longer than for existing CDMA implementations.

## 5 Conclusion

The developed spline-wavelet robust bent code scheme provides higher robust parameters in case of implementation in communication systems and also gives a better level of protection from side-channels attacks in comparison with existing solutions. But the time spent on encoding information with a robust code is much longer than with linear codes. Therefore, the implementation of robust codes depends on security requirements.

**Acknowledgements** This work was supported by the Ministry of Science and Higher Education of the Russian Federation 'Goszadanie' № 075-01024-21-02 from 29.09.2021 (project FSEE-2021-0015)

# References

1. Cramer R, Fehr S, Padro C (2013) Algebraic manipulation detection codes. Sci China Math 56(7):1349–1358. https://doi.org/10.1007/s11425-013-4654-5
2. Karpovsky MG, Kulikowski K, Wang Z (2007) Robust error detection in communication and computation channels. In: Keynote paper, international workshop on spectral techniques
3. Levina A, Taranov S (2015) Spline-wavelet robust code under nonuniform codeword distribution. In: 3rd international IEEE computer, communication, control and information technology. IEEE
4. Levina A, Taranov S (2015) Algorithms of constructing linear and robust codes based on wavelet decomposition and its application. In: El Hajji S, Nitaj A, Carlet C, Souidi EM (eds) Codes, Cryptology, and Information Security. Springer, Cham, pp 247–258. https://doi.org/10.1007/978-3-319-18681-8_20
5. Levina AB, Taranov SV (2015) Second-order spline-wavelet robust code under nonuniform codeword distribution. Procedia Comput Sci 62:297–302
6. Levina AB, Taranov SV (2017) Creation of codes based on wavelet transformation and its application in ADV612 chips. Int J Wavelets Multiresol Inf Process 15(2):1750014
7. Carlet C (2007) Boolean functions for cryptography and error correcting codes. In: Hammer P, Crama Y (eds) Chapter of the monograph «Boolean methods and models». Cambridge University Press, Cambridge
8. Schanes K, Dobernig K, Gozet B (2018) Food waste matters – a systemic review of household food waste practice and their policy implications. J Clean Prod 182:1978–1991
9. Alla L, Gleb R, Igor Z (2019) Spline-wavelet bent robust codes. In: Proceedings of the federated conference on computer science and information systems, pp 227–230
10. Tokareva N (2015) Bent functions: results and applications to cryptography
11. Claude C, Sihem M (2015) Four decades of research on bent functions. In: Designs, codes and cryptography, p 78. https://doi.org/10.1007/s10623-015-0145-8
12. Fekri F, Mersereau RM, Schafer RW (1999) Theory of wavelet transform over finite fields. IEEE Int Conf Acoust Speech Signal Process 3:1213–1216
13. Dem'yanovich KY (2008) Minimal splines and wavelets. Vestnik SPSU 41(2):88–101
14. Fekri F, McLaughlin SW, Mersereau RM, Schafer RW (1999) Double circulant self-dual codes using finite-field wavelet transforms. In: Fossorier M, Imai H, Lin S, Poli A (eds) Applied algebra, algebraic algorithms and error-correcting codes. Springer, Heidelberg, pp 355–363. https://doi.org/10.1007/3-540-46796-3_35
15. Munjal M, Chawla P, Radaur J (2014) A review comparison of different spreading codes for DS CDMA. Int J Sci Res 2:995–999

# On the Possibility of Increasing the Accuracy of Computations in the Floating-Point Format with Multiple Exponents

Shamil Otsokov and Aleksei Mishin

**Abstract** The paper proposes a format for the representation of numbers with multiple exponents, it is shown that it provides higher accuracy of calculations than the traditional floating-point format with the same length of floating-point coefficient with numbers close to zero and expands the numerical representation range. Formulas for the quantity of presentable numbers in the format with multiple exponent and floating-point coefficient are obtained.

**Keywords** Subnormal numbers · Floating point number · Multiple exponents · Machine epsilon · Distribution floating point number

## 1 Introduction

It is known that subnormal numbers are a subset of denormalized numbers. Any nonzero number in absolute magnitude less than the positive smallest normal number is subnormal. Due to subnormal numbers, addition and subtraction of floating-point numbers will never result in a complete loss of significant digits, i.e. two adjacent floating-point numbers will have a non-zero difference in the floating-point format. Arithmetic operations with subnormal numbers should be supported in floating-point coprocessors in accordance with the IEEE 754 floating-point standard [1, 8]. Some hardware manufacturers do not support working with subnormal numbers at the hardware level, only at the software level, which significantly reduces the speed of such calculations. However, the speed of calculations with subnormal numbers remains significantly lower than with normalized numbers even with hardware implementation (6 times slower) [2–4]. In this connection, some processor models provide the ability to round subnormal numbers to zero [5].

S. Otsokov (✉) · A. Mishin
National Research University "Moscow Power Engineering Institute", Moscow, Russia
e-mail: OtsokovShA@mpei.ru

A. Mishin
e-mail: MishinAlA@mpei.ru

© The Author(s), under exclusive license to Springer Nature Switzerland AG 2022
A. Tchernykh et al. (eds.), *Mathematics and its Applications in New Computer Systems*,
Lecture Notes in Networks and Systems 424,
https://doi.org/10.1007/978-3-030-97020-8_44

487

The purpose of this article is to investigate the floating-point format with a double exponent in order to exclude operations with subnormal numbers that are performed slower than with normalized numbers.

The following paragraph of this article sets out the format for representing numbers with a double exponent and algorithms for performing arithmetic operations for high-precision calculations with numbers close to zero [11–14].

Let us consider this format in the next paragraph.

## 2 Format for the Representation of Floating-Point Numbers with Multiple Exponents

Any floating-point number in the binary base notation [9]

$$A = \pm m \cdot 2^e \tag{1}$$

where

$m-$ the floating-point coefficient in the $q$ base number representation system, $1 \le m < 2$

$e-$ exponent part, $|e| \le e_{\max}$.

Let us define some parameters of the floating-point format for numbers of the form (1)

Let $n-$ be the number of digits in the fractional part of the floating-point coefficient.

Maximum positive presentable normalized floating-point number $A_{\max}$

$$A\,\max = 2^{e_{\max}} \cdot \left(1 - 2^{-n}\right) \tag{2}$$

The smallest such number greater than one is determined by the formula [1]:

$$(1.00...01)_2 = 1 + 2^{-n} \tag{3}$$

Machine epsilon $\varepsilon$ is the difference between one and the next larger number [1]:

$$\varepsilon = (0.00...01)_2 = 2^{-n} \tag{4}$$

For a number of the form (1), we define $ulp(x)$ for positive $x$ as the difference between $x$ and the next larger floating-point number, for negative $ulp(x)$ - as the difference between $x$ and the next smaller floating-point number.

$$ulp = (0.00\ldots01)_2 \cdot 2^e = 2^{-n} \cdot 2^e$$

The minimal positive presentable normalized floating-point number of the form (1) is equal to

$$A \min = 2^{-e_{\max}} \tag{5}$$

The number of positive presentable normalized floating-point numbers is calculated as the product of all possible floating-point coefficients in binary notation for each value of the exponents from $-e_{\max}$ to $e_{\max}$.

$$(2 \cdot e_{\max} + 1) \cdot 2^n = e_{\max} \cdot 2^{n+1} + 2^n \tag{6}$$

Let us consider the following floating-point format with multiple exponent

$$A = \pm m_1 \cdot q_1^{e_1} \cdot q_2^{e_2} \cdot \ldots \cdot q_m^{e_m} \tag{7}$$

where
$m_1$ — the floating-point coefficients of a number in the $q_1$ base number representation system satisfying the normalization condition
$1 \leq m_1 < q_1$,
$e_1$ — exponent part, $|e_1| \leq e_{\max}$.
$e_2$ — exponent part, $|e_2| \leq e_{\max}$,
$e_m$ — exponent part, $|e_m| \leq e_{\max}$.
$q_2 = q_1^{k_1}$, $k_1 \in N$,
$q_3 = q_2^{k_2}$, $k_2 \in N$

$$q_m = q_{m-1}^{k_{m-1}}, k_{m-1} \in N$$

Let the length of the fractional part of the floating-point $m_1$ be equal to $n$. Let us consider some parameters for numbers of the form (7)
Maximum positive presentable normalized floating-point number

$$A\max_1 = q_1^{e_{\max}} \cdot q_2^{e_{\max}} \cdot \ldots \cdot q_m^{e_{\max}} \cdot (1 - q_1^{-n}) \tag{8}$$

Obviously, $A \max_1 > A \max$ when $q_1 \geq 2$.
The smallest presentable number in the format (7) greater than one is equal to

$$(1.00 \ldots 01)_2 = 1 + q_1^{-n}$$

Machine epsilon $\varepsilon$ is the difference between one and the next larger number defined as follows:

$$\varepsilon = (0.00 \ldots 01)_2 = q_1^{-n}$$

For numbers of the form (7), we define $ulp(x)$.

$$ulp = \varepsilon \cdot q_1^{e1} \cdot q_2^{e2} \upsilon$$

The difference between a positive number of the form (7) and its nearest larger number is:

$$(0.00\ldots01)_2 \cdot q_1^{e1} \cdot q_2^{e2} \cdot \ldots \cdot q_m^{em} = q_1^{-n} \cdot q_1^{e1} \cdot q_2^{e2} \cdot \ldots \cdot q_m^{em}$$
$$= q_1^{-n+e1} \cdot q_2^{e2} \cdot \ldots \cdot q_m^{em}$$

$$(9)$$

The minimal positive presentable normalized floating-point number of the form (7) is equal to

$$A\min_1 = q_1^{-e_{max}} \cdot q_2^{-e_{max}} \cdot \ldots \cdot q_m^{-e_{max}}$$    (10)

Obviously, $A\min_1 < A\min$ when $q_1 \geq 2$.

and from the inequation of the relatively maximal values it follows that the range of representation of numbers with multiple exponents is higher than the range of traditional floating-point numbers.

Let us estimate the quantity of presentable floating-point numbers with traditional and multiple exponents.

## 3   Estimating the Quantity of Presentable Numbers with Multiple Exponent

The quantity of different presentable floating-point numbers within the representation range affects the accuracy of calculations, the more floating-point numbers, the higher the accuracy and the smaller the rounding error. This quantity depends on the length of the floating-point coefficient and the range of changes in the floating-point number exponents.

The quantity of positive presentable normalized floating-point numbers of the form (1) is not difficult to calculate and is equal to

$$N = (2 \cdot e_{max} + 1) \cdot 2^n = e_{max} \cdot 2^{n+1} + 2^n$$    (11)

Since the quantity of possible floating-point coefficient values in the binary number system is $2^n$, and the exponent varies from $-e_{max}$ to $e_{max}$., the plus one term is added to take into account the exponent equal to zero.

The quantity of negative floating-point numbers of the form (1) coincides with the positive ones. Let us consider an estimate for the quantity of different presentable positive normalized floating-point numbers $N_1$ of the form (8).

$$N_1 < (2 \cdot e_{max} + 1)^m \cdot q_1^n$$

This estimate does not determine the exact value for the quantity of different floating-point numbers of the form (7), because there are duplicates among the numbers.

We substitute $q_2$ in (7) to obtain

$$m_1 \cdot q_1^{e1} \cdot q_2^{e2} \cdot \ldots \cdot q_m^{em}$$
$$= m_1 \cdot q_1^{e1} \cdot (q_1^{k1})^{e2} \cdot \ldots \cdot (q_{m-1}^{km-1})^{em}$$
$$= m_1 \cdot q_1^{e1} \cdot (q_1^{k1})^{e2} \cdot (q_1^{k1 \cdot k2})^{e3} \cdot \ldots \cdot (q_1^{k1 \cdot k2 \cdot \ldots \cdot km-1})^{em}$$

Then, for example, repeated are the numbers of the form:
$m_1 \cdot q_1^k (q_1^k)^0, m_1 \cdot q_1^0 (q_1^k)^1$, which match when
$e_1 = k, e_2 = 0$ and $e_1 = 0, e_2 = 1$ if $k < e_{max}$.
The following conclusion is true.

## 4 Results

### 4.1 Quantity of Different Floating-Point Numbers

**Statement 1**
The quantity of different floating-point numbers of the form (7) is equal to

$$N_1 = 2 \cdot e_{max} \cdot (1 + k_1 + k_1 k_2 + k_1 k_2 k_2 + \ldots + k_1 k_2 \cdot \ldots \cdot k_{m-1}) \cdot q_1^n$$

**Proof** It is not difficult to see that the estimation of the quantity of different integers of the form (7) comes down to the estimation of different values of the expression:

$$y(e_1, e_2, \ldots, em) = e_1 + k_1 e_2 + k_1 k_2 e_3 + \ldots + k_1 k_2 \cdot \ldots \cdot k_{m-1} e_m,$$

$$\partial de$$

$$|e_1| \leq e_{max},$$

$$|e_2| \leq e_{max},$$

$$\ldots$$

$$|e_m| \leq e_{max},$$

$$k_i < e_{max}$$

$$(12)$$

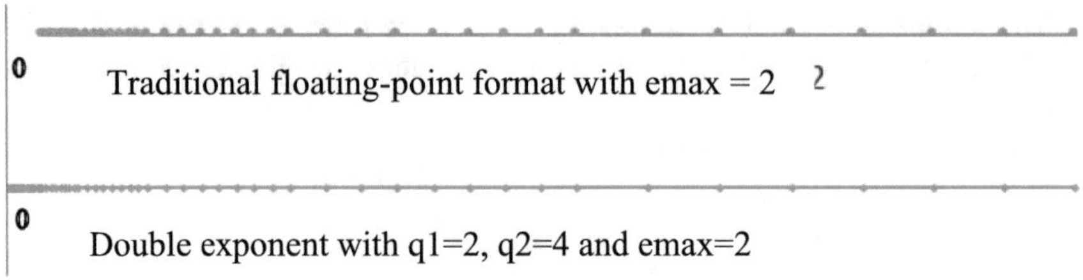

**Fig. 1** Distribution of traditional floating-point and double-exponential numbers

Let us estimate the values of the above expression (12)

$$|y(e_1, e_2, ..., em)| \leq e_{max} + k_1 e_{max} + k_1 k_2 e_{max} + ... + k_1 k_2 \cdot ... \cdot k_{m-1} e_{max}$$
$$= e_{max} \cdot (1 + k_1 + k_1 k_2 + k_1 k_2 k_2 + ... + k_1 k_2 \cdot ... \cdot k_{m-1}) \leq e_{max} \cdot k_1 k_2 \cdot ... \cdot k_m$$

That is, the maximum positive value that the function (12) can take is equal to:

$$K = e_{max} \cdot (1 + k_1 + k_1 k_2 + k_1 k_2 k_2 + ... + k_1 k_2 \cdot ... \cdot k_{m-1})$$

Since there are a total of $K$ different positive values of the function (12), the total quantity of different numbers that the function can take (12) is

$$N_1 = 2 \cdot e_{max} \cdot (1 + k_1 + k_1 k_2 + k_1 k_2 k_2 + ... + k_1 k_2 \cdot ... \cdot k_{m-1}) \cdot q_1^n$$

Q.E.D

Figure 1 shows that the distribution density of floating-point numbers with a double exponent having $q_1 = 2, q_2 = 4, k_1 = 2, m = 2$ near zero is higher than that of traditional floating-point numbers. The upper distribution corresponds to traditional normalized floating-point numbers, the minimal of which is 0.25. The lower distribution corresponds to normalized numbers with a double exponent, the minimal of which is 0.015625.

To represent numbers less than 0.25 in the floating-point format, subnormal numbers are used, which increase the accuracy of calculations with small numbers. The minimal number in this case is

$$A \min Sub = (0.00...01)_2 \cdot 2^{-e_{max}} = 2^{-n} . 2^{-e_{max}}$$
$$= 2^{-n-e_{max}} \tag{13}$$

Let's compare the minimal positive presentable normalized floating-point number with multiple exponent (10) with the expression (13). According to (10)

$$Amin_1 = q_1^{-e_{max} \cdot (1 + k_1 + k_1 k_2 + k_1 k_2 k_2 + ... + k_1 k_2 \cdot ... \cdot k_{m-1})}$$

For $q_1 = 2$:

$Amin_1 \le A \min Sub,$

$2^{-e_{max} \cdot (1+k_1+k_1k_2+k_1k_2k_2+...+k_1k_2 \cdot ... \cdot k_{m-1})} \le 2^{-n-e_{max}},$

$-e_{max} \cdot (1 + k_1 + k_1k_2 + k_1k_2k_2 + ... + k_1k_2 \cdot ... \cdot k_{m-1}) \le -n - e_{max},$   (14)

$e_{max} \cdot (1 + k_1 + k_1k_2 + k_1k_2k_2 + ... + k_1k_2 \cdot ... \cdot k_{m-1}) \ge n + e_{max},$

$e_{max} (1 + k_1 + k_1k_2 + k_1k_2k_2 + ... + k_1k_2 \cdot ... \cdot k_{m-1}) \ge n$

Thus, when condition (14) is met, the minimal presentable number with a multiple exponent for which the first exponent is binary is less than the minimal presentable subnormal number.

In particular, for $e_{max} = 2$, n = 3, $q_1 = 2$, $q_2 = 4$, $k = 2$

$$Amin_1 = 0.015625$$
$$A \min Sub = 0.03125$$

In the range from 0 to 1, there are 48 normalized floating-point numbers with a double exponent and 16 traditional normalized floating-point numbers and 25 subnormal numbers. In the range from 1 to 7.5, the quantity of numbers in both formats is the same.

It also follows from the condition (14) that the introduction of an additional exponent and the use of a format with multiple exponents makes it possible to abandon the use of subnormal numbers, arithmetic operations with which are difficult for processor manufacturers to support.

Let us consider the rules for performing arithmetic operations.

Let two numbers of the form (7) be given, consider the rules of multiplication and division:

$$A = m_{A_1} \cdot q_1^{eA_1} \cdot q_2^{eA_2} \cdot ... \cdot q_m^{eA_m},$$   (15)

$$B = m_{B_1} \cdot q_1^{eB_1} \cdot q_2^{eB_2} \cdot ... \cdot q_m^{eB_m},$$   (16)

## 4.2  Rules of Multiplication and Division

Multiplication and division of these numbers is performed according to the rules of multiplication and division of floating-point numbers, i.e.

$$A \cdot B = m_{A_1} \cdot m_{B_1} \cdot q_1^{eA_1+eB_1} \cdot q_2^{eA_2+eB_2} \cdot ... \cdot q_m^{eA_m+eB_m}$$

The division is performed in the same way, i.e. the formula is valid:

$$A/B = (m_{A_1}/m_{B_1}) \cdot q_1^{eA_1-eB_1} \cdot q_2^{eA_2-eB_2} \cdot ... \cdot q_m^{eA_m-eB_m}$$

Consider the rules for adding and subtracting numbers of the form (7)

## 4.3 Rules of Addition and Subtraction

The rules for addition and subtraction are similar to the addition of floating-point numbers, only the alignment of the exponents is performed by the set of exponents. For the case of $m = 2$, these rules have the form:

1)  Alignment of exponents $eA_1, eA_2$ and $eB_1, eB_2$

      1.1.    The number exponents are compared $eA_1, eB_1$

      1.2.    If they are not equal, the floating-point coefficient of the number with the smaller exponent is shifted to the right by the difference of the larger and smaller exponents from $eA_1, eB_1$.

      1.3.    The number exponents are compared $eA_2, eB_2$

      1.4.    If they are not equal, the floating-point coefficient of the number with the smaller exponent is shifted to the right by the difference of the larger and smaller exponents from $eA_2, eB_2$

Let $eA_1 > eB_1, eA_2 > eB_2$, then

The floating-point coefficient is set in the $q_1$ base number representation system and the floating-point coefficient shift to the right by $s$ digits in paragraph 1.2 is equivalent to dividing by $q_1^s$.

Let us consider the method of shifting, if division by $q_2^s$ is required, when aligning the exponents in paragraph 1.4. Let the floating-point coefficient be given in binary notation and have the form

$$m_A = a_1 \cdot q_1^{-1} + \ldots + a_{n-1} \cdot q_1^{-(n-1)},$$

$$m_A \cdot q_2^{-s} = \left(q_1^k\right)^{-s}\left(a_1 \cdot q_1^{-1} + \ldots + a_{n-1} \cdot q_1^{-(n-1)}\right)$$

$$= q_1^{-ks} \cdot \left(a_1 \cdot q_1^{-1} + \ldots + a_{n-1} \cdot q_1^{-(n-1)}\right)$$

Then dividing by $q_1^{ks}$ is equivalent to shifting the floating-point coefficient to the right by $k \cdot s$ digits.

In the next paragraph, we will consider the results of experimental studies of the proposed format with multiple exponents.

## 4.4 Results of Experimental Research

Numerical experiments were carried out to find the value of the following expression using traditional floating-point arithmetic and multiple exponent arithmetic with ($m = 2$):

$$S = t \cdot \sum_{i=1}^{10} \frac{1}{i \cdot t}, \tag{17}$$

where $t$ is some constant.

In expression (17), the introduction of the constant s was due to the fact that arithmetic operations were performed with numbers close to zero.

Floating-point calculations were performed in binary notation for the length of the floating-point coefficient $n = 6$ and for various values of $t$.

For calculations with a double exponent, the following parameters were assumed:

$$q_1 = 2, k = 2, q_2 = q_1^2 = 4, n = 6, e_{max} = 10$$

The same range of representation of floating-point and double-exponent numbers from $-2016$ to $2016$ was used. At the same time, floating-point numbers in this range of a given length were 1407, and numbers with a double exponent in the same range with the same floating-point coefficient length were 2624. There were 1217 more numbers with a double exponent, and they were in the ranges $[-9.313\text{E}-10, -3.051\text{E}-05]$ and $[9.313\text{E}-10, 3.051\text{E}-05]$. The minimal positive presentable number in the double exponent format was $9.313225746154785\text{e}-10$, and the floating-point number was $3.0517578125\text{e}-05$.

Figure 2 below shows the dependence of the relative error of this expression calculation on the parameter $t$ for calculations with multiple exponent (FloatDoubleExponent) ($m = 2$) and for floating-point calculations (Float) with the same floating-point coefficient length.

It can be seen from the graph in Fig. 2 that as the parameter $t$ increases, the relative error for calculations of the value of the expression S in floating-point format

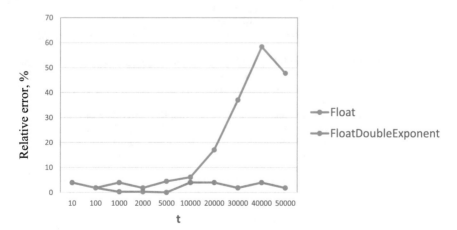

**Fig. 2** The relative error of expression S calculation

increases, which is explained by the fact that there are more presentable numbers close to zero in the format with multiple exponents.

## 5 Conclusion

This article shows that the floating-point format with multiple exponents enables to represent a larger quantity of numbers close to zero and get more accurate results compared to the traditional floating-point format when performing arithmetic operations with such numbers. In addition, the proposed format enables to abandon the use of subnormal numbers, hardware and software support for which is difficult to implement. A promising direction for further research is the choice of the optimal number of exponents.

## References

1. Overton ML (2001) Numerical computing with IEEE floating point arithmetic. Soc Indust Appl Math 104 p
2. Dooley I, Kale L (2006) Quantifying the interference caused by subnormal floating-point values. In: Proceedings of the workshop on operating system interference in high performance applications
3. Fog IA (2011) Instruction tables: lists of instruction latencies, throughputs and micro-operation breakdowns for Intel, AMD and VIA CPUs. Copenhagen University College of Engineering
4. Marc A, David K, Keaton M, Ranjit J, Sorin L, Hovav S (2015) On subnormal floating point and abnormal timing. In: IEEE symposium on security and privacy
5. Schwarz E, Schmookler M, Trong SD (2003) Hardware implementations of denormalized numbers. In: Proceedings 16th IEEE symposium on computer arithmetic (Arith16). 16th IEEE symposium on computer arithmetic. IEEE Computer Society, pp 104–111
6. Bailey DH, Borwein JM (2015) High-precision arithmetic in mathematical physics. Mathematics 3:337–367
7. Bailey DH, Barrio R, Borwein JM (2012) High-precision computation: Mathematical physics and dynamics. Appl Math Comput 218(20):10106–10121
8. Muller J-M, Brisebarre N, de Dinechin F, Jeannerod C-P, Lef'evre V, Melquiond G, Revol N, Stehl'e D, Torres S (2010) Handbook of floating-point arithmetic. Birkhauser, Boston, 572 p
9. Israel K (2002) Computer arithmetic algorithms, 2nd edn. University of Massachusetts
10. Fousse L, Hanrot G, Lefèvre V, Pélissier P, Zimmermann P (2007) MPFR: a multiple-precision binary floating-point library with correct rounding. ACM Trans Math Softw 33(2):13
11. MParithm - package for high precision computation (2015) www.wolfgang-ehrhardt.de/mp_intro.html
12. GNU Scientific Library 2.5 released (2018) https://savannah.gnu.org/forum/forum.php?forum_id=9175
13. Operations with multi-bit real numbers of the ZReal type. http://ishodniki.ru/list/index.php?action=name&show=pascal-math&cat=11
14. Otsokov ShA, Magomedov ShG (2020) Using of redundant signed-digit numeral system for accelerating and improving the accuracy of computer floating-point calculations. Int J Adv Comput Sci Appl 11(9)

# Method of Building a Scalable Neural Control System for Underwater Robotic Complexes

**I. M. Dantsevich**⬤, **M. L. Somko**⬤, **E. P. Khaleeva**⬤, and **M. N. Lyutikova**⬤

**Abstract** The article examines the method of building a neural control system of underwater robotic complexes under the conditions of impossibility of physical expansion of data exchange channels. The method is based on the use of wavelet transformation functions possessing a compact spectrum (corresponding to Riesz basis conditions). A tabular type control system consists of neurons formed according to the principles of spectral density maximum limit of a control signal power, which fits to the spectral radius of the normalized evaluation unit. The decomposition coefficients of control signal allow synthesizing broadcast commands on the on-board control complex, based on the reference points of control tables. The complex forms a self-learning neural network adapting to the verbal reactions of a pilot's who controls the complex or specified control criteria under the fully automatic complex dynamics mode.

**Keywords** Daubechies wavelet · Remotely operated vehicle (ROV) · Propulsor · Control system · Numerical methods · Positioning · Propulsor power · Mathematical model · Transformation matrix · Command signal · Transfer coefficients

## 1 Introduction

Management of an underwater complex comprising an underwater vehicle, a garage-sinker, a set of working tools on the vehicle board is usually performed by two specialists. One of them is a pilot of a ROV and the other is an engineer of underwater technical work [1, 2].

I. M. Dantsevich · M. L. Somko · E. P. Khaleeva · M. N. Lyutikova (✉)
Admiral Ushakov Maritime State University, Novorossiysk, Russian Federation
e-mail: mnlyutikova@mail.ru

M. L. Somko
e-mail: zur_mga@nsma.ru

© The Author(s), under exclusive license to Springer Nature Switzerland AG 2022    497
A. Tchernykh et al. (eds.), *Mathematics and its Applications in New Computer Systems*,
Lecture Notes in Networks and Systems 424,
https://doi.org/10.1007/978-3-030-97020-8_45

The ROV carrier has zero floatability, and, therefore, does not have counterweights of onboard manipulators, which movement is able to change the device primary balance.

The control dynamics requires certain skills; a sudden change in propulsor power can cause turbidity that complicates the use of CCTV monitoring and increases the work value. Based on these and other factors (undercurrents, water density in different layers, etc.), the process of work is constantly requiring to adjust the power and steepness, to change the modes of underwater lighting and ROV position while studying various objects [2].

It is necessary to adjust the propellers parameters to maintain desired operating parameters in automatic control mode, for example, movement over an underwater pipeline or cable. In this case, the usual adjustable regulators are not sufficient. The problem is partially solved by the use of training simulators of underwater technical work and staff training [3].

## 2 Materials and Methods

A multilayer control system is formed from the sets of control functions in the longitudinal-transverse and vertical subspaces forming the vectors of control signals:

$$\nabla = 2MS - R, \tag{1}$$

where $M$– correlation matrix of the input signal, $S$– matrix of weight coefficients, $R$ – vector column for the counts of intercorrelation function of input parameters and useful signal.

The problem of control function classifying is defined as a fuzzy problem of classification, clustering and control function approximation.

The control tables are made according to control vectors U1, U2, U3 and U4, respectively, "forward–backward", "left–right" turns, "left–right" shift and "emersion-dive". Control instructions transmission with the use of fiber-optic interfaces during full-duplex data transmission and control instructions requires the application of special measures to facilitate the traffic [4]. The first reason is that the control signals decoding must ensure real-time operation. The second one is that video streaming, implementing the ROV dynamics visual feedback, should be provided without delays, dangerous owning to the loss of visual contact with moving objects and objects located on the bottom surface [5].

As practice shows, the implementation of these requirements does not always allow obtaining a satisfactory result, and it is often difficult to achieve good ROV dynamics for these reasons.

Applying the theory of neurons, we define a neuron weight function as:

$$\tilde{S}(\xi) = m(\xi/2)\tilde{S}(\xi/2), \tag{2}$$

where $m(\xi) = \frac{1}{\sqrt{2}} \sum_{n \in S} h_n e^{2\pi i n \xi}$ transformation filter.

The neuron output signal:

$$y(\xi) = sign\left(\sum_{i=1}^{N} \tilde{S}_i(\xi) X_i(\xi) - b\right), \tag{3}$$

where: $\xi$– iteration step, $X_i(\xi)$– set of input parameters values at $\xi$ iteration step, $b$– shift.

The ROV dynamics consists of control system high-speed reactions, such as a change in the scale of input counts, as well as steepness depending on a signal rise time.

The application of fast Fourier transform algorithms to the evaluation of control signals spectrum leads to the loss of information on the control signal phase. In this aspect, the adaptation algorithm is most effective in the basis of orthonormal bursts (wavelets) [4–6, 11].

In this case, let us consider two scaling functions:

$$\tilde{S}(\xi) = m\left(\frac{\xi}{2}\right) \tilde{S}\left(\frac{\xi}{2}\right);$$

$$\tilde{D}(\xi) = \tilde{m}(\xi/2)\tilde{D}(\xi/2). \tag{4}$$

It is possible to write the following data for the single period scaling functions:

$$S = \sum_{n \in Z} h_n S_{1n};$$

$$D = \sum_{n \in Z} h_n D_{1n}. \tag{5}$$

Taking into account the fact that the characteristic of propulsor power (steepness) is convenient of a linear form:

$$D = Te^{T-1} \tag{6}$$

where $T = \tilde{S}_n/\tilde{S}_{n+1}$, and (2.6) is the following equation solution:

$$\frac{dD(\xi)}{d\xi} = e^{S_n/S_{n+1}}. \tag{7}$$

A neuron self-organization is possible by forming function thresholding. The biorthogonal system of bursts makes it possible to implement both the lower threshold (soft thresholding) and the upper one (hard thresholding).

Let us suppose that autocorrelation function $R^{D,N}$ is from scaling function $\varphi^{D,N}$, and

$$d_N(k) = D_N(k) = d^k \cos(\omega k + \phi). \tag{8}$$

Using Euler's formula,

$$D_N(k) = d^k \cos(\omega k + \varphi) = d^k \frac{e^{-\omega i(k+\varphi)} + e^{\omega i(k+\varphi)}}{2}, \tag{9}$$

we suppose $D(z) = \frac{1}{1-d_z{-1}}$ definitely

$$D(z) = \sum_{i=1}^{N} \frac{\cos\varphi_i - d_i \cos(\omega_i - \varphi_i)z^{-1}}{1 - 2d_i \cos(\omega_i)\, z^{-1} + z^{-2}}. \tag{10}$$

The expression (10) corresponds to a discrete decaying sinusoid where the spectrum of scaling function is compact.

The adaptation of operating frequency of control function (1) is the process of calculating the gradient corresponding to the product of intercorrelation function and control signal input sequence. A discrete convolution of two sets is the sum of feature-based products of intercorrelation functions and input sequences:

$$R = D * X. \tag{11}$$

The matrix D (convolution matrix):

$$D = \begin{bmatrix}
d(0) & 0 & 0 & \cdots & 0 \\
d(1) & d(0) & 0 & \cdots & 0 \\
d(2) & d(1) & d(0) & \cdots & 0 \\
\vdots & \vdots & \vdots & \ddots & \vdots \\
d(N) & d(N-1) & d(N-2) & \cdots & d(0) \\
d(N+1) & d(N) & d(N-1) & \cdots & d(1) \\
\vdots & \vdots & \vdots & \ddots & \vdots \\
d(M) & d(M-1) & d(M-2) & \cdots & d(M-N) \\
0 & d(M) & d(M-1) & \cdots & d(M-N+1) \\
0 & 0 & d(M) & \cdots & d(M-N+2) \\
\vdots & \vdots & \vdots & \ddots & \vdots \\
0 & 0 & 0 & 0 & d(M)
\end{bmatrix}. \tag{12}$$

## 3   Results

ROV semi-automatic control mode is provided by using the pilot's joystick for the original control signal decomposition $W(nT) = w_0 \cdot (2^0 + 2^1 + 2^2 + 2^3) = 1$, given that $w_0 = 0,066$MC, for the 4th order decomposition:

$$W(4T) = 0,066 \cdot 1 + 0.066 \cdot 2 + 0.066 \cdot 4 + 0.066 \cdot 8 \approx 1,$$

with uniform acceleration with a pilot's maximum possible verbal reaction (decomposition in the orthogonal basis of control function input).

Expansion of decomposition points adds the decomposition coefficients with the multiple value of $2^n$. The use of basic decomposition functions in Daubeche filters [14] allows implementing a real-time algorithm.

Let us consider the results obtained using the example of joystick control. The control signal input implementation along with the system noise is shown in the Fig. 1.

The analysis of input implementation in the example is performed in Daubechies analyzing filter of the 4th order, as shown in the Fig. 2. The decomposition of input implementation with the specified thresholding parameters allows constructing the control matrix spectrum, which elements are spectral densities of the original implementation signal.

The obtained estimate of input signal spectrum allows us to determine the power of a group of propulsors and desirable bias in the control system.

The implementation in question is described by the object equation:

$$\dot{D}(t) = P(D(t), t) - C_{\alpha\beta}(t)V(t); \tag{13}$$

where:

$$P(D(t), t) = \sum_{i=0}^{s} P_i D_{n,i}(t); \quad D_{n,i}(t) = \binom{s}{i} t(1-t)^{s-1}; \quad \binom{s}{i} = \frac{s!}{i!(s-i)!};$$
$$\tag{14}$$

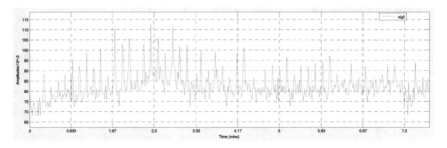

**Fig. 1** Input implementation of a control signal with noise

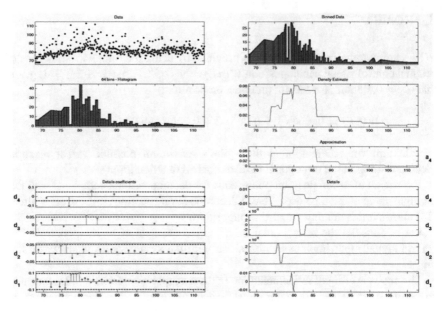

**Fig. 2** Analysis of input implementation by means of Wavelet Toolbox package in the Matlab package [13]

We group the coefficients by 4 interpolation points and we express it in a matrix form:

$$P(D(t), t) = \begin{bmatrix} t^3 & t^2 & t & 1 \end{bmatrix} \cdot \begin{bmatrix} -1 & 3 & -3 & 1 \\ 3 & -6 & 3 & 0 \\ -3 & 3 & 0 & 0 \\ 1 & 0 & 0 & 0 \end{bmatrix} \cdot \begin{bmatrix} P_{x1} \\ P_{x2} \\ P_{x3} \\ P_{x4} \end{bmatrix}; \qquad (15)$$

- $P(D(t), t)$ – the operator built according to (1), power $P(t)$ according to control dynamics $D(t)$;
- $C_{\alpha\beta}(t)$ – the matrix of ROV resistance coefficients according to the angle of attack $\alpha$ and yaw angle $\beta$;
- $V(t)$ – matrix of ROV speed dynamics;
- $\begin{matrix} P_{x1} \\ P_{x2} \\ P_{x3} \\ P_{x4} \end{matrix}$ – power values based on 4 approximation points of control vector according to control variant points U1.

Lagrange criterion correlates the functional with the conditions [8–10]:

$$\frac{\partial Lag}{\partial \widehat{F}^D} = 0, \quad \frac{\partial Lag}{\partial \widehat{F}^B} = 0, \quad \frac{\partial Lag}{\partial \Lambda} = 0; \qquad (16)$$

where $\widehat{\partial F}^{D}$, $\widehat{\partial F}^{W}$ - estimates of spectral operators of propulsion-steering unit control and actual forces creating stops and moments.

According to Lagrange criterion, two of four estimates, $d_1$ and $d_4$, correspond to the condition in the observed implementation; the estimates $d_2$ and $d_3$ have a minimum threshold of comparability, determining the minimum value of discrepancy coefficient with control signal specified value. The spectral characteristic of control signal is shown in the Fig. 3.

The power values of propulsion-steering units (movement along the power characteristic approximated by 4 points $P_{x\Sigma}$) for observed dynamics, taking into account the integration time value $t = 0,066c$:

$$P(D(t), t) = 81.47 \cdot d_1 \cdot P_{x1} + 0.028 \cdot d_2 \cdot P_{x4}$$
$$= 81.47 \cdot 0,097 \cdot P_{x1} + 0.028 \cdot 0,049 \cdot P_{x4} = 7.9 \cdot P_{x1} + 0.001 \cdot P_{x4}. \tag{17}$$

The scale changing of coefficients conversion of propulsion-steering unit power is performed by increasing the dimension of Daubechies filters, as shown in a program window in the Fig. 3.

As a result of thresholding, it is possible to reduce threshold in the window of the 4th order transformation of Matlab program, (the coefficient at $d_2, d_3, d_4$ have the dimension order $10^{-2}$), they are excluded from implementation as having a noise character.

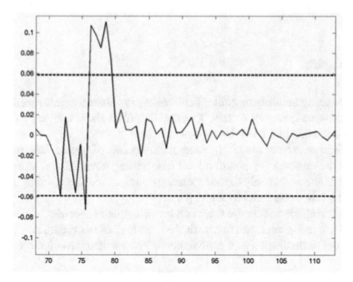

**Fig. 3** Building of control input signal spectrum based on the estimates of spectral density of original random signal by means of Wavelet Toolbox package in Matlab package, on abscissa axis are the signal samples, on ordinate axis is the control signal value [13]

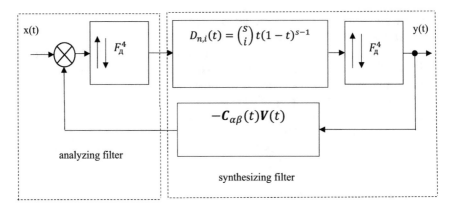

**Fig. 4** A block diagram of an adaptive controller with a multilayer self-learning neural network

Figure 4 shows a block diagram of a self-learning neural network. Based on the decomposition coefficients, the 4th-order Daubechies analyzing filter forms the nodal points of network characteristics tied to the values of propulsion-steering unit stops obtained during the system calibration in an experimental pool. The coefficients corresponding to the criteria of spectral estimates (16) are transmitted directly to the ROV.

According to the coefficients obtained on ROV board, an interpolated control function is synthesized on the basis of control tables. The decisive feedback (Fig. 4) makes it possible to isolate a perturbation component caused by medium impact on the underwater vehicle in a multiphase (multi-speed) control system.

## 4 Discussion

The idea of using an analyzing filter bank makes it possible to implement a multiphase (multi-speed) control system. Thresholding from above restricts the group of interpolation points of the tabular control system.

The reference control model by interpolation points of tabular description form (similar to the one on ROV board) allows transmitting not all signal traffic through the control channels, but only the interpolation points, according to which the control function curve is constructed by nodal points.

The analyzing filter structure works on the principle of recording the next count of original control implementation with the "erasing" of the rightmost count if the calculation of normalized signal convolution is greater than the condition:

$$0,707 \leq \widehat{F}^D \leq 1 \tag{18}$$

Then a new wavelet (filter) is interpolated by increasing the frequency (reducing the integration time by 0.066 ms), if less, then the frequency decreases by the same amount. The control neuron synthesis and its "collapse" is provided by threading from above (hard threading) and from below (soft threading) of signals according to their spectral estimates.

## 5 Conclusion

The proposed method of ROV control system synthesis using thresholding functions based on spectral estimates of analyzing and synthesizing filters assumes a reduction in control signal traffic transmission with a multilayer structure of adaptive algorithm.

This method is relevant when organizing a feedback with the use of a visual control channel where, besides the control commands, transmission and processing of video traffic from the ROV board take place, and two data streams superimposition will lead to time delays in control commands and video transmission from the ROV.

The neural network formed according to the estimates of control signals and feedback builds an intelligent control network with ROV dynamic parameters "observation" reducing the amount of information transmitted.

## References

1. Yuh J, Ura T, Bekey GA (eds) (2012) Underwater robots. Springer
2. Kheckert EV, Dantsevich IM, Liutikova MN, Khaleeva EP (2021) The technology of geophysical vessel control for the study of the world ocean. IOP Conf Ser Earth Environ Sci 872(1):012001
3. Liutikova MN, Dantsevich IM, Pankina SI (eds) (2021) The intelligent underwater laboratory. In: 1st international conference on marine geology and engineering. IOP conference series: earth and environmental science. IOP Publishing, Novorossiysk
4. Choi SK, Yuh J, Keevil N (1993) Design of omni-directional underwater robotic vehicle. In: Proceedings of OCEANS 1993. IEEE, pp I192–I197
5. Dantsevich I, Zviagintsev N, Tarasenko A (2012) Control of unmanned underwater vehicles: monograph
6. Valueva MV, Nagornov NN, Liakhov PA, Valuev GV, Cherviakov NI (2020) Application of the residue number system to reduce hardware costs of the convolutional neural network implementation. Math Comput Simul 177:232–243
7. Ma Y, Ma S, Wu Y, Pei X, Gorb SN, Wang Z, Zhou F (2018) Remote control over underwater dynamic attachment/detachment and locomotion. Adv Mater 30(30):1801595
8. Seto ML (ed) (2012) Marine robot autonomy. Springer, Canada
9. Wang YH, Wang SX, Xie CG (2007) Dynamic analysis and system design on an underwater glider propelled by temperature difference energy. J Tianjin Univ 40(2):133–138
10. Liao Y, Wang L, Li Y, Li Y, Jiang Q (2016) The intelligent control system and experiments for an unmanned wave glider. PLoS ONE 11(12):e0168792
11. Dantsevich IM, Liutikova MN, Novikov AY, Osmukha SA (2020) Analysis of a nonlinear system dynamics in the Morlet wavelet basis. IOP Conf Ser Mater Sci Eng 873(1):012035

12. Wang X, Yao X, Zhang L (2020) Path planning under constraints and path following control of autonomous underwater vehicle with dynamical uncertainties and wave disturbances. J Intell Rob Syst 99(3):891–908
13. Trauth MH, Gebbers R, Marwan N, Sillmann E (2007) MATLAB recipes for earth sciences, vol 34. Springer, Berlin
14. Cherviakov NI, Kondrashov IuV (2009) Development of methods of fast wavelet transformation with the help of Daubechies filters. Econ Inf 12(15):70

# Analytical Review of Methods for Detection, Localization and Error Correction in the Residue Number System

**A. Gladkov⑩, N. Gladkova⑩, and N. Kucherov⑩**

**Abstract** In the modern world, one of the main goods is information. In this regard, two main tasks arise: increasing the data transfer rate and checking the correctness of the transmitted data, and if errors are found, correcting them. One of the ways to speed up the processing and transmission of data is to use a Residue Number System (RNS) to represent them. In this paper, we have considered some methods for detecting and correcting errors in the data presented in the RNS. Two effective methods for finding and correcting errors are the projection method and the syndrome method. The article analyzes these two methods for excess RNS. The effectiveness of each of them for detecting, localizing and correcting errors is shown using specific examples. As a result of the considered example, conclusions were drawn about the effectiveness of the methods of the syndrome and projections. It was shown that with minimal redundancy, a high speed of the methods was obtained. It is noted in the work that for the projection method only 9 projections can be used to correct errors.

**Keywords** Error correction · Residue number system · Projection method · Syndrome method

## 1 Introduction

To solve the problems of correct data presentation during transmission and storage, error correction codes are used. Such codes are an important part of computing. As a rule, the detection of error correction in a redundant RNS (RRNS) is performed in three sequential stages: checking for errors, identifying erroneous digits of the remainder, and error correction. Existing multiple error detection and correction

A. Gladkov (✉) · N. Kucherov
North-Caucasus Federal University, Stavropol, Russia
e-mail: agladkov@ncfu.ru

N. Kucherov
e-mail: nkucherov@ncfu.ru

N. Gladkova
NCFU, North Caucasian Center for Mathematical Research, Stavropol, Russia

algorithms use three different methods to determine erroneous remainder digits. These are consistency checks using the syndrome [1, 2], reconstruction of numbers using the Chinese Remainder Theorem (CRT) [3, 4] and the module projection [2, 5]. These algorithms require recursive and iterative computations, comparisons that do not lend themselves to parallel hardware implementation.

## 2 Projection Method

Let us consider an approach to the detection and correction of multiple errors in the RRNS [6]. The bits of the RRNS residuals are divided into three groups so that any combination of errors can be uniquely identified by one of the seven error location categories. From the obtained representation of the remainder, three syndromes are deduced to detect up to 2t and correct up to t errors of the remainder digits, where the number of redundant modules is 2t. Compared to existing algorithms, the proposed algorithm is simpler and has a very low complexity of error decoding. The syndrome computation method calculates the difference between the received residual digits and the residual digits in extended bases. Syndromes $|\Delta|_{m_i}$ for $i = 1, 2, \ldots, k$, $k + 1, \ldots, k + 2t$ are grouped into the following sets:

$$S_1 \equiv \left\{ |\Delta|_{m_{k+1}}, |\Delta|_{m_{k+2}}, \ldots, |\Delta|_{m_{k+2t}} \right\}$$

$$S_2 \equiv \left\{ |\Delta|_{m_{k-t}}, |\Delta|_{m_{k-t+1}}, \ldots, |\Delta|_{m_{k+t-1}} \right\}$$

$$S_3 \equiv \left\{ |\Delta|_{m_{k-2t-1}}, |\Delta|_{m_{k-2t}}, \ldots, |\Delta|_{m_{k-2}} \right\}$$

$$\vdots$$

$$S_{[(k+t-1)/(t+1)]+1} \equiv \left\{ |\Delta|_{m_1}, |\Delta|_{m_2}, \ldots, |\Delta|_{m_{2t}} \right\} \tag{1}$$

where $k$ – is the number of information modules.

The location of the error of the received digits of the remainder is determined by $S_i$ for $i = 1, 2, \ldots, [(k + t - 1)/(t + 1)] + 1$, provided that the number of nonzero elements is less than or equal to t. An error-free remainder representation is then constructed by replacing the erroneous remainder digits with the corresponding base extended remainder digits. The complexity of this algorithm depends on the number of test sets and the number of syndromes calculated for each set.

**Example 1** Let's build a projection for the code (8,4)-RRNS.

| 12 | | | | | | | |
|----|----|----|----|----|----|----|----|
| 13 | 23 | | | | | | |
| 14 | 24 | 34 | | | | | |
| 15 | 25 | 35 | 45 | | | | |
| 16 | 26 | 36 | 46 | 56 | | | |
| 17 | 27 | 37 | 47 | 57 | 67 | | |
| 18 | 28 | 38 | 48 | 58 | 68 | 78 | |
| 19 | 29 | 39 | 49 | 59 | 69 | 79 | 89 |

According $t \leq \left[\frac{(n-k)}{2}\right]$ *or* $t \leq [r/2]$ the maximum error correction factor is $t = \frac{(8-4)}{2} = 2$ for a given code.

Taking into account that $DC = (\{x_1, x_2\}, \{x_3, x_4\}, \ldots, \{x_{n-1}, x_n\})$, we have $DC = (\{x_1, x_2\}, \{x_3, x_4\}, \{x_5, x_6\}, \{x_7, x_8\}, )$

Index $=$ choose$(\frac{n}{2}, t)$ we have Index $= (4,2) =$ $(\{1, 2\}, \{1, 3\}, \{1, 4\}, \{2, 3\}, \{2, 4\}, \{3, 4\})$, the number of modular protrusions is $C_4^2$ we have:

$$V = (\{x_1, x_2\} \cup \{x_3, x_4\}, \{x_1, x_2\} \cup \{x_5, x_6\}, \{x_1, x_2\} \cup \{x_7, x_8\}, \{x_3, x_4\}$$
$$\cup\{x_5, x_6\}, \{x_3, x_4\} \cup \{x_7, x_8\}, \{x_5, x_6\} \cup \{x_7, x_8\})$$
$$= (\{x_1, x_2, x_3, x_4\}, \{x_1, x_2, x_5, x_6\}, \{x_1, x_2, x_7, x_8\}, \{x_3, x_4, x_5, x_6\},$$
$$\{x_3, x_4, x_7, x_8\}, \{x_5, x_6, x_7, x_8\})$$

Using the brute force method, we find the number of errors and deleting them:

1) 1234: (12,13,14,23,24,34) 6 bugs fit
2) 1256: (12,15,16,25,26,56) 5 errors are suitable, since 12 have already used
3) 1278: (12,17,18,27,28,78) 5 errors are suitable, since 12 have already used
4) 3456: (34,35,36,45,46,56) 4 errors are suitable, since 34 and 54 have already used
5) 3489: (34,38,39,48,49,89) 5 errors are suitable, since 34 have already used
6) 1279: (12,17,19,27,29,79) 3 errors are suitable, since 12, 17 and 27 have already used
7) 3457: (34,35,37,45,47,57) 3 errors are suitable, since 34, 35 and 45 have already used
8) 5689: (56,58,59,68,69,89) 4 errors are suitable, since 56 and 89 have already used
9) 1267: (12,16,17,26,27,67) 1 error is suitable, since 12,16,17,26 and 27 have already been used

## 3 Syndrome Method

The method of syndromic decoding is based on the method of extending the RNS radix system. Its main essence is as follows:

Let, as a result of calculations in the RNS, a certain number $A = (a_1, a_2, \ldots, a_{n+r})$ is obtained. To determine the correctness of the number A, it is necessary to determine the values of the residuals $|A|_{p_{n+1}}, |A|_{p_{n+2}}, \ldots, |A|_{p_{n+r}}$ by control grounds, which are an extension of the Mixed Radix Conversion (MRC) base system.

To do this, you need to use the known residuals $a_1, a_2, \ldots, a_{n+r}$ and the corresponding decomposition matrix of orthogonal bases, calculated by the formula

$$A = a_1 + a_2 p_1 + a_3 p_1 p_2 + \cdots + a_n p_1 p_2 \cdots p_n,$$

by converting the number. A into MRC by the method of joint use of MRC and MRC, calculate $a_1, a_2, \ldots, a_{n+r}$. In order for the number A to lie in the initial range $[0, P]$, by the condition

$$a_{n+1} = a_{n+2} = \cdots = a_{n+r} = 0,$$

the values $a_{n+1}, a_{n+2}, \ldots, a_{n+r}$ must be equal to zero, otherwise the value of the number A is out of the dynamic range. This fact is used to find $|A|_{p_{n+1}}, |A|_{p_{n+2}}, \ldots, |A|_{p_{n+r}}$.

Then you should compare the residuals $a_{n+1}, a_{n+2}, \ldots, a_{n+r}$ with the values $|A|_{p_{n+1}}, |A|_{p_{n+2}}, \ldots, |A|_{p_{n+r}}$ in the same modules, but formed already in the course of calculations on the input data, similar to those that resulted in the number A. Comparison of residuals by control bases you can do them by subtracting:

$$\delta_i = \left| \alpha_{n+i} - |A|_{p+i} \right|_{p_{n+i}}, \text{ i.e. } i = 1, \ldots, r. \tag{2}$$

The numbers $\delta_i$ are called error syndromes, i.e. they allow you to determine the correctness of the calculation result in the RNS. Assuming that no more than l residuals turned out to be distorted in the resulting number A, for a code with a minimum distance $d_{min} = 2l + 1$, the following properties of error syndromes can be formulated:

- If all the values of the syndromes are equal to zero: $\delta_1 = \delta_2 = \cdots = \delta_i = 0$, then no error occurred, and vector A is a code word.
- If j values of syndromes $(1 \leq j \leq l)$ are not equal to zero, then errors have occurred in the corresponding residuals by control units, while the vector A is a codeword.
- If more than l values of syndromes are not equal to zero, then at least a single error has occurred in vector $A$.

Based on the numerical values of error syndromes, error constants are formed in such a way that when they are added to the information digits of the controlled number $A$ the error in the number is eliminated. In the case when an error was made due to an excess discharge, to correct it, you need to subtract the value of the error from the discharge in which it was made.

Further, on the basis of the considered examples, the processes of error correction on informational and redundant bases by the method of syndromic decoding will be presented.

## 4  Examples of Practical Application of the Syndrome Decoding Method

Let the RNS base system be given: $p_1 = 2$, $p_2 = 3$, $p_3 = 5$, $p_4 = 7$, $p_5 = 11$, and $p_1$, $p_2$, $p_3$ – are information modules, and $p_4$, $p_5$ – are control modules. Then the full range $P = 2 \cdot 3 \cdot 5 \cdot 7 \cdot 11 = 2310$ and the orthogonal bases of the system:

$$B_1 = 1155, \ B_2 = 1540, \ B_3 = 1386, \ B_4 = 330, \ B_5 = 210$$

In this case, the decomposition matrix of orthogonal bases in the MRC will have the form:

$$\begin{pmatrix} 1 & 1 & 2 & 3 & 5 \\ 0 & 2 & 1 & 2 & 7 \\ 0 & 0 & 1 & 4 & 6 \\ 0 & 0 & 0 & 4 & 1 \\ 0 & 0 & 0 & 0 & 1 \end{pmatrix}$$

Errors in the RNS are characterized by the depth $\Delta_{\text{огл}}$, determined by the value of the base $p_i$. The depth of error is the difference between a correct and a corrupted digit in base $p_i$. Therefore, depending on whether $(\Delta_+)$ is added or $(\Delta_-)$ is subtracted, the error value will be considered based on an example.

Let the initial number be $A = 17 = (1, 2, 2, 3, 6)$. Let us assume that an information error has occurred with the addition of the error value. The wrong number $\overline{A}$ has the form: $\overline{A} = (1, 2, 4, 3, 6)$. Let's analyze this number. For this, we use the translation of the number $\overline{A}$ from RNS to MRC, performed using the base extension method.

We define error syndromes using the formula (2)

$$\delta_1 = |3 - 1|_7 = 2, \ \delta_2 = |6 - 7|_{11} = 10$$

We find the base and depth of the error according to [1], while using the values opposite to the error syndromes, since the error occurred with the addition of the error value.

For base $p_1$:

$$P_1 = \frac{P}{p_1} = \frac{30}{2} = 15$$

$$|\lambda_{11}|_7 = \left|-\delta_1 \cdot P_1^{-1}\right|_7 = \left|-\frac{2}{15}\right|_7 = |-2 \cdot 1|_7 = 5$$

$$|\lambda_{12}|_{11} = \left|-\delta_2 \cdot P_1^{-1}\right|_{11} = \left|-\frac{10}{15}\right|_{11} = \left|-\frac{2}{3}\right| = |-2 \cdot 4|_{11} = 3$$

$$5 \neq 3$$

For base $p_2$:

$$P_2 = \frac{P}{p_2} = \frac{30}{3} = 10$$

$$|\lambda_{21}|_7 = \left|-\delta_1 \cdot P_2^{-1}\right|_7 = \left|-\frac{2}{10}\right|_7 = \left|-\frac{1}{5}\right| = |-1 \cdot 3|_7 = 4$$

$$|\lambda_{22}|_{11} = \left|-\delta_2 \cdot P_2^{-1}\right|_{11} = \left|-\frac{10}{10}\right|_{11} = |-1| = 10$$

$$4 \neq 10$$

For base $p_2$:

$$P_3 = \frac{P}{p_3} = \frac{30}{5} = 6$$

$$|\lambda_{31}|_7 = \left|-\delta_1 \cdot P_3^{-1}\right|_7 = \left|-\frac{2}{6}\right|_7 = \left|-\frac{1}{3}\right| = |-1 \cdot 5|_7 = 2$$

$$|\lambda_{32}|_{11} = \left|-\delta_2 \cdot P_3^{-1}\right|_{11} = \left|-\frac{10}{6}\right|_{11} = \left|-\frac{5}{3}\right| = |-5 \cdot 4| = 2$$

$$2 = 2$$

Therefore, the error occurred at the base $p_3$, while the magnitude of the error is $\Delta_+ = |r \cdot P_3|_{p_3} = |2 - 6|_5 = 2$. Since the error occurred with the addition of the error value, then to correct it, the same error value must be subtracted

$$(0, 0, 0) - (0, 0, 2) = (0, 0, 3)$$

According to the results obtained, error syndromes $\delta_1 = 2$ and $\delta_2 = 10$ correspond to the error constant $(0, 0, 3)$

Using the result obtained, you can correct the error in the number:

$$(1, 2, 4, 3, 6) - (0, 0, 2, 0, 0) = (1, 2, 2, 3, 6)$$

Arguing similarly if the error is on an informational basis with the subtraction of the error value. In this case, to correct it, the same value must be added.

Based on the results obtained in the example, we can conclude that each error constant corresponds to two sets of error syndromes values, depending on whether the error value is added or subtracted. Moreover, the error constants are universal in the sense that they do not depend on the number, but only on the basis of the system.

Error constants for RNS with bases $p_1 = 2$, $p_2 = 3$, $p_3 = 5$, $p_4 = 7$, $p_5 = 11$, can be written in the table

| Radical error $p_1$ | $\delta_1$ | $\delta_2$ | Radical error $p_2$ | $\delta_1$ | $\delta_2$ | Radical error $p_3$ | $\delta_1$ | $\delta_2$ |
|---|---|---|---|---|---|---|---|---|
| | $\Delta_+$ | $\Delta_-$ | | $\Delta_+$ | $\Delta_-$ | | $\Delta_+$ | $\Delta_-$ |
| (1, 0, 0) | 6, 7 | 1, 4 | (0, 1, 0) | 1, 2 | 3, 10 | (0, 0, 1) | 6, 6 | 4, 9 |
| | | | (0, 2, 0) | 4, 1 | 6, 9 | (0, 0, 2) | 5, 1 | 3, 4 |
| | | | | | | (0, 0, 3) | 2, 10 | 4, 7 |
| | | | | | | (0, 0, 4) | 3, 2 | 1, 5 |

# 5 Conclusion

The article examines the problem of reducing the computational complexity of algorithms for detecting, localizing and correcting errors in RNS. It is shown that the universal projection method has exponential computational complexity depending on the number of errors to be corrected. Reducing the computational complexity of the projection method can be achieved by using the maximum likelihood method, which allows you to replace the overflow criterion with the Hamming distance criterion. However, using the Hamming distance allows you to reduce the number of computed errors by a factor of two or more, only for codes that correct one error in the RRNS. If the error correction code in RRNS corrects two or more errors, then the maximum likelihood method asymptotically converges to the classical projection method.

An alternative way to solve the problem is to use the syndrome method. The main disadvantage of the syndrome method in RNS is the large error tables. Various techniques are used to reduce the size of tables, but at the same time, the size of the table can be reduced only if the number of control bases is two more than the number of working bases, then by calculating three syndromes, you can localize and correct the error.

In the future, it is planned to eliminate the problem of the projection method by using the approaches from the projection method.

**Acknowledgements** The responded study was funded by Russian Federation President Grant MK-24.2020.9.

# References

1. Yau SS, Liu YC (1973) Error correction in redundant residue number systems. IEEE Trans Comput 100(1):5–11
2. Sun JD, Krishna H (1992) A coding theory approach to error control in redundant residue number systems. II. Multiple Error detection and correction. IEEE Trans Circuit Syst II: Analog Digital Signal Process **39**(1):18–34
3. Mandelbaum DM (1976) On a class of arithmetic codes and a decoding algorithm proof: if t 5 r, then M $/C mod M, has exactly t nonzero. IEEE Trans Inf Theory
4. Goldreich O, Ron D, Sudan M (1999). Chinese remaindering with errors. In Proceedings of the thirty-first annual ACM symposium on theory of computing. pp 1330–1338
5. Goh VT, Siddiqi MU (2008) Multiple error detection and correction based on redundant residue number systems. IEEE Trans Commun 56(3):325–330
6. Tay TF, Chang CH (2015) A non-iterative multiple residue digit error detection and correction algorithm in RRNS. IEEE Trans Comput 65(2):396–408

# A High-Order Difference Scheme for the Diffusion Equation of Multi-term and Distributed Orders

**A. Alikhanov , A. Apekov , and C. Huang**

**Abstract** A difference schemes of a higher order of approximation for the time-fractional diffusion equation of multi-term and distributed orders are constructed on the basis of the L2 type formula. The stability and convergence of the proposed difference schemes is proved by the method of energy inequalities.

**Keywords** Time-fractional diffusion equation · Finite difference method · Stability · Convergence

## 1 Introduction

Differential equations with fractional order derivatives provide a powerful mathematical tool for accurate and realistic description of physical and chemical processes proceeding in media with fractal geometry [1, 3–6]. It is known that the order of a fractional derivative depends on the fractal dimension of medium [7, 8]. It is therefore reasonable to construct mathematical models based on partial differential equations with the distributed order derivatives [1, 9–11]. Analytical methods for solving such equations are scarcely effective, so that the development of the corresponding numerical methods is very important.

The initial-boundary-value problems for the generalized multi-term time fractional diffusion equation over an open bounded domain $G \times (0, T)$, $G \in \mathbb{R}^n$ were

A. Alikhanov (✉) · A. Apekov
North Caucasus Center for Mathematical Research, North-Caucasus Federal University,
355017 Stavropol, Russia
e-mail: aaalikhanov@gmail.com

A. Apekov
e-mail: amapekov@ncfu.ru

C. Huang
School of Mathematics and Statistics, Huazhong University of Science and Technology,
Wuhan 430074, China
e-mail: cmhuang@mail.ntut.edu.tw

considered [12]. Multi-term linear and non-linear diffusion-wave equations of fractional order were solved in [13] using the Adomian decomposition method. Applications of the homotopy analysis and new modified homotopy perturbation methods to solutions of multi-term linear and nonlinear diffusion-wave equations of fractional order are discussed in [14, 15].

Using the energy inequality method, a priori estimates for the solution of the Dirichlet, Robin and nonlocal boundary value problems for the fractional, variable and distributed orders diffusion equation with Caputo fractional derivative have been obtained [16–22].

In [21] a super-convergence point is found for the interpolation approximation of the linear combination of multi-term fractional derivatives. The derived numerical differentiation formula can achieve not less than the second order accuracy. Then the formula is used to numerically solve the time multi-term and distributed-order fractional sub-diffusion equations.

In the present paper we use $L2$ type difference analog of the fractional Caputo derivative with the order of approximation $\mathcal{O}(\tau^{3-\alpha})$ for each $\in (0, 1)$. Difference schemes of the second order of approximation in space and the $(3 - \alpha_0)$th-order in time for the diffusion equation of multi-term and distributed orders with variable coefficients are built. By means of the method of energy inequalities, the stability and convergence of these schemes are proven. Numerical computations of some test problems confirming reliability of the obtained results are implemented.

## 2  Materials and Methods

### 2.1  The L2 Type Fractional Numerical Differentiation Formula

Let us consider the uniform grid $\overline{\omega}_\tau = \{t_j = j\tau,\ j = 0, 1, \ldots, M;\ T = \tau M\}$. For the Caputo fractional derivative of the order $\alpha_r$, $0 < \alpha_r < 1$, of the function $u(t) \in C^3[0, T]$ at the fixed point $t_{j+1}$, $j \in \{1, 2, \ldots, M - 1\}$ the following equalities are valid

$$\partial_{0t_{j+1}}^{\alpha_r} u(t) = \frac{1}{\Gamma(1 - \alpha_r)} \int_0^{t_{j+1}} \frac{u'^{(\eta)} d\eta}{\left(t_{j+1} - \eta\right)^{\alpha_r}}$$

$$= \frac{1}{\Gamma(1 - \alpha_r)} \int_0^{t_2} \frac{u'^{(\eta)} d\eta}{\left(t_{j+1} - \eta\right)^{\alpha_r}} + \frac{1}{\Gamma(1 - \alpha_r)} \sum_{s=2}^{j} \int_{t_s}^{t_{s+1}} \frac{u'^{(\eta)} d\eta}{\left(t_{j+1} - \eta\right)^{\alpha_r}}, \tag{1}$$

where

$$\partial_{0t}^{\alpha_r} u(t) = \frac{1}{\Gamma(1 - \alpha_r)} \int_0^t \frac{u'^{(\eta)} d\eta}{(t - \eta)^{\alpha_r}}$$

is the Caputo fractional derivative of order $\alpha_r \in (0, 1)$.

As shown in the paper [23] on each interval $\left[t_s, t_{s+1}\right]$ ($1 \leq s \leq j$), applying the quadratic interpolation $\Pi_{2,s}u(t)$ of $u(t)$ that uses three points $(t_{s-1}, u(t_{s-1}))$, $(t_s, u(t_s))$ and $(t_{s+1}, u(t_{s+1}))$, we arrive at L2 type difference analogue of Caputo fractional derivative

$$\Delta^{\alpha_r}_{0t_{j+1}} u = \frac{\tau^{1-\alpha_r}}{\Gamma(2-\alpha_r)} \sum_{s=0}^{j} c^{(\alpha_r)}_{j-s} u_{t,s}, \tag{2}$$

where

$$a^{(\alpha_r)}_l = (l+1)^{1-\alpha_r} - l^{1-\alpha_r},$$

$$b^{(\alpha_r)}_l = \frac{1}{2-\alpha_r}\left[(l+1)^{2-\alpha_r} - l^{2-\alpha_r}\right] - \frac{1}{2}\left[(l+1)^{1-\alpha_r} + l^{1-\alpha_r}\right], \quad l \geq 0,$$

for $j = 1$

$$c^{(\alpha_r)}_s = \begin{cases} a^{(\alpha_r)}_0 + b^{(\alpha_r)}_0 + b^{(\alpha_r)}_1, & s = 0, \\ a^{(\alpha_r)}_1 - b^{(\alpha_r)}_1 - b^{(\alpha_r)}_0, & s = 1, \end{cases} \tag{3}$$

for $j = 2$

$$c^{(\alpha_r)}_s = \begin{cases} a^{(\alpha_r)}_0 + b^{(\alpha_r)}_0, & s = 0, \\ a^{(\alpha_r)}_1 + b^{(\alpha_r)}_1 + b^{(\alpha_r)}_2 - b^{(\alpha_r)}_0, & s = 1, \\ a^{(\alpha_r)}_2 - b^{(\alpha_r)}_2 - b^{(\alpha_r)}_1, & s = 2, \end{cases} \tag{4}$$

and for $j \geq 3$,

$$c^{(\alpha_r)}_s = \begin{cases} a^{(\alpha_r)}_0 + b^{(\alpha_r)}_0, & s = 0, \\ a^{(\alpha_r)}_s + b^{(\alpha_r)}_s - b^{(\alpha_r)}_{s-1}, & 1 \leq s \leq j-2, \\ a^{(\alpha_r)}_{j-1} + b^{(\alpha_r)}_{j-1} + b^{(\alpha_r)}_j - b^{(\alpha_r)}_{j-2}, & s = j-1, \\ a^{(\alpha_r)}_j - b^{(\alpha_r)}_j - b^{(\alpha_r)}_{j-1}, & s = j. \end{cases} \tag{5}$$

**Lemma 1** [23] *For any $\alpha \in (0, 1)$, $j = 1, 2, \ldots, M-1$ and $u(t) \in C^3\left[0, t_{j+1}\right]$*

$$\left| \partial^{\alpha}_{0t_{j+1}} u - \Delta^{\alpha}_{0t_{j+1}} u \right| = \mathcal{O}(\tau^{3-\alpha}). \tag{6}$$

For $j = 1$ we have

$$c^{(\alpha)}_0 = \frac{2+\alpha}{2^{\alpha}(2-\alpha)}, \quad c^{(\alpha)}_1 = \frac{2-3\alpha}{2^{\alpha}(2-\alpha)}, \quad c^{(\alpha)}_0 + 3c^{(\alpha)}_1 = \frac{2^{3-\alpha}(1-\alpha)}{2-\alpha} > 0.$$

For $j \geq 2$, the next lemma shows properties of the coefficient $c_s^{(\alpha)}$ defined in (4) and (5).

**Lemma 2** [23] *For any* $\alpha \in (0, 1)$ *and* $c_s^{(\alpha)}$ $(0 \leq s \leq j, j \geq 2)$ *the following inequalities are valid*

$$\frac{11}{16} \cdot \frac{1-\alpha}{(j+1)^\alpha} < c_j^{(\alpha)} < \frac{1-\alpha}{j^\alpha}, \tag{7}$$

$$c_0^{(\alpha)} > c_2^{(\alpha)} > c_3^{(\alpha)} > \ldots > c_{j-2}^{(\alpha)} > c_{j-1}^{(\alpha)} > c_j^{(\alpha)}, \tag{8}$$

$$c_0^{(\alpha)} + 3c_1^{(\alpha)} - 4c_2^{(\alpha)} > 0. \tag{9}$$

**Lemma 3** [23] *For any real constants* $c_0, c_1$ *such that* $c_0 \geq \max\{c_1, -3c_1\}$, *and* $\{v_j\}_{j=0}^{j=M}$ *the following inequality holds*

$$v_{j+1}\big(c_0 v_{j+1} - (c_0 - c_1)v_j - c_1 v_{j-1}\big) \geq E_{j+1} - E_j, \quad j = 1, \ldots, M-1, \tag{10}$$

where

$$E_j = \left(\frac{1}{2}\sqrt{\frac{c_0 - c_1}{2}} + \frac{1}{2}\sqrt{\frac{c_0 + 3c_1}{2}}\right)^2 v_j^2$$

$$+ \left(\sqrt{\frac{c_0 - c_1}{2}}v_j - \left(\frac{1}{2}\sqrt{\frac{c_0 - c_1}{2}} + \frac{1}{2}\sqrt{\frac{c_0 + 3c_1}{2}}\right)v_{j-1}\right)^2,$$

$$j = 1, 2, \ldots, M.$$

**Lemma 4** [23] *For any function* $v(t)$ *defined on the grid* $\overline{\omega}_\tau$ *one has the inequality*

$$v_{j+1}\Delta_{0t_{j+1}}^\alpha v \geq \frac{\tau^{-\alpha}}{\Gamma(2-\alpha)}\big(E_{j+1} - E_j\big) + \frac{1}{2}\overline{\Delta}_{0t_{j+1}}^\alpha v^2 = \frac{\tau^{-\alpha}}{\Gamma(2-\alpha)}\big(\mathcal{E}_{j+1} - \mathcal{E}_j\big) - \frac{\tau^{-\alpha}}{2\Gamma(2-\alpha)}\overline{c}_j^{(\alpha)} v_0^2, \tag{11}$$

where

$$\overline{\Delta}_{0t_{j+1}}^\alpha v = \frac{\tau^{-\alpha}}{\Gamma(2-\alpha)}\sum_{s=0}^{j}\overline{c}_{j-s}^{(\alpha)}(v_{s+1} - v_s), \quad j = 1, 2, \ldots, M,$$

$$\overline{c}_0^{(\alpha)} = c_2^{(\alpha)}, \ \overline{c}_1^{(\alpha)} = c_2^{(\alpha)}, \ \overline{c}_s^{(\alpha)} = c_s^{(\alpha)}, \quad s = 2, 3, \ldots, j,$$

$$\text{for} \quad j = 1, 2, 3, \ldots, M, \quad E_j = E_j\big(c_0^{(\alpha)} - c_2^{(\alpha)}, c_1^{(\alpha)} - c_2^{(\alpha)}\big),$$

$$\mathcal{E}_j = E_j + \frac{1}{2} \sum_{s=0}^{j-1} \overline{c}_{j-1-s}^{(\alpha)} v_{s+1}^2.$$

## 2.2 A Difference Scheme for the Multi-term Time-Fractional Diffusion Equation

In this section we consider the multi-term time fractional diffusion equation with variable coefficients

$$\partial_{0t}^{\alpha_0} u + \sum_{r=1}^{m} w_r \partial_{0t}^{\alpha_r} u = \frac{\partial}{\partial x}\left(k(x,t)\frac{\partial u}{\partial x}\right) \tag{12}$$
$$-q(x,t)u + f(x,t), \quad 0 < x < l, \quad 0 < t \le T,$$

$$u(0,t) = 0, \ u(l,t) = 0, \ 0 \le t \le T, \ u(x,0) = u_0(x), \ 0 \le x \le l, \tag{13}$$

where $0 < \alpha_m < \ldots < \alpha_0 < 1$, $w_r \ge 0$, $k(x,t) \ge c_1 > 0$, $q(x,t) \ge 0$.

In [21] for such an equation a difference scheme of a higher order of approximation using the $L2\text{-}1_\sigma$ - formula was constructed and its stability and convergence were proved for sufficiently small values of $\tau$. The difference analogue of the high-order of approximation of the Caputo fractional derivative proposed in this paper can be used to solve problem (12) and (13) without any restrictions on the grid steps. In addition, the approximation order of the difference scheme is higher than $L2\text{-}1_\sigma$ - type difference scheme.

Let the following difference scheme be put into a correspondence with differential problem (12) and (13)

$$\Delta_{0t_{j+1}}^{\alpha} y_i + \sum_{r=1}^{m} w_r \Delta_{0t_{j+1}}^{\alpha_r} y_i = \left(a y_{\bar{x}}^{j+1}\right)_{x,i} \tag{14}$$
$$-d y_i^{j+1} + \varphi_i^{j+1}, \quad i = 1, 2, \ldots, N-1, \quad j = 1, 2, \ldots, M-1,$$

$$y(0,t) = 0, \ y(l,t) = 0, \ t \in \overline{\omega}_\tau, \ y(x,0) = u_0(x), \ x \in \overline{\omega}_h, \tag{15}$$

where $y_{\bar{x},i} = (y_i - y_{i-1})/h$, $y_{x,i} = (y_{i+1} - y_i)/h$, $a_i^{j+1} = k\left(x_{i-\frac{1}{2}}, t_{j+1}\right)$, $d_i^{j+1} = q(x_i, t_{j+1})$, $\varphi_i^{j+1} = f(x_i, t_{j+1})$.

The difference scheme (14) and (15) has the approximation order $\mathcal{O}(h^2 + \tau^{3-\alpha_0})$. We assume that the solution $y_i^1$ and its difference derivative $y_{\bar{x},i}^1$ are found with the order of accuracy $\mathcal{O}(h^2 + \tau^{3-\alpha_0})$. For example, we can use $L1$-formula [19] and solve problem (12) and (13) on the time layer $[0, \tau]$ with step $\tau_1 = \mathcal{O}\left(\tau^{\frac{3-\alpha_0}{2-\alpha_0}}\right)$.

## 2.3  Stability and Convergence

**Theorem 1** *The difference scheme* (14) *and* (15) *is unconditionally stable and its solution satisfies the following a priori estimate:*

$$
\sum_{j=1}^{M-1}\left(\|y^{j+1}\|_0^2 + \|y_{\bar{x}}^{j+1}\|_0^2\right)
$$

$$
\tau \le M_1\left(\|y^1\|_0^2 + \|y_{\bar{x}}^1\|_0^2 + \|y^0\|_0^2 + \|y_{\bar{x}}^0\|_0^2 + \sum_{j=1}^{M-1}\|\varphi^{j+1}\|_0^2\tau\right),
$$

(16)

where $M_1 > 0$ is a known number independent of $h$ and $\tau$.

**Proof** Taking the inner product of the Eq. (14) with $y^{j+1}$, we have

$$
\left(y^{j+1}, \Delta_{0t_{j+1}}^{\alpha_0} y\right) + \sum_{r=1}^{m} w_r\left(y^{j+1}, \Delta_{0t_{j+1}}^{\alpha_r} y\right)
$$

$$
-\left(y^{j+1}, \left(ay_{\bar{x}}^{j+1}\right)_x\right) + \left(dy^{j+1}, y^{j+1}\right) = \left(y^{j+1}, \varphi^{j+1}\right).
$$

(17)

Using Lemma 4, we obtain

$$
\left(y^{j+1}, \Delta_{0t_{j+1}}^{\alpha_r} y\right) \ge \frac{\tau^{-\alpha_r}}{\Gamma(2-\alpha_r)}\left(E_{j+1}^{(\alpha_r)} - E_j^{(\alpha_r)}\right) + \frac{1}{2}\overline{\Delta}_{0t_{j+1}}^{\alpha_r}\|y\|_0^2
$$

$$
= \frac{\tau^{-\alpha_r}}{\Gamma(2-\alpha_r)}\left(\mathcal{E}_{j+1}^{(\alpha_r)} - \mathcal{E}_j^{(\alpha_r)}\right)
$$

$$
- \frac{\tau^{-\alpha_r}}{2\Gamma(2-\alpha_r)}\overline{c}_j^{(\alpha_r)}\|y^0\|_0^2, \quad j = 1, 2, \ldots, M-1, \ r = 0, 1, \ldots, m,
$$

where

$$
E_j^{(\alpha)} = \left(\frac{1}{2}\sqrt{\frac{c_0^{(\alpha)} - c_1^{(\alpha)}}{2}} + \frac{1}{2}\sqrt{\frac{c_0^{(\alpha)} + 3c_1^{(\alpha)} - 4c_2^{(\alpha)}}{2}}\right)^2 \|y^j\|_0^2
$$

$$
+ \left(\sqrt{\frac{c_0^{(\alpha)} - c_1^{(\alpha)}}{2}}\|y^j\|_0 - \left(\frac{1}{2}\sqrt{\frac{c_0^{(\alpha)} - c_1^{(\alpha)}}{2}} + \frac{1}{2}\sqrt{\frac{c_0^{(\alpha)} + 3c_1^{(\alpha)} - 4c_2^{(\alpha)}}{2}}\right)\|y^{j-1}\|_0\right)^2.
$$

$$
\mathcal{E}_j^{(\alpha)} = E_j^{(\alpha)} + \frac{1}{2}\sum_{s=0}^{j-1}\overline{c}_{j-1-s}^{(\alpha)}\|y^{s+1}\|_0^2.
$$

From (17), using that

$$-\left(y^{j+1}, \left(ay_{\bar{x}}^{j+1}\right)_x\right) + \left(dy^{j+1}, y^{j+1}\right) \geq c_1 \|y_{\bar{x}}^{j+1}\|_0^2,$$

$$\left(y^{j+1}, \varphi^{j+1}\right) \leq \frac{c_1}{l^2}\|y^{j+1}\|_0^2 + \frac{l^2}{4c_1}\|\varphi^{j+1}\|_0^2 \leq \frac{c_1}{2}\|y_{\bar{x}}^{j+1}\|_0^2 + \frac{l^2}{4c_1}\|\varphi^{j+1}\|_0^2,$$

one obtains the inequality

$$\frac{\tau^{-\alpha_0}}{\Gamma(2-\alpha_0)}\left(\mathcal{E}_{j+1}^{(\alpha_0)} - \mathcal{E}_j^{(\alpha_0)}\right) + \sum_{r=1}^{m} w_r \frac{\tau^{-\alpha_r}}{\Gamma(2-\alpha_r)}\left(\mathcal{E}_{j+1}^{(\alpha_r)} - \mathcal{E}_j^{(\alpha_r)}\right) + \frac{c_1}{2}\|y_{\bar{x}}^{j+1}\|_0^2$$

$$\leq \frac{l^2}{4c_1}\|\varphi^{j+1}\|_0^2 + \left(\frac{\tau^{-\alpha_0}}{2\Gamma(2-\alpha_0)}\bar{c}_j^{(\alpha_0)} + \sum_{r=1}^{m} w_r \frac{\tau^{-\alpha_r}}{2\Gamma(2-\alpha_r)}\bar{c}_j^{(\alpha_r)}\right)\|y^0\|_0^2. \quad (18)$$

Multiplying inequality (18) by $\tau$ and summing the resulting relation over $j$ from 1 to $M - 1$ and taking into account inequality (7), one obtains a priori estimate (16).

The stability and convergence of the difference scheme (14) and (15) follow from the a priori estimate (16).

## 2.4 A Difference Scheme for the Time-Fractional Diffusion Equation of Distributed Order

In this section we consider the time fractional diffusion equation of distributed order with variable coefficients

$$\int_0^1 w(\alpha)\partial_{0t}^\alpha u\, d\alpha = \frac{\partial}{\partial x}\left(k(x, t)\frac{\partial u}{\partial x}\right) - q(x, t)u + f(x, t), \quad 0 < x < l, \ 0 < t \leq T,$$

$$(19)$$

$$u(0, t) = 0, \ u(l, t) = 0, \ 0 \leq t \leq T, \ u(x, 0) = u_0(x), \ 0 \leq x \leq l, \quad (20)$$

To the differential problem (19) and (20) we assign the following difference scheme

$$\sum_{r=0}^{m} w_r h_\alpha \Delta_{0t_{j+1}}^{\alpha_r} y_i = \left(ay_{\bar{x}}^{j+1}\right)_{x,i}$$

$$(21)$$

$$-dy_i^{j+1} + \varphi_i^{j+1}, \ i = 1, 2, \ldots, N - 1, \ j = 1, 2, \ldots, M - 1,$$

$$y(0, t) = 0, \ y(l, t) = 0, \ t \in \bar{\omega}_\tau, \ y(x, 0) = u_0(x), \ x \in \bar{\omega}_h, \quad (22)$$

where $a_i^{j+1} = k\left(x_{i-\frac{1}{2}}, t_{j+1}\right)$, $d_i^{j+1} = q\left(x_i, t_{j+1}\right)$, $\varphi_i^{j+1} = f\left(x_i, t_{j+1}\right)$,

$$\int_0^1 w(\alpha)\partial_{0t}^\alpha u d\alpha = \sum_{r=0}^m w_r h_\alpha \Delta_{0t_{j+1}}^{\alpha_r} u_i + O\left(h_\alpha^4\right)$$

is a numerical integration over the variable $\alpha$ according to Simpson's rule with a step $h_\alpha = 1/m$.

The difference scheme (21) and (22) has the approximation order $O\left(h^2 + \tau^2 + h_\alpha^4\right)$.

We assume that the solution $y_i^1$ and its difference derivative $y_{\bar{x},i}^1$ are found with the order of accuracy $O\left(h^2 + \tau^2 + h_\alpha^4\right)$. For example, we can use $L1$-formula [19] and solve problem (19) and (20) on the time layer $[0, \tau]$ with step $\tau_1 = O\left(\tau^2\right)$.

## 2.5 Stability and Convergence

**Theorem 2** *The difference scheme* (21) *and* (22) *is unconditionally stable and its solution satisfies the following a priori estimate:*

$$\sum_{j=1}^{M-1}\left(\|y^{j+1}\|_0^2 + \|y_{\bar{x}}^{j+1}]\|_0^2\right)$$

$$\tau \leq M_2\left(\|y^1\|_0^2 + \|y_{\bar{x}}^1]\|_0^2 + \|y^0\|_0^2 + \|y_{\bar{x}}^0]\|_0^2 + \sum_{j=1}^{M-1}\|\varphi^{j+1}\|_0^2\tau\right), \tag{23}$$

where $M_2 > 0$ is a known number independent of $h$ and $\tau$.

**Proof** The proof is similar to the proof of Theorem 1. □

The stability and convergence of the difference scheme (21) and (22) follow from the a priori estimate (23).

## 3  Conclusion

In the current paper a new difference schemes of the second approximation order in space and the $3 - \alpha_0$ approximation order in time for the time-fractional diffusion equation of multi-term and distributed orders with variable coefficients are constructed. The stability and convergence of these schemes with the rate equal to the order of the approximation error are proved.

**Acknowledgements** The first and second authors acknowledge the financial support from the North-Caucasus Center for Mathematical Research under agreement no. 075-02-2022-892 with the

Ministry of Science and Higher Education of the Russian Federation. The first and third authors also jointly funded by RFBR (no. 20-51-53007) and NSFC (no. 12011530058)

# References

1. Nakhushev AM (2003) Fractional calculus and its applications, FIZMATLIT, Moscow, (in Russian)
2. Oldham KB, Spanier J (1974) The fractional calculus. Academic Press, New York
3. Podlubny I (1999) Fractional differential equations. Academic Press, San Diego
4. Hilfer R (2000) Applications of fractional calculus in physics. World Scientific, Singapore
5. Kilbas AA, Srivastava HM, Trujillo JJ (2006) Theory and applications of fractional differential equation. Elsevier, Amsterdam
6. Uchaikin VV (2008) Method of fractional derivatives. Artishok, Ul'janovsk
7. Kobelev VL, Kobelev YL, Romanov EP (1998) Non-debye relaxation and diffusion in fractal space. Dokl Akad Nauk 361:755–758
8. Kobelev VL, Kobelev YL, Romanov EP (1999) Self-maintained processes in the case of nonlinear fractal diffusion. Dokl Akad Nauk 369:332–333
9. Lorenzo CF, Hartley TT (2002) Variable order and distributed order fractional operators. Nonlin Dynam 29:57–98
10. Jiao Z, Chen Y, Podlubny I (2012) Distributed-order dynamic systems. Stability, simulation, applications and perspectives. Springer, London
11. Luchko Y (2009) Boundary value problems for the generalized time-fractional diffusion equation of distributed order. Fract Calc Appl Anal 12:409–422
12. Luchko Y (2011) Initial-boundary-value problems for the generalized multi-term time-fractional diffusion equation. J Math Anal Appl 374:538–548
13. Daftardar-Gejji V, Bhalekar S (2008) Solving multi-term linear and non-linear diffusion-wave equations of fractional order by Adomian decomposition method. Appl Math Comput 202:113–120
14. Jafari H, Golbabai A, Seifi S, Sayevand K (2010) Homotopy analysis method for solving multi-term linear and nonlinear diffusion–wave equations of fractional order. Comput Math Appl 59:1337–1344
15. Jafari H, Aminataei A (2011) An algorithm for solving multi-term diffusion-wave equations of fractional order. Comput Math Appl T 62:1091–1097
16. Alikhanov AA (2010) A priori estimates for solutions of boundary value problems for fractional-order equations. Differ Equ 46:660–666
17. Alikhanov AA (2015) Stability and convergence of difference schemes approximating a two-parameter nonlocal boundary value problem for time-fractional diffusion equation. Comput Math Model 26:252–272
18. Alikhanov AA (2012) Boundary value problems for the diffusion equation of the variable order in differential and difference settings. Appl Math Comput 219:3938–3946
19. Alikhanov AA (2015) Numerical methods of solutions of boundary value problems for the multi-term variable-distributed order diffusion equation. Appl Math Comput 268:12–22
20. Alikhanov AA (2015) A new deference scheme for the time fractional diffusion equation. J Comput Phys 280:424–438
21. Gao G-H, Alikhanov AA, Sun Z-Z (2017) The temporal second order difference schemes based on the interpolation approximation for solving the time multiterm and distributed-order fractional sub-diffusion equations. J Sci Comput 73:93–121
22. Du R, Alikhanov AA, Sun Z-Z (2020) Temporal second order difference schemes for the multi-dimensional variable-order time fractional sub-diffusion equations. Comput Math Appl 79:2952–2972
23. Alikhanov AA, Huang C (2021) A high-order L2 type difference scheme for the time-fractional diffusion equation. Appl Numer Math 172:546–565

# A Locally One-Dimensional Difference Scheme for a Multidimensional Integro-Differential Equation of Parabolic Type of General Form

**Z. V. Beshtokova** ⓘ

**Abstract** The first boundary value problem for a multidimensional integro-differential equation of parabolic type of general form with variable coefficients is investigated. To solve numerically the multidimensional problem, a locally one-dimensional difference scheme is constructed, the essence of the idea of which is to reduce the transition from layer to layer to sequential solving of a number of one-dimensional problems in each of the coordinate directions. It is shown that the approximation error for the locally one-dimensional scheme is $O(|h|^2 + \tau^2)$, where $|h|^2 = h_1^2, h_2^2, ..., h_p^2$. Using the method of energy inequalities in the $L_2-$norm, for the solution of a locally one-dimensional difference scheme, an a priori estimate is obtained. The obtained estimate implies uniqueness, stability with respect to the right-hand side and initial data, as well as the convergence of the solution of the difference problem to the solution of the original differential problem with a rate equal to the approximation error. In the two-dimensional case (for p = 2), an algorithm for finding the approximate solution of the problem under consideration is constructed and numerical calculations of test examples are carried out, illustrating the theoretical calculations obtained in the work.

**Keywords** First initial-boundary value problem · Locally one-dimensional scheme · A priori estimate · Difference scheme · Parabolic equation · Integro-differential equation · Equation with memory

## 1 Introduction

The work is devoted to the construction of a locally one-dimensional (economical) difference scheme for the numerical solution of the first initial-boundary value problem for a multidimensional integro-differential equation in partial derivatives of parabolic type of general form, the main idea of which is to reduce the transition from layer to layer to the sequential solution of a number of one-dimensional problems

Z. V. Beshtokova (✉)
Institute of Applied Mathematics and Automation, Kabardino-Balkar Research Center of Russian Academy of Sciences, Nalchik, Russia
e-mail: zarabaeva@yandex.ru

© The Author(s), under exclusive license to Springer Nature Switzerland AG 2022          525
A. Tchernykh et al. (eds.), *Mathematics and its Applications in New Computer Systems*,
Lecture Notes in Networks and Systems 424,
https://doi.org/10.1007/978-3-030-97020-8_48

in each of the coordinate directions. Moreover, although each of the intermediate problems may not approximate the original differential problem, in the aggregate and in special norms such an approximation takes place. These methods are called splitting methods, which were developed in the works of Douglas J., Peaceman D.W., Rachford H.H. [1, 2], N.N. Yanenko [3], A.A. Samarsky [4–8], G.I. Marchuk [9], E.G. Dyakonova [10, 11] and others.

The works are devoted to the construction of locally one-dimensional schemes for the numerical solution of multidimensional parabolic equations: [6–8, 12–17]

Thus, in [6], in an arbitrary domain G, a locally one-dimensional scheme is considered for solving linear and quasilinear parabolic equations. The stability of the difference scheme with respect to the right-hand side, boundary and initial data is proved, as well as convergence with a rate $O(h^2 + \tau)$. In [7], locally one-dimensional difference schemes are considered on arbitrary «nonuniform grids» for linear and quasilinear equations of parabolic type with «heat conductivity coefficient» $k_\alpha = k_\alpha(x, t, u)$ depending on the «temperature» $u = u(x, t)$. These schemes converge on arbitrary non-uniform grids $\omega_h$. In [8], locally one-dimensional difference schemes were studied for hyperbolic equations in an arbitrary domain G. These schemes converge on arbitrary nonuniform grids $\omega_h$.

In the work [12] locally-one-dimensional difference schemes for the fractional diffusion equation in multidimensional domains are considered. Stability and convergence of locally one-dimensional schemes for this equation are proved. In [13] for a fractional diffusion equation with Robin boundary conditions, locally one-dimensional difference schemes are considered and their stability and convergence are proved.

Work [14] is devoted to the construction of locally one-dimensional schemes for a parabolic equation with a nonstationary boundary condition, when a concentrated heat capacity of a certain quantity is placed on the boundary of the domain. Using the maximum principle, an a priori estimate is obtained in the uniform metric, which implies the convergence of the difference scheme on a cubic grid. In [15] locally one-dimensional difference schemes are considered as applied to a fractional diffusion equation with variable coefficients in a domain of complex geometry. They are proved to be stable and uniformly convergent for the problem under study.

This work is a continuation of the author's series of works [16, 17] devoted to the study of local and nonlocal boundary value problems for multidimensional parabolic equations.

## 2 Materials and Methods

### 2.1 Problem statement

In a cylinder $\overline{Q}_T = \overline{G} \times [0, T]$, the base of which is a $p$-dimensional rectangular parallelepiped $\overline{G} = \{x = (x_1, x_2, ..., x_p) : 0 \leq x_\alpha \leq l_\alpha, \alpha = 1, 2, ..., p\}$ with boundary $\Gamma$, $\overline{G} = G \cup \Gamma$, consider the problem

$$\frac{\partial u}{\partial t} = Lu + f(x, t), \quad (x, t) \in Q_T, \tag{1}$$

$$u|_\Gamma = 0, \quad 0 \le t \le T, \tag{2}$$

$$u(x, 0) = u_0(x), \quad x \in \overline{G}, \tag{3}$$

where $Lu = \sum\limits_{\alpha=1}^{p} L_\alpha u,$

$$L_\alpha u = \frac{\partial}{\partial x_\alpha}\left(k_\alpha(x, t)\frac{\partial u}{\partial x_\alpha}\right) + r_\alpha(x, t)\frac{\partial u}{\partial x_\alpha} - q_\alpha(x, t)u - \frac{1}{p}\int\limits_0^t K(x, t, \tau)u(x, \tau)d\tau,$$

$$0 < c_0 \le k_\alpha(x, t) \le c_1, |K(x, t, \tau)|, |r_\alpha|, |q_\alpha| \le c_2,$$

$$k_\alpha(x, t) \in C^{3,1}(Q_T), \quad K(x, t, \tau), \quad r_\alpha(x, t), \quad q_\alpha(x, t) \in C^{2,1}(Q_T),$$

$c_0, c_1, c_2$—положительные постоянные, $\alpha = 1, 2, ..., p, 0 < \tau < t$.

In what follows, $M_i, i = 1, 2, ...,$ denote positive constants depending only on the input data of the problem under consideration.

## 2.2 Locally One-Dimensional Scheme

On the segment $[0, T]$ we introduce a uniform grid $\overline{\omega}_\tau = \{t_j = j\tau, = 0, 1, ..., j_0\}$ with a step $\tau = T/j_0$. We divide each interval $(t_j, t_{j+1})$ into $p$ parts by points $t_{j+\frac{\alpha}{p}} = t_j + \tau\frac{\alpha}{p}, \alpha = 1, 2, ..., p$ and denote by $\Delta_\alpha = (t_{j+\frac{\alpha-1}{p}}, t_{j+\frac{\alpha}{p}}]$.

We choose the spatial grid to be uniform in each direction $Ox_\alpha$ with a step $h_\alpha = \frac{l_\alpha}{N_\alpha}, \alpha = 1, 2, ..., p$ :

$$\omega_h = \prod\limits_{\alpha=1}^{p} \omega_{h_\alpha}, \omega_{h_\alpha} = \{x_\alpha^{(i_\alpha)} = i_\alpha h_\alpha : i_\alpha = 1, ..., N_\alpha - 1, \alpha = 1, 2, ..., p\}.$$

Equation (1) can be rewritten as

$$\sum\limits_{\alpha=1}^{p} \Re_\alpha u = 0, \Re_\alpha u = \frac{1}{p}\frac{\partial u}{\partial t} - L_\alpha u - f_\alpha, \sum\limits_{\alpha=1}^{p} f_\alpha = f.$$

where $f_\alpha(x, t)$, $(\alpha = 1, 2, ..., p)$— arbitrary functions with the same smoothness as $f(x, t)$ and satisfying the normalization condition $\sum_{\alpha=1}^{p} f_\alpha = f$.

On each half-interval $\Delta_\alpha$, $\alpha = 1, 2, ..., p$ we will successively solve the problems

$$\mathfrak{R}_\alpha \vartheta_{(\alpha)} = \frac{1}{p} \frac{\partial \vartheta_{(\alpha)}}{\partial t} - L_\alpha \vartheta_{(\alpha)} - f_\alpha = 0, x \in G, t \in \Delta_\alpha, \tag{4}$$

$$\begin{cases} \vartheta_{(\alpha)} = 0, x_\alpha = 0, \\ \vartheta_{(\alpha)} = 0, x_\alpha = l_\alpha, \end{cases} \tag{5}$$

assuming at the same time [20, стр. 522]

$$\vartheta_{(1)}(x, 0) = u_0(x), \vartheta_{(1)}(x, t_j) = \vartheta_{(p)}(x, t_j), j = 1, 2, ...,$$

$$\vartheta_{(\alpha)}(x, t_{j+\frac{\alpha-1}{p}}) = \vartheta_{(\alpha-1)}(x, t_{j+\frac{\alpha-1}{p}}), j = 2, 3, ..., p.$$

Similarly [20, p. 401], we obtain for the last equation of number $\alpha$ a monotonic scheme of the second order of approximation in $h_\alpha$. To do this, we consider the last equation for a fixed $\alpha$ with a perturbed operator $\tilde{L}_\alpha$ :

$$\frac{1}{p} \frac{\partial \vartheta_{(\alpha)}}{\partial t} = \tilde{L}_\alpha \vartheta_{(\alpha)} + f_{(\alpha)}, t \in \Delta_\alpha, \alpha = 1, 2, ..., p, \tag{6}$$

where

$$\tilde{L}_\alpha \vartheta_{(\alpha)} = \chi_\alpha \frac{\partial}{\partial x_\alpha} \left( k_\alpha(x, t) \frac{\partial \vartheta_{(\alpha)}}{\partial x_\alpha} \right) + r_\alpha(x, t) \frac{\partial \vartheta_{(\alpha)}}{\partial x_\alpha} - q_\alpha(x, t) \vartheta_{(\alpha)}$$

$$- \frac{1}{p} \int_0^t K(x, t, \tau) \vartheta_{(\alpha)}(x, \tau) d\tau,$$

$\varkappa_\alpha = \frac{1}{1+R_\alpha}$, $R_\alpha = \frac{0.5 h_\alpha |r_\alpha|}{k_\alpha}$ — is the difference Reynolds number.
We approximate each Eq. (6) of number $\alpha$ by an implicit scheme on the half-interval $\Delta_\alpha$, then we obtain a chain of p one-dimensional difference equations:

$$\frac{y^{j+\frac{\alpha}{p}} - y^{j+\frac{\alpha-1}{p}}}{\tau} = \tilde{\Lambda}_\alpha y^{j+\frac{\alpha}{p}} + \varphi_\alpha^{j+\frac{\alpha}{p}}, \alpha = 1, 2, ..., p, \tag{7}$$

$$y^{j+\frac{\alpha}{p}}|_{\gamma_{h,\alpha}} = 0, \tag{8}$$

$$y(x, 0) = u_0(x), \tag{9}$$

where

$$\tilde{\Lambda}_\alpha y^{j+\frac{\alpha}{p}} = \varkappa_\alpha (a_\alpha y_{\bar{x}_\alpha}^{j+\frac{\alpha}{p}})_{x_\alpha} + b_\alpha^+ a_\alpha^{(+1_\alpha)} y_{x_\alpha}^{j+\frac{\alpha}{p}} + b_\alpha^- a_\alpha y_{\bar{x}_\alpha}^{j+\frac{\alpha}{p}} - d_\alpha y^{j+\frac{\alpha}{p}}$$

$$-\frac{1}{p}\sum\nolimits_{j'=0}^{j} K\left(x_1, x_2, \ldots, x_p; t_j, t_{j'}\right) y\left(x, t^{j'+\frac{\alpha}{p}}\right)\tau,$$

$$r_\alpha^+ = 0.5(r_\alpha + |r_\alpha|) \geq 0, r_\alpha^- = 0.5(r_\alpha - |r_\alpha|) \leq 0, b_\alpha^+ = \frac{r_\alpha^+}{k_\alpha}, b_\alpha^- = \frac{r_\alpha^-}{k_\alpha}, \bar{t} = t^{j+\frac{1}{2}},$$

$$r_\alpha = r_\alpha^+ + r_\alpha^-, a^{(1_\alpha)} = a_{i_\alpha+1,\alpha} = k_\alpha(x_{i_\alpha-1/2}, \bar{t}), \varphi_\alpha^{j+\frac{\alpha}{p}} = f_\alpha(x, \bar{t}), d_\alpha = q_\alpha(x_{i_\alpha}, \bar{t}),$$

$$x^{(-0.5\alpha)} = (x_1, \ldots, x_{\alpha-1}, x_\alpha - 0.5h_\alpha, x_{\alpha+1}, \ldots, x_p), \gamma_{h,\alpha}-\text{set of boundary nodes in}$$
the $x_\alpha$ direction.

## 2.3 Approximation Error for a Locally One-Dimensional Scheme

The characteristic of the accuracy of a locally one-dimensional scheme is the difference $z^{j+\frac{\alpha}{p}} = y^{j+\frac{\alpha}{p}} - u^{j+\frac{\alpha}{p}}$, where $u^{j+\frac{\alpha}{p}}$ — be the solution of problem (1)–(2). Substituting $y^{j+\frac{\alpha}{p}} = z^{j+\frac{\alpha}{p}} + u^{j+\frac{\alpha}{p}}$ into the difference problem (7)–(9), we obtain the following problem for the error $z^{j+\frac{\alpha}{p}}$ :

$$\frac{z^{j+\frac{\alpha}{p}} - z^{j+\frac{\alpha-1}{p}}}{\tau} = \tilde{\Lambda}_\alpha z^{j+\frac{\alpha}{p}} + \tilde{\psi}_\alpha^{j+\frac{\alpha}{p}}, \tag{10}$$

$$z^{j+\frac{\alpha}{p}} = 0 \text{ if } x \in \gamma_{h,\alpha}, \ z(x,0) = 0. \tag{11}$$

Denoting by

$$\psi_\alpha^o = \left(L_\alpha u + f_\alpha - \frac{1}{p}\frac{\partial u}{\partial t}\right)^{j+1/2},$$

and noticing that $\sum_{\alpha=1}^{p}\psi_\alpha^o = 0$, if $\sum_{\alpha=1}^{p} f_\alpha = f$, we represent $\psi_\alpha^{j+\frac{\alpha}{p}}$ in the form $\psi_\alpha^{j+\frac{\alpha}{p}} = \psi_\alpha^o + \psi_\alpha^*$ :

$$\psi_\alpha^{j+\frac{\alpha}{p}} = \tilde{\Lambda}_\alpha u^{j+\frac{\alpha}{p}} + \varphi^{j+\frac{\alpha}{p}} - \frac{u^{j+\frac{\alpha}{p}} - u^{j+\frac{\alpha-1}{p}}}{\tau} + \psi_\alpha^o - \psi_\alpha^o = \left(\tilde{\Lambda}_\alpha u^{j+\frac{\alpha}{p}} - L_\alpha u^{j+\frac{1}{2}}\right)$$

$$+ \left(\varphi_\alpha^{j+\frac{\alpha}{p}} - f_\alpha^{j+\frac{1}{2}}\right) - \left(\frac{u^{j+\frac{\alpha}{p}} - u^{j+\frac{\alpha-1}{p}}}{\tau} - \frac{1}{p}\left(\frac{\partial u}{\partial t}\right)^{j+1/2}\right) + \psi_\alpha^o = \psi_\alpha^o + \psi_\alpha^*.$$

Obviously $\psi_\alpha^* = O(h_\alpha^2 + \tau), \psi_\alpha^o = O(1),$

$$\sum_{\alpha=1}^{p} \psi_{\alpha}^{j+\frac{\alpha}{p}} = \sum_{\alpha=1}^{p} \psi_{\alpha}^{o} + \sum_{\alpha=1}^{p} \psi_{\alpha}^{*} = O(|h|^2 + \tau), |h|^2 = h_1^2 + h_2^2 + \dots + h_p^2.$$

## 2.4 Stability of a Locally One-Dimensional Scheme

Let us multiply Eq. (7) scalarly by $y^{(\alpha)} = y^{j+\frac{\alpha}{p}}$ :

$$\left(\frac{1}{p}y_{\bar{t}}^{(\alpha)}, y^{(\alpha)}\right) - \left(\tilde{\Lambda}_{\alpha}y^{(\alpha)}, y^{(\alpha)}\right) = \left(\varphi_{\alpha}, y^{(\alpha)}\right), \tag{12}$$

where

$$\frac{1}{p}y_{\bar{t}}^{(\alpha)} = \frac{y^{j+\frac{\alpha}{p}} - y^{j+\frac{\alpha-1}{p}}}{\tau}, \ (u,v)_{\alpha} = \sum_{i_\alpha=1}^{N_\alpha-1} u_{i_\alpha} v_{i_\alpha} h_\alpha, \ \|y^{(\alpha)}\|_{L_2(\alpha)}^2 = \sum_{i_\alpha=1}^{N_\alpha-1} y^2 h_\alpha,$$

$$(u,v) = \sum_{x\in\omega_h} uvH, \ H = \prod_{\alpha=1}^{p} h_\alpha, \ \|y^{(\alpha)}\|_{L_2(\omega_h)}^2 = \sum_{i_\beta \neq i_\alpha} \|y^{(\alpha)}\|_{L_2(\alpha)}^2 H/h_\alpha.$$

Transforming each term in identity (12) using the Cauchy inequality with $\varepsilon$, the Cauchy-Bunyakovsky-Schwarz inequality, lemma 1 [18], and, choosing $h \leq h_0 = \min\left\{\frac{2k_\alpha}{|r_\alpha|}, \frac{1}{c_3}\right\}$, we get

$$\left(\|y^{(\alpha)}\|_{L_2(\omega_h)}^2\right)_{\bar{t}} + \|y_{\bar{x}_\alpha}^{(\alpha)}\|_{L_2(\omega_h)}^2 \leq M_1 \sum_{j'=0}^{j} \tau\|y\left(x, t^{j'+\frac{\alpha}{p}}\right)\|_{L_2(\omega_h)}^2 + M_2\|y^{j+\frac{\alpha}{p}}\|_{L_2(\omega_h)}^2$$

$$+M_3\|\varphi^{j+\frac{\alpha}{p}}\|_{L_2(\omega_h)}^2. \tag{13}$$

Let us summarize (13) first over $\alpha$ from 1 to p:

$$\left(\|y\|_{L_2(\omega_h)}^2\right)_{\bar{t}} + \sum_{\alpha=1}^{p} \|y_{\bar{x}_\alpha}^{j+\frac{\alpha}{p}}\|_{L_2(\omega_h)}^2 \leq M_1 \sum_{\alpha=1}^{p} \sum_{j'=0}^{j} \|y\left(x, t^{j'+\frac{\alpha}{p}}\right)\|_{L_2(\omega_h)}^2 \tau$$

$$+M_2 \sum_{\alpha=1}^{p} \|y^{j+\frac{\alpha}{p}}\|_{L_2(\omega_h)}^2 + M_3 \sum_{\alpha=1}^{p} \|\varphi^{j+\frac{\alpha}{p}}\|_{L_2(\omega_h)}^2,$$

and then over $j'$ from 0 to $j$:

$$\|y^{j+1}\|^2_{L_2(\omega_h)} + \sum_{j'=0}^{j} \tau \sum_{\alpha=1}^{p} \|y_{\overline{x}_\alpha}^{j'+\frac{\alpha}{p}}\|^2_{L_2(\omega_h)} \le M_1 \sum_{j'=0}^{j} \tau \sum_{\alpha=1}^{p} \sum_{s=0}^{j'} \tau \|y\left(x, t^{s+\frac{\alpha}{p}}\right)\|^2_{L_2(\omega_h)}$$

$$+M_2 \sum_{j'=0}^{j} \tau \sum_{\alpha=1}^{p} \|y^{j'+\frac{\alpha}{p}}\|^2_{L_2(\omega_h)} + M_4 \left( \sum_{j'=0}^{j} \tau \sum_{\alpha=1}^{p} \|\varphi^{j'+\frac{\alpha}{p}}\|^2_{L_2(\omega_h)} + \|y^0\|^2_{L_2(\omega_h)} \right)$$

or

$$\|y^{j+1}\|^2_{L_2(\omega_h)} + \sum_{j'=0}^{j} \sum_{\alpha=1}^{p} \|y_{\overline{x}_\alpha}^{j'+\frac{\alpha}{p}}]\|^2_{L_2(\omega_h)} \le M_5 \sum_{j'=0}^{j} \tau \sum_{\alpha=1}^{p} \|y\left(x, t^{j'+\frac{\alpha}{p}}\right)\|^2_{L_2(\omega_h)}$$

$$+M_6 \left( \sum_{j'=0}^{j} \tau \sum_{\alpha=1}^{p} \|\varphi^{j'+\frac{\alpha}{p}}\|^2_{L_2(\omega_h)} + \|y^0\|^2_{L_2(\omega_h)} \right) \tag{14}$$

From (14), we have

$$\|y^{j+1}\|^2_{L_2(\omega_h)} \le M_5 \sum_{j'=0}^{j} \tau \sum_{\alpha=1}^{p} \|y^{j'+\frac{\alpha}{p}}\|^2_{L_2(\omega_h)} + M_6 F^j, \tag{15}$$

where $F^j = \sum_{j'=0}^{j} \tau \sum_{\alpha=1}^{p} \|\varphi^{j'+\frac{\alpha}{p}}\|^2_{L_2(\omega_h)} + \|y^0\|^2_{L_2(\omega_h)}$, $M_5 = TM_1 + M_2$.
Let us show that the following inequality holds

$$\max_{1 \le \alpha \le p} \|y^{j+\frac{\alpha}{p}}\|^2_{L_2(\omega_h)} \le \nu_1 \sum_{j'=0}^{j-1} \tau \max_{1 \le \alpha \le p} \|y^{j'+\frac{\alpha}{p}}\|^2_{L_2(\omega_h)} + \nu_2 F^j, \tag{16}$$

where $\nu_1, \nu_2-$ are known positive constants.
To this end, we rewrite inequality (13) in the following form

$$\|y^{j+\frac{\alpha}{p}}\|^2_{L_2(\omega_h)} \le \|y^{j+\frac{\alpha-1}{p}}\|^2_{L_2(\omega_h)} + M_1 \tau \sum_{j'=0}^{j} \tau \|y^{j'+\frac{\alpha}{p}}\|^2_{L_2(\omega_h)}$$

$$+M_2 \tau \|y^{j+\frac{\alpha}{p}}\|^2_{L_2(\omega_h)} + M_3 \tau \|\varphi^{j+\frac{\alpha}{p}}\|^2_{L_2(\omega_h)}. \tag{17}$$

Summing (17) over $\alpha'$ from 1 to $\alpha$, then we get

$$\|y^{j+\frac{\alpha}{p}}\|^2_{L_2(\omega_h)} \le \|y^j\|^2_{L_2(\omega_h)} + M_1 \sum_{\alpha'=1}^{\alpha} \tau \sum_{j'=0}^{j} \tau \|y^{j'+\frac{\alpha'}{p}}\|^2_{L_2(\omega_h)}$$

$$+M_2\tau \sum_{\alpha'=1}^{\alpha} \|y^{j+\frac{\alpha'}{p}}\|^2_{L_2(\omega_h)} + M_3\tau \sum_{\alpha'=1}^{\alpha} \|\varphi^{j+\frac{\alpha'}{p}}\|^2_{L_2(\omega_h)} \le \|y^j\|^2_{L_2(\omega_h)}$$

$$+M_1 \sum_{\alpha=1}^{p} \tau \sum_{j'=0}^{j} \tau \|y^{j'+\frac{\alpha}{p}}\|^2_{L_2(\omega_h)} + M_2\tau \sum_{\alpha=1}^{p} \|y^{j+\frac{\alpha}{p}}\|^2_{L_2(\omega_h)}$$

$$+M_3\tau \sum_{\alpha=1}^{p} \|\varphi^{j+\frac{\alpha}{p}}\|^2_{L_2(\omega_h)}. \tag{18}$$

Without loss of generality, we can assume that

$$\max_{1\le\alpha'\le p} \|y^{j+\frac{\alpha'}{p}}\|^2_{L_2(\omega_h)} = \|y^{j+\frac{\alpha}{p}}\|^2_{L_2(\omega_h)},$$

otherwise (17) will be summed up to such a value of $\alpha$ that $\|y^{j+\frac{\alpha}{p}}\|^2_{L_2(\omega_h)}$ reaches its maximum value for a fixed j. Then (18) can be rewritten as

$$\max_{1\le\alpha\le p} \|y^{j+\frac{\alpha}{p}}\|^2_{L_2(\omega_h)} \le \|y^j\|^2_{L_2(\omega_h)} + pM_2\tau \max_{1\le\alpha\le p} \|y^{j+\frac{\alpha}{p}}\|^2_{L_2(\omega_h)}$$

$$+pM_1T \sum_{j'=0}^{j} \max_{1\le\alpha\le p} \|y^{j'+\frac{\alpha}{p}}\|^2_{L_2(\omega_h)}\tau + M_3\tau \sum_{\alpha=1}^{p} \|\varphi^{j+\frac{\alpha}{p}}\|^2_{L_2(\omega_h)}. \tag{19}$$

We rewrite (19) once again in the following form

$$(1 - pM_2\tau) \max_{1\le\alpha\le p} \|y^{j+\frac{\alpha}{p}}\|^2_{L_2(\omega_h)} \le pM_1T \sum_{j'=0}^{j} \max_{1\le\alpha\le p} \|y^{j'+\frac{\alpha}{p}}\|^2_{L_2(\omega_h)}\tau$$

$$+\|y^j\|^2_{L_2(\omega_h)} + M_3\tau \sum_{\alpha=1}^{p} \|\varphi^{j+\frac{\alpha}{p}}\|^2_{L_2(\omega_h)}. \tag{20}$$

Choosing $\tau \le \tau_0 = \frac{1}{2pM_2}$, from (20), we find

$$\max_{1\le\alpha\le p} \|y^{j+\frac{\alpha}{p}}\|^2_{L_2(\omega_h)} \le M_7 \sum_{j'=0}^{j} \max_{1\le\alpha\le p} \|y^{j'+\frac{\alpha}{p}}\|^2_{L_2(\omega_h)}\tau + M_8\overline{F}^j, \tag{21}$$

where $\overline{F}^j = \|y^j\|^2_{L_2(\omega_h)} + \tau \sum_{\alpha=1}^{p} \|\varphi^{j+\frac{\alpha}{p}}\|^2_{L_2(\omega_h)}$.

Based on Lemma 4 [19, p. 171], from (21), we obtain the a priori estimate

$$\max_{1 \leq \alpha \leq p} \|y^{j+\frac{\alpha}{p}}\|^2_{L_2(\omega_h)} \leq M_9 \|y^j\|^2_{L_2(\omega_h)} + \tau M_{10} \sum_{\alpha=1}^{p} \|\varphi^{j+\frac{\alpha}{p}}\|^2_{L_2(\omega_h)}. \qquad (22)$$

Since it follows from (15) that

$$\|y^j\|^2_{L_2(\omega_h)} \leq M_5 \sum_{j'=0}^{j-1} \tau \max_{1 \leq \alpha \leq p} \|y^{j'+\frac{\alpha}{p}}\|^2_{L_2(\omega_h)} + M_6 F^j,$$

then, from (22), we have

$$\max_{1 \leq \alpha \leq p} \|y^{j+\frac{\alpha}{p}}\|^2_{L_2(\omega_h)} \leq \nu_1 \sum_{j'=0}^{j-1} \tau \max_{1 \leq \alpha \leq p} \|y^{j'+\frac{\alpha}{p}}\|^2_{L_2(\omega_h)} + \nu_2 F^j.$$

Introducing the notation $g_{j+1} = \max_{1 \leq \alpha \leq p} \|y^{j+\frac{\alpha}{p}}\|^2_{L_2(\omega_h)}$, the last inequality can be rewritten as follows:

$$g_{j+1} \leq \nu_1 \sum_{k=1}^{j} \tau g_k + \nu_2 F^j, \qquad (23)$$

where $\nu_1$, $\nu_2-$ are known positive constants.

Applying Lemma 4 [19, p. 171] to (23), from (14), we obtain the a priori estimate

$$\|y^{j+1}\|^2_{L_2(\omega_h)} + \sum_{j'=0}^{j} \tau \sum_{\alpha=1}^{p} \|y_{\bar{x}_\alpha}^{j'+\frac{\alpha}{p}}\|^2_{L_2(\omega_h)}$$

$$\leq M \left( \sum_{j=0}^{j} \tau \sum_{\alpha=1}^{p} \|\varphi^{j'+\frac{\alpha}{p}}\|^2_{L_2(\omega_h)} + \|y^0\|^2_{L_2(\omega_h)} \right), \qquad (24)$$

where $M = const > 0$ does not depend on $h_\alpha$ or $\tau$.

## 2.5 Convergence of a Locally One-Dimensional Scheme

By analogy with [20], we represent the solution of the problem for the error $z_{(\alpha)} = z^{j+\frac{\alpha}{p}}$ as the sum $z_{(\alpha)} = \upsilon_{(\alpha)} + \eta_{(\alpha)}$, where $\eta_{(\alpha)}$ is determined by the conditions

$$\frac{\eta_{(\alpha)} - \eta_{(\alpha-1)}}{\tau} = \psi_\alpha^o, x \in \omega_h + \gamma_\alpha, \alpha = 1, 2, ..., p, \qquad (25)$$

$$\eta(x, 0) = 0.$$

From (25), it follows that $\eta^{j+1} = \eta_{(p)} = \eta^j + \tau(\psi_1^o + \psi_2^o + ... + \psi_p^o) = \eta^j = ... = \eta^0 = 0$. For $\eta^\alpha = \tau(\psi_1^o + \psi_2^o + ... + \psi_\alpha^o) = -\tau(\psi_{\alpha+1}^o + ... + \psi_p^o) = O(\tau)$.
The function $\upsilon_{(\alpha)}$ is determined by the conditions

$$\frac{\upsilon_{(\alpha)} - \upsilon_{(\alpha-1)}}{\tau} = \Lambda_\alpha \upsilon_{(\alpha)} + \widetilde{\psi}_\alpha, x \in \omega_h, \alpha = 1, 2, ..., p, \tag{26}$$

$$\upsilon_{(\alpha)} = -\eta_\alpha, x_\alpha \in \gamma_{h,\alpha}, \upsilon(x, 0) = 0, \widetilde{\psi}_\alpha = \psi_\alpha^* + \Lambda_\alpha \eta_{(\alpha)}.$$

If there exist continuous in the closed domain $\overline{Q}_T$ derivatives $\frac{\partial^4 u}{\partial x_\alpha^2 \partial x_\beta^2}$, $\alpha \neq \beta$, then $\Lambda_\alpha \eta_{(\alpha)} = -\tau \Lambda_\alpha(\psi_{\alpha+1}^o + ... + \psi_p^o) = O(\tau)$.
We estimate the solution of problem (26) using (26).
Since $\eta^j = 0$, $\eta_{(\alpha)} = O(\tau)$, $\|z^j\| \leq \|\upsilon^j\|$, then it follows from estimate (24)

## 3 Results

So, the following are true.
   **Theorem 1.** The locally one-dimensional scheme (7)–(9) is stable with respect to the right-hand side and initial data, so estimate (24) is valid for the solution of the difference problem (7)–(9) with $h \leq h_0, \tau \leq \tau_0$.
   **Theorem 2.** Let problem (1) –(3) have a unique solution u (x, t) continuous in $\overline{Q}_T$ and there exist derivatives also continuous in $\overline{Q}_T$

$$\frac{\partial^2 u}{\partial t^2}, \frac{\partial^4 u}{\partial x_\alpha^2 \partial x_\beta^2}, \frac{\partial^3 u}{\partial x_\alpha^2 \partial t}, \frac{\partial^2 f}{\partial x_\alpha^2}, \alpha = 1, 2, ..., p, \alpha \neq \beta,$$

then the locally one-dimensional scheme (7)–(9) converges to the solution of the differential problem (1) –(3) with the rate $O(|h|^2 + \tau)$, so that for sufficiently small h, $\tau$ the following estimate is valid

$$\|y^{j+1} - u^{j+1}\|_1 \leq M(|h|^2 + \tau), 0 < h \leq h_0, 0 < \tau \leq \tau_0,$$

where $\|z^{j+1}\|_1 = \left( \|z^{j+1}\|_{L_2(\omega_h)}^2 + \sum_{j'=0}^j \tau \sum_{\alpha=1}^p \|z_{\overline{x}_\alpha}^{j'+\frac{\alpha}{p}}\|_{L_2(\omega_h)}^2 \right)^{\frac{1}{2}}$, $|h|^2 = h_1^2 + h_2^2 + ... + h_p^2$.

**Numerical Experiment**
The coefficients of the equation and boundary conditions of problem (1)–(4) are selected so that the exact solution of the problem is the function $u(x, t) = t^3(x_1^4 - l_1 x_1^3)(x_2^4 - l_2 x_2^3)$.
   Below in Tables 1, 2, we present the maximum value of the error ($z = y - u$) and the computational order of convergence (CO) in the norms $\| \cdot \|_{L_2(\omega_{h\tau})}$ and $\| \cdot \|_{C(\omega_{h\tau})}$,

**Table 1** Change of the error in the norm $\| \cdot \|_{L_2(w_{h\tau})}$ while the mesh size is decreasing: $t = 1$, when $\overline{h} = h_1 = h_2 = \sqrt{\tau}$

| $\overline{h}$ | Maximum value of the error | CO in $\| \cdot \|_{L_2(w_{h\tau})}$ |
|---|---|---|
| 1/20 | 0.001857970 | |
| 1/40 | 0.000593163 | 1.6472 |
| 1/80 | 0.000162875 | 1.8647 |

**Table 2** Change of the error in the norm $\| \cdot \|_{C(w_{h\tau})}$ while the mesh size is decreasing: $t = 1$, when $\overline{h} = h_1 = h_2 = \sqrt{\tau}$

| $\overline{h}$ | Maximum value of the error | CO in $\| \cdot \|_{C(w_{h\tau})}$ |
|---|---|---|
| 1/20 | 0.007803440 | |
| 1/40 | 0.002881834 | 1.4371 |
| 1/80 | 0.000856583 | 1.7503 |

where $\|y\|_{C(w_{h\tau})} = \max_{(x_i,t_j)\in w_{h\tau}} |y|$, when $\overline{h} = h_1 = h_2 = \sqrt{\tau}$, while the mesh size is decreasing. The error is being reduced in accordance with the order of approximation $O(h^2 + (\sqrt{\tau})^2)$.

## 4 Conclusion

The Dirichlet problem is studied for a multidimensional integro-differential equation of parabolic type of general form with variable coefficients. A locally one-dimensional difference scheme is constructed. Using the method of energy inequalities, an a priori estimate is obtained for the solution of a locally one-dimensional difference scheme. The obtained estimate implies the uniqueness, stability, and also the convergence of the solution of the difference problem to the solution of the original differential problem with a rate equal to the approximation error. Numerical calculations of test examples are carried out to illustrate the theoretical calculations obtained in the work.

## References

1. Douglas J, Rachford HH (1956) On the numerical solution of heat conduction problems in two and three space variables. Trans Amer Math Soc 82(2):421–439
2. Peaceman DW, Rachford HH (1955) The numerical solution of parabolic and elliptic differential equations. J Industr Math Soc 3(1):28–41
3. Yanenko NN (1963) On the convergence of the splitting method for the heat conductivity equation with variable coefficients. USSR Computat Math Math Phys 2(5):1094–1100. https://doi.org/10.1016/0041-5553(63)90516-0
4. Samarskii AA (1963) Homogeneous difference schemes on non-uniform nets for equations of parabolic type. USSR Comput Math Math Phys 3(2):351–393
5. Samarskii AA (1980) Some problems of the theory of differential equations. Differ Uravn 16(11):1925–1935

6. Samarskii AA (1963) On an economical difference method for the solution of a multi-dimensional parabolic equation in an arbitrary region. USSR Comput Math Math Phys 2(5):894–926

7. Samarskii AA (1963) Local one dimensional difference schemes on non-uniform nets. USSR Comput Math Math Phys 3(3):572–619

8. Samarskii AA (1964) Local one-dimensional difference schemes for multi-dimensional hyperbolic equations in an arbitrary region. USSR Comput Math Math Phys 4(4):21–35

9. Marchuk GI (1995) Splitting-up methods for non-stationary problems. Comput Math Math Phys 35(6):667–671

10. D'yakonov EG (1962) Difference schemes with a splitting operator for nonstationary equations. Dokl Akad Nauk SSSR 144(1):29–32

11. D'yakonov EG (1963) Difference schemes with a "disintegrating" operator for multidimensional problems. USSR Computat Math Math Phys 2(4):581–607. https://doi.org/10.1016/0041-5553(63)90531-7

12. Lafisheva MM, Shhanukov-Lafishev MKH (2008) Locally one-dimensional difference schemes for the fractional order diffusion equation. Comput Math Math Phys 48(10):1875–1884

13. Bazzaev AK, Shkhanukov-Lafishev MKh (2010) A locally one-dimensional scheme for a fractional-order diffusion equation with boundary conditions of the third kind. Comput Math Math Phys 50(7):1141–1149

14. Shkhanukov MKh, Lafisheva MM, Nakhusheva FM, Mambetova AB (2013) The locally-one-dimensional scheme for the equation of heat conductivity with the concentrated thermal capacity. Vladikavkaz. Mat. Zh. 15(4):58–64

15. Bazzaev AK, Shkhanukov-Lafishev MKh (2016) Locally one-dimensional schemes for the diffusion equation with a fractional time derivative in an arbitrary domain. Comput Math Math Phys 56(1):106–115

16. Beshtokova ZV, Shkhanukov-Lafishev MK (2018) Locally one-dimensional difference scheme for the third boundary value problem for a parabolic equation of the general form with a nonlocal source. Differ Eq 54:870–880

17. Beshtokova ZV, Lafisheva MM, Shkhanukov-Lafishev MKh (2018) Locally one-dimensional difference schemes for parabolic equations in media possessing memory. Comput Math Math Phys. 58(9):1477–1488

18. Andreev VB (1968) The convergence of difference schemes which approximate the second and third boundary value problems for elliptic equations. USSR Comput Math Math Phys 8(6):44–62

19. Samarskii AA, Gulin AV (1973) Stability of difference schemes. Nauka, Moscow

20. Samarskiy AA (1983) Teoriya raznostnykh skhem. [Theory of difference schemes]. Nauka, Moscow

# Solution for Inverse Boundary Value Problems on the Power of a Concentrated Charge in a Mathematical Model of Subsidence Soils Compaction

**Elena O. Tarasenko**[ID]**, Andrey V. Gladkov**[ID]**, and Natalia A. Gladkova**[ID]

**Abstract** The soils of the North Caucasus contain a large share of subsidence loess, which accounts for about 80% of the entire area. Compaction of subsidence soils through deep explosions appears a relevant issue when designing and constructing buildings and other structures. The analysis of the experimental data obtained through explosions shows that it can be a reasonable idea to employ methods of differential calculus to carry out mathematical modeling of this process. Diffusion equations featuring a high degree of reliability can describe a mathematical model of subsidence loess compaction by deep explosions.

While describing changes in the compacted soil density, in the case involving a concentrated explosive charge, a function was used, which is a Gaussian function of the compacted soil density distribution. This article offers a look at the equations presented by the boundary value problem with a system of initial and boundary conditions, aiming to solve the inverse applied problem of restoring the power of a concentrated explosive charge. There have been analytical and numerical solutions of the problem obtained. A computational experiment has been held to calculate the power of a concentrated charge relying on a developed program. An analysis of the literature focusing on this issue reveals lack of elaboration of the issue as well as lack of a mathematical description for loess soil compaction problems.

**Keywords** Subsidence soils compaction · Loess · Mathematical modeling · Boundary value problems with initial and boundary conditions

## 1 Introduction

The area of the North Caucasus contains abundant subsidence loess, which accounts for about 80% of the entire territory. From a structural point of view, loess is rich in

E. O. Tarasenko · A. V. Gladkov (✉) · N. A. Gladkova
North-Caucasus Center for Mathematical Research, NCFU, Stavropol, Russia
e-mail: agladkov@ncfu.ru

E. O. Tarasenko
e-mail: etarasenko@ncfu.ru

© The Author(s), under exclusive license to Springer Nature Switzerland AG 2022     537
A. Tchernykh et al. (eds.), *Mathematics and its Applications in New Computer Systems*,
Lecture Notes in Networks and Systems 424,
https://doi.org/10.1007/978-3-030-97020-8_49

macropores, which allow water to penetrate deep into the soil. Further on, a rapid soaking of the loess occurs, and there appear uneven deformations manifested as subsidence. Compaction of subsidence soils is an urgent issue that has to be taken into account when dealing with the design and construction of buildings and other structures [2, 3, 10, 11].

## 2 Materials and Methods

### 2.1 Mathematical Modeling for Solving Inverse Problems of Charge Power in Subsidence Soils

Here we will consider the construction of analytical solutions to some boundary value problems with set initial and boundary conditions within a mathematical model of subsidence soil compaction. In this article, we will offer a development for a solution to boundary value problems aimed at restoring the power of a concentrated explosive charge.

Within the mathematical model of subsidence loess compaction by deep explosions, the change in the compacted soil density value can be described by a diffusion equation of the following type [5, 12]

$$\frac{\partial q}{\partial t} + U\frac{\partial q}{\partial x} = \frac{\partial}{\partial x}K_x\frac{\partial q}{\partial x} + \frac{\partial}{\partial y}K_y\frac{\partial q}{\partial y} + \frac{\partial}{\partial z}K_z\frac{\partial q}{\partial z} \tag{1}$$

with the following initial condition

$$q(0, x, y, z) = Q\delta(x)\delta(y)\delta(z - H), \tag{2}$$

where $Q = const > 0$, $\delta(x) - \delta$ is a Dirac function.

Let us assume that $U, K_x, K_y, K_z$ are continuous functions regarding the argument $z$ : $U = U(z)$, $K_x = K_x(z)$, $K_y = K_y(z)$, $K_z = K_z(z)$.

In order to describe changes in the compacted soil density, apart from Eq. (2), in case of a concentrated explosive charge, function [12] can be employed

$$q_1(t, x, y, z) = \frac{Q}{(2\pi)^{3/2}\sigma_x(t)\sigma_y(t)\sigma_y(t)} \exp\left\{-\left[\frac{(x - \overline{U}t)^2}{2\sigma_x^2(t)} + \frac{y^2}{2\sigma_y^2(t)} + \frac{(z - H)^2}{2\sigma_z^2(t)}\right]\right\}, \tag{3}$$

which is the Gaussian function of the compacted soil density distribution, and here $\sigma_x^2(t)$, $\sigma_y^2(t)$, $\sigma_z^2(t)$ are dispersion coordinate changes of gas atoms at $t$ time along the $Ox, Oy, Oz$ axes, so respectively $(\sigma_x^2(t), \sigma_y^2(t), \sigma_z^2(t)$ are continuous functions of

$t$ argument, $t \geq 0$, $\overline{U} = const$), $\overline{U}$ the average velocity of gas atom transfer along the $Ox$ axis.

The compacted soil density values $q_1$, obtained through (3), pretty much match the practical data [4, 7].

At $t \to \infty$ we will have [8]

$$\frac{\sigma_x^2(t)}{t} \to \sigma_x^2, \quad \frac{\sigma_y^2(t)}{t} \to \sigma_y^2, \quad \frac{\sigma_z^2(t)}{t} \to \sigma_z^2,$$

where $\sigma_x^2 > 0$, $\sigma_y^2 > 0$, $\sigma_z^2 > 0$ are certain constants. Given this, at sufficiently large $t$ we will implement linear transformations of the dispersion coordinate soil changes by approximate equations:

$$\sigma_x^2(t) = \sigma_x^2 \cdot t, \ \sigma_y^2(t) = \sigma_y^2 \cdot t, \ \sigma_z^2(t) = \sigma_z^2 \cdot t. \tag{4}$$

The compacted soil density $q_1(t, x, y, z)$, following (3), given the validity of linear transformations (4) will satisfy the equation

$$\frac{\partial q_1}{\partial t} + \overline{U}\frac{\partial q_1}{\partial x} - \frac{1}{2}\sigma_x^2\frac{\partial^2 q_1}{\partial x^2} - \frac{1}{2}\sigma_y^2\frac{\partial^2 q_1}{\partial y^2} - \frac{1}{2}\sigma_z^2\frac{\partial^2 q_1}{\partial z^2} = 0$$

and initial condition (2). Transforming Eq. (1) will produce

$$\frac{\partial q}{\partial t} + \overline{U}\frac{\partial q}{\partial x} - \frac{1}{2}\sigma_x^2\frac{\partial^2 q}{\partial x^2} - \frac{1}{2}\sigma_y^2\frac{\partial^2 q}{\partial y^2} - \frac{1}{2}\sigma_z^2\frac{\partial^2 q}{\partial z^2} + (U(z) - \overline{U})\frac{\partial q}{\partial x}$$
$$= \frac{\partial}{\partial x}\left(K_x(z) - \frac{1}{2}\sigma_x^2\right)\frac{\partial q}{\partial x} + \frac{\partial}{\partial y}\left(K_y(z) - \frac{1}{2}\sigma_y^2\right)\frac{\partial q}{\partial y} + \frac{\partial}{\partial z}\left(K_z(z) - \frac{1}{2}\sigma_z^2\right)\frac{\partial q}{\partial z}. \tag{5}$$

It is obvious that the less the differences

$$(U(z) - \overline{U}), \ \left(K_x(z) - \frac{1}{2}\sigma_x^2\right), \ \left(K_y(z) - \frac{1}{2}\sigma_y^2\right), \ \left(K_z(z) - \frac{1}{2}\sigma_z^2\right)$$

deviate from 0, the less will $q_1$, obtained through formula (3), deviate from the exact solution $q$ (valid for cases where it continuously depends on the coefficients of Eq. (1)). For $\overline{U}$, $\sigma_x^2$, $\sigma_y^2$, $\sigma_z^2$ to have the smallest deviation within the $[0, h]$ interval respectively from $U(z)$, $K_x(z)$, $K_y(z)$, $K_z(z)$, where $h$ is the thickness of the compacted subsidence soil, it is enough to have the following conditions met:

$$\int_0^h (U(z) - \overline{U})^2 dz \to \min_{\overline{U}},$$

$$\int_0^h \left( K_x(z) - \frac{1}{2}\sigma_x^2 \right)^2 dz \to \min_{\sigma_x^2},$$

$$\int_0^h \left( K_y(z) - \frac{1}{2}\sigma_y^2 \right)^2 dz \to \min_{\sigma_y^2},$$

$$\int_0^h \left( K_z(z) - \frac{1}{2}\sigma_z^2 \right)^2 dz \to \min_{\sigma_z^2}, \qquad (6)$$

and then $\overline{U}$, $\sigma_x^2$, $\sigma_y^2$, $\sigma_z^2$, meeting conditions (6), will appear as:

$$\overline{U} = \frac{1}{2} \int_0^h U(z)dz,$$

$$\sigma_x^2 = \frac{2}{h} \int_0^h K_x(z)dz,$$

$$\sigma_y^2 = \frac{2}{h} \int_0^h K_y(z)dz,$$

$$\sigma_z^2 = \frac{2}{h} \int_0^h K_z(z)dz. \qquad (7)$$

Therefore, function $q_1(t, x, y, z)$, obtained based on formula (3), where $\sigma_x(t)$, $\sigma_y(t)$, $\sigma_z(t)$, are identified from formulae (4), (7), is to be taken as an approximate solution of mathematical problem (1), (2).

Relying on formula (3) can produce expressions for calculating the average values of the compacted soil density from the concentrated charge at the surface discharge $z = 0$ [1, 9]

$$\left. \frac{\partial q(t, x, y, z)}{\partial z} \right|_{z=0} = 0 \qquad (8)$$

or, in the conditions of compacting subsidence soil

$$q(t, x, y, 0) = 0. \qquad (9)$$

They will, respectively, appear as follows:

$$q_2(t, x, y, z) = \frac{Q}{(2\pi)^{3/2}\sigma_x\sigma_y\sigma_z t^3} \exp\left\{-\left(\frac{(x - \overline{U}t)^2}{2\sigma_x^2 t} + \frac{y^2}{2\sigma_y^2 t}\right)\right\}$$

$$\times \left[\exp\left\{-\frac{(z - H)^2}{2\sigma_z^2 t}\right\} + \exp\left\{-\frac{(z + H)^2}{2\sigma_z^2 t}\right\}\right], \tag{10}$$

$$q_3(t, x, y, z) = \frac{Q}{(2\pi)^{3/2}\sigma_x\sigma_y\sigma_z t^3} \exp\left\{-\left(\frac{(x - \overline{U}t)^2}{2\sigma_x^2 t} + \frac{y^2}{2\sigma_y^2 t}\right)\right\}$$

$$\times \left[\exp\left\{-\frac{(z - H)^2}{2\sigma_z^2 t}\right\} - \exp\left\{-\frac{(z + H)^2}{2\sigma_z^2 t}\right\}\right], \tag{11}$$

Further, we will integrate equalities (10), (11) by $t$ within the interval of 0 to $\infty$, obtain the calculated expressions for $q_2(x, y, z)$, $q_3(x, y, z)$ from a stationary concentrated explosive charge located at the $(0, 0, H)$ point:

$$q_i(x, y, z) = \int\limits_0^\infty q_i(t, x, y, z)dt, \ i = 2, 3,$$

where $H$ is the depth of placing the charge in the compacted soil.

Formulae (10) and (11) can be used to solve inverse problems within mathematical models of subsidence soil compaction through deep concentrated explosions.

## 2.2 Formulation of Inverse Problems on Charge Power Recovery

Assume there are known average values of the density $q_2(x, y, z)$ и $q_3(x, y, z)$ of the soil compacted after a deep explosion with a concentrated charge under conditions of surface emission and complete absorption of gas by the surrounding soil (loess compaction implemented); the depth of the explosive is set as H; there are also $\sigma_x^2(t)$, $\sigma_y^2(t)$, $\sigma_z^2(t)$ known, which are dispersion coordinate changes of gas atoms at $t$ time along the $Ox$, $Oy$, $Oz$ axes, respectively.

Identify the unknown concentrated charge power or the amount of gas ejected by the charge of atoms $Q$.

## 2.3 Analytical Solution for Inverse Problems of Subsidence Soils Compaction

We will present the solutions of inverse problems on the concentrated charge power of the mathematical model of subsidence soil compaction by deep explosions in an analytical way. Let us consider two cases where there is a surface discharge and loess compaction performed, during soil compaction through the said method.

1. If the considered system features typical anisotropy properties (provided gas atoms are reflected completely from the surrounding subsidence soil – surface discharge), the desired $Q$ root of Eq. (10) will assume the following analytical representation

$$Q = \frac{q_2(t, \ x, \ y, \ z)(2\pi)^{3/2}\sigma_x\sigma_y\sigma_z t^3}{\exp\left\{-\left(\frac{(x-\bar{U}t)^2}{2\sigma_x^2 t} + \frac{y^2}{2\sigma_y^2 t}\right)\right\}\left[\exp\left\{-\frac{(z-H)^2}{2\sigma_z^2 t}\right\} + \exp\left\{-\frac{(z+H)^2}{2\sigma_z^2 t}\right\}\right]}. \tag{12}$$

If the analyzed system has typical isotropy properties under the conditions of surface ejection and assuming that $\sigma_z = \sigma_y = \sigma_x$, then the desired $Q$ root of Eq. (10) will take the analytical representation as follows below

$$Q = \frac{q_2(t, \ x, \ y, \ z)(2\pi)^{3/2}\sigma_z^3 t^3}{\exp\left\{-\left(\frac{(x-\bar{U}t)^2}{2\sigma_z^2 t} + \frac{y^2}{2\sigma_z^2 t}\right)\right\}\left[\exp\left\{-\frac{(z-H)^2}{2\sigma_z^2 t}\right\} + \exp\left\{-\frac{(z+H)^2}{2\sigma_z^2 t}\right\}\right]}. \tag{13}$$

2. If the system has characteristic anisotropy properties (in case gas atoms are completely absorbed by the surrounding subsidence soil – subsidence loess compaction), then the desired $Q$ root of Eq. (11) will take the following analytical representation

$$Q = \frac{q_3(t, \ x, \ y, \ z)(2\pi)^{3/2}\sigma_x\sigma_y\sigma_z t^3}{\exp\left\{-\left(\frac{(x-\bar{U}t)^2}{2\sigma_x^2 t} + \frac{y^2}{2\sigma_y^2 t}\right)\right\}\left[\exp\left\{-\frac{(z-H)^2}{2\sigma_z^2 t}\right\} - \exp\left\{-\frac{(z+H)^2}{2\sigma_z^2 t}\right\}\right]}. \tag{14}$$

In the event the system features characteristic isotropy properties under the conditions of subsidence loess compaction, and assuming that $\sigma_z = \sigma_y = \sigma_x$, then the desired $Q$ root of Eq. (11) will assume an analytical representation of

$$Q = \frac{q_3(t, \ x, \ y, \ z)(2\pi)^{3/2}\sigma_z^3 t^3}{\exp\left\{-\left(\frac{(x-\bar{U}t)^2}{2\sigma_z^2 t} + \frac{y^2}{2\sigma_z^2 t}\right)\right\}\left[\exp\left\{-\frac{(z-H)^2}{2\sigma_z^2 t}\right\} - \exp\left\{-\frac{(z+H)^2}{2\sigma_z^2 t}\right\}\right]}. \tag{15}$$

## 3 Results

**Software Implementation of the Analytical Solution for the Inverse Problem**
For the above-proposed solutions of inverse problems, while performing the mathematical modeling of a deep explosion, we will carry out their practical implementation at a civil construction facility, thus aiming to exclude loess subsidence.

The calculations are performed involving the experimental data from the work by Galai B.F. [7], which contain details regarding hydro-explosive compaction of a 20-m subsidence layer at the facility of *Dormitory for 203 persons at the Prikumsky Plastics Plant, Neighborhood 7*. Table 1 [6, 7] shows the soil density and humidity indicators following a deep explosion with a concentrated charge. The horizontal gas propagation parameter $\vec{u} = 0\,m/s$ (vector along the axis of the abscissa $Ox$). The system under consideration features typical isotropy properties ($\sigma_z = \sigma_y = \sigma_x$). The compaction is performed under conditions of complete absorption of gas by the surrounding soil. The depth of placing the explosive is $H = 6$ m.

Find the $Q$ power of the concentrated charge.

Taking into account that the system possesses characteristic isotropy properties, we will solve the problem relying on formula (13). The dispersion coordinate changes of gas atoms at $t$ time along the $Oz$ axis are to be determined by the soil skeleton density (Table 1) $\sigma_z \in (1, 61;\ 1, 65)$.

To solve the problem, a software program in the C# programming language was developed. Figure 1 shows the result of running the software. When placing the explosive to a depth of 6 m on a $5 \times 5$ grid, an average charge power value of 5 kg was obtained.

Analyzing the outcome suggests that the proposed mathematical model matches experimental data and can be used in planning the production of robots for subsidence loess compaction by deep explosions.

**Table 1** Indicators of soil density and moisture [7]

| Depth, $m$ | Moisture, % | Skeleton density, $g/cm^3$ |
|---|---|---|
| 3.0 | 13.4 | 1.80 |
| 4.0 | 14.2 | 1.77 |
| 4.5 | 18.1 | 1.65 |
| 5.0 | 13.8–14.2 | 1.62–1.70 |
| 6.0 | 12.1–14.1 | 1.61–1.65 |
| 7.4 | 14.8–15.0 | 1.61–1.65 |
| 8.7 | 17.3–17.4 | 1.61–1.62 |
| 9.1 | 15.7–17.0 | 1.63–1.75 |
| 10.0 | 17.4 | 1.75 |
| 11.0 | 15.3–16.5 | 1.61–1.68 |
| 11.3 | 14.7 | 1.64 |

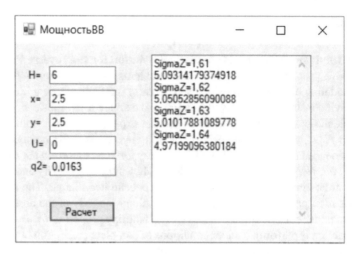

**Fig. 1** Outcome of running the developed software

# 4 Conclusion

Mathematical modeling of subsidence soils compaction by deep explosions entails inverse problems. The solution for some of them is offered above.

There have been analytical solutions constructed for (12), (13) and (14), (15) of the boundary value problem for restoring the power of a concentrated explosive charge in systems with typical features of anisotropy and isotropy, under the conditions of surface ejection, as well as subsidence loess compaction.

Based on the obtained analytical results, there has been practical implementation offered for the solution of the inverse problem at a civil construction facility. The mathematical calculations point at the feasibility of the proposed approach to estimating the power of a concentrated explosive charge, natural loess compaction. A numerical solution has been obtained for the inverse applied problem.

**Acknowledgements** The authors herewith express their gratitude to Professor Boris Fedorovich Galai (Doctor of Geological and Mineralogical Sciences, Department of Construction, North-Caucasus Federal University) for his assistance in conducting this research.

**Funding** This work has been supported by the North-Caucasus Center for Mathematical Research subject to Agreement #075-02-2021-1749 with the Ministry of Science and Higher Education of the Russian Federation.

# References

1. Amšiejus J, Dirgėlienė N (2007) Probabilistic assessment of soil shear strength parameters using triaxial test result. Baltic J Road Bridge Eng 2(3):125–131

2. Tsukamoto Y, Ishihara K (2010) Analysis on settlement of soil deposits following liquefaction during earthquakes. Soils Found 50(3):399–441
3. Ivanov PL (1966) Compaction of Cohesionless Soils by Explosives. In: Proceedings of the YI international conference on soil mechanics and foundation engineering, monreal, vol. 3, pp 352–354
4. Ishihara K (2009) New challenges in Geotechnique for ground hazards due to intensely strong earthquake shaking. Geotech Geol Earthquake Eng 11:91–114
5. Tarasenko EO, Gladkov AV (2021) Statistical modeling of the economic efficiency of blasting operations for compaction of subsidence soils by the method of random optimization. Rev Appl Ind Math 28:1–5
6. Sviderskikh AV, Noskov IV (2019) Analysis of changes in the loess soils properties depending on porosity. Newslett Euras Sci 11(2):1–7
7. Galai BF (2016) Manual on subsidence loess soil compaction of by deep explosions in the North-Caucasus (survey, design, work) Servisshkola, NCFU, Stavropol, Russia
8. Denisov AM (1994) Introduction to the theory of inverse problems. Moscow State University Publishing House, Moscow, Russia
9. Petrakov AA, Prokopov AYu, Petrakova NA, Panasyuk MD (2021) Interpretation of soil strength properties for numerical studies. Newslett Tula State Univ 1:225–236
10. Pantiushina EV (2011) Loess soils and engineering methods for eliminating their subsidence properties. Polzunovsky Newslett 1:127–130
11. Gridnevsky AV, Prokopov AYu (2019) Natural and man-made conditions for formation of flooding of inter-girder spaces of Rostov-on-Don. Newslett Tula State Univ Earth Sci 2:26–37
12. Tarasenko EO, Tarasenko VS, Gladkov AV (2019) Mathematical modeling of subsidence loess soil compaction in the North Caucasus by deep explosions. Newslett Tomsk Polytech Univ Georesour Eng 330(11):94–101

# Dynamic Model of Inter-Industry Balance, the Problem of Its Digitizing and Simulation of Structural Transient Processes

Gudieva Natalia👁 and Toroptsev Evgeny👁

**Abstract** This paper represents the authors' concept and mathematical support for investigation of structural transient processes in macroeconomics, their statistical research base in the field of inter-industry analysis, structural forecasting, sustainability, economic dynamics and economic growth. The mathematical research is formalized by Cauchy problem for ordinary differential equations. In addition to these areas of analysis, our work gives new opportunities to the investigation of economic systems inter-industry persistence. The proposed analysis of transient processes in macroeconomics has no analogues in economic research and it is intended to the results development and evaluation in the implemented economic policy. Based on official statistics data, the article solves the problem of digitizing the dynamic model of inter-industry balance, presented in the form of the system of differential equations. Originally, this model has been transferred from a set of "purely theoretical constructions" to the class of computable ones, i.e., econometrically (quantitatively, numerically) defined models. Our research has revealed the advantages of calculation of matrix exponent and its integral in analysis and forecasting tasks over the other numerical methods. The proposed method makes it possible to construct effectively a difference scheme for integration with any step of observation of the economy aptitude for expanded reproduction. An important task was to ensure computational aptitude for the author's theory of structural/proper dynamic features of the economy as a system. It became necessary to develop methods for generating our own statistical database of computational studies. It would be impossible to quantify the main dynamic inter-industry model without this problem solution.

**Keywords** Cauchy problem · Differential equation · Macroeconomics · Inter-industry analysis · Structural forecasting · Economic system · Neurocryptography · Inter-industry balance

G. Natalia (✉)
North-Caucasus Federal University, Stavropol, Russia
e-mail: gudieva82@bk.ru

T. Evgeny
North Caucasus Center for Mathematical Research, North-Caucasus Federal University, 355017 Stavropol, Russia

© The Author(s), under exclusive license to Springer Nature Switzerland AG 2022    547
A. Tchernykh et al. (eds.), *Mathematics and its Applications in New Computer Systems*,
Lecture Notes in Networks and Systems 424,
https://doi.org/10.1007/978-3-030-97020-8_50

# 1 Introduction

A great number native and foreign scientists' works has been devoted to the investigation of transient processes [1, 2] and economic dynamics [3, 4]. The results of these studies have been published on periodicals pages [5, 6] for a long time, a lot of monographic generalizations, dissertations [7, 8] and university textbooks on economics are dedicated to them. The presented set of references contains the examples and approaches to the analysis of transient processes in economy and, of course, it is still to be completed. Actually, these processes are constantly occurring in economy, connected both with various kinds of reforms, and due to the fact that, originally, the economic systems are unbalanced [9, 10].Their structure is also in a permanent transient process, "mastering" the results of scientific and technological progress by means of technologies development [11], extension of goods and services range and quality [12] and introduction of new forms of management of production processes [13].

# 2 Materials and Methods

The use of structural models of economy is necessary for analysis and visualization of structural transient processes. The dynamic inter-industry balance (DIB) is one of the major leaders among them which is formalized in the form of Cauchy problem for a system of differential and algebraic equations represented as follows:

$$(E - A)X(t) - BpX(t) = Y(t), X(0) = X_0 \tag{1}$$

where $p = \frac{d}{dt}$ – a time differentiation symbol, A – a matrix of direct material cost coefficients, B – a matrix of inter-industry persistence, Y(t) – a vector of final demand (consumption), X(t) – a vector of gross output by type of foreign economic activity (FEA) [26], E – a unity matrix, and the vector BpX(p) characterizes the dynamics of accumulation/reduction of all "capital" types in the context of (FEA).

The Eq. (1) was first derived by V.V. Leontev in 1952, and a contemporary Russian reader most likely knows his book "Economic essays. Theories, research, facts and politics" [14]. The matrixes and vectors included in (1) are also characterized there. And although the author refers to the matrix B of dimension $(100 \times 100)$ for the American economy, neither Russian nor Soviet statistics have ever been engaged in its development [15].

This circumstance made it impossible to digitize the model (1) directly, i.e., it was impractical to classify it as a computable (practical) model. The development of B would make it possible to use the numerical analysis of Cauchy problem for ordinary differential equations. A researcher only needs to set the initial/boundary conditions and start the integration process.

The experience of our calculations, which elements are graphically presented in the Fig. 1, allows us to assert that the model (1) provides the possibility of forecasting gross foreign trade output by FEA for medium-term with accuracy of 10% at the worst. Taking into account all possible sources of errors (unconformity of the model to the real economy, statistical data errors, numerical method inaccuracy, rounding inaccuracy in computer calculations), such result should be considered satisfactory.

The Rosstat data in the specified figure correspond in nominal terms to five sections of chosen FEA. At the accelerating forecasting stage the applied approximation reflects the monotonous growth of nominal gross production in Russia when considering its dynamics from year to year. At the same time, a monthly output explication shows oscillatory movements better than an annual one. This is also recorded by Rosstat data.

The vector of gross output for 2011 is chosen as the initial conditions $X_0$. The dependencies shown in the Fig. 1 are used to digitize our model and contain the forecasting results. The model is continuous, so it allows determining the gross output for any value of $t$.

Let us pay attention to the fact that according to the data in the Fig. 1 and using standard forecasting methods for time series, it is possible to obtain only the persistent forecast with one or another rising dynamics of the graphs. These data, demonstrating monotonous dynamics, lead to monotonous forecasts. Calculated indicators of proper

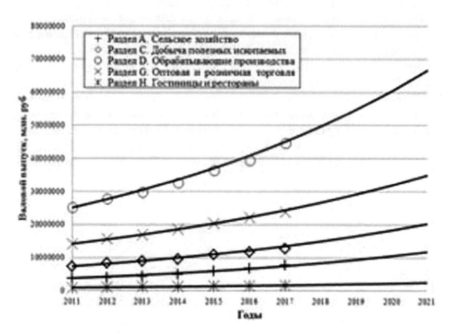

**Fig. 1** An example of forecasting gross production by foreign economic activity (Rosstat data and own calculations)

dynamic properties (PDP) [16], allow visualizing the oscillatory motion components, and also require information minimum according to Rosstat data.

If the model (1) is transformed into an ordinary Cauchy form, and the integrated pairs are detected by means of direct calculation in the spectrum of the model state matrix, then the integration (1) will visualize deviations and allow assessing the impact of the mentioned pairs on the medium-term economic perspective. In general, the analysis of the discussed model allows solving a wide range of problems of oscillatory and aperiodic static stability of economy as a system. It also assesses numerically its structural aptitude (or unavailability) for expanded reproduction (i.e., for the economic growth).

The foregoing, from a new point of view, allows us to investigate the problem of economic system persistence and obtain the instrumental methods of numerical analysis. At the same time, it should be remembered that structural persistence is "responsible" for the transient processes quality, including systemic stability of economy and economic growth. Usually, the difference analogs of the model (1) are used for $\Delta t = 1$, i.e., one year. Then the elements of the persistent matrix B are numerically equal to the mid-year incremental capital output ratio.

The calculation experience for transient processes allows us to distinguish three types of economic systems that differ in their dynamic properties. The Figs. 2, 3 and 4 qualitatively show different types of transient processes, depending on the proper dynamic properties (PDP) of economic system models.

The transient processes shown in the Fig. 2.a demonstrate a decline in gross production for all types of FEA. In a closed conservative system, this decline reaches zero, which in the classic sense indicates the simulated object stability.

Due to the absence of forces interfering with the process, i.e., idealized macrosystem destruction, the process is stable and easily achievable. From a mathematical point of view, the proper dynamic properties (PDP) are characterized by a negative spectrum of eigenvalues, leading the trajectories to zero (to a production stop), regardless of the eigenvectors. A return to growth is associated with investment flows from the outside when it is impossible by a simple redistribution of declining resources between economic agents within the system. These investments should be

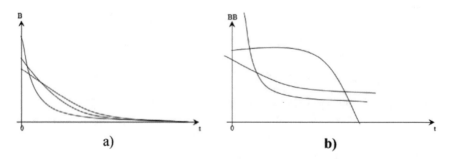

a)                                                      b)

**Fig. 2  a** Gross production decline **b** The growth of some and the decline of other foreign economic activity

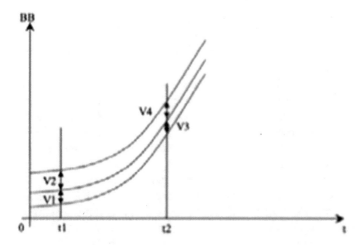

**Fig. 3** Trunk gross output growth

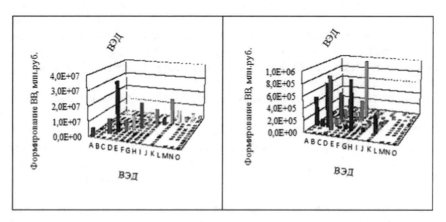

**Fig. 4** Qualitative representation of the table of formation of output of goods and services (Rosstat data)

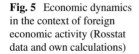

**Fig. 5** Economic dynamics in the context of foreign economic activity (Rosstat data and own calculations)

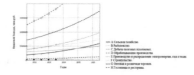

directed in accordance with certain proportions as external influences on the system, otherwise the conditions and effects in the Fig. 2.b may occur.

The economic situation where a negative growth in the output of some types of FEA and a positive growth of the others is possible is described by an alternating spectrum of eigenvalues, their weak observability and, accordingly, weak controllability from various external influences.

Let us consider the models of macroeconomic systems where the entire spectrum of eigenvalues of state matrices is positive (Fig. 3). If this growth is proportional according to the formula $\frac{V1}{V2} = \frac{V3}{V4}$, so such growth is called trunk or balanced.

In practice, this process is determined by management circuit prioritization in relation to various industries, which are, for example, locomotives ones for the others and donors for the budget.

The problem of applying differential equations to dynamic models of inter-industry balance (MIB) is directly related to the lack of initial data, usually presented in a matrix form. The official statistics develops a table of direct material costs based on two tables related to resources and use of goods and services. The rest of data is available only in the form of vectors, what is obviously insufficient to digitize the system of differential equations describing the dynamics of main indicators of MIB.

The tables on output formation of goods and services presented, for example, in the papers published by Rosstat "National Russian Accounts in … years." are available with a time lag of one year. They provide the original material for the model digitizing and have predominance of the main diagonal over non-diagonal elements, i.e., profile products production. The FEA always dominate. This phenomenon is qualitatively shown in the Fig. 4.

We admit that the gross output vector is determined by summing the row elements of discussed table.

The following data components for solving our problem are the ratios of inter-mediate consumption and added value for FEA. It is known, the gross value added (GVA) can be used to calculate the gross domestic product (GDP). With an error acceptable for our investigation, we can assume that the GDP structure corresponds to the GVA one.

If we had a matrix for intermediate consumption formation represented as, and a dimensionally equal matrix for added value formation $MVA$, then a static MIB would allow us to write the following equation:

$$PM^T = MIC + MVA, \tag{2}$$

where $PM$ – transposed matrix of goods and services output (of gross output – production matrix), $T$ – transposing symbol. Here and further, several consecutive capital letters mean one mathematical object, and not a product of variables. Mathematically, the formula (2) implements the combination of matrixes of the I and III quadrants of the MIB, but statistics do not allow doing it "head-on" for already mentioned reason, providing the relevant vectors instead of $MIC$ and $MVA$: $ICV$ (intermediate consumption vector) and $VAV$ (value added vector). Then the output vector:

$$X = ICV + VAV, \tag{3}$$

Now let us do the following: we represent $PM$ matrix as two terms, as in the formula (2), but taking into account the proportions given by the vectors from (3):

$$PM^T = diag\left((PM^T)^{-1} \times ICV\right) \times PM^T + diag\left((PM^T)^{-1} \times VAV\right) \times PM^T.$$
(4)

Two terms on the right side of equality (4), similarly to (2), are the matrices $MIC$ and $MVA$, respectively.

The transition from financial flows to cost coefficients involves consideration and explanation of the following presentation, recorded on the basis of (2):

$$PM^T \times diag\left(\frac{1}{X}\right) \times = MIC \times diag\left(\frac{1}{X}\right) \times X + MVA \times diag\left(\frac{1}{X}\right) \times X.$$
(5)

It can be seen from (5) that the diagonal matrix product $diag\left(\frac{1}{X}\right)$ by the vector $X$ gives a unit vector. The matrix multiplication $PM^T$ by a column-vector with single elements implements a row-by-row integration operation, which results in a gross output vector $X$. It means, the gross output is to the left of the equal sign in (5). The product $PM^T \times diag\left(\frac{1}{X}\right) = BM$ gives a matrix (let it be called a basic matrix), which being multiplied by the vector $X$ gives the same vector $X$, i.e., $X$ is an eigenvector of the matrix $BM$, and it corresponds to an eigenvalue equal to one. The basic matrix explicitly takes into account industry and inter-industry interactions in the economy, which are initially "collapsed" in the proportions of the output vector X.

It is obvious that the Eq. (5) is a variant of the static MIB – a specific type of Leontev's dual price model. In fact, the Eq. (5) can be rewritten according to the columns of the MIB scheme as follows:

$$BM \times X = A^T \times X + MVA' \times X, \quad MVA' = MVA \times diag\left(\frac{1}{X}\right)$$

$$X = BM^{-1} \times A^T \times X + BM^{-1} \times MVA' \times X,$$

$$X = A^T \times X + VAV$$
(6)

As can be seen from the last Eq. (6), the information about material costs matrix structure is still in the first term of the right part, while the structure of gross value added is still "collapsed" in the vector $VAV$.

Let us pay attention to the basic matrix obtained from the data of 2015. The starting year in our research was 2011. This year, obvious impact of the global crisis of 2009–2010 on the Russian economy has faded away. So, our economy structure has been recovering up to the pre-crisis state from the starting year, increasing the gross production rates. The gross production structure that was developed by 2015 is presented in the Table 1.

**Table 1** Basic matrix ($BM$) of the Russian economy for 2015

|   | A | B | C | D | E | F | G | H | I | J | K | L | M | N | O |
|---|---|---|---|---|---|---|---|---|---|---|---|---|---|---|---|
| A | 0.91 | 0 | 0 | 0.014 | 0 | 0.001 | 0.001 | 0.002 | 0 | 0 | 0.001 | 0 | 0 | 0 | 0.001 |
| B | 0 | 0.635 | 0 | 0.003 | 0 | 0 | 0 | 0 | 0.001 | 0 | 0 | 0 | 0 | 0 | 0 |
| C | 0 | 0 | 0.908 | 0.021 | 0.002 | 0.009 | 0.002 | 0.003 | 0.003 | 0 | 0.004 | 0 | 0 | 0.001 | 0.001 |
| D | 0.005 | 0.01 | 0.083 | 0.943 | 0.015 | 0.007 | 0.025 | 0.009 | 0.009 | 0 | 0.016 | 0 | 0 | 0.001 | 0.004 |
| E | 0 | 0 | 0 | 0.001 | 0.959 | 0.005 | 0.002 | 0.001 | 0.001 | 0 | 0.003 | 0 | 0 | 0 | 0.056 |
| F | 0 | 0 | 0.001 | 0.002 | 0 | 0.975 | 0.003 | 0.001 | 0.003 | 0 | 0.003 | 0 | 0 | 0 | 0.002 |
| G | 0 | 0 | 0.001 | 0.023 | 0 | 0.002 | 0.879 | 0.02 | 0.029 | 0.002 | 0.065 | 0 | 0 | 0 | 0.003 |
| H | 0 | 0 | 0 | 0 | 0 | 0 | 0 | 0.979 | 0 | 0 | 0.001 | 0 | 0 | 0 | 0.002 |
| I | 0 | 0 | 0.001 | 0.001 | 0.003 | 0.005 | 0.003 | 0.007 | 0.971 | 0.002 | 0.006 | 0 | 0 | 0.001 | 0.009 |
| J | 0 | 0 | 0 | 0 | 0 | 0 | 0 | 0 | 0 | 0.998 | 0 | 0 | 0 | 0 | 0 |
| K | 0 | 0 | 0.002 | 0.008 | 0.005 | 0.006 | 0.004 | 0.008 | 0.003 | 0.001 | 0.968 | 0 | 0.004 | 0.002 | 0.005 |
| L | 0 | 0 | 0 | 0 | 0 | 0 | 0 | 0 | 0 | 0 | 0.002 | 0.995 | 0 | 0 | 0 |
| M | 0 | 0 | 0 | 0 | 0 | 0 | 0 | 0.001 | 0 | 0 | 0 | 0 | 0.998 | 0 | 0 |
| N | 0 | 0 | 0 | 0 | 0 | 0 | 0 | 0.001 | 0 | 0 | 0 | 0 | 0 | 0.998 | 0 |
| O | 0 | 0 | 0 | 0 | 0.002 | 0 | 0 | 0.005 | 0 | 0 | 0.002 | 0 | 0.001 | 0 | 0.972 |

**Table 2** Classification of foreign economic activity in the SNA

| | FEA |
|---|---|
| A | 1. Agriculture, hunting and forestry |
| B | 2. Fishing, fish farming |
| C | 3. Mining |
| D | 4. Manufacturing industries |
| E | 5. Production and distribution of electricity, gas and water |
| F | 6. Construction |
| G | 7. Wholesale and retail trading; repair of motor vehicles, motorcycles, household goods and personal items |
| H | 8. Hotels and restaurants |
| I | 9. Transport and communications |
| J | 10. Financial activities |
| K | 11. Real estate transactions, rental and provision of services |
| L | 12. State administration and military security; social insurance |
| M | 13. Education |
| N | 14. Healthcare and social services provision |
| O | 15. Provision of other public, social and personal services |
| P | 16. Household activities |

The matrix is calculated according to the statistical collection "National Russian Accounts in 2011–2016", rows and columns names correspond to the names of VEA of the same collection (Table 2).

We have come up with the basic matrix in order to show the diagonal dominance of its elements indicating the linearity of its dynamics, at least at this level of model detailing, and the weakness of inter-industry relations Russian economy. The zeros in the given matrix indicate the absence of mutual financial flows, i.e., the absence of inter-industry relations. Theoretically, the dynamic model of MIB can be represented as follows:

$$BM \times X = A \times X + B \frac{dX}{dt} + L \times X, \tag{7}$$

where $L$ – a matrix of labor costs provided by the households. The matrix elements are determined, for example, by norms of consumption and labor intensiveness in various VEA. It is in general terms and considering that $L$ takes into account different demand components for non-productive consumption in the government sector and export–import balance as well.

The Eq. (7) is transformed into an ordinary Cauchy form for the initial conditions of the year 2011:

$$B \frac{dX}{dt} = B^{-1} \times (BM - A - L) \times X, \, X(0) = X_{2011}, \tag{8}$$

what through the matrix exponent can be seen as follows:

$$BX(t) = \exp(Gt) \times X(0).X(0) = X_{2011}, G = B^{-1} \times (BM - A - L), \quad (9)$$

where $\exp(Gt)$ – the matrix exponent.

The Eqs. (8) and (9) will provide us with variants of the matrix B. Gross output in nominal terms for the period of 2011–2015 is represented by the corresponding icons in the Fig. 3. It also shows theoretical curves that properly correlate with the statistical data. These curves are the exponents obtained for the diagonal matrix G, which elements were determined by the formulas:

$$G = diag(\lambda_j), \lambda_j = \frac{\ln\left(\frac{x_{j,2015}}{x_{j,2011}}\right)}{2015 - 2011}, \quad (10)$$

Now let us pay attention to what follows from the Eq. (6):

$$MVA' = BM - A^T, \quad (11)$$

We can see from (11) that the $MVA$ matrix is already computable by means of two ways. Let us now assume for certainty that half of the value added is spent on satisfying the aggregate final demand, i.e., $L = (1/2)MVA$.

The following is obvious:

$$G = B^{-1}\left(BM - A - \frac{1}{2}MVA\right) = B^{-1}\left(BM - A - \frac{1}{2}(BM - A^T)\right)$$
$$= B^{-1}\left(\frac{1}{2}(BM + A^T) - A\right) \quad (12)$$

Now, it is not difficult to calculate the matrix $B$ for the case under consideration:

$$B = \left(\frac{1}{2}(BM + A^T) - A\right) \times G^{-1} \quad (13)$$

## 3  Results

The research results for the matrix $B$ correspond to our assumptions presented in the Table 3. Proportion setting does not change the integral curves trajectory, since the matrix of the closed system $G$ remains unchanged and determines the transition process characteristics. The variations of elements $B$ are compensated by the variations in final and capital demand.

Thus, the dynamic MIB model digitization can be considered completed. It is a variant of a fundamental scientific problem solving what opens up broad prospects

**Table 3** The matrix $B$ model (1)

|   | A | B | C | D | E | F | G | H | I | J | K | L | M | N | O |
|---|---|---|---|---|---|---|---|---|---|---|---|---|---|---|---|
| A | 2.248 | 0.001 |  | 0.039 | 0.001 | 0.004 | 0.002 | 0.005 | 0.001 |  | 0.002 |  |  |  | 0.002 |
| B |  | 1.497 |  | 0.012 |  |  | 0.001 |  | 0.002 |  |  |  |  |  |  |
| C |  | 0.000 | 3.226 | 0.040 | 0.010 | 0.050 | 0.008 | 0.011 | 0.011 |  | 0.012 |  |  | 0.003 | 0.002 |
| D | 0.006 | 0.009 | 0.123 | 1.260 | 0.037 | 0.016 | 0.037 | 0.013 | 0.013 |  | 0.021 |  |  | 0.002 | 0.005 |
| E |  |  |  | 0.001 | 2.728 | 0.012 | 0.003 | 0.001 | 0.001 |  | 0.004 |  |  |  | 0.085 |
| F |  |  | 0.002 | 0.004 | 0.002 | 3.738 | 0.008 | 0.003 | 0.009 |  | 0.008 |  |  |  | 0.005 |
| G | 0.001 |  | 0.005 | 0.072 | 0.001 | 0.012 | 2.997 | 0.064 | 0.098 | 0.006 | 0.196 |  |  | 0.001 | 0.009 |
| H |  |  |  |  |  |  | 0.001 | 2.299 |  |  | 0.002 |  |  |  | 0.005 |
| I |  |  | 0.002 | 0.004 | 0.013 | 0.020 | 0.008 | 0.016 | 2.584 | 0.005 | 0.015 |  | 0.001 | 0.002 | 0.022 |
| J |  |  |  |  |  |  |  |  |  | 3.483 | 0.002 |  |  |  |  |
| K | 0.001 | 0.001 | 0.008 | 0.031 | 0.032 | 0.034 | 0.015 | 0.032 | 0.012 | 0.002 | 3.465 |  | 0.022 | 0.006 | 0.019 |
| L |  |  |  |  |  |  |  |  |  |  | 0.008 | 3.843 |  |  |  |
| M |  |  |  |  |  |  |  | 0.003 |  |  |  |  | 6.363 | 0.000 | 0.001 |
| N |  |  |  |  |  |  |  | 0.005 |  |  |  |  |  | 2.941 | 0.001 |
| O |  |  |  |  | 0.009 | 0.001 |  | 0.016 | 0.001 |  | 0.004 |  | 0.003 |  | 2.892 |

for using dynamic models of MIB in the form of systems of ordinary differential equations.

It would be appropriate to admit that the unfilled positions in the Table 3 do not mean zeros, the values that could stand there are negligible in every sense and, therefore, insignificant for analysis. We also inform that we have removed from consideration Households as they contained only one diagonal element in the Output Formation Table.

Let us pay attention to numerical analysis methods of model (1) and fairly refer it to the computable type. At the same time, we will focus on the study of transient processes and obtaining related assessments of structural economic dynamics. We present the following Eq. (1) in the ordinary form:

$$pX(t) = DX(t) + C(t), X(0) = X_0, \tag{14}$$

where the construction $D = B^{-1}(E - A)$ is a structural matrix of economy, and the vector $C(t) = -B^{-1}Y(t)$ is defined as one that sets disturbances and determines the forced mode of economy functioning (disturbing vector).

The course of mathematical analysis provides the data that the complete integral of a linear inhomogeneous differential Eq. (14) is equal to the general solution of corresponding homogeneous equation $X_{CB}(t)$, its free component, and the partial solution $X_B(t)$ of the inhomogeneous equation obtained under the condition $t \rightarrow \infty$ and called the forced or steady-state mode of the system operation. The forced component is caused by the structure of a final demand and inter-industry persistence, and the whole output represents the sum of:

$$X(t) = X_B(t) + X_{CB}(t) \tag{15}$$

Our long-term experience with the model (1) indicates that the problem (14) analytical solution and computational process subsequent design are suitable while using the matrix exponent and its integral. Then free components solution (the solution of the corresponding homogeneous equation at $C = 0$) has the following form:

$$X_{CB}(t) = e^{Dt} \cdot X(0) \tag{16}$$

The system solution under the influence of compelling forces ($C \neq 0$) will be represented as follows:

$$X(t) = e^{Dt} \cdot F(t), \tag{17}$$

where $F(t)$ – an intermediate matrix function of our model.

We differentiate (17) by time in order to remove this function from consideration and obtain:

$$\frac{dX(t)}{dt} = De^{Dt}F(t) + e^{Dt}\frac{dF(t)}{dt}, \tag{18}$$

or

$$\frac{dX(t)}{dt} = DX(t) + e^{Dt}\frac{dF(t)}{dt}, \tag{19}$$

It follows from expression (7) that

$$C(t) = e^{Dt}\frac{dF(t)}{dt}, \tag{20}$$

When constructing a computational algorithm, we initially assume that $C = const$ and consider the necessary transformations, and then consider the general case when $C(t)$, i.e., it is a time function.

If the expression (20) is multiplied on the right and left sides by the matrix exponent $e^{-Dt}$, then we get the following integral in order to remove the function $F(t)$ from further consideration:

$$F(t) = \int_{-\infty}^{t} e^{-D\Theta}d\Theta \cdot C = \int_{-\infty}^{0} e^{-D\Theta} \cdot C + \int_{0}^{t} e^{-D\Theta}d\Theta \cdot C, \tag{21}$$

By inserting (21) in (17), we obtain:

$$X(t) = e^{Dt}\int_{-\infty}^{0} e^{-D\Theta}d\Theta \cdot C + \int_{0}^{t} e^{-D\Theta}d\Theta \cdot C, \tag{22}$$

And inserting $t = 0$ in (22) we define that vector of initial conditions:

$$X(0) = \int_{-\infty}^{0} e^{-D\Theta}d\Theta \cdot C, \tag{23}$$

Taking into account (23), we represent the formula for our model analytical solution:

$$X(t) = e^{Dt}X(0) + \int_{0}^{t} e^{D\tau}d\tau \cdot C. \tag{24}$$

The direct use of formula (23) requires the calculation of the matrix exponent and its integral. We will not investigate the possibilities based on the application of Lagrange-Sylvester formula [17], as well as on the representation of this exponent by a power series from the structural matrix $D$ [18]. These two fundamental monographs, which still make a decisive contribution to the training of world-class algebraists, allow us to do this on our own.

Our research experience shows that it is advisable to construct a difference scheme for numerical integration based on the following uniformly converging series:

$$Xe^{Dt} = E + \sum_{k=1}^{\infty} \frac{D^k t^k}{k!}. \tag{25}$$

$$\int_0^t e^{D\tau} d\tau = t + \sum_{k=1}^{\infty} \frac{D^k t^{k+1}}{(k+1)!}. \tag{26}$$

The well-known (25) and (26) disadvantage is a rapid increase of the number of sums terms for achieving the necessary account accuracy at any large $t$ and it is overcome as follows.

The constructed difference scheme gives the solutions that are on the grid with the desired step $H$ and approximate continuous and exact solutions (23) of our model depending on the value of $H$. Randomly the set step for solution $H$ observation generates a discrete series of integration variable values $t_n = nH$, $n = 0, 1, 2, \ldots$. Then for sequentially calculated values of gross outputs $X_n = X(t_n)$, $X_{n+1} = X(t_{n+1})$ we present the equations:

$$X_{n+1} = e^{D(t_n + H)} X_0 + \int_0^{t_{n+1}} e^{D\tau} d\tau \cdot C, \tag{27}$$

$$X_n = e^{Dt_n} X_0 + \int_0^{t_n} e^{D\tau} d\tau \cdot C. \tag{28}$$

Now, if we multiply (25) by $e^{DH}$ and (26) subtract this result, we obtain the following equation for the model (23) solution in discrete time with a randomly set step $H$ of its observation:

$$X_{n+1} = e^{DH} X_n + \int_0^H e^{D\tau} d\tau \cdot C. \tag{29}$$

As we can see, the solutions (29) do not contain an integration variable $t$ ($t$ is determined by the step number $H$) and can be obtained with high computational efficiency if the exponent of the structural matrix of the economy and this exponent integral are preliminary calculated. It is also seen that the solutions (29) for the zero matrix $D$ are easily obtained by the quadrature formula of the mean rectangles.

So, we need to calculate two objects: $e^{DH}$ and $\int_0^H e^{D\tau} d\tau$. We will use the technique first presented by Professor of Leningrad Polytechnic Institute Yu. V. Rakitskii in (31), and introduce such small parameter (initial step) of the computational procedure as follows:

$$h = H/2^N, \tag{30}$$

where $N$ – an integer chosen in such a way that it would be possible to calculate $e^{Dh}$ for a small step $h$ determined by (30) with a small number of terms in the formula (25). The selection of the number $N$ regulates the method error. It can always be reduced to an acceptable level. After that, the recurrent formula is the following:

$$\varphi_{k+1} = \varphi_k \cdot \varphi_k = \varphi_k^2, \tag{31}$$

where $\varphi_k = e^{2^k Dh}$, $k = 0, 1, \ldots, N$; $\varphi_0 = e^{Dh}$; $\varphi_N = e^{2^N Dh} = e^{DH}$, allows us obtaining the desired exponent from the structural matrix.

A similar recurrent formula is constructed for the integral of this matrix exponent. We assume that

$$\Phi_k = \int_0^{2^k h} e^{D\tau} d\tau, k = 0, 1, \ldots, N, \tag{32}$$

and obtain from (32)

$$\Phi_1 = \int_0^{2h} e^{D\tau} d\tau = \int_0^h e^{D\tau} d\tau + \int_h^{2h} e^{D\tau} d\tau = \int_0^h e^{D\tau} d\tau + \int_0^h e^{D\tau} d\tau \cdot e^{Dh}$$
$$= \int_0^h e^{D\tau} d\tau (E + e^{Dh}) = \Phi_0(E + \varphi_0), \tag{33}$$

and then, by generalizing (33), we have

$$\Phi_{k+1} = \Phi_k(E + \phi_k). \tag{34}$$

The formulas (32)-(34) allow us to see the necessary integral.

The comparison of series (25) and (26) demonstrate $\varphi_k = E + D\Phi_k$ and allow us transforming the recurrent formula (33) to the form (35):

$$\varphi_{k+1} = \Phi_k(2E + D\Phi_k), \tag{35}$$

what helps to calculate $e^{DH}$ и $\int_0^H e^{D\tau} d\tau$ independently on each other, i.e. the matrixes $\Phi_k$ and $\varphi_k$ are calculated independently/

The mentioned above allows us to write down the numerical method working formula:

$$X_{n+1} = \overline{\varphi}_N X_n + \overline{\Phi}_N \cdot C, \tag{36}$$

where the dashes above the matrixes mean their approximate calculation with an accuracy regulated by the choice of an integer $N$ and the number of terms of the sums in (25) and (26). The presented recurrence relations (31), (34) or (35) allow us to observe consistently the solution at points $H, 2 \cdot H, 3 \cdot H, \ldots$ according to the formula (24.1).

Instead of matrixes $\Phi_k$ it is possible to calculate the vector recursively:

$$g_k = \int_0^{2^k h} d^{D\tau} d\tau \cdot C = \Phi_k \cdot C, k = 0, 1, \ldots, N, \tag{37}$$

what reduces the algorithm computational complexity. The use of (37) may be important for modern laptops with MIB dimensions of more than 100 differential equations and solving optimization problems. Since the matrixes $\Phi_k$ and $(E + \varphi_k)$ are permuted during multiplication, we obtain the following recurrent formula for the vector $g_{k+1}$ calculation:

$$g_{k+1} = (E + \varphi_k)g_k, k = 1, \overline{N} - 1 . \tag{38}$$

The formula (38) is used in the computational algorithm together with (29), and then the method working formula will take the form:

$$X_{n+1} = \overline{\varphi}_N X_n + \overline{g}_N . \tag{39}$$

If by selecting the number $N$ we achieve a small number of terms of the sums in (25) and (26) with sufficient accuracy for calculating the matrixes $\varphi_0$ and $\Phi_0$, the problem of developing a computational procedure at $C = const$ is solved.

Sticking at the question of the accuracy of the computational procedure in our today's review, we will leave aside the problem of the algorithm stability until the planned consideration of the rigidity phenomenon in dynamic models of MIB. The problem of numerical analysis stability becomes of great importance precisely in those conditions of functioning when rapidly transient processes occur in the monetary sector of the economy in the background of the slower ones.

This problem is complicated by the need of the models digitizing and calculations implementation based on Rosstat data containing difficult-to-estimate errors. Here we assume that a table of forecast values of FEA output is made on the basis of the observation step of the solution $H$ for a set of consecutive values of the integration variable $t$.

It means that the step $H$ must be chosen from the condition of correct observation of the components dynamics of the vector $X(t)$. According to the data of the table, such an approximating function should be constructed, which will meet the needs of economic analysis and management of the development process with the necessary accuracy on the medium-term horizon.

Let us consider what the use of the recurrent scheme (29) practically provides. We assume that the computational process is constructed using Taylor formula, which in our case supplies with the necessary integration accuracy on the segment $[0, T]$. Let the step $h$ h is generated to meet this requirement and four expansion terms from Taylor series are used, which gives the following working formula:

$$X_{n+1} = \left( E + hD + \frac{d^2 D^2}{2} + \frac{h^3 D^3}{6} + \frac{h^4 D^4}{24} \right) X_n + h \left( E + \frac{hD}{2} + \frac{h^2 D^2}{6} + \frac{h^3 D^3}{24} \right)$$

$$C = \overline{\varphi}_0 X_n + \Phi_0 C = \overline{\varphi}_0 X_n + \overline{g}_0$$

$$\tag{40}$$

The formula (40) represents the well-known Runge–Kutta method of the 4th degree with solution approximation by Taylor polynomial of the 4th degree at step $h$. $h$. On the other hand, the difference Eq. (29) generates a competitive with the (40) scheme:

$$X(mH + H) = \overline{\varphi}_N X(mH) + \overline{g}_N, \, m = 0, 1, \ldots, H = 2^N h. \tag{41}$$

If we assume that $\frac{T}{h}$ is an integer then the method (40) implements $\frac{T}{h}$ multiplying a matrix by a vector, and method (41) requires only such multiplications $\frac{T}{H} = 2^{-N} T/h$ while ensuring the same accuracy. Moreover, $n$ – the dimension of the model (14) and additionally the method (51) implement $(n - 1)N$ of matrixes $\varphi_k$ multiplication from (31) for $\overline{\varphi}_N$ and we obtain the same number of matrix–vector multiplications in accordance with (38) to form $\overline{g}_N$.

The presented basic computational costs of the methods suggest that for rather large $\frac{T}{h}$ (it is guaranteed in the presence of fast-transition processes in the economy), the method (41) has greater computational efficiency compared to the standard Runge–Kutta method (40). At the same time, it is not rare for a researcher to be interested only in the final integration (only one endpoint when $H = T$.) Nothing prevents us from doing it and applying formula (41) only once. There are no restrictions on the step of solution observation $H$. It can be chosen randomly, regardless to the change in the components of the gross output vector. The required accuracy of the solution is provided by the choice of the initial step $h$.

At a given step of solution observation $H$, an integer $N$ is determined so that the value $h = H/2^N$ provides the necessary integration accuracy. The same "necessary accuracy" is achieved if we focus on the inequality from [19]:

$$h \leq \frac{k_H}{||D||}, \tag{42}$$

where $k_H$ – a constant, and $||D||$ –Euclidean norm of the structural matrix. The experience of calculations suggests that the value $k_H$ for the rigid models (13)-(14) can be assumed as $\sim 0, 1 \ldots 0, 2$ due to the need to pass the so-called boundary layer and this constant can be increased fifty times for non-rigid models [20–39].

Having decided on the initial step $h$, we calculate $\overline{\Phi}_0$, and $\overline{\varphi}_0$ according to the formulas:

$$\overline{\varphi}_0 = e^{Dh} \approx E + Dh + \ldots + \frac{D^v h^v}{v!}, \tag{43}$$

$$\overline{\Phi}_0 = \int_0^h e^{D\tau} d\tau \approx h \left( E + \frac{hD}{2} + \ldots + \frac{h^{v-1} D^{v-1}}{v!} \right), \tag{44}$$

at low values of $v$, instead of (43) and (44) it is possible to use such fractional-rational approximations of these matrixes as [18]:

$$\overline{\varphi}_0 = E = D\Phi_0 \approx \left( E - \frac{hD}{2} + \frac{h^2D^2}{12} \right)^{-1} \left( E + \frac{hD}{2} + \frac{h^2D^2}{12} \right), \quad (45)$$

$$\overline{\Phi}_0 = \int_0^h e^{D\tau} d\tau \approx h \left( E - \frac{hD}{2} + \frac{h^2D^2}{12} \right)^{-1}. \quad (46)$$

Now we calculate:

$$\overline{g}_0 = \overline{\Phi}_0 C \approx \left( E - \frac{hD}{2} + \frac{h^2D^2}{12} \right)^{-1} \cdot hC. \quad (47)$$

Then $\overline{\varphi}_N$ and $\overline{g}_N$ are calculated by the formulas (31) and (38). Finally, we obtain an approximate solution of the MIB model at the next step of observing the solution by the formula (41). So, (43), (44), (31), (28) and (41) mentioned in the order of their use represent the group of working formulas of the method.

There is only one comment should be said regarding the integration of dynamic MIBs of the form (13) - (14). The estimate (40) associated with the choice of the initial step it can provide the underestimated values, which will lead to a critical accumulation of computational error, which is known even from the educational literature. To overcome this negativity, it is necessary to increase the constant $k_H$ up to the value when the quotient (40) becomes approximately equal to $\left( \max_i |\lambda_i| \right)^{-1}$ where $\lambda_i$ are the structural matrix eigenvalues. In this case, Runge rule allows us to check the integration accuracy of the model. Two integration options are necessary for it: one with the initially specified value $k_H$ and the other with twice smaller value. In the second case the number of steps in the recurrent formulas (31) and (38) increases by one.

The considered algorithm for integrating dynamic MIBs with small and obvious changes can be used to solve the same problem with variable final demand. The model (2) will take the form:

$$pX(t) = DX(t) + C(t), X(t_0) = X_0. \quad (48)$$

We keep all designations and variables the same, and we take approximately constant and equal $C(t_n + H/2)$ the components of the final demand vector $C(t)$ to build a computational algorithm on the segment $[t_n, t_n + H]$. Under these assumptions, the algorithm modifications are performed easier and more obviously. The temporal dynamics of the final demand allows us to assume that its values of $C(t_n + \tau)$ are sufficiently representable by power polynomials. The model (48) slightly changes its form:

$$pX(t_n + \tau)DX(t_n + \tau) + C\left( t_n = \frac{H}{2} \right), \tau \in [0, H]. \quad (49)$$

Then the difference Eq. (31) takes a new form:

$$X_{n+1} = e^{DH} X_n + \int_0^H e^{D\tau} d\tau \cdot C\left(t_n + \frac{H}{2}\right). \tag{50}$$

If we consider the vector $C(t)$ zero then free components (50) accurately describe the model (49) solution and at $C(t) \neq 0$ the Eq. (50) provides the solutions $X_{n+1}$, approximating the model solutions (49) in the points $t_{n+1}$.

The requirements for choosing the observation step the solution $H$ are obvious. It must be selected in accordance with the change speed of the vector-function of the final demand $C(t)$. Here we should admit that the final demand forecasting is a task that has been successfully solved by economists for a long time.

The changes of the vector $C\left(t_n + \frac{H}{2}\right)$ from step to step are of necessary consideration in a computational procedure. So, if the matrix $\overline{\varphi}_N$ and the vector $\overline{g}_N$ were constructed according to the formula (31) and (38) respectively for the constant vector $C(t)$, the last vector is not calculated in this case. The matrix $\Phi_N$ is constructed instead of it according to the formulas (34) or (35), and the working formula of the method takes the following form:

$$pX(t_n + \tau)DX(t_n + \tau) + C\left(t_n = \frac{H}{2}\right), \tau \in [0, H]. \tag{51}$$

Finally, let us consider the following non-stationary dynamic model of MIB

$$pX(t_n + \tau)DX(t_n + \tau) + C\left(t_n = \frac{H}{2}\right), \tau \in [0, H], \tag{52}$$

which takes into account the dependence of the elements of the structural matrix $D$ on the time. But it only can be done if these elements dynamics is predictable. Then the following restrictions are considered for the system (52):

$$X(t_0) = X_0, \|D(t) - D_n\| \leq \varepsilon_n, t_n \leq t \leq t_n + H_n, D_n = const, \tag{53}$$

the corresponding models (52) refer to the restrictions (53) and the difference equation will be the following

$$X_{n+1} = e^{D_n H_n} + \int_0^{H_n} e^{D_n \tau} d\tau \cdot C\left(t_n + \frac{H_n}{2}\right), \tag{54}$$

allowing to take into account changes in all coefficients at each step of integration.

It is obvious that the more detailed the model we strive to build, the more questions arise about its digitization. A researcher has to find a compromise between the statistical capabilities of the source data and the forecasting horizon.

The difference correlation (54) requires a greater number of arithmetic operations than those discussed above. The reason lays in the need of multiple multiplication of square matrices, which is comparable in complexity to the inversion of the matrix by Gauss method. Recently, the amount of calculations to account the variable coefficients at each step of integration was recognized rather large. At present, modern laptops are successfully coping with this task, so the problem has lost its urgency.

## 4 Conclusion

As a conclusion, we point that mathematical procedures for the investigation of structural transient processes based on the dynamic MIB model are proposed and justified in this paper. The algorithms define a sequence of consistently performed calculations, and could be easily formalized on a computer. The structural dynamics analysis methods based on the calculation of the matrix exponent and its integral claim to be a part of a complex mathematical support for solving problems of sustainability, economic dynamics and growth, the quality of transient process in macroeconomics.

This article has launched the development of the author's statistical research base in the field of structural forecasting, stability of economic dynamics and economic growth:

1. For the first time, a matrix of inter-industry persistence has been obtained, which allows applying the ordinary differential equations of dynamic motion in order to predict the dynamics of gross production. Digitization of the model is performed according to Rosstat data based on obvious mathematical and economic assumptions. It moves the model (1) from a set of "purely theoretical constructions" into a set of computable or practical ones.
2. For the first time, a basic task has been obtained which will make it possible to determine the necessary resources to achieve the required structural changes in the economy.

**Acknowledgements** The article was prepared with the financial support of the Russian Foundation for Basic Research. Grant № 20-010-00084A "Mathematical modeling of stability and macroeconomic dynamics".

## References

1. Lyubushin NP, Babicheva NE, Konyshkov AS (2017) Sustainable development: assessment, analysis, forecasting. Econ Anal Theory Pract 16(12):471
2. Kim KA (2011) Transient processes are a prerequisite for the development of the economic system. KRSU Bull 11(2):135
3. Svetunkov SG, Abdullaev IS (2009) Economic dynamics and production functions. Bulletin of the Orenburg State University 5
4. Svetunkov SG, Abdullaev IS (2010) Comparative analysis of production functions in models of economic dynamics. Petersburg State University of Economics, Bulletin of the St, p 5

5. Lyubushin NP, Babicheva NE, Usachev DG, Shustova MN (2015) Genesis of the concept "sustainable development of economic systems of various hierarchical levels." Region Econ Theory Pract 48:423
6. Tupchienko VA (2013). Scientific and technological progress and its impact on economic growth. Econ Anal Theory Pract 24(327)
7. Almon K (2012) The art of economic modeling/Otv. ed. MN Uzyakov. M, MAX press
8. Svetunkov S (2012) Complex-valued modeling in economics and finance. Springer Science & Business Media. https://doi.org/10.1007/978-1-4614-5876-0
9. Glazyev S (2020) Economy of the future. Does Russia have a chance? Litres
10. Posamantir EI (2014) Computable general equilibrium of the economy and transport. Transport in a dynamic intersectoral balance. M, PoliPrint Service
11. Brignolfson M (2019) Brignolfson E., McAfee E. Machine, platform, crowd. Our digital future. M, Mann, Ivanov and Ferber
12. Pogosov IA, Sokolovskaya EA (2015) Factors of long-term economic growth: the ratio of capital and labor in the growth of the gross income of the economy, the number of employees and labor productivity. Forecast Probl 6:18–30
13. Pogosov IA (2015) Factors of long-term economic growth: scientific and technological progress and capital intensity of production. Forecasti Prob 5:423–433
14. Leontiev V (1990) Economic essays. Theory, research, facts and politics: Per. from English M, Politizdat, 264
15. Suvorov NV, Treshchina SV, Balashova EE, Davidkova OB, Zenkova, GV (2015) The role of the technological factor in the development of the Russian economy: the results of predictive and analytical studies. Scientific works: Institute for Economic Forecasting RAS 13:8–75
16. Toroptsev EL, Marakhovskii AS, Duszynski R (2019) Theoretical framework for a set of equilibrium and input-output models development. Econ Anal Theory Pract. https://doi.org/10.24891/EA.18.3.427
17. Ashimov AA, Aisakova BA, Alshanov RA, Borovskiy YV, Borovskiy NY, Novikov DA, Sultanov BT (2014) Parametric regulation of economic growth based on nonautonomous computable general equilibrium models. Autom Remote Control 75(6):1041–1054
18. Valieva OV (2019) Global value chains: new rules and institutions. In: Institutional transformation of the economy: resources and institutions (ITERI-2019), pp 25–26
19. Suvorov NV, Treshchina SV, Beletsky YuV, Balashova EE (2017) Balance and factor models as a tool for analyzing and forecasting the structure of the economy. Scientific works, Institute for Economic Forecasting RAS 15
20. Shirov AA, Sayapova AR, Yantovsky AA (2015) Integrated intersectoral balance as an element of analysis and forecasting of relations in the post-Soviet space. Forecast Prob 1:11–21
21. Shirov AA, Yantovskii AA (2017) RIM interindustry macroeconomic model: development of instruments under current economic conditions. Stud Russ Econ Dev 28(3):241–252
22. Almon K (2012) The art of economic modeling. INP RAS
23. Almon C, Grassini M (1980) The Changing Structure of Employment in Italy
24. Ivanter VV, Belousov DR, Blohin AA (2017) Structural and Investment Policy for Sustainable Growth and Modernization of the Economy. Scientific report. Institute of Economic Forecasting Russian Academy of Sciences 34
25. Shirov AA (2016) From a crisis of financing mechanisms to sustainable economic growth. Stud Russ Econ Dev 27(4):359–366
26. Malkina MYu, Ovcharov AO (2019) Assessment of financial instability of economic systems: a variety of methods and models. Econ Anal Theory Pract 18(7):1273–1294
27. Uzyakova ES, Uzyakov RM (2018) Analysis of the impact of scientific and technological development on growth using the intersectoral balance toolkit. Forecast problems 6 (171)
28. Kuznetsov SYu, Piontkovsky DI, Sokolov DD, Starchikova OS (2016) Empirical comparison of mathematical methods for constructing time series of the input-output table system. Econ J Higher School Econ 20(4):711–730
29. Egiev SK (2016) Uncertainty shocks and short-term fluctuations in the Russian economy. Econ Horiz 6:49–57

30. Pankova SV, Popov VV (2016) Modeling the influence of the dynamics of the structure of mutual trade on the development of the economic potential of the Russian Federation. Econ Anal Theory Pract 12(459):4–13
31. Kuznetsov SA, Tikhobaev VM (2019) An iterative approach to the calculation of inter-industry production volumes and the development of an alternative concept of the national economy. Econ Anal Theory Pract 18(6):1166–1180
32. De Vries GJ, Erumban AA, Timmer MP, Voskoboynikov I, Wu HX (2012) Deconstructing the BRICs: structural transformation and aggregate productivity growth. J Comp Econ 40(2):211–227
33. Fagiolo G, Roventini A (2016) Macroeconomic policy in DSGE and agent-based models redux: New developments and challenges ahead. SSRN 2763735
34. Almon C (2018) Why are input-output tables important? Stud Russ Econ Dev 29(6):584–587
35. Suslov VI, Domozhirov DA, Ibragimov NM, Kostin VS, Melnikova LV, Tsyplakov AA (2016) Agent-based multiregional input-output model of the Russian economy. Ekonomika i matematicheskie metody= Econ Math Methods 52(1):112–131
36. Domozhirov DA, Ibragimov NM, Melnikova LV, Tsyplakov AA (2017). Integration of input–output approach into agent-based modeling. Part 1. Methodological principles. World of economics and management/Vestnik NSU. Series Soc Econ Sci 17(1):86–99
37. Domozhirov DA, Ibragimov NM, Melnikova LM (2016) Integration of input–output approach into agent-based modeling: part 2. Interregional analysis in an artificial economy. World Econ Manag 2017(2):15–25
38. Dolyatovsky VA, Grethko MV (2018) Modeling of mechanisms of behavior of economic agents. Econ Anal Theory Pract 17(10):1835–1848
39. Safiullin MR, Elshin LA, Prygunova MI (2016) Methodological approaches to forecasting the mid-term cycles of economic systems with the predominant type of administrative-command control. J Econ Econ Educ Res 17:277

# Anthropomorphic Model of States of Subjects of Critical Information Infrastructure Under Destructive Influences

**E. A. Maksimova⊙, M. A. Lapina⊙, V. G. Lapin⊙, and A. M. Rusakov**

**Abstract** Critical information infrastructure (CII) is an object of protection, the problem of ensuring the security of which is one of the priorities at the international level. Approaches to solving this problem in each country are determined independently, taking into account the peculiarities of the construction and development of CII. In the Russian Federation, at the legislative level, the issue was updated in 2017. However, at the methodological level, the approach used to ensure the safety of the RF CII is not based on a systematic approach. This leads to gross errors and inaccuracies in the course of making management decisions, therefore, to an increase in information security risks. When considering the subject of CII as a system, it becomes necessary to study inter-object relationships as sources of destructive influences that can lead to the effect of infrastructural destructives, i.e., to the self-destruction of infrastructure. For this, at the initial stage, it is proposed to build an anthropomorphic model of the states of the CII subjects. In the course of working with this model, it is possible to predict the development of the situation of self-destruction of the infrastructure of the CII subject in conditions of uncertainty.

**Keywords** Information security · Subject of critical information infrastructure · Anthropomorphism · Destructive influences · Inter-object relationships · Static model · Cognitive map

E. A. Maksimova · A. M. Rusakov
FSBEI HE "MIREA - Russian Technological University", Moscow 119454, Russian Federation
e-mail: maksimova@mirea.ru

M. A. Lapina (✉)
North Caucasus Federal University, Stavropol 355000, Russian Federation
e-mail: mlapina@ncfu.ru

V. G. Lapin
Stavropol Regional Clinical Advisory and Diagnostic Center, Stavropol 355000, Russian Federation

# 1   Introduction

Key changes in the scientific, technological, and economic spheres of life, the priority of using information technologies have dramatically changed the relationship between infrastructures. The result is the creation of more interconnected and complex infrastructures with, as a rule, greater centralization of management. This situation is accompanied by the actualization of tasks related to the identification and analysis of interdependencies between them [1, 2].

Critical Information Infrastructure (CII) is one of the types of infrastructure. The security of the CII RF today is one of the "pain points" in the field of information security (IS), affecting almost all spheres of social life. Its support is carried out on a regulatory basis in dynamics, that is, with the constant introduction of new regulations and methodological documents. Nevertheless, the main problem, in the author's opinion, is associated with the lack of a systematic approach as a methodological basis for the development of regulatory requirements for the development of CII.

# 2   Formulation of the Problem

A regulatory feature and a hypothetical contradiction of the subject area is the introduction of the actual concept of "CII subject". So, according to the Federal Law of the Russian Federation of July 26, 2017, No. 187-FZ "On the Security of Critical Information Infrastructure in the Russian Federation", the subject of CII is understood as a legal entity - the owner of CII facilities. CII objects - information systems, information and telecommunication networks, automated control systems - are independent units functioning within the CII. Here the concept of "the subject of CII" is not presented as a system.

It is important to note that at the regulatory level, the need to provide information on the relationships between CII objects when providing data for categorization is indicated (Decree of the Government of the Russian Federation of February 8, 2018 No. 127 "On approval of the Rules for categorizing objects of critical information infrastructure of the Russian Federation, as well as a list of indicators of criteria for the significance of objects critical information infrastructure of the Russian Federation and their values"). However, methodological support for its implementation has not yet been developed. The subject of CII (and CII as a whole) is not considered as a system in the documents of IS regulators.

In addition, the very concept of CII is not unambiguous from the point of view of a systems approach. On the one hand, CII can be considered as a set of objects, and then a horizontal hierarchy is observed in its structure (by types of objects). On the other hand, as a set of subjects; then a vertical hierarchical structure can be traced. The indicated contradiction leads to significant errors in assessing the safety and efficiency of the functioning of the CII at all levels, to gross errors in the course of making managerial decisions, therefore, to an increase in the risks of it's IS.

## 3  Conditions for Building a Model of States of CII Subjects

It is proposed to use a systematic approach as a methodological basis for the study, i.e. we will consider the subject of CII as a system of interacting objects, as well as the means of their interconnection, which are in the ownership of this subject. Based on the specifics of the CII architecture, we can talk about several variants of "spatial slices", within which it is possible to build a model of states of CII subjects. For more concretization, let us define as a working cut the level of CII subjects, as structural components and independent system units of CII.

Within the framework of the declared provisions, it is possible to solve the problem of studying the relationships in the CII as a system. However, they represent a complex subject for the task of analysis; its solution at the initial stage is possible by designating the dimensions of the taxonomy that create their main aspects. For example, these are types of interconnections, characteristics of infrastructure, infrastructure environment, binding and response behavior, types of infrastructural effects leading to destructive (destructive) impacts.

## 4  Classification of Types of Inter-Object Connections on the Subject of CII

In the course of the study, the subject of CII is considered as a complex structured system that functions under the influence of a large number of factors. The method of cognitive modeling and the scenario approach [3–7] (which will be used in the development of a model of states of CII subjects under destructive influences in a static mode) has been identified as the most optimal method for working with these systems.

To form a classification sensitivity, from the point of view of considering the full range of options for inter-object interaction on the subject of CII, an anthropomorphic approach was used, which was first applied in the research of prof. M.V. Buinevich and K.E. Izrailov [8, 9], where they identified 9 anthropomorphic types of vulnerability interactions: obligatory and optional symbiosis, commensalism, parasitism, commensalism, parasitism, predation, neutralism, amensalism, allelopathy, and competition.

This classification, in the author's opinion, is the most "sensitive" and most fully reflects all kinds of inter-object interaction in CII. The use of this approach makes it possible to substantiate the synergistic effect arising in the subject of CII and to determine the moment of the appearance of the "bifurcation point" [10].

Let us present the possible forms of symbiosis and antibiosis concerning the study of CII at the level of interaction between its elements. So, under obligate symbiosis, we mean a form of symbiosis in which CII objects cannot function without each other. A kind of obligate symbiosis is:

- mutualism is a form of obligatory mutually beneficial functioning of two or more CII objects, in which mutually beneficial assistance is necessary since the objects are interdependent;
- facultative symbiosis (protocooperation) - a form of symbiosis, when coexistence for CII objects is beneficial, but not necessary;
- commensalism - a form of symbiosis in which one CII object benefits from the relationship, and the other receives neither benefit nor harm;
- neutralism - a form of symbiosis in which the objects of CII do not have a mutual effect.
- Let's designate the forms of implementation of antibiosis at the CII:
- amensalism - a form of antibiosis, in which one CII object negatively affects another, but itself does not experience either negative or positive influence;
- allelopathy - a form of antibiosis, in which the objects of CII have a mutually harmful effect on each other, due to their functional parameters;
- competition is a form of antibiosis, in which two objects of CII are inherently infrastructural "enemies".

## 5   Modes of Functioning of the Subject of CII

The construction and study of models of states of CII subjects under destructive influences can be considered in static and dynamic modes; these modes are determined based on the infrastructural mechanics of the CII subject.

So, in a static mode, the functioning of the CII subject S occurs with an established infrastructure: changes in the states of the CII subject are carried out with a constant quantitative/qualitative composition at the infrastructural level, that is, provided that the following system characteristics remain unchanged in time:

- $N_s\,(O_i)$ – number of KII facilities:

$$\frac{dN_S(O_i)}{dt} = 0, \tag{1}$$

- $SYS_S(O_i, O_j)$ – system of inter-object relationships at the level of the considered subject of CII:

$$\frac{dSYS_S(O_i, O_j)}{dt} = 0, \tag{2}$$

- $SYS_{<S_k,S_l>}(O_i, O_j)$ – system of inter-object relationships at the level of inter-subjective interaction:

$$\frac{dSYS_{<S_k,S_l>}(O_i, O_j)}{dt} = 0. \tag{3}$$

Thus, the static model of the CII subject is described by a system of differential equations:

$$\{\frac{dN_s(O_i)}{dt} = 0, \frac{dSYS_S(O_i, O_j)}{dt} = 0, \frac{dSYS_{<S_k, S_l>}(O_i, O_j)}{dt} = 0. \quad (4)$$

The subject of CII as a system will switch to a dynamic mode as soon as condition (1) is violated. In this case, conditions (2) and (3) can either be violated or remain in their original state.

When modeling the states of CII subjects under destructive influences in a static mode (hereinafter referred to as a static model), the main problematic issue will be determined by the study of the types of inter-object connections (MOS) that affect the state of the system. As an indicator of the latter, we take the functionality of the CII subject F (S), by which we mean an integrative indicator reflecting the possibility of the CII subject, as a system, implementing its functions. In greater detail, similarly, we introduce the concept of the functionality of the CII -F (O_i) object. In the context of building a static model, this indicator is defined as the concept of a fuzzy cognitive model.

# 6   Methodology and Discussion

To build a static model, based on the introduced definitions of the types of MOS on the subject of CII, we will present the basic second-order models; the number of objects - structural elements of CII subjects, determines the order of the model.

To develop basic models, we will build adjacency matrices of inter-object influences. Since in the basic models 2 objects are considered in the structure of the CII subject, the matrices will have the second order. At the intersection of the rows and columns of the matrices, there will be quantitative values of the degrees of inter-object influence Rij - the conditional value of the influence of the object Oi on the object Oj.

Considering that MOS can be bipolar, the degree of inter objective influence has a "+" sign with a positive impact, i.e. when an increase in the value of the concept F (Oi) leads to an increase in the value of the concept F (Oj), the "−" sign - otherwise. To obtain a quantitative assessment and the convenience of working with experts, a scale of correspondence between the qualitative and quantitative assessments of the impact of MOS on the CII object was formed, which also works at the level of the bipolarity of concepts: "very weak" - 0.1; "Moderate" - 0.3; "Significant" - 0.5; "Strong" - 0.7; "Very strong" - 1.0.

The values of the elements of the basic MOC adjacency matrices are set according to the following rule:

$$R_{ij} = \{0, if i = j R_{ij}, if i \neq j,$$

*where* $R_{ij} \in [-1, +1]$.

We will consider the static model in two approximations; in both, the structure of the system remains unchanged and basic models are investigated. The first, to define the type of ISM as a destructive impact of an infrastructural nature. In the second - on the subject of determining the type of inter-object communication as a destructive impact of an infrastructural nature during the activation of objects. Here, the activation of objects is understood as a situation in which a change in the strength of the object's influence is made.

The second-order static model in the first approximation is presented in the form of a cognitive model for assessing the mutual complexity in the subject of CII. Depending on the type of MOS, where the type influences the presence of the corresponding connection in the cognitive map, all basic models can be represented by three concepts (vertices):

- "Assessment of the functionality of the subject of CII" F (S) (target vertex);
- O1M, O2M - the influence of the object O1 and O2, respectively, for a model of type "M" (M is a conditional number)

## 7  Experiment

To identify the types of MOS as destructive impacts of infrastructural.

Raker, experimental studies were carried out with basic static models. For the development and research of a static model in the first approximation, the notation and tools of the Mind Modeler system were used [11]. During the experiment, the structure of the CII subject remained unchanged; only the type of connection and the quantitative assessment of the degree of influence of the concepts - F (O1M) and F (O2M).

The statement of the problem for the experimental study of the functionality of subjects of CII with synchronous/asynchronous changes in object functionality is presented in Table 1.

We will consider the simulation results through the meaning of functionality, which is a bipolar sigmoid (Fig. 1).

The results of the graphical analysis of the data obtained made it possible to obtain an assessment of the change in the functionality of the CII subject from the values of concepts (object functionality) for each type of IOC (Table 2) and to establish a number of interesting dependencies.

First, the functionality of the CII subject changes with a synchronous change in the influence of objects for all types of IOC (dependence I). Secondly, the functionality of the CII subject changes with an asynchronous change in the functionality of objects for all types of MOS: (F (O1M) $\equiv$ const, F (O2M) - varies) (dependence II). Thirdly, the functionality of the CII subject changes with an asynchronous change in the influence of objects only for MOS such as "Commensalism" and "Amensalism": (F (O2M) $\equiv$ const, F (O1M) - varies) (dependence III). And fourthly, the functionality of the O1M object changes with an asynchronous change in the influence of objects

**Table 1** Statement of the problem for the experimental study of the functionality of the subject of CII

| Initial data | | Changing object functionality | |
|---|---|---|---|
| **Basic MOS Matrix** | **Cognitive model** | **synchronous** | **asynchronous** |

**1. Obligate symbiosis**

|  | $F(O_1 1)$ | $F(O_2 1)$ |
|---|---|---|
| $F(O_1 1)$ | 0 | +1 |
| $F(O_2 1)$ | +1 | 0 |

Cognitive model: Функциональность субъекта КИИ (1); $O_1 1$, $O_2 1$

synchronous: $F(O_1 1) = F(O_2 1) = \underline{-1.0; +0.9}$, Step = 0.1

asynchronous: $F(O_1 1) \neq F(O_2 1) \Rightarrow (F(O_1 1) \equiv const) \Rightarrow F(O_2 1) = \underline{-1.0; +0.9}$, Step = 0.1

**2. Optional symbiosis**

|  | $F(O_1 2)$ | $F(O_2 2)$ |
|---|---|---|
| $F(O_1 2)$ | 0 | +0.3 |
| $F(O_2 2)$ | +0.3 | 0 |

Cognitive model: Функциональность субъекта КИИ (2); $O_1 2$, $O_2 2$

synchronous: $F(O_1 2) = F(O_2 2) = \underline{-1.0; +1.0}$, $F(O_1 2) \neq 0.3$, Step = 0.1

asynchronous: $F(O21) \neq F(O22)$: $(F(O21) \equiv const) \Rightarrow (F(O22) = \underline{-1.0; +1.0}$, $F(O_1 2) \neq 0.3$, Step = 0.1

**3. Commensalism**

|  | $F(O_1 3)$ | $F(O_2 3)$ |
|---|---|---|
| $F(O_1 3)$ | 0 | 0 |
| $F(O_2 3)$ | +0,5 | 0 |

Cognitive model: Функциональность субъекта КИИ (3); $O_1 3$, $O_2 3$

synchronous: $F(O_1 3) = F(O_2 3) = \underline{-1.0; +1.0}$, Step = 0.1

asynchronous: $F(O_1 3) \neq F(O_2 3)$: 1) $(F(O_1 3) \equiv const) \Rightarrow (F(O_2 3) = \underline{-1.0; +1.0}$, Step = 0.1); 2) $(F(O_2 3) \equiv const) \Rightarrow (F(O_1 3) = \underline{-1.0; +1.0}$, Step = 0.1)

**4. Neutralism**

|  | $F(O_1 4)$ | $F(O_2 4)$ |
|---|---|---|
| $F(O_1 4)$ | 0 | 0 |
| $F(O_2 4)$ | 0 | 0 |

Cognitive model: Функциональность субъекта КИИ (4); $O_1 4$, $O_2 4$

synchronous: $F(O_1 4) = F(O_2 4) = \underline{-1.0; +1.0}$, Step = 0.1

asynchronous: $F(O_1 4) \neq F(O_2 4)$: $(F(O_1 4) \equiv const) \Rightarrow F(O_2 4) = \underline{-1.0; +1.0}$, Step = 0,1)

**5. Amensalism**

|  | $F(O_1 5)$ | $F(O_2 5)$ |
|---|---|---|
| $F(O_1 5)$ | 0 | 0 |
| $F(O_2 5)$ | −0,5 | 0 |

Cognitive model: Функциональность субъекта КИИ (5); $O_1 5$, $O_2 5$

synchronous: $F(O_1 5) = F(O_2 5) = \underline{-1.0; +1.0}$, Step = 0.1

asynchronous: $F(O_1 5) \neq F(O_2 5)$: 1) $(F(O_1 5) \equiv const) \Rightarrow (F(O_2 5) = \underline{-1.0; +1.0}, F(O_2 5) \neq -0.5$, Step = 0.); 2) $(F(O_2 5) \equiv const) \Rightarrow (F(O_1 5) = \underline{-1.0; +1.0}$, Step = 0.1)

**6. Allelopathy**

|  | $F(O_1 6)$ | $F(O_2 6)$ |
|---|---|---|
| $F(O_1 6)$ | 0 | −0,7 |
| $F(O_2 6)$ | −0,7 | 0 |

Cognitive model: Функциональность субъекта КИИ (6); $O_1 6$, $O_2 6$

synchronous: $F(O_1 6) = F(O_2 6) = \underline{-1.0; +1.0}$, $F(O_2 6) \neq -0.7$, Step = 0.1

asynchronous: $F(O_1 6) \neq F(O_2 6)$: $(F(O_1 6) \equiv const) \Rightarrow F(O_2 6) = \underline{-1.0; +1.0}$, Step = 0.1

**7. Competition**

|  | $F(O_1 7)$ | $F(O_2 7)$ |
|---|---|---|
| $F(O_1 7)$ | 0 | −1 |
| $F(O_2 7)$ | −1 | 0 |

Cognitive model: Функциональность субъекта КИИ (7); $O_1 7$, $O_2 7$

synchronous: $F(O_1 7) = F(O_2 7) = \underline{-0.9; 0.9}$, Step = 0.1

asynchronous: $F(O_1 7) \neq F(O_2 7)$: $(F(O_1 7) \equiv const) \Rightarrow F(O_2 7) = \underline{-1.0; +1.0}$, Step = 0.1

for all types of MOS, except for "Neutralism" $F(O1M) \equiv const$, $F(O2M)$ - varies) (dependency IV).

Thus, the functionality of the CII subject changes either with a synchronous change in the object functionality, or with a change in the object functionality of one of the structural elements. In case of MOS of the type "Facultative symbiosis" and "Allelopathy" for any variants of changes in the object functionality of structural

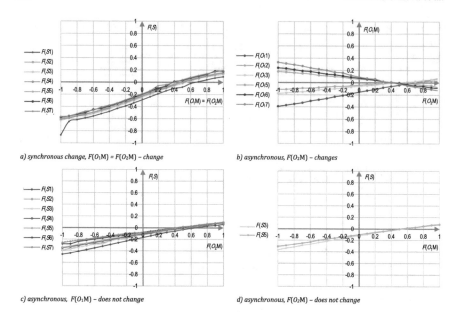

a) synchronous change, $F(O_1M) = F(O_2M)$ – change

b) asynchronous, $F(O_2M)$ – changes

c) asynchronous, $F(O_1M)$ – does not change

d) asynchronous, $F(O_2M)$ – does not change

**Fig. 1** CII-subject functionality changes during synchronous/asynchronous changes in object functionality

**Table 2** Assessment of changes in functionality from concept values

| № p/p | MOS type | Dependence | | | |
|---|---|---|---|---|---|
| | | I | II | III | IV |
| 1 | Obligate symbiosis | Yes | Yes | No | Yes |
| 2 | Optional symbiosis | Yes | Yes | No | Yes |
| 3 | Commensalism | Yes | Yes | Yes | Yes |
| 4 | Neutralism | Yes | Yes | No | No |
| 5 | Amensalism | Yes | Yes | Yes | Yes |
| 6 | Allelopathy | Yes | Yes | No | Yes |
| 7 | Competition | yes | Yes | No | yes |

elements, a change in the reaction of the functionality of the CII subject occurs; in addition, a change in the functionality of one of the objects entails a change in the functionality of the other object. If the influence of object A changes, then with MOS of all types, except for "Neutralism", there is a change in the functionality of object B.

Since a change in object functionality affects the functionality of the subject, it is necessary to determine the type and strength of this influence. Based on the results of the analysis of the data obtained during the experiment, a table of the dynamics of changes in functionality at the level of the subject of CII was formed (Table 3).

**Table 3** Dynamics of changes in functionality at the CII-subject level

| № p/p | MOS type | I | | II | | III | | | IV | | |
|---|---|---|---|---|---|---|---|---|---|---|---|
| | | $F(O_{ij}M)$ | $F(S)$ | $F(O_{ij}M)$ | $F(S)$ | $F(O_1M)$ | $F(O_2M)$ | $R(S)$ | $F(O_1M)$ | $F(O_2M)$ | $F(S)$ |
| 1 | Obligate symbiosis | ← | ← | ← | ← | ← | – | – | – | ← | ← |
| 2 | Optional symbiosis | ← | ← | ← | ← | ← | – | – | – | ← | ← |
| 3 | Commensalism | ← | ← | ← | ← | ← | – | ← | – | ← | ← |
| 4 | Neutralism | ← | ← | ← | ← | ← | – | – | – | ← | – |
| 5 | Amensalism | ← | ← | ← | ← | ← | – | ← | – | ← | → |
| 6 | Allelopathy | ← | ← | ← | ← | ← | – | – | – | ← | → |
| 7 | Competition | ← | ← | ← | ← | ← | – | – | – | ← | → |

**Table 4** Indicators of the models of interobject connections as destructive-forming

| № p/p | MOS type | Limit values of the reaction of activated objects during synchronous changes in object functionality | |
|---|---|---|---|
| 1 | Obligate symbiosis | [−1, −0,5] | [Very strong, substantial] |
| 2 | Optional symbiosis | [−1, −0,5] | [Very strong, substantial] |
| 3 | Commensalism | [−1, −0,6] | [Very strong, strong] |
| 4 | Neutralism | [−1, −0,6] | [Very strong, strong] |
| 5 | Amensalism | [−1, −0,7] | [Very strong, strong] |
| 6 | Allelopathy | [−1, −0,7] | [Very strong, strong] |
| 7 | Competition | [−1, −0,8] | [Very strong] |

To define the types of MOS as a destructive generators, we will consider those that, with certain changes in the values of the influence of objects, will lead to a negative effect for the subjective functionality at the level [−1, 0.5], that is, at the level of a very "strong ... significant" influence. Table 4 shows the types and states of MOS obtained experimentally, which can be considered as destructive-generating at the level of the library of basic models.

It is important to note that in models of the anabiotic type, an asynchronous change in object functionality leads to a decrease in the functionality of the target. At the level of basic models, this cannot be considered as a destructive-forming effect, but it can become such with an increase in the infrastructural complexity of the CII subject.

Thus, in the static mode of the functioning of the subject, the CII will be destructively forming MOS of all types only with a synchronous change in the object functionality of structural elements within the limits of the limiting values of the change in the influence of the influencing object.

# 8  Conclusion

In the course of the study, a model of states of CII subjects under destructive influences in a static mode was developed, based on the developed library of basic adjacency matrices and anthropomorphic classification of the types of MOS. An experimental study of this model made it possible to identify the following conditions for the manifestation of intersubject connections as destructively forming: the type of intersubject communication, the variant of changing the object functionality (synchronous/asynchronous), and the value of the influence of the influencing object.

It has been experimentally proved that under certain conditions, both a point change in the functionality of the subject of CII and an integrative change in the functionality of the subject of CII and the functionality of the object of influence can be traced.

Since the influence of MOS can be cumulative, therefore, in the subject of CII, in the presence of the corresponding types of MOS as destructive-forming, the effect of infrastructural destructives may manifest over time, i.e. self-destruction of the subject of CII as a system.

The presented static model does not consider the situation of infrastructural dynamics of the subject of CII, which manifests itself in violation of condition (1) and, as a consequence, violation of conditions (2) and (3). To build a dynamic model, it is necessary to use, for example, logical-probabilistic modeling with the construction of models of different levels of complexity, allowing, among other things, to adjust the categories of importance of CII objects at all stages of the life cycle of the CII subject [12–14]. In this situation, the palette of destructive influences will expand due to the description of possible errors of an infrastructure nature at different stages of the life cycle of the CII subject.

# References

1. Maksimova EA (2021) "Smart decisions" in development of a model for protecting information of a subject of critical information infrastructure. Lect Notes Netw Syst 155:1213–1221. https://doi.org/10.1007/978-3-030-59126-7_132
2. Maksimova EA, Baranov VV (2021) Predicting destructive malicious impacts on the subject of critical information infrastructure. In: Singh PK, Veselov G, Vyatkin V, Pljonkin A, Dodero JM, Kumar Y (eds) Futuristic trends in network and communication technologies: third international conference, FTNCT 2020, Taganrog, Russia, October 14–16, 2020, revised selected papers, part I. Springer, Singapore, pp 88–99. https://doi.org/10.1007/978-981-16-1480-4_8
3. Azhmukhamedov IM (2014) Management of poorly formalized socio-technical systems based on fuzzy cognitive modeling (on the example of integrated information security systems). Dissertation for the degree of Doctor of Technical Sciences, Astrakhan.
4. Sadovnikova NP, Zhidkova NP (2012) The choice of strategies for territorial development based on cognitive analysis and scenario modeling. Internet Bull VolgGASU 7(21):4
5. Sadovnikova NP (2011) Application of the cognitive modeling for analysis of the ecological and economical efficiency of the urban planning project. Internet-Vestnik VolgGASU Ser Civil Eng Inf 5(14). Access mode: www.vestnik.vgasu.ru
6. Roberts FC (1986) Discrete mathematical models with applications to social, biological and environmental problems: per. from English. Nauka, Moscow, 496p
7. Obiedat M, Samarasinghe S (2013) Fuzzy representation and aggregation of fuzzy cognitive maps. In: 20th international congress on modeling and simulation, Adelaide, Australia, pp 690–694. www.mssanz.org.au/modsim2013
8. Buinevich MV (2019) Anthropomorphic approach to describing the interaction of vulnerabilities in the program code. Part 1. Types of interactions. In: Buinevich MV, Izrailov KE (eds) Information security. Inside, vol 5 no 89. pp 78–85
9. Buinevich MV (2019) Anthropomorphic approach to describing the interaction of vulnerabilities in the program code. Part 2. Metrics of vulnerabilities. In: Buinevich MV, Izrailov KE (eds) Information security. Inside, vol 6, no 90, pp 61–65

10. Veselov GYe (2017) The theory of hierarchical control of complex systems: a synergetic approach. In: Veselov GE (ed) VIII All-Russian scientific conference "System synthesis and applied synergetics": collection of scientific papers, 18–20 September 2017. Southern Federal University, Nizhniy Arkhyz, pp 23–43
11. http://www.mentalmodeler.org/
12. Ryabinin IA (2010) Logical-probabilistic analysis and its modern capabilities. Interdiscip Sci Appl J Biosphere 2(1):23–28
13. Alekseev VV (2009) Development of logical and probabilistic models, methods and algorithms for risk and efficiency management in structurally complex systems Abstract of the thesis for the degree of candidate of technical sciences Specialty 05.13.18 - Mathematical modeling, numerical methods and program complexes, St. Petersburg
14. Mozhaev AS, Gromov VN (2000) Theoretical foundations of the general logical-probabilistic method of automated system modeling. SPb. VITU, 144p

Printed in the United States
by Baker & Taylor Publisher Services